From
NATURAL PHILOSOPHY
to the
SCIENCES

From
NATURAL PHILOSOPHY
to the
SCIENCES

Writing the History of
Nineteenth-Century Science

❦

Edited by
DAVID CAHAN

THE UNIVERSITY OF CHICAGO PRESS
CHICAGO AND LONDON

DAVID CAHAN is Charles Bessey Professor of History at the University of Nebraska. He is the author of *An Institute for an Empire: The Physikalisch-Technische Reichsanstalt 1871–1918* and coauthor of *Science at the American Frontier: A Biography of DeWitt Bristol Brace.* He is also the editor of Hermann von Helmholtz's *Science and Culture: Popular and Philosophical Essays,* published by the University of Chicago Press.

The University of Chicago Press, Chicago 60637
The University of Chicago Press, Ltd., London
© 2003 by The University of Chicago
All rights reserved. Published 2003
Printed in the United States of America
12 11 10 09 08 07 06 05 04 03   5 4 3 2 1

ISBN (cloth): 0-226-08927-4
ISBN (paper): 0-226-08928-2

Library of Congress Cataloging-in-Publication Data

From natural philosophy to the sciences : writing the history of nineteenth-century science / edited by David Cahan.
    p.   cm.
    Includes bibliographical references and index.
    ISBN 0-226-08927-4 (alk. paper) — ISBN 0-226-08928-2 (pbk. : alk. paper)
    1. Science—History—19th century   I. Cahan, David.
    Q125 .F695   2003
    509'.034—dc21

2002155951

Contents

ACKNOWLEDGMENTS

SUSAN ABRAMS, executive editor at the University of Chicago Press, has supported the publication of this volume from its early stages, and it is a pleasure to thank her for her imaginative and wise counsel. I also thank the volume's copyeditor, Joel Score, for his precision and skill in preparing the manuscript for publication.

DAVID CAHAN

CONTRIBUTORS

BERNADETTE BENSAUDE-VINCENT is professor of the history and philosophy of science at the Université de Paris X. She is the author of *Lavoisier: Mémoires d'une révolution* (1993); *Dans le laboratoire de Lavoisier, livre illustré* (1993); and, with Isabelle Stengers, *A History of Chemistry* (1997). In addition, she has edited *Eloge du mixte: Matériaux nouveaux et philosophie ancienne* (1998); with Ferdinando Abbri, *Lavoisier in European Context: Negotiating a New Language for Chemistry* (1995); and, with A. Lundgren et al., *Communicating Chemistry: Textbooks and Their Audiences* (2000).

JED Z. BUCHWALD is Dreyfuss Professor of the History of Science at the California Institute of Technology. He is the author of *From Maxwell to Microphysics: Aspects of Electromagnetic Theory in the Last Quarter of the Nineteenth Century* (1985); *The Rise of the Wave Theory of Light: Optical Theory and Experiment in the Early Nineteenth Century* (1989); and *The Creation of Scientific Effects: Heinrich Hertz and Electric Waves* (1994). He is also the coeditor of *Archive for History of Exact Sciences*.

DAVID CAHAN is Charles Bessey Professor of History at the University of Nebraska. He is the author of *An Institute for an Empire: The Physikalisch-Technische Reichsanstalt, 1871–1918* (1989) and, with M. Eugene Rudd, *Science at the American Frontier: A Biography of DeWitt Bristol Brace* (2000). He is also the editor of *Letters of Hermann von Helmholtz to His Parents: The Medical Education of a German Scientist, 1837–1846* (1993); *Hermann von Helmholtz and the Foundations of Nineteenth-Century Science* (1993); and a collection of Helmholtz's essays, *Science and Culture: Popular and Philosophical Essays* (1995).

JOSEPH DAUBEN, a former editor of *Historia Mathematica*, is professor of history and history of science in the Department of History, Lehman College, City University of New York, and the Ph.D. Program in History, the Graduate Center, CUNY. He is the author of *Georg Cantor: His Mathematics and Philosophy of the Infinite* (1979) and *Abraham Robinson: The Creation of Nonstandard Analysis, A Personal and Mathematical Odyssey* (1995). He is also the editor of

History of Mathematics from Antiquity to the Present (1985; rev. CD-ROM ed. 2000) and, with Christoph J. Scriba, *Writing the History of Mathematics: Its Historical Development* (2002).

Frederick Gregory is professor of history of science in the Department of History at the University of Florida. His *Scientific Materialism in Nineteenth-Century Germany* (1977) was followed by another study of German science in its cultural context, *Nature Lost? Natural Science and the German Theological Traditions of the Nineteenth Century* (1992). He is the editor of the English translation of J. F. Fries's 1805 work, *Wissen, Glaube und Ahndung,* which appeared in 1989 as *Knowledge, Belief, and Aesthetic Sense.* He has focused most recently on the history of science and religion.

Michael Hagner is a senior fellow at the Max-Planck-Institut für Wissenschaftsgeschichte in Berlin. He is the author of *Homo Cerebralis: Der Wandel vom Seelenorgan zum Gehirn* (1997; English trans. 2004), has edited *Ecce Cortex: Beiträge zur Geschichte des modernen Gehirns* (1999) and *Ansichten der Wissenschaftsgeschichte* (2001), and has coedited an issue of *Science in Context* (2001). He is currently working on a book on elite brain research in the nineteenth and twentieth centuries.

Sungook Hong is associate professor at the Institute for the History and Philosophy of Science and Technology, University of Toronto. He works on the history of nineteenth-century physics, the historical interaction between science and technology, and historiographical issues in the history of science and technology. He is the author of *Wireless: From Marconi's Black-box to the Audion* (2001).

David R. Oldroyd is Honorary Visiting Professor at the School of History and Philosophy of Science at the University of New South Wales. His principal publications in the history of the geosciences are *The Highlands Controversy: Constructing Geological Knowledge through Fieldwork in Nineteenth-Century Britain* (1990), *Thinking about the Earth: A History of Ideas in Geology* (1996), *Sciences of the Earth: Studies in the History of Mineralogy and Geology* (1998), and *Earth, Water, Ice, and Fire: Two Hundred Years of Geological Research in the English Lake District* (2002).

Theodore M. Porter is professor of history of science in the Department of History, UCLA. His books include *The Rise of Statistical Thinking, 1820–1900* (1986) and *Trust in Numbers: The Pursuit of Objectivity in Science and Public Life* (1995). He is a coauthor of *The Empire of Chance: How Probability Changed Science and Everyday Life* (1989) and coeditor, with Dorothy Ross, of *The Cam-*

bridge History of Science, vol. 7, *Modern Social Sciences* (2002). His latest work concerns the early career of the statistician Karl Pearson.

ROBERT J. RICHARDS is professor at the University of Chicago in the departments of History, Philosophy, and Psychology and director of the Fishbein Center for the History of Science and Medicine. He is the author of *Darwin and the Emergence of Evolutionary Theories of Mind and Behavior* (1987; winner of the Pfizer Prize in the history of science), *The Meaning of Evolution: The Morphological Construction and Ideological Reconstruction of Darwin's Theory* (1992), and *The Romantic Conception of Life: Science and Philosophy in the Age of Goethe* (2002).

ULRICH WENGENROTH is professor of the history of technology at the Munich Center for the History of Science and Technology. He has published a number of studies concerning the history of technology and business history in nineteenth- and twentieth-century Europe. Among his books are *Enterprise and Technology: The German and British Steel Industries, 1865–1895* (1994) and, as editor, *Technik und Wirtschaft* (1993). He is currently working on forms of technological knowledge and directing a research group of the Federal Ministry of Science and Education on innovation systems and the culture of innovation in Germany.

From
NATURAL PHILOSOPHY
to the
SCIENCES

One

LOOKING AT NINETEENTH-CENTURY SCIENCE: AN INTRODUCTION

David Cahan

❧

DURING THE PAST half century the history of nineteenth-century science has attracted enormous interest from a wide variety of scholars. Apart from historians of science proper, historians of technology, medicine, religion, general intellectual and cultural history, literature, art, music, and socioeconomic and political life have turned to the study of nineteenth-century science, both for its intrinsic interest and for its importance for their own fields of scholarship. The study of nineteenth-century science is flourishing.

In the course of the long nineteenth century—here understood (without any political implications) to have begun in the late eighteenth century and ended in the early twentieth—the scientific enterprise underwent enormous and unprecedented intellectual and social changes. Developments in the sciences during this period arguably equaled or exceeded those in natural philosophy during the Scientific Revolution of the sixteenth and seventeenth centuries, and in virtually every respect, be it intellectual range, theory formation, empirical results, or instrumentation. Moreover, the sciences underwent unprecedented institutional growth and had a large role in reshaping society—just as society helped reshape them. That is why scholars have occasionally referred to this later period (or a portion of it) as "the Second Scientific Revolution."[1] Yet the label has never really stuck or been much used, and perhaps for good reason: "revolution" seems a dubious term for a set of intellectually and socially diverse events that occurred in several different countries over a period of a century or more. Indeed, scholars of the early modern Scientific Revolution have recently raised doubts about the meaning and aptness of the label even for that period.[2] Notwithstanding the untold number of

---

1. Kuhn, "Function of Measurement," 220; Mendelsohn, "Emergence of Science," 4; and idem, "Context of Nineteenth-Century Science," xvii. See also Cunningham and Williams, "Decentring the 'Big Picture.'" Brush, *History of Modern Science,* applies the label to the period from 1800 to 1950.

2. Lindberg and Westman, *Reappraisals of the Scientific Revolution.*

landmark, innovative results during both periods, it seems more fitting to speak of an evolution of the sciences than of a revolution.

By the early 1700s, when the Scientific Revolution had largely run its course, "science" still meant natural philosophy—taking that term, as does this volume's title, to include natural history. Natural philosophy had by then shed its Aristotelian metaphysics, rejected occult qualities in explanation, adopted new standards of evidence and experiment, created entirely new sorts of instrumentation, and generally incorporated new concepts and results. This was especially the case in the exact sciences of astronomy, mechanics, and optics. Yet "science" was just beginning to assume its modern meaning and scope; this understanding emerged more definitively between the late Enlightenment and the early twentieth century. The reconceptualization and quantitative explosion of scientific knowledge; the new institutional and social structures of the sciences; the enthusiastic applications of the sciences and their practical relevance to medicine, technology, and industry; and the implications for religion and literature—these and other transformations occurred during the nineteenth century and marked a new epoch in science.

As the contributors to this volume articulate in some detail, it was in the nineteenth century that the modern disciplines of chemistry, physics, mathematics, biology, and the earth sciences, as well as the social sciences, assumed their more or less contemporary form and simultaneously reshaped the institutional landscape of science. New terms like "biology" and "physics," and "biologist" and "physicist," were created to describe the new disciplines and their practitioners, just as the more general term "scientist" was created to reflect a new general social category. Longstanding designations like "mathematician," "astronomer," and "chemist" now took on new, more narrowly defined meanings. Certainly by the final third of the nineteenth century, one could speak legitimately, that is, in a modern sense, of "science," "scientists," and the disciplines of science. These new labels and categories reflected the fact that science had both delimited itself more fully from philosophy, theology, and other types of traditional learning and culture and differentiated itself internally into increasingly specialized regions of knowledge. At the same time, new institutions, such as specialized societies and institutes, were created, and the notion of a "scientific community" appeared. Moreover, interactions between and among the sciences and other aspects of culture, the economy, the state, and society in general became more significant. In many minds "the nineteenth century" and "science" became synonymous with "progress."

Historians of science have, of course, long been aware of this transformation. During the past half century, as I have already suggested, scholarship in the history of nineteenth-century science has produced an enormous increase in knowledge about individual nineteenth-century scientists, conceptual and theoretical changes, experimental practices, scientific institutions, and the social, political, and cultural contexts and implications of nineteenth-century science. To give a quantitative measure to the point, the *Isis Current Bibliography* for 1999 lists over 750 books and articles published in the previous year or so dealing with nineteenth-century science and related matters of medicine and technology.[3]

This half century of intense scholarship has created an enticing intellectual opportunity. We now have a far better historical basis upon which to construct a sophisticated vision and analysis of the structure, growth, dynamics, and cultural context of nineteenth-century science as a whole. How should we explain the large-scale transformation of natural philosophy into the sciences during the long nineteenth century? What were the principal causes at play? Did these vary over time and place, and if so, how? What were the relationships between science as a whole and other aspects of culture? The essays in this volume are meant, in part, to stimulate scholars to think about formulating responses to these, the big questions.

In previous generations, three remarkable scholars offered quite different interpretations of the overall pattern of nineteenth-century science. The first and most extensive account was that of John Theodore Merz, an Englishman who received a rigorous education in Germany in mathematics, physics, philosophy, and theology, but who subsequently earned his living in the chemical and, especially, the electrical manufacturing business in Britain. Merz boldly led the way with the publication of his four-volume work, *A History of European Thought in the Nineteenth Century* (1904–12). He sought unity both within nineteenth-century science proper and in its relationship to nineteenth-century thought in general. He did so in part by portraying what he called the "scientific spirits" of France, Germany, and England as well as the various "views of nature": the astronomical, atomic, kinetic, physical, morphological, genetic, vitalistic, psychophysical, and the statistical. (He also sketched the development of mathematics during the century).[4] Furthermore, he sought to unify science and philosophy.[5] His overall aim for these intellectually ambitious and broad volumes was "to contribute something towards a unification of thought."[6] While one may be skeptical about a historiographical approach based on "spirits" and "views," one may nonetheless admire the account that emerges from Merz's classificatory vision.

---

3. Similar figures hold for other years: the *Isis Critical Bibliography* for 1989 reported about 800 books and articles and that for 1979 about 680. Neu, ed., *Isis Current Bibliography of the History of Science and Its Cultural Influences 1999*, 129–69; Neu, ed., *Isis Current Bibliography of the History of Science and Its Cultural Influences 1989*, 111–49; and Neu, ed., *Isis Current Bibliography of the History of Science and Its Cultural Influences 1979*, 108–41.

4. Merz, *History of European Thought*, vols. 1 and 2.

5. Ibid., vols. 3 and 4.

6. Ibid., 1: v.

The second account issued from John Desmond Bernal, also an En-glishman, but one who was an academic crystallographer and who wrote in the early 1950s in a broadly Marxist vein. Bernal argued that the development of science in the nineteenth century, as in others, correlated closely with de-velopments in the social and economic worlds. "The phases of the evolution of modern science," he declared, "mark the successive crises of capitalist economy." For Bernal, industrial capitalism was the economic dynamic be-hind nineteenth-century science.[7] But Bernal's correct recognition of the importance of economic development for the general material and social de-velopment of the sciences devolved into an overly simplistic causal relation-ship between capitalism and science.

The third account came from Joseph Ben-David, an Israeli-American historical sociologist who developed his model of the sociology of science during the 1960s and 1970s, evincing, in contrast to Bernal, a decidedly pro-capitalist, free-enterprise spirit. Ben-David portrayed science's development, including that during the nineteenth century, largely in terms of "the scien-tific role" and competition among scientists and their potential state patrons.[8] His model of the social evolution of science and its increasing relevance to so-ciety, from ancient Greece to contemporary America, served him best for the nineteenth and twentieth centuries.[9] Here, too, while one may generally agree with Ben-David's emphasis on competition and the scientist's role, one may nonetheless notice that other causal elements came into play, and that the role of the individual scientist underwent many historical permutations.

Whatever shortcomings scholars today might find with these perspectives, the fact remains that Merz, Bernal, and Ben-David sought to provide a sense of the unity of nineteenth-century science.[10] Though one-sided and some-times short on empirical detail, their accounts remain intellectually provoca-tive. While contemporary historians of nineteenth-century science frequently employ one or more of the general intellectual, economic, or sociological ori-entations used by Merz, Bernal, or Ben-David, they have largely confined their work to well-circumscribed, chronologically limited studies. It is time, I

7. Bernal, *Science in History,* 1:4. Bernal's analysis of nineteenth-century science is presented in two volumes of his four-volume work *Science in History* (vol. 2, *The Scientific and Industrial Revolu-tions,* and vol. 4, *The Social Sciences: Conclusion),* and in his *Science and Industry in the Nineteenth Cen-tury.*

8. Ben-David, *Scientist's Role in Society,* esp. chaps. 6–8. For other, related works by Ben-David, see the bibliography.

9. Kuhn, "Scientific Growth: Reflections on Ben-David's 'Scientific Role,' " provides an excel-lent appreciation and critique of Ben-David.

10. Sharlin, *Convergent Century,* offers an unconvincing account for unification. Knight, *Age of Science,* and Russell, *Science and Social Change,* offer good, modern introductory overviews of nineteenth-century science.

would suggest, for scholars to consider attempting a new, broad, and synthetic interpretation of the development of nineteenth-century science as a whole, and to do so on the basis of the innumerable microstudies of science that have appeared during the past half century.

From Natural Philosophy to the Sciences: Writing the History of Nineteenth-Century Science addresses this state of scholarly affairs. Its overall objective is twofold: First, to present historiographical analyses of work done by scholars of nineteenth-century science, emphasizing studies of the disciplinary formation of the sciences and the sciences' relationships with other parts of culture and society. These analyses aim to highlight the broader intellectual, socioeconomic, political, and cultural contexts within which nineteenth-century science developed. Second, to pose questions for future scholarship that will lead to a broader understanding of nineteenth-century science as a whole, as well as to further specialized studies.

Each essay seeks to give a thematic historiographical analysis of the most important problems, intellectual traditions, literature, methods, modes of explanation, and so on in a given field of scholarship. The huge number of scholarly books and articles published to date and the limited space available mean that no contributor could possibly assess all the literature in her or his field. Instead, the authors have focused on what they consider to be the most important studies in their respective areas, making judgments as to which are historiographically most thought-provoking and instructive. At the same time, the volume's extensive bibliography lists many of the outstanding and influential contributions to the history of nineteenth-century science. The volume thus largely eschews giving historical accounts of nineteenth-century science, seeking instead to direct attention to the best writing about nineteenth-century science.

The contributors have naturally sought not to beg historiographical questions. They have challenged explanatory assumptions, whether implicit or explicit, and criticized the misuse of supposed terms of explanation, for example, "Darwinism" and "science-technology push." They have raised the often-neglected but essential issue of periodization, which includes judgments about the important "turning points" in a field and the degree of unity of the field. Each contributor has employed the expression "nineteenth-century science" in the way that he or she thought most appropriate for his or her essay. Developments before 1800 were obviously so crucial for the later course of some sciences—chemistry and geology, for example—that the historiographer must reach back into the late eighteenth century in order to present an adequate understanding of those disciplines' overall development. In other fields—"physics," for example—one can more reasonably focus on the deep, path-setting intellectual and institutional changes that occurred between 1800 and 1900. Moreover, each essay has sought to be as comparative and interna-

tionally oriented as possible and to point to tendentious, nationalist biases in
its field, without, however, neglecting the fact that some sciences proceeded
more fully or more rapidly in one or a few countries during the period in question. The contributions to this volume by several European scholars may help
avert what is all too often an excessively American or British orientation. Furthermore, the classification of six of the essays—those on biology, the earth
sciences, mathematics, physics, chemistry, and the social sciences—according
to modern discipline is intended solely as an organizational device and is most
certainly not meant to prejudice the important and central historiographical
issue concerning the origin and formation of the modern, specialized scientific disciplines. Finally, I regret that I was unable to secure essays on astronomy,
on the philosophical foundations of science, and on science and literature,
and I note that the recent rise of interest in such topics as the study of museums, zoos and botanical gardens, environmentalism, science in the non-
Western world, and science and colonialism will in due course require
additional historiographical work.

From Natural Philosophy to the Sciences also aims to take its place among recent volumes devoted to historiographical analyses of the history of science.
For the early modern period, there is David Lindberg and Robert S. Westman's recent edited set of essays, *Reappraisals of the Scientific Revolution,* as
well as H. Floris Cohen's *The Scientific Revolution: A Historiographical Inquiry.*
For Enlightenment science, there is the older but still useful set of essays
edited by G. S. Rousseau and Roy Porter, *The Ferment of Knowledge: Studies in
the Historiography of Eighteenth-Century Science.* And although we still lack a
comprehensive historiographical study of twentieth-century science, we do
have Sally Gregory Kohlstedt and Margaret W. Rossiter's important edited
volume of essays on twentieth-century American science, *Historical Writings
on American Science.* The present volume on nineteenth-century science thus
is intended to fill an essential gap in the historiography of the history of science, complementing these and other works. It aims to encourage historiographical thought and debate among a broad spectrum of scholars, and to
help the history of science move toward higher levels of analysis.

THE ESSAYS in this volume suggest three very broad themes or contentions
about nineteenth-century science.

First, as already indicated, the very character of "science" changed during
the course of the nineteenth century. A number of the essays explicitly discuss the transformation of natural philosophy and natural history into a set of
well-defined, specialized scientific disciplines. In his essay on mathematics,
Joseph Dauben argues that during the course of the century mathematics
transformed itself from a field broadly devoted to applied mathematics (to

use the modern term) into an autonomous discipline concerned with its own logical foundations and proofs. Its concerns expanded beyond the mathematical dimension of other subjects (e.g., astronomy and geography); at the same time, however, it continued to contribute to many of the new scientific disciplines. Jed Buchwald and Sungook Hong, in their essay on physics, argue that the very meaning of "physics" changed during the course of the century. In taking up the issue of the transformation of natural philosophy in the eighteenth century into physics in the nineteenth, they point especially to the mathematization of physics, the changing relationship of theory and experiment, and the interaction between physics and technology. In their view, modern scholarship indicates a three-stage periodization for nineteenth-century physics: By around 1830, it had become a discipline; between then and the 1860s it became institutionalized in the universities, proved itself useful to certain industries, and developed some overarching principles (e.g., conservation of energy). Thereafter, as physics laboratories and institutes became commonplace, continuum physics developed in the form of electromagnetism and, toward the end of the century, microphysics began to emerge. Buchwald and Hong see a historiographical shift from a concern with high theory to a concern with low theory, instruments, and experimentation. They examine the complex relations of theory and experiment by looking at recent studies concerned with such leading figures as Michael Faraday, William Thomson, James Clerk Maxwell, the "Maxwellians," and Hermann von Helmholtz.

In his historiographical treatment of nineteenth-century biology, Robert J. Richards argues that "evolutionary theory has been the obsession of the discipline." That obsession—otherwise known as the Darwin industry—began in 1959, the centennial of Charles Darwin's *Origin of Species*. It changed dramatically in the mid- to late 1970s as scholars began reporting on their examinations of the Darwin manuscripts at Cambridge University. And it reached an inflection point in 1982, the centennial of Darwin's death. Richards argues that a number of Darwin's interpreters have sought to view him in terms of twentieth-century biology or have seen his thought as the scientific embodiment of political liberalism or a full interpenetration of the social and the scientific, inspired and conditioned by his ambient socioeconomic and political world. Still others, Richards further argues, have sought to capture the personal and nonpersonal sources of Darwin's thought without any simplistic reduction.

Beyond evolutionary theory, Richards points out that the historiography of nineteenth-century biology has also been concerned with social Darwinism and evolutionary ethics, biology and religion, biology and literature, morphology and romantic biology, neurophysiology, genetics, and cell theory. He urges scholars to devote more attention to all of these fields, noting that they too (and not just evolutionary theory) helped constitute the discipline of biology that emerged after 1800. To do so would be to help put evolutionary theory in its proper historical place.

The formation and transformation of new disciplines was perhaps most dramatic in the cases of the earth and the social sciences. As David Oldroyd argues, the discipline of the earth sciences simply did not exist before about 1800, and through much of the nineteenth century a sense of disciplinary unity was largely absent. During the course of the century, geology, geomagnetical studies, and exploration and topographical mapping, to name only three major areas, did much to help shape the earth sciences into a discipline. Historians of the earth sciences, Oldroyd says, are largely divided between scientist-historians and historians of geology, the former mostly concerned with internalist issues and the latter with a wider range of social and intellectual issues that bear on scientific developments. Though this distinction is weakening, it continues to shape (and sometimes plague) the field. Yet Oldroyd argues that the fault lies not only with modern historians. Led by geology, the nineteenth-century earth sciences were themselves in what Oldroyd calls a "reconnaissance" mode. They explored and mapped the world (more precisely, central and western Europe and America), greatly increasing the amount of empirical data. Yet they could offer no unifying theory. Oldroyd argues that scholars are not even agreed as to when geology began, other than to point to the late eighteenth, early nineteenth centuries when the work of various miners, surveyors, and engineers, and the theories of James Hutton, Abraham Werner, and Georges Cuvier, quickened interest. The issue of periodization remains particularly vexing with the earth sciences.

Theodore Porter makes a similar point for the history and historiography of the social sciences. Like Oldroyd, Porter sees the social sciences as being in search of order. He argues that, while their roots go back to the late Enlightenment (if not antiquity), their birth as organized disciplines did not occur before about 1890. Yet their identity cannot be reduced to the establishment of university disciplines, since they originated in different intellectual traditions, administrative and political practices, nationalities, and cultures and were closely tied to a wide variety of social issues and discourses. As a result, the social scientific disciplines were diverse and disunified. Porter argues that defining social science became a problematic only around 1890; it is one that today's historians have yet to resolve satisfactorily, not least because many of them continue to view the formation of the social scientific disciplines in the nineteenth century through the "lenses of the modern disciplines." The only historiographic agreement, he argues, is that some sort of "landmark shift" happened "between about 1870 and 1890," a shift associated with such luminaries as Alfred Marshall, Wilhelm Wundt, Sigmund Freud, Emile Durkheim, Max Weber, and Franz Boas. Yet Porter maintains that this shift was as much institutional as intellectual, as much American as European, and was intimately tied to larger socioeconomic, political, and cultural issues. Porter's essay is thus concerned with a variety of issues: "the invention of 'social science' in the early nineteenth century, and its bearing on the contemporaneous reshaping of 'science' as a privileged category"; the retrospective reshaping of nineteenth-century social science into separate disciplinary histories; the linking of the natural and the social sciences, especially biology and statistics; the political setting of social science and its use as a tool of reform and administration; and, finally, the issue of social science's periodization and its setting within a broad swath of socioeconomic, political, and cultural history.

The second theme identifiable in several of these essays concerns the social reconfiguration of the sciences during the nineteenth century. In my own essay, on scientific institutions and communities, I argue that there was no identifiable scientific community before the early nineteenth century. Although a limited number of scientific institutions and specialized societies had emerged by or during the late Enlightenment, modern scholarship shows that it was only after about 1830 that scientific communities began to take meaningful shape and to function on a national and international scale. To help trace the emergence and growth of scientific communities from the late eighteenth century onwards, I review empirical studies of the major scientific institutions and communities of nineteenth-century Britain, France, Germany, and the United States. In particular, I discuss studies on the demography of nineteenth-century scientific institutions and communities, on the (enlarged) role of universities as the setting for much of the scientific enterprise, on new types of nonacademic institutions, and on national academies of science and scientific societies, and I urge that much more work be done on each of these areas. I also argue that there emerged an imagined community of science, which was composed of real, particular individual institutions and social structures that effectively functioned together to constitute the nineteenth-century scientific community.

The sense of belonging to a national or an international scientific community was very much based on scientific disciplines and their specialized associations, and these in turn were due to new or expanded intellectual developments. As several of the contributors to this volume emphasize, there was a dialectical relationship between intellectual developments and the surrounding social and cultural worlds. Dauben, for example, remarks on the rise of mathematics journals and institutions, as well as on the changing roles of women in mathematics and on relations with society. In her essay on the historiography of chemistry, Bernadette Bensaude-Vincent argues for increased attention to the cultural history of science in order to understand scientific creativity and development. (As does Richards, who puts forth a theoretical schema for doing so.) Bensaude-Vincent is explicitly concerned with the problematic presentist historiographical outlook of all too many historians of science, many of whom are professional scientists keenly interested in exploring the history of their own disciplines. She argues that this outlook has led much of the historiography of nineteenth-century chemistry to focus on "identifying the prejudices or obstacles that delayed the acceptance of novelties." While she acknowledges that this approach brings several advantages—for example, it is useful in identifying key discoveries, in providing periodization, in assemblying sourcebooks, and in locating heroes—she argues that it has also led to an uncritical acceptance of "the historiographical categories forged by chemists." Dauben, Richards, Oldroyd, and Porter say much the same thing with regard to their respective disciplines. Bensaude-Vincent further maintains that contemporary historiography needs (and seeks) to revise "the entire issue of the reception or diffusion of scientific innovations." Toward that end, she analyzes the historiographical roles and uses of the Chemical Revolution, biography, chemical atomism, the legends of chemistry, disciplinary identity and boundaries, research schools, professionalization, textbooks, and applied chemistry, among other themes. Throughout her analysis she advocates a "multifaceted cultural approach," one that seeks to critically assess individuals, concepts, practices, events, instrumentation, societies, and all other aspects associated with nineteenth-century chemistry. She argues that there has been too much heroization of chemists and not enough study of their real, daily lives as set within their local and national contexts.

All of these contributors argue that, if we are to appreciate more fully the development of the sciences during the nineteenth century, then historians need to incorporate much more social and cultural history into their studies, both in order to understand the general context in which the sciences as a whole functioned and to understand the more direct bearing that cultural context had on particular individuals, concepts, and disciplines. To move in this direction, we need more biographical studies that ground the development of science in the lives of concrete, individual scientists, examining these lives in a critical way, without either heroizing or diabolizing them, and placing them within their appropriate cultural contexts.

Finally, a third theme addressed by a number of contributors concerns the applications of science or its increased interrelationships with other aspects of culture or society. While developments in medicine, technology, and religion constitute obvious candidates here, modern historiography also points to other interrelationships involving the natural and social sciences.

Medicine and technology, as Michael Hagner and Ulrich Wengenroth argue, respectively, saw the contents and methods of their domains transformed, in part through the introduction of science. But for both this introduction and its meaning are highly problematic. Hagner maintains that while "scientific medicine" became the hallmark of nineteenth-century medicine (as well as its twentieth-century successor), there is no unequivocal understanding as to precisely what that designation meant. His essay explores the wide variety of interpretations that "scientific medicine" has engendered: from the philosophical, the cultural, and the anthropological to perspectives stressing experimental physiology, pathology, and the therapeutic foundations of medicine as science and as therapy, not to mention the medical education of doctors and the scientific work done in hospitals. He points out that while the history of experimental medicine has recently attracted much attention, there has been too little investigation of the different types of experiment in medicine and their significance, a point that Buchwald and Hong also note in their discussion of the historiography of physics. Hagner further finds that the history of medicine as the history of ideas has been largely thrown overboard and that "interest in the relationship between philosophy and medicine seems to have disappeared," a point similarly noted by Bensaude-Vincent in her discussion of the historiography of chemistry. In general, Hagner sees a lack of theory and methodology in the historiography of medicine; above all, he argues, there is no grand narrative in the history of medicine. Finally, like Richards on biology and Bensaude-Vincent on chemistry, Hagner notes the need for further biographical studies of medical practitioners.

Wengenroth, for his part, argues that one of the central questions in the historiography of nineteenth-century technology concerns the sources of innovation: to what extent was it due to science, and just how dependent were science and technology on one another? The old positivist school of the 1950s and 1960s regarded industrial technology as applied science and saw a one-way, linear "science-technology push." Wengenroth reports that contemporary historiography rejects this viewpoint: historians of technology have found little evidence that science played much of a role in the Industrial Revolution, and indeed, the most sophisticated technology was itself of only minor importance to industrial development during the first two-thirds or so of the nineteenth century. Starting in the 1870s, Wengenroth maintains, engineering schools and institutes of technology, especially in Germany, started to make a difference in accelerating industrial change and growth. Academic science and industrial technology began to create a common language, as well as engineering theory, which allowed for "intensified exchanges between the two fields." It was only then, he finds, that "a still quite limited number of science-based industries in chemistry, electrical engineering, optics, and mechanical engineering eventually provided the empirical background for evaluating the contribution of science to technology and industry at the end of the nineteenth century," and helped give the means for the Continent (above all, Germany) and America to catch up with Britain industrially. The resulting relationship, the historiography shows, was dynamically interactive, not mono-causally linear. To describe it, Wengenroth argues for a triangular model connecting science, technology, and industry, with "multidirectional pushes and pulls to and from each of the three elements."

Frederick Gregory argues that there have been essentially five stages of historiographical development in the representation of relations between science and religion during the nineteenth century. In the first, a series of writers prior to World War I used (or abused) that relation in the service of a larger agenda: to promote or criticize an interpretation of an alleged warfare between science and religion. The main issues here concerned the age of the earth and the origins of the solar system and of species. In the second stage (1919–39), the historiography moved beyond "warfare" in an attempt to reintroduce teleology into science. Moreover, scholars began to stress that both "science" and "religion" were ever-changing historical phenomena. The third stage, which developed after World War II, was characterized by broad visions and was due principally to a fuller understanding of the social and religious context of geology's development, the history of evolution (the Darwin industry), and (thanks especially to Thomas Kuhn) the relevance of the social history of science. The fourth stage, in the 1970s and 1980s, was marked by the rise of the social history of the relations of science and religion. "The field," Gregory writes, "became much more open than ever before to investigations of all sorts, especially those that uncovered social attitudes about science that reflected widely shared sentiments." It was now that the Darwin industry began to have its effect on thinking about the relations of science and religion, and it was now that discontent appeared over the notion of a conflict between the two. Indeed, the conflict interpretation now became suspect. By contrast, the fifth and current stage, which Gregory argues began in the 1990s, is characterized by more scholars looking to cultural and biographical studies for enlightenment about the relationship.

The essays in this volume demonstrate that the individual achievements of scholars of the history of nineteenth-century science during the past generation have been extraordinary. Yet they also reveal that the historiography of nineteenth-century science is still very much a work in progress. No single scholar or viewpoint has yet been able to provide a broad and intellectually satisfying account of the dynamics behind the developments of any single science, let alone of the sciences as a whole. Contemporary scholarship has yet to find its Merz, Bernal, or Ben-David. Though some of us remain skeptical of the approaches of these three bold and innovative scholars, none doubts that the attempt to achieve a broad understanding of the transformation of natural philosophy into the sciences and of the general dynamic underlying scientific development during the nineteenth century is a task eminently worth pursuing. It is hoped that the essays in this volume will encourage readers in their own attempts to come to terms with nineteenth-century science.

ACKNOWLEDGMENTS

I thank Susan Abrams, Jean Axelrad Cahan, Frederick Gregory, and Theodore Porter for their helpful comments on earlier versions of this introduction.

Two

BIOLOGY

ROBERT J. RICHARDS

✺

HISTORIANS, the good ones, mark a century by intellectual and social boundaries rather than by the turn of the calendar page. Only through fortuitous accident might occasions of consequence occur at the very beginning of a century. Imaginative historians do tend, however, to invest a date like 1800 with powers that attract events of significance. It is thus both fortunate and condign that "biology" came to linguistic and conceptual birth with the new century. In 1800, Karl Friedrich Burdach, a romantic naturalist, suggested that his coinage *Biologie* be used to indicate the study of human beings from a morphological, physiological, and psychological perspective.[1] Many other neologisms of the period (and Burdach issued quite a few) were stillborn or survived only for a short while. *Biologie,* though, fit the time, and with slight adjustment received its modern meaning two years later at the hands of the *Naturphilosoph* Gottfried Reinhold Treviranus. In his multivolume treatise *Biologie, oder Philosophie der lebenden Natur* (1802–22), Treviranus announced: "The objects of our research will be the different forms and manifestations of life, the conditions and laws under which these phenomena occur, and the causes through which they have been effected. The science that concerns itself with these objects we will indicate by the name biology *[Biologie]* or the doctrine of life *[Lebenslehre]*."[2] Jean-Baptiste de Lamarck, also in 1802, employed the term with comparable intention.[3] In the work of both of these biologists, the word became immediately associated with the theory of the transmutation of species—a new term in recognition of the new laws of life.

1. Though the term "biology" came to bear its familiar meaning at the beginning of the nineteenth century, it had a long gestation. Michael Christoph Hanov, a disciple of Christian Wolff, used the word in the title of his *Philosophiae naturalis* (1766). In this volume, biology seems to have designated zoology. (I am grateful to Peter McLaughlin for this reference.) In 1797, Theodor Georg Roose employed the word once in a book on *Lebenskraft.* In the preface, he referred to his own tract as a "sketch of a biology." See Roose, *Grundzüge der Lehre von der Lebenskraft,* 1. For Burdach's usage, see his *Propädeutik zum Studium der gesammten Heilkunst,* 62. The linguistic priorities are discussed in Jahn et al., *Geschichte der Biologie,* 283–89.

2. Treviranus, *Biologie,* 1:4.

3. Lamarck first used the term *Biologie* in his *Hydrogéologie* (1802) to distinguish that part of terrestrial physics dealing with living creatures. See Grassé, "'La Biologie.'"

16

Treviranus thought the progressive deposition of fossils evinced a modification of species over time. And Lamarck, in the very year of 1800, declared, in his "Discours d'Ouverture," that diverse environmental influences would cause creatures to adopt new habits that could alter anatomical parts, which themselves would become heritable, thus progressively modifying species. Biology, as it came to birth at the beginning of the nineteenth century, had evolutionary theory within its genetic depths. After midcentury, of course, biological study would explode, with a prodigious outpouring of evolutionary and counterevolutionary literature. Though the history of science exhibits no radical discontinuities (of the sort Foucault or Kuhn have imagined), evolutionary theory did quickly form into an enormous and powerful force, disrupting everything within its conceptual territory. This surge of evolutionary thought has endlessly fascinated historians of the nineteenth century, and they have devoted more pages to its study than to any other subject falling under the rubric of biology.

Between 1795 and 1800, the German Romantic movement took shape through the literary, philosophic, and scientific efforts of a select band of individuals resident in and around Jena, that small university outpost near Weimar. Its developmental ideal of *Bildung* (formation), which organized thought in biology, literature, and personal culture, readied the soil in Germany for the reception of evolutionary seeds blown over from France in the early part of the nineteenth century and the more fruitful germinations from England in the later years. The conceptual ground for the Romantic movement was prepared by the literary and historical researches of the brothers Friedrich and Wilhelm Schlegel, by the poetry and iconic personality of Friedrich von Hardenberg (Novalis), by the idealistic philosophy and personal magnetism of Friedrich Schelling, and by the dynamic art and science of Johann Wolfgang von Goethe. In 1797, Schelling's *Ideen zu einer Philosophie der Natur* appeared (giving the name *Naturphilosophie* its particular contours), and then in rapid succession his *Von der Weltseele* (1798) and *System der transscendentalen Idealismus* (1800). These books provided philosophical guidance for numerous works of biological importance that would penetrate far into the decades of the new century—for instance, Goethe's own collection of tracts, *Zur Morphologie* (1817–24), the many studies in physiology and zoology of Treviranus and Johann Christian Reil, and the morphological researches of Burdach, Lorenz Oken, Carl Gustav Carus, and ultimately Richard Owen. The Romantic movement also gave focus to the scientific vision of Alexander von Humboldt, who rashly but systematically conducted the kind of autoexperimentation in electrophysiology that would insinuate the self into the biology of the new century. In 1799, Humboldt sailed for the Americas, where he would spend five years exploring the geological and biological features of the New World and, not incidentally, creating a scientific persona that would come to epitomize, for the first half of the nineteenth century, the natural-scientific researcher. Humboldt recounted his extraordinary journey in a multivolume work, *Travels to the Equinoctial Regions of the New Continent* (1818–29).[4] The book inspired Charles Darwin and Ernst Haeckel to embark on comparable voyages of adventure and research. The conceptual, moral, and aesthetic tides of the Romantic movement would wash through the century, cresting in the evolutionary theories of Darwin and Haeckel.

The conceit that the nineteenth century was "Darwin's century" carries more significance than the immediately obvious. It also portends an alteration in historiographic practice. In *On the Origin of Species* (1859), Darwin proposed that the study of living nature would assume a new meaning when undertaken from a historical vantage:

> When we no longer look at an organic being as a savage looks at a ship, as at something wholly beyond his comprehension; when we regard every production of nature as one which has had a history . . . how far more interesting, I speak from experience, will the study of natural history become.[5]

The same might be said of the history of biology. The practice of the historiography of science, and that of biology in particular, has gradually moved from a concentration on the logical skeleton of theory—say, in the still quite useful *History of Biology* (1920–24) by Erik Nordenskiöld or in the more recent and even more useful *Geschichte der Biologie*, largely by Ilse Jahn (1998)[6]—to an examination of the full, fleshy creature. This has happened when more austere intellectual history has recovered its cultural context, when the theories that Darwin, Mendel, Haeckel, Galton, and Pasteur advanced have been understood as the products of multiple forces operative on the minds and hearts of such scientists. For the historian, this requires an imaginative and empirically inspired return to the past to catch the now dead theories when they were full of life. Sometimes, of course, the resources for recovering that context are meager and the best the historian can do is lay out the skeleton. But the full pleasures of the dance with the past can only be had when those con-

4. The volumes were originally published in French, with German and English translations quickly following. Darwin took the first two volumes with him on the *Beagle*, and he acquired others when he returned. See Humboldt and Bonpland, *Personal Narrative*. A convenient edition is the new German translation: Humboldt, *Reise in die Äquinoktial-Gegenden des Neuen Kontinents*.

5. Darwin, *On the Origin of Species*, 485–86.

6. Jahn et al., *Geschichte der Biologie*, is a magnificent general history of biology, with authoritative thematic essays addressing particular time periods and a collection of some 1,600 brief biographies. The essays on the nineteenth century treat of such topics as biology in the Goethe era, *Naturphilosophie*, botany, zoology, Darwin and evolutionary theory, theories of heredity, methodology, institutions, and developmental and comparative physiology. Nordenskiöld, *History of Biology*, includes brief intellectual biographies of the biologists considered, with emphasis on their major contributions.

structive forces have been reconstituted so that the companion offers a lively step and a knowing smile.

The works of the historians I will discuss rarely meet the ideal of a fully reconstructed cultural history of biology. Some provide merely the bare bones of a theory and neglect its author, who becomes only a name for a given set of ideas. Others produce a flabby creature that lacks the stiff structures of science—much about politics and social status, little about the hard elements of biological theory and practice. Some few historians do, however, articulate the bones, presenting the remains of past biology in vivid poses with the imaginative skill of the artist, making it spring to life once again. I will have more to say about the ideal of cultural history of biology in the last section of this essay.

In discussing nineteenth-century historiography of biology, I cite articles only occasionally, since their number is uncountable and space is finite. The medium of expression for most historians has been the extended monograph, and that genre certainly has had the principal role in shaping the field. I have chosen books that I believe have been of major importance and added a few others for contrast. Evolutionary theory has been the obsession of the discipline, so the largest fraction of works I will discuss reflects that concentration. Evolutionary biology, then, will be my starting point. Thereafter, I turn to social Darwinism and evolutionary ethics (section 2), biology and religion (section 3), biology and literature (section 4), morphology and romantic biology (section 5), neurophysiology (section 6), genetics and cell theory (section 7), and biography (section 8). In the last section of this essay, I sketch two contrasting modes in history of biology, intellectual history and cultural history.

1. Evolutionary Biology

During the last four decades studies in the history of nineteenth-century biology have proliferated, moving the history of science community beyond the bounds established by histories of the physical sciences, which dominated the previous half century. The occasion for the transformation was the celebration, in 1959, of the centenary of Darwin's *Origin of Species*. The commemoration stimulated the publication of several books whose oppositional considerations suggested a quite unsettled view of the scientific status of evolutionary theory and its underlying metaphysics and thereby made poignant the very nature of scientific theory itself. Loren Eiseley, in his *Darwin's Century* (1958), presented, in highly literate and sometimes elegant prose, the character of Darwin's accomplishment, the reactions of his contemporaries, and the prospects for the future. In that latter consideration, Eiseley seemed to take away what he had so felicitously offered in the first part of his book: he attempted to free human beings (by historical argument) from the biological determinism assumed by Darwin's theory. Elaborating some considerations of Alfred Russel Wallace, and unaware of the latter's dalliance with spiritualism, Eiseley declared: "The mind of man, by indetermination, by the power of choice and cultural communication, by the great powers of thought, is on the verge of escape from the blind control of that deterministic world with which the Darwinists had unconsciously shackled man."[7]

Eiseley's history gave vent to his distrust of the underlying metaphysics of Darwin's theory; and so he found in all corners of the Englishman's science subtle deficiencies, rough edges, and misaligned ideas that indicated the whole would eventually come clanking and sputtering to a halt. In a subsequent volume, *Darwin and the Mysterious Mr. X* (1979), Eiseley, upon due reflection, had decided that Darwin did not even deserve the attributions of originality initially conceded to him. Eiseley now maintained that Edward Blyth, an obscure naturalist, had formulated the fundamental Darwinian concepts—variation, struggle for existence, natural and sexual selection—already in 1835, and that Darwin had tacitly appropriated them as his own. John Greene, while not suggesting the kind of fraud that Eiseley eventually did, nonetheless maintained that Darwin's fundamental ideas had been anticipated by an obscure physician, William Wells, in 1818. Greene's *Death of Adam* (1959) dissolved Darwin's genius into the musings of his predecessors. Greene likewise found the metaphysics of Darwinism distasteful, as he later made clear in his *Science, Ideology, and World View* (1981).

The attitudes of Eiseley and Greene found their complement in the work of Gertrude Himmelfarb. In her compelling, if irritating, study *Darwin and the Darwinian Revolution* (1959), she argued that the scientific core of Darwin's theory sank into confusion, while the dogmatic shell might be retrieved by sly Marxists. In this respect, she prophesied correctly if obliquely, since Marxist historians, of considerably more benign character than this Cold Warrior envisioned, have seized upon evolutionary theory as a subject for social analysis. Stephen Jay Gould, Robert Young, and Adrian Desmond, whose works I will more thoroughly discuss below, have each detected varying aspects of the theory to be generated by political and social assumptions. The studies of Eiseley and Himmelfarb gained force when the philosopher Karl Popper (1974) set his own small bomb under Darwin's theory. He argued that because natural selection could not predict new variations and new species, it could not, for reasons of logical symmetry, explain their origin. Further, he construed the theory as simply a tautology: the fit survive, and we know they are fit because they survive. The theory, he concluded, failed as science but thrived happily as metaphysics.[8]

7. Eiseley, *Darwin's Century*, 350.

8. Popper, "Darwinism as a Metaphysical Research Programme." See also Ruse, "Karl Popper and Evolutionary Biology."

These initial studies of the origins of evolutionary thought brought a counterreaction from historically minded biologists such as Ernst Mayr. Mayr began writing historical essays during the 1960s and 1970s; his work culminated in the 1982 book *The Growth of Biological Thought,* two-thirds of whose almost one thousand pages he devoted to evolution and genetics. His history fashioned Darwin into the very model of the biological scientist, and its trajectory had a definite end, namely the vindication of Darwinian theory against the likes of Eiseley, Greene, Himmelfarb, and Popper. The model of the proper evolutionist, though, ill-suited Herbert Spencer—at least, in Mayr's estimation. He devoted only three paragraphs to Spencer, who, after Darwin, was certainly the most influential nineteenth-century English evolutionist. Mayr thought "it would be quite justifiable to ignore Spencer totally in a history of biological ideas because his positive contributions were nil."[9] This attitude, needless to say, poorly comported with that of the younger, professionally trained historians whose interests became trapped in the tangle of evolution, politics, and social relationships. Like Brer Rabbit, they loved the brier patch, where the likes of Spencer could be found. But alas, poor Spencer still awaits the monograph that will show exactly what it was about his philosophy and science that captivated intellects of power and influence during the late nineteenth century.[10]

Another scientist turned historian who began writing in the wake of the Darwinian centennial is Michael Ghiselin. His *Triumph of the Darwinian Method* (1969) provided a literate public, especially scientists, with a general introduction to Darwin's thought. But Ghiselin also found in Darwin's work those singular features that raised it above even very clever science, something that anointed Darwin's ideas as scientific touchstones, contact with which could reveal the many other claims made by biologists to be gold or dross. When that special aspect of Darwin's thought was revealed, however, the expectant reader met disappointment. In Ghiselin's estimation, what made Darwin's method triumphant turned out to be its putative hypothetico-deductive character. In other words, Darwin's method was just what the logical empiricists took to be the technique of all good science, and Darwinian theory was, after all, good science—therefore, hypothetico-deductive. This was a Darwin the logical empiricists could learn to love.

9. Mayr, *Growth of Biological Thought,* 386. Mayr has been often charged with being whiggish in his history because of its orthogenetic trajectory. In "When Is Historiography Whiggish?" he defends himself, arguing that a "developmental history" of evolutionary thought might selectively ignore those who made no contribution to contemporary theory. Mayr forgets, however, that consequent conceptual development is shaped by the intellectual environment, even when that environment contains the thought of figures who make "no contribution" to subsequently triumphant theory.

10. J. D. Y. Peel's *Herbert Spencer: The Evolution of a Sociologist* is helpful but does not consider the breath of Spencer's work or the depth of his achievement, especially in biology.

David Hull and Michael Ruse, two leading philosophers of biology, have, from the very beginnings of their careers, made special study of the history of evolutionary theory.[11] In *Darwin and His Critics* (1973), Hull collected early contemporary reviews of the *Origin of Species*—those of J. D. Hooker, Adam Sedgwick, Richard Owen, and others. He prefaced the collection with a series of essays that treated topics relevant to evolutionary controversies (e.g., inductive method, occult qualities, teleology, essences). Like Ghiselin, Hull strove to make Darwin's thought look respectable to logical-empiricist eyes (though in more recent work, Hull has thrown sand in those very eyes).[12] Ruse had a similar goal, pursing it through such books as *The Darwinian Revolution* (1979), *Taking Darwin Seriously* (1986), *Evolutionary Naturalism* (1995), and *Monad to Man* (1996). In this last, Ruse surveyed the development of evolutionary thought in the nineteenth and twentieth centuries, anchoring that thought in Darwin's accomplishment. The book, rich from archival digging, posed several interesting questions, of which two stand out: Does Darwin's theory intrinsically imply biological progress? and What was the professional status of the theory prior to the synthesis of evolution and genetics during the 1930s and 1940s? Ruse handled these questions deftly and almost persuasively. He argued that notions of progress clung to Darwin's theory like barnacles to a ship—inevitable attachments if one plied the waters of the mid-nineteenth century but eliminable with enough analytical scraping. He also maintained that, because of such accretions, scientists like Thomas Henry Huxley might take evolutionary theory out on a pleasure cruise, as something to entertain the masses, but would never seek to introduce it into professional work. Ruse improbably concluded that evolutionary theory did not become a respectable scientific subject in the professional literature (at least in the English-speaking world) until the synthesis of evolutionary theory and genetics undertaken by J. B. S. Haldane, R. A. Fisher, and Sewall Wright during the 1930s and 1940s.

In the mid- to late 1970s, the scholarship on Darwin changed decidedly. Historians began making pilgrimages to Cambridge, where a huge trove of unpublished manuscripts and letters lay buried beneath vague catalog titles (e.g., in the "Black Box"). Historians looked first to the material pertaining to the young Darwin and the formulation of his fundamental ideas. Here was empirical work that would help settle, among other questions, Darwin's originality. Howard Gruber used Darwin's early notebooks, especially those devoted to questions of human evolution, to uncover the particular nature of Darwin's genius. Gruber's *Darwin on Man* (1974) brought Piagetian psychology to the reconstruction of Darwin's theory of species change and, most in-

---

11. Hull and Ruse are also the editors of *The Philosophy of Biology*, the most important and comprehensive collection of philosophical articles dealing with aspects of evolutionary theory.

12. See, e.g., Hull, *Science as a Process*.

terestingly, the evolution of human moral and intellectual traits. Gruber conceived the field of Darwin's genius not as flashing with the brilliance that Huxley manifested, but more as a landscape slowly evolving, one in which underground forces inexorably push up towering mountains with dramatic vistas. Indeed, not Piagetian stages but Lyellian gradualism served as the implicit model.

Edward Manier also sustained claims to Darwin's inventiveness but in relation to another stratum of thought, the philosophical. Manier explored Darwin's "virtual" interactions with a group of social, political, and philosophical writers whom he dubbed Darwin's "cultural circle." *The Young Darwin and His Cultural Circle* (1978) depicted Darwin in dialogue, via their books and papers, with the likes of Charles Lyell, Jean-Baptiste de Lamarck, James Mackintosh, William Whewell, Thomas Malthus, David Hume, Dugald Stewart, and others of somewhat narrower fame. The young, philosophically curious naturalist held up his end of the conversation by annotating their books and jotting reactions in his notebooks. Manier, using this archival material, argued that Darwin's creativity lay both in slowly formulating synthetic notions out of the metaphysical and ethical ideas of his circle and in the ways he wove those notions through his biological theories. These philosophical threads, as Manier perceptively observed, were not fashioned from brittle materialism and mechanism, but from a supple Scottish realism and Wordsworthian romanticism. Moreover, the argumentative structure of Darwin's early evolutionary theory could hardly be called hypothetico-deductive, since virtually nothing of the theory could be tested in the way that scheme demanded. Manier offered an important corrective to the accounts of those historians and philosophers reading Darwin's ideas off the surface of the *Origin of Species*.

During this same period, historians David Kohn and Dov Ospovat, whose arguments would set the agenda for much subsequent scholarship, also made generous and insightful use of Darwin's unpublished papers and letters. Kohn argued in his "Theories to Work By" (1980) that Darwin inched his way toward natural selection, trying out a variety of hypotheses for the production of transformations, each of which he then dropped into his toolbox of auxiliary aids. Darwin's work on these initial hypotheses, according to Kohn, prepared him to see the significance of Malthus's observation about the tremendous reproductive capacities of organisms, namely, that with many more organisms produced than could survive, those that by chance had some advantage within their particular environments would, in competition with others, be more likely to reach reproductive age and pass on their advantageous traits, which had gradually to transform species.

Ospovat, in his *Development of Darwin's Theory* (1981), constructed and judiciously sustained an important thesis regarding Darwin's development. He maintained that Darwin, throughout his early notebooks and essays (1837–44), retained an assumption from his days as a student of William Paley. Paley and other natural theologians had held that biological adaptations were perfect, since authored by an all-perfect Creator. Darwin, as Ospovat pictured him, substituted evolutionary processes for the hand of the divine but still thought of the products as being perfect. During the early 1850s, though, Darwin began to adjust his theory, finally giving up the idea of perfect adaptations. Manier, Kohn, and Ospovat indicated by their histories what real advantage could be gained in understanding the origins and significance of Darwin's thought by examining his unpublished papers and letters.

Just as the anniversary of the *Origin of Species* stimulated the first wave of Darwinian scholarship, so the anniversary of Darwin's death produced a second, but more historiographically forceful response. The build-up began in 1982 with a conference at the Florence Center for the History and Philosophy of Science and crested with the 1985 publication of *The Darwinian Heritage*, edited by Kohn, which included the papers presented there as well as many others. Virtually all the major Darwinian scholars of the time contributed to the volume, which set many of the questions for subsequent studies, not only in the history of evolutionary theory but also in its philosophy. Gruber and Silvan Schweber depicted the immediate and wider context of Darwin's theorizing; Phillip Sloan examined Darwin's very early work on invertebrates, with telling results; Frank Sulloway subjected Darwin's *Beagle* writings to the kind of statistical content analysis that would later prove an obsession for him; Kohn and Jonathan Hodge focused on the notebooks and pieced together the immediate origins of the theory; and Malcolm Kottler followed the debates between Darwin and Wallace over the origin of interspecies sterility and the nature of sexual selection. Peter Bowler, Paul Weindling, and others provided comparative analyses of the reception of Darwinism in different national cultures. The more philosophical aspects of Darwin's ideas were considered by John Beatty, Elliott Sober, and Hull. Finally, Kottler constructed a comprehensive bibliography. The book demonstrated to the larger community how history of science might become more deeply satisfying—for the historian, almost sanctifying—when augmented by hard archival work. It became one of the great products of the "Darwin industry," which has been belching smoke ever since.

Not every historian of Darwinism has the sooty look of a laborer in the archives. Peter Bowler has kept his Irish tweeds neat, confining his efforts to published works. He must, though, be reckoned one of the captains of the Darwin industry, so great has been his output and influence. He has probably devoted more pages than any other historian to the reconstruction of Darwin's theory and those of the many other evolutionists writing in the wake of the *Origin of Species*. In *Evolution: The History of an Idea* (1984), *The Eclipse of*

Darwinism (1983), *The Non-Darwinian Revolution* (1988), and other books Bowler has brought to his reconstructions a particular thesis, which he shares with Ruse, Gould, and Mayr. He has insisted that natural selection forms the essence of Darwinism.[13] Thus we should not mistake, for example, Haeckel's work as furthering the Darwinian revolution, since it gave minimal attention to natural selection. Bowler is quite convinced that Haeckel's use of Lamarckian notions, his progressivism, and his theory of recapitulation completely separated his biology from Darwin's. This tendency among historians like Bowler (as well as Ruse and Gould) to distance Haeckel from the Darwinian tradition derives, I suspect, more from Haeckel's perceived connections with the Nazis than from the theoretical markers of Lamarckism, progressivism, and recapitulation. I have argued in *The Meaning of Evolution* (1992) that the heart of Darwin's theory, from its inception through its mature development, beat precisely to progressivist and recapitulationist rhythms. It is less controversial to observe, though not often remarked, that Darwin always allowed for acquired characteristics to be inherited and to serve as traits upon which natural selection might operate. Bowler's interesting and wide-ranging studies have, by contrast, striven to preserve Darwin as an authentic scientist (by our lights), one untainted with the kind of scientific and moral corruptions associated with someone like Haeckel.

Darwin, unique among figures in the history of nineteenth-century science, has drawn scientists to the history of their discipline. The best-known of those so drawn is Stephen Jay Gould, who started writing on the history of biology in the early 1970s. His succulently readable essays began appearing in the magazine *Natural History* and have been subsequently collected in a number of plump little volumes, beginning with *Ever Since Darwin* (1977) and trailing off into books bearing rather more kitschy titles: *The Flamingo's Smile* (1985), *Bully for Brontosaurus* (1991), *Eight Little Piggies* (1993), *Dinosaur in a Haystack* (1995), and *Leonardo's Mountain of Clams and the Diet of Worms* (1999). Gould has been prodigiously productive; his many other books dealing with aspects of the history of evolution, genetics, and geology include *Ontogeny and Phylogeny* (1977), *The Mismeasure of Man* (1981), *Time's Arrow, Time's Cycle* (1987), and *Wonderful Life* (1989). In most of these, Gould has found a fascinating hook that quickly lands the reader. And while professional historians might cavil at some of his claims, they are undoubtedly beneficiaries of the large readership he has cultivated for the history of biology. Gould writes, as he himself has frequently avowed, as a Marxist historian. His Marxism,

13. Jean Gayon shares the view that natural selection virtually defines Darwinism. In the best philosophical study to date, *Darwinism's Struggle for Survival*, he traces the fortunes of the idea of natural selection from Darwin through the modern synthesis to Richard Lewontin and Mottoo Kimura.

though, is decently attired in a J. Press work shirt. He exposes the ideological taints—usually racial and class biases—that tincture much of nineteenth- and twentieth-century biology; but he knows that good science, at its heart, remains pure. From his historical analyses, he catalogs (rather often) what he takes to be the lessons of evolutionary history: that evolution is contingent and nonprogressive; that human beings form one species; that we should not assume given traits have arisen as specific adaptations; and that, consequently, human mental activity betrays no lingering inclinations, acquired from our Pleistocene ancestors, toward specific behaviors. Ironically, such a list of purity rules would cast Darwin himself from the ranks of the saved.

Ghiselin, Mayr, Ruse, Bowler, and Gould represent one strong interpretative wing of Darwinian studies. In their hands Darwin's theory has been molded to late-twentieth-century specifications. They implicitly regard scientific theories as abstract entities that can be differently instantiated in the nineteenth century or today, while exhibiting the same essential features. For instance, Ruse, Bowler, and Gould have consistently interpreted Darwin's theory as having a logic and evidentiary base that renders evolutionary progress impossible. Lamarckians and Spencerians might have fabricated theories of progressive evolution (which sanctioned racist ideologies), but Darwin, they believe, rejected the idea of biological progress—or at least his theory did. Darwin the historical individual might actually have succumbed to the idea of progress, but somehow he constructed a theory that remained scientifically pure—that is, as pure as we like them today. By separating Darwin's actual words and logic, as expressed in the *Origin of Species* and the *Descent of Man,* from the theory as an abstract entity, these writers have attempted to keep the essences of Darwinism inviolate, free from what they take to be nonscientific corruption.[14]

Another wing of Darwinian scholarship, the left wing, has understood Darwin's theory, as well as earlier and later evolutionary constructions, to be saturated with social and political features, stains that sink right to the core of Darwinian thought.[15] This is actually a quite traditional way to understand Darwinism. In the 1920s Erik Nordenskiöld argued, in his comprehensive *History of Biology,* that Darwin's romantic conceptions had been scientifically refuted long ago, even though they retained cultural momentum. Political

14. I have discussed the misadventures of philosophers and scientists who hold Darwin up to the mirror and see a fellow bedecked in Land's End apparel, ready to travel to Woods Hole in the spring. See my "The Epistemology of Historical Interpretation: Progressivity and Recapitulation in Darwin's Theory."

15. As suggested above, Gould professes a moderate Marxism, arguing that the pith of Darwinian thought remains objectively valid, while the skin (e.g., notions of progress) yielded to the kind of social and political influences to which Darwin, as an upper-middle-class Whig, was subject.

liberalism, he urged, had initially given evolutionary theory its push: "From the beginning Darwin's theory was an obvious ally to liberalism; it was at once a means of elevating the doctrine of free competition, which has been one of the most vital corner-stones of the movement of progress, to the rank of natural law, and similarly the leading principle of liberalism, progress, was confirmed by the new theory."[16] This political interpretation of evolutionary theory has been sustained more recently by such historians as Robert Young and Adrian Desmond. Young published a series of essays, beginning in the late 1960s and continuing through the early 1980s, that attempted to place the development of Darwin's ideas, as well as those of Spencer and other early evolutionists, in a common intellectual and social context. The earlier of these essays, several of which he collected in *Darwin's Metaphor* (1985), pulled from a variety of printed sources—the *Westminster Review,* the *Edinburgh Review,* the *Quarterly Review,* and other Victorian periodicals, as well as from monographs in political economy and social theory. These essays were measured and convincing, complex and insightful, unlike Young's later essays, which sometimes became mired in a creed that removed them from the category of history to that of polemic. In an essay published in *The Darwinian Heritage,* Young urged that

> once it is granted natural and theological conceptions are, in significant ways, projections of social ones, then important aspects of all of the Darwinian debate are social ones, and the distinction between Darwinism and Social Darwinism is one of level and scope, not of what is social and what is asocial. ... The point I'm making is that biological ideas have to be seen as constituted by, evoked by, and following an agenda set by, larger social forces that determine the tempo, the mode, the mood, and the meaning of nature.[17]

If biological theory has its fundamental meanings determined by social beliefs, by ideology, then the question becomes which ideology do you prefer, not which theory is the most coherent and supported by the most evidence. Young became fully persuaded that recent applications of evolutionary theory to explore animal and human social traits, especially in the work of sociobiologists, were driven by a pernicious ideology; his later essays fire the claxon to sound the appropriate warnings. Despite the polemical cast of Young's recent work (or maybe because of it), he remains a thinker of power and urgency.

Adrian Desmond shared Young's sense of the interpenetration of the social and the scientific, and portrayed that sense in a series of monographs written with verve and panache. In *Archetypes and Ancestors* (1985), he examined the Huxley-Owen debates and detected beneath the scientific surface,

16. Nordenskiöld, *History of Biology,* 477.
17. Young, "Darwinism Is Social," 610, 622. See also Young, "Malthus and the Evolutionists."

scarred as it was by considerable acrimony, an ideological divide separating the rising professionals of strong materialistic bent from the establishment and church-supported idealists. Desmond continued his digging into the substructure of scientific debate in *The Politics of Evolution* (1989), which attempted to show how radical reformers in the London medical scene, during the early part of the nineteenth century, had embraced evolutionary theory as part of an effort to overcome the elite political and scientific hegemony of orthodox institutions, such as the Royal College of Surgeons. He argued that this context made Darwin hesitate to publish a theory that had been associated with "dissenting or atheistic lowlife, with activists campaigning against the 'fornicating' Church, with teachers in court for their politics, with men who despised the 'political archbishops' and their corporation 'toads.'" Desmond imagined that Darwin, who sought to articulate "a Malthusian science for the rising industrial-professional middle classes," was frozen with political fear and so delayed publication of his work for some twenty years.[18] Desmond and Young thus interpreted evolutionary science as deriving its force and danger from the political message ticking away in the carved hollow of its theory.

After the Darwin centennial, interest spilled over—though only slightly—to other evolutionists, especially to his predecessors Jean-Baptiste de Lamarck and Robert Chambers. Lamarck had been pictured either as merely a precursor of Darwin or as one whose theory of evolution, in the words of Charles Coulston Gillispie, "belongs to the contracting and self-defeating history of subjective science."[19] Richard Burkhardt rectified these contorted approaches by a thoroughgoing contextualization of Lamarck's thought in his *The Spirit of System: Lamarck and Evolutionary Biology* (1977). Burkhardt showed how Lamarck's evolutionary ideas fit into his other concerns, in physics, chemistry, meteorology, and geology. By the standards of his time, Lamarck proved no more subjective in his science than his opponents. Burkhardt set a high standard for other historians—he had consulted the important archives, he wrote clearly and forcefully, and he was interested in Lamarck's theory for its own sake, judging it by relevant criteria. He also sought to answer the historian's usual kind of question: How did Lamarck arrive at his evolutionary theory? Answer: As a conchologist aware of the similarities between fossil shells and those of living species, and as a geologist seeking to avoid Cuverian theories of geological catastrophe and animal extinction, he sought refuge in the idea of a gradual mutability of species over long periods of time. Pietro Corsi, in his finely wrought *The Age of Lamarck* (1989), broadened the contextualizing of Lamarck's theory by considering in detail those ideas of his antagonists

18. Desmond, *Politics of Evolution*, 413, 410.
19. Gillispie, "Lamarck and Darwin," 286.

(e.g., Cuvier) and friendly associates (e.g., Geoffroy Saint-Hilaire and Bory de Saint-Vincent). In the new edition of his book, *Lamarck: Genèse et enjeux du transformisme, 1770–1830* (2001), Corsi adds a list of students who attended Lamarck's lectures and includes extracts from his notebooks—a considerable resource for those interested in transformation theory before Darwin.

Robert Chambers's anonymously published *Vestiges of the Natural History of Creation* (1844) set the Victorian community in some modest ferment. Its lively and imaginative style captured the fancy of a large reading public; but its amateur science enraged the likes of Adam Sedgwick and Thomas Henry Huxley (before the latter became a convert to evolutionary theory) and led Darwin to fear that it had poisoned the waters for any transmutational proposals. James Secord has done more than any recent historian to bring Chambers's work back into view. He edited a reprint (1994) of the first edition of the *Vestiges* along with its sequel, and traced the reactions of various segments of British society to the book in his expansive volume *Victorian Sensation* (2001). Secord's history displays great scholarly force and subversive intent. He presumes that the entire meaning of Chambers's book was furnished by its particular readers, as if the author's conceptions were sands reshaped by the tides of readers' political, social, and religious concerns. This interpretive principle allows him to propose that Darwin's accomplishment has no more claim on genius than Chambers's. Secord thus turns the historical advantage to the underlaborers in science, such as Chambers—a move consonant with the constructionist movement still alive in history of science, if gasping for breath. Like the work of Young and Desmond, that of Secord cannot be ignored, even if its epistemological assumptions collapse under the weight of his own authorial acumen.

My own study of Darwinism—*Darwin and the Emergence of Evolutionary Theories of Mind and Behavior* (1987)—initially focused on the impact of evolutionary considerations on the understanding of psychological life, from animal instinct to human mind and moral behavior. I examined Darwin's evolutionary constructions of instinct, mind, and morality, and then traced their operations in the works of a range of biologists and psychologists— Herbert Spencer, George Romanes, Conway Lloyd Morgan, James Mark Baldwin, and William James, through to John B. Watson, Konrad Lorenz, and Edward O. Wilson. I argued that the earlier studies of Young and Desmond, which certainly made vividly compelling reading, were shackled to the a priori conviction that science must be a surrogate for politics and social philosophy. I attempted what I thought a more empirical strategy, focusing on individual scientists—their education, experience, and psychological dispositions—to discover whence their scientific theories emerged and what the several forces that shaped their thought might be. The forces were usually multiple, with, perhaps, a more powerful impetus coming now from the political side, now from the philosophical, now from the religious, and always against the inertia of the scientific. Desmond and Young usually examined the external context of ideas first, then moved inward to characterize the mind of the scientist. And since they began with the a priori assumption that the political-social context was the most important, they came easily to regard the science as stuck in politics like a fly in molasses—and they themselves had a great taste for molasses. If one began, however, with the individual mind—working out the formative experiences, examining the books read, assessing the interests that moved the soul—then one might more adequately determine what features of the external environment had the most purchase on the scientist. Darwin may have grown up in a political and social context of individualistic utilitarianism, but his biology of moral behavior turned out to be authentically altruistic and expressly antiutilitarian. One simply could not predict in advance what the most powerful forces shaping the science might be. By focusing on the application of biological theory to mind and behavior, many of those various currents could be seen welling to large and more identifiable proportions.

2. Social Darwinism and Evolutionary Ethics

The rise of sociobiology, following in the turmoil created by Edward O. Wilson's *Sociobiology* (1975), has undoubtedly served as stimulus for many of the new books devoted to social Darwinism, especially in the different national contexts of Britain, France, Germany, and America.[20] The connection between *Darwinismus* and Nazism has often been served up as a cautionary tale, with Haeckel as the favorite target for making the moral explicit. In recent works, Haeckel has been frequently indicted for the supposed transmutation of evolutionary theory into something quite sinister. Jürgen Sandmann's *Der Bruch mit der humanitären Tradition* (1990) argues that Darwin's and Haeckel's biologizing of ethics, grounding ethical behavior in animal instinct, had undermined the philosophical presumption that human beings were capable of moral, free choice. This naturalizing of the ethical bond had the effect, so Sandmann believes, of justifying the racial-hygienic measures of the Third Reich. The historical inquisition of Haeckel began with Daniel Gasman's *Scientific Origins of National Socialism* in 1971. Gould echoed Gasman's argument in *Ontogeny and Phylogeny* (1977), and Gasman himself reprised it in *Haeckel's Monism and the Birth of Fascist Ideology* (1998). Haeckel and the rise of Nazi

---

20. See the following excellent studies: Jones, *Social Darwinism and English Thought;* L. Clark, *Social Darwinism in France;* Bannister, *Social Darwinism: Science and Myth in Anglo-American Social Thought;* Degler, *In Search of Human Nature;* Pittenger, *American Socialists and Evolutionary Thought;* and Weikart, *Socialist Darwinism.*

biology have received much more nuanced, complex, and ultimately more satisfying treatment in Paul Weindling's comprehensive and extensively documented study *Health, Race and German Politics between National Unification and Nazism, 1870–1945* (1989). Weindling finds traces of Haeckelian monism in the thought of both Nazi biologists and their opponents. He points out, for instance, that it was precisely the more scientifically grounded ethics of Haeckel that gave comfort to those Weimar physicians who espoused greater sexual freedom for women and homosexuals. Nazi theories of racial hygiene had many more sources than Haeckel, whose views could be recruited by conservatives and liberals alike.

The historical interpretation of evolutionary theory in a hard political and moral light, with Haeckel garishly illuminated, raises three questions that authors like Sandmann, Gasman, and Gould usually do not tackle. First, since Haeckel died in 1919, in what sense can the rise of Nazi science be morally attributed to him? Second, and more generally, by what historiographic and moral criteria can we assign guilt to historical figures? And finally, did the very structure of evolutionary theory lead to reasonable expectations of considerable differences in character among human groups? Virtually every nineteenth-century biologist harbored, by our twenty-first-century lights, racist beliefs. In the cases of Darwin, Spencer, and Haeckel, one has to ask what were the standards of their time and place, and how deviant—if at all— were their opinions? This does not settle the question, of course, since whole groups can indulge in immoral behavior. We must further ask: Could the scientists have believed differently, or was the presumption of the inferiority of other ethnic groups too entrenched to be reasonably overcome? Since virtually every evolutionist of the last century (and well into the twentieth) conceived evolutionary theory as charting a progressive development of creatures, the establishment of racial hierarchies among humans seems hardly surprising or morally indictable. The judgment about the moral deficiencies of nineteenth-century scientists like Spencer or Haeckel has come, in the books just mentioned, much too swiftly and unproblematically.

The broader ethical implications of evolutionary theory (beyond the preoccupation with Nazi appropriations) have received renewed attention both from philosophers and historians. On the historical side two books corner the market: Paul Farber's *The Temptations of Evolutionary Ethics* (1994) and the collection *Biology and the Foundation of Ethics* (1999), edited by Jane Maienschein and Michael Ruse. Farber examines theories of evolutionary ethics, its promoters and critics, from Darwin, Spencer, and Thomas Henry Huxley through John Dewey, Julian Huxley, and C. H. Waddington. He contends that the various efforts at constructing an evolutionary ethics (including the contemporary work of Wilson, Ruse, Richard Alexander, and myself) have suffered from oversimplification and a lack of rational justification for the values that evolution might promote—in short, that the proponents commit the supposed naturalistic fallacy. But what Farber and other critics fail to come to terms with is that human beings are natural and that, consequently, their modes of action must also be natural. We are not denigrated by the recognition of our status as biological creatures. Our biology is, rather, elevated by that recognition.

The unique collection by Maienschein and Ruse contains historical essays by some dozen authors, who direct their efforts at the biological ethics of Aristotle, the British moralists, Lamarck, Darwin, German *Naturphilosophen*, Friedrich Nietzsche, Julian Huxley, George Gaylord Simpson, early twentieth-century geneticists, Nazi biologists, and several contemporary biophilosophers. The collection, of course, lacks the unity of Farber's monograph, but the several authors treat their respective materials in much greater philosophical and historical depth.

3. Biology and Religion

The relationship between science and religion has been glossed almost entirely in terms of religious sects' and individuals' reaction to and accommodation of evolutionary theory.[21] Prior to Darwin's *Origin of Species,* a biological scientist did not need to segregate his religious from his scientific beliefs. The biologists of the early nineteenth century, particularly in England and France, could admire in organisms the intricate handiwork of the Creator. In Germany, though, the transformation of God into nature had been gaining apace, especially in the work of the Romantics and Goethe, both of whom found philosophic refuge in Spinoza's formula *Deus sive Natura*. Darwin himself became a legatee of this identification, as I will mention in more detail below. The publication of the *Origin*, however, radically altered the comity achieved between science and religion in the early part of the century. The first wave of reaction followed immediately on the book's appearance, though, by the time of Darwin's death in 1882, various kinds of reconciliation had been made on both sides. The second wave—this of gigantic proportions—swelled as the result of the polemical cast of Haeckel's books and essays in popular science. At every turn, he spiked the tentacles of organized religion, which he saw slithering up from Rome and out of the churches of northern Germany, threatening to strangle empirical science and liberal government. He preached the sheer incompatibility of religious superstition and scientific reason. The public conflict between science and religion, which still

---

21. For a full treatment of this topic, see Frederick Gregory's essay "Science and Religion" in this volume.

echoes loudly in our day, can be attributed in large measure to the volcanic impact of Haeckel.

To understand the nature of the antagonism, one can do hardly better than to consult a scholarly work written at the end of the nineteenth century, Andrew Dickson White's two-volume *A History of the Warfare of Science with Theology in Christendom* (1896). The book ranges widely (from ancient religions to Galileo to German form criticism), but White's first chapter sketches out the principal conflict: that between a dogmatic interpretation of scripture and a scientific interpretation of life. "Darwin's *Origin of Species*," he wrote, "had come into the theological world like a plough into an ant-hill. Everywhere those thus rudely awakened from their old comfort and repose had swarmed forth angry and confused."[22] His solution to the conflict was simple: theologians had to abandon a literal interpretation of scripture, since the truth of evolutionary theory had been well established by Darwin and Haeckel. Historians who have subsequently written on the relationship of religion to biology in the nineteenth century have refrained from making such recommendations explicitly, though their own religious beliefs have colored their accounts, sometimes in marked ways.

Neal Gillespie was the first to devote a monograph, *Charles Darwin and the Problem of Creation* (1979), to examining the theological context of the *Origin* and the trajectory of Darwin's own religious beliefs. Though he spoke the new language of Kuhn and Foucault, he reached conclusions that had already been broached in the literature: during the *Beagle* voyage, Darwin had given up special creation as incompatible with the new positivistic approach to nature that he had come to endorse under the influence of Lyell; he yet retained a casual deism, according to which God had established universal laws that fully accounted for all natural occurrences. Darwin's own belief in Christianity had, as he himself admitted in his autobiography, slowly slipped away—and, it seems, without causing any of the psychological trauma that beset other eminent Victorians.[23] Gillespie thus has Darwin retaining theism—albeit greatly attenuated in later life—while giving up dogmatic religion altogether. This general picture has not changed in more recent literature, though now most would take Darwin at his word in the *Autobiography* when he averred that Huxley's term "agnostic" best suited him.[24] What is missing from Gillespie, and most other literature on Darwin, however, is a

22. White, *History of the Warfare*, 1:70.

23. See Darwin, *Autobiography*, 87.

24. Bernard Lightman's highly instructive *Origins of Agnosticism* shows how pervasive was the attitude in the second half of the nineteenth century. He attributes the fundamental epistemology of agnosticism to Kant and traces his influence (by way of Henry Mansel) on such Victorians as Thomas Henry Huxley, John Morley, William Kingdon Clifford, and Leslie Stephen.

careful examination of the way in which various arguments and conceits in the *Origin* had been molded by deeply embedded habits of thought developed in his reading of the natural theological literature, and the way in which the Romantic conception of nature came to reside in the *Origin of Species* and *Descent of Man* (1871).

Most of the literature on science and religion has focused on the impact of Darwinian theory on Protestants. John Greene, in the wake of the *Origin* anniversary, published three lectures—*Darwin and the Modern World View* (1961)—in which he lightly sketched the reaction to Darwinism of scripture scholars and those theologians still working in a natural theological tradition. James Moore describes, in much greater depth, the ways mainline Protestant thinkers in Britain and America came to terms with Darwinism. His *Post-Darwinian Controversies* (1979) defends the novel proposition that the more religiously orthodox individuals could adjust to Darwin's theory, since their views were more consonant with those of the Darwin who once studied for the ministry, while the more liberal thinkers were likely to succumb to non-Darwinian evolutionary theory. This latter group, according to Moore, had so identified nature and nature's God that any mechanical explanation of the sort Darwin proposed simply could not work. Moore uses cognitive-dissonance theory to argue his thesis, which in several instances produces that very phenomenon in the reader. Nonetheless, he offers the most probing analysis extant of British Protestant accommodations to Darwinism. Jon Roberts focuses on the American Protestant reaction, without attempting to sustain any overarching thesis (and remaining absolutely mute about Moore's position—indeed, strangely not even mentioning him in a book that makes extensive use of the secondary literature). His *Darwinism and the Divine in America* (1988) is stronger than Moore's on the theological niceties, but simply classifies all evolutionists as more or less Darwinians. Roberts recognizes that not all evolutionary thinkers might so casually be classified but claims his theologians were not all that careful either. Together, Moore's and Roberts's books provide a detailed examination of British and American Protestant reactions to evolutionary theory. What is yet wanting, though, are sustained examinations of Catholic and Jewish reactions.[25]

4. Biology and Literature

Some scholars have found the literary values of Darwin's work as compelling as its moral, political, and theological values. The literary interest in Darwin—and he is virtually the only nineteenth-century scientific figure, save

25. There are some brief indications of the latter responses in Swetlitz, "Response of American Reform Rabbis."

Freud, to attract more than passing attention—forms part of a larger concern, expressed during the last two decades, with the rhetoric of science. Literary critics like Gillian Beer and George Levine move easily from Charles Dickens's *Bleak House* and George Eliot's *Silas Marner* to Darwin's *Origin of Species*. Beer and Levine in particular have opened up Darwin's texts for a kind of examination not previously considered by historians more attentive to the surface logic and evidentiary support for evolutionary theory. Beer's *Darwin's Plots* (1983) and Levine's *Darwin and the Novelists* (1988) explore in fine detail the metaphorical structure of the *Origin,* as well as the resonance of Darwin's ideas in the fiction of Eliot, Dickens, and other Victorian writers. Investigations of this sort, when done well, do reveal more of the deep structure of Darwin's thought than might be supposed. Beer and Levine, though, have written as literary critics more than as literary historians. What is missing from their work is an effort to trace back a metaphor, a turn of phrase, or an imaginative trope to Darwin's notebooks, essays, and letters, in order to catch these figures as they first emerge. To attempt simply to locate Darwin's language within the confines of, say, the *Origin of Species,* does not fully illuminate the deeper meaning his particular linguistic configurations convey. An archaeology of Darwin's texts is required. When, for example, describing the operations of natural selection in the *Origin,* Darwin freighted his tropes with moral implications that were formed in his early essays, implications that became sedimented in the deep structure of the *Origin.* Tracing back the metaphor of the "selector" to its source in those early essays reveals, for instance, that Darwin originally pictured nature as a being that acts intelligently and with moral concern for her creatures. This is a nature that first appeared in Darwin's *Journal of Researches of H.M.S. Beagle* (1839), which took as a model Humboldt's *Travels to the Equinoctial Regions of the New Continent.* Recognition of these connections must readjust the usual understanding, which portrays Darwinian nature as mechanical and intrinsically devoid of moral and aesthetic value.[26]

Helmut Müller-Sievers has intentions similar to those of Beer and Levine but orchestrates his study with a bit of a twist and a postmodern beat. Instead of examining the metaphors in a biological text, he investigates the impact of a biological idea—that of epigenesis—as it has been used metaphorically in texts of philosophy and literature. His *Self-Generation: Biology, Philosophy, and Literature Around 1800* (1997) is loosely historical but falls more within the genre of critical theory, the assumptions and approaches of which often make historians queasy. Müller-Sievers, however, knows rather well the history of the

---

26. I have attempted to excavate the moral depths of Darwin's tropes in "Darwin's Romantic Biology." Manier had already detected the deeper moral structure of Darwin's conception of nature in his *Young Darwin.*

concept of epigenesis (i.e., the idea that the fetus begins in a homogeneous state and gradually achieves more articulate structure). And he is at home in the philosophical works of Kant, Fichte, and Schelling, his main concerns. His remarks consequently illuminate some of the darker areas in the philosophy of the period, where the metaphor of epigenesis had been lurking. Shadows lengthen, however, when he uses the flickering candle of this metaphor to explore the literature of Goethe and Beaumarchais. Despite liabilities, the literary and philosophical investigations of Beer, Levine, and Müller-Sievers do provoke the historian to expand the horizons of more traditional history of science. And they frequently write with an imaginative verve that accelerates their ideas past concepts slogging in the diligent prose of most historians.

5. Morphology and Romantic Biology

The impetus provided by evolutionary studies has carried over into associated areas of the history of biology. Morphology as a science of animal form came to birth at about the same time as fully articulated descent theories. Goethe initiated the study of underlying structures of plants and animals and gave currency to the term *Morphologie*.[27] Many subsequent researchers moved Goethe's studies forward (often along divergent paths) during the early part of the nineteenth century, including such naturalists as Karl Burdach, Carl Friedrich Kielmeyer, Georges Cuvier, Karl Ernst von Baer, Carl Gustav Carus, Lorenz Oken, Louis Agassiz, and Richard Owen. Later in the century, Darwin, with only dim awareness of their origin, and Haeckel, with devotional recognition of their author, turned Goethe's morphological ideas about unity of type to evolutionary advantage. The most powerful general study of these many individuals is still E. S. Russell's classic *Form and Function* (1916). Russell—familiar with a wide range of literature, especially in nineteenth-century German and French periodicals—reconstructed the extensive debates among zoologists as to the relative importance of form and function in understanding the design of animals. The cardinal dispute can be simply put: Did form determine function or function form? Russell himself sided with those who favored function as the determining factor. Like Cuvier, he thought function held the key to the shape and arrangement of animal parts. He regarded as insufficient the morphological theories of the evolutionists, who also sought, of course, to explain the character of structures as the result of function. According to Russell, Darwin did not understand Cuvier's dynamic view of the

27. Burdach seems to have been the first to use the term in print. See Burdach, *Propädeutik*, 62. See also Schmid, "Über die Herkunft der Ausdrücke Morphologie und Biologie." Goethe may still have the honor of having formulated the term. In September 1796 he entered the word in his diary, and in November used it, in a letter to his friend the poet Friedrich Schiller, to describe the study of forms in nature (organic and inorganic). See, Goethe, *Die Schriften*, 9b:88, 90.

correlation of parts and Haeckel focused on homologies to construct dubious phylogenetic systems. These were the views of a late-nineteenth century biologist not fully adjusted to the new evolutionary dispensation; Russell's deeply learned analyses, nonetheless, provide a wealth of insights into the literature of nineteenth-century morphology.

Recent historiography has accorded the most attention to Cuvier, von Baer, and Owen. William Coleman (1964) and Dorinda Outram (1984) have composed brief intellectual biographies of Cuvier, while Martin Rudwick has translated large portions of Cuvier's work on fossils and geology in his *Georges Cuvier, Fossil Bones, and Geological Catastrophes* (1997). Rudwick prefaces the various translations with brief, luminous essays putting Cuvier's ideas into context.[28] Toby Appel has adroitly traced the confrontation between Cuvier and Geoffroy Saint-Hilaire in her multifaceted and culturally rich *Cuvier-Geoffroy Debate* (1984). Carl Friedrich Kielmeyer, who had a significant influence on thinkers from Schelling to Goethe, wrote little in his lifetime and has enjoyed only passing attention since. Kai Torsten Kanz, however, has done much to alleviate this historical neglect with his facsimile reproduction of and introduction to Kielmeyer's most famous *Rede*, his *Ueber die Verhältnisse der organischen Kräfte* ([1795] 1993), as well as with his edited collection on Kielmeyer, *Philosophie des organischen in der Goethezeit* (1994). Carus and Oken have yet to find extensive treatment in recent historical literature.[29] Agassiz and Owen, though, have been well served in, respectively, Mary Winsor's *Reading the Shape of Nature* (1991) and Nicolaas Rupke's *Richard Owen: Victorian Naturalist* (1994). Winsor details the morphological views of Agassiz in the first part of her book and then, with more enthusiasm, turns to his founding of the Museum of Comparative Zoology at Harvard and its fate in the latter part of the century. Rupke competently situates Owen within the social, political, and institutional contexts of mid-nineteenth-century Britain. Philip Rehbock's *Philosophical Naturalists* (1983) sketches the theories of several of Owen's contemporaries, who espoused a British version of transcendentalism: e.g., Robert Knox, Martin Barry, William Carpenter, and Edward Forbes (all also vividly discussed in Desmond's *Politics of Evolution*).

Von Baer became a major theoretical force in both German and British morphology in the first half of the century, and his work forms the focal point of Timothy Lenoir's often-cited monograph, *The Strategy of Life* (1982). Lenoir maintains that several early-nineteenth-century German morphologists and physiologists—Reil, Kielmeyer, Johann Friedrich Blumenbach, Johann Frie-

---

28. Rudwick's *Meaning of Fossils* is an indispensable aid in sorting out the history of fossils from the seventeenth to the nineteenth centuries.

29. An older book on Carus, done in East Germany (without entailing much of the usual ideology), still can be recommended: Genschorek, *Carl Gustav Carus*.

drich Meckel, Friedrich Tiedemann, and particularly von Baer—had adopted Kant's principle of teleo-mechanism. According to Kant, any proper explanation in science must employ only mechanical and mathematical laws. Yet he admitted that human understanding ineluctably had to construe biological organisms teleologically, that is, *as if* they had been intelligently constructed according to a plan. But in Kant's view, this teleological assumption could only be a heuristic principle of method, not a constitutive principle of natural operation. Hence teleological notions could only be used methodologically, never determinatively. Lenoir argues that the aforementioned German biologists endorsed Kant's proposal, since it allowed them to explore organisms teleologically, without stepping into the muck of Romantic *Naturphilosophie,* which rather mystically attributed final causes to nature herself. Lenoir's interpretation of early German biology has found acceptance with several authors writing on this period—e.g., James Larson in his *Interpreting Nature* (1994). Lenoir's thesis, though, tacitly assumes that the zoologists just mentioned both clearly understood Kant's critical philosophy and acquiesced in its chief conclusion, namely, that study of organic nature, which necessarily depended on teleological considerations, could not become authentic science *(Naturwissenschaft).* This seems to me an unwarranted assumption on both counts. A careful examination of the thinkers in question leads, I believe, to a very different conclusion, namely that they understood nature to be teleologically structured intrinsically and that they believed their approaches to biology met the criteria of *Naturwissenschaft.*[30] Lenoir's book provides yet a further demonstration of the significant value of a history clearly organized with a thesis vigorously advanced—even if subsequent research might reject the thesis.

A steadier view of German morphology of the early nineteenth century would detect, if one looks in the right direction, its umbilical connections to the Romantic movement. The work of such thinkers as Reil, Kielmeyer, Burdach, Oken, Humboldt, and even Johannes Müller have filial ties to *Naturphilosophie,* which historians are now recovering.[31] Like Lenoir, Mayr rejects the possibility of any formative impact of Romantic *Naturphilosophie* on the main currents of nineteenth-century biology. He regards the work of the central scientific figures of the movement, Schelling and Oken, as "fantastic if not ludicrous."[32] This, however, is the judgment of a twentieth-century biologist

30. I have argued against Lenoir's interpretation in "Kant and Blumenbach on the *Bildungstrieb.*"

31. The most problematic of this lot is Müller. But recent scholarship has argued that Schelling, for instance, had a significant impact on Müller's conception of the science of physiology and its practice. See Tsouyopoulos, "Schellings Naturphilosophie," and Gregory, "Hat Müller die Naturphilosohie wirklich aufgegeben?," both in Hagner and Wahrig-Schmidt, *Johannes Müller.* For an account of the impact of Romanticism on Reil, see my "Rhapsodies on a Cat-Piano."

32. Mayr, *Growth of Biological Thought,* 388.

rather than a sympathetic historian of the nineteenth century. Against such assumption, I have attempted a revaluation of the relationship of Romantic *Naturphilosophie* to the main currents of nineteenth-century biology in my *Romantic Conception of Life* (2002).

Lynn Nyhart and Peter Bowler have each authored excellent histories that treat of morphology in the last half of the nineteenth century.[33] In her *Biology Takes Form* (1995), Nyhart traces the development of German morphology, especially Carl Gegenbaur's and Ernst Haeckel's evolutionary applications of the discipline, against the background of its shifting place within the German university system. She provides the deepest analysis of Gegenbaur's science now available, situating it with the work of Haeckel during their years together at Jena. Nyhart makes extensive use of archival documentation to get a more exact profile of the university context of morphology. Entrance to the archives has become the rite of passage for younger historians of science, who, as is the case with Nyhart, are being trained almost exclusively now in history departments instead of philosophy or science departments. Of comparable value is *Life's Splendid Drama* (1998), Bowler's comprehensive study of morphology in late Victorian England. Bowler has not, typically, sullied his hands with crumbling archival documents, but he has read extensively and intelligently in the published literature. In his earlier books, he contended that Darwinian thought, just after Darwin's death, had been eclipsed by non-Darwinian evolutionary schemes (i.e., those that eliminated or greatly reduced the role of natural selection) and that it only reemerged with the Modern Synthesis of the 1930s and 1940s. In *Life's Splendid Drama*, he rejects this earlier picture. His research has now convinced him that evolutionary morphology, with its concern to establish systematic genealogies, preoccupied Darwinians at the turn of the century and that this work also established the ground for the synthesis to come. The force of Bowler's analyses—which spreads over the work of biologists like Darwin, Thomas Henry Huxley, Francis Balfour, Edward Drinker Cope, Henry Fairfield Osborn, E. Ray Lankester, Anton Dohrn, and Haeckel—shakes the supports of Ruse's contrasting thesis that evolutionary studies did not become "professionalized" until the 1930s and 1940s.

6. Neurophysiology

Neurophysiology in the nineteenth century advanced rapidly as dissectional research, aided by developments of cell theory and ever-better microscopes, joined with animal experimentation and human clinical observation. At the beginning of the century hypotheses about nerve conduction via electrical

---

33. One can follow the subsequent fate of morphology in America via Ronald Rainger's interesting study *An Agenda for Antiquity*.

fluid (stemming from the researches of Luigi Galvani and Alessandro Volta) produced tremendous excitement,[34] leading Alexander von Humboldt to undertake extensive electroneural experiments, many done on his own body, with attendant puffs of burning flesh.[35] Attempts to derive a functional understanding of the central nervous system received impetus from the phrenological ideas of Franz Joseph Gall, who sought to locate psychological and behavioral traits in specific areas of the brain. The assumption of cerebral localization of sensory and motor functions became a dominant theme of neurophysiological investigations throughout the century. After midcentury, evolutionary theory suggested how specific brain regions acquired their functions.

The history of brain research has received some attention in the recent historical literature, though still the best study may be Robert Young's *Mind, Brain, and Adaptation in the Nineteenth Century* (1970). Young traced the debates about cerebral localization from Gall, Müller, Pierre Flourens, and Alexander Bain through to Spencer, Pierre Paul Broca, Hughlings Jackson, Gustav Fritsch, Eduard Hitzig, and David Ferrier. Young did not attempt to locate his actors very deeply within their cultural context, but he did lay out their views in a straightforward, confident way, taking each of his subjects through their major works on the brain and relating their views to one another. His approach to individual thinkers can be contrasted to that of Edwin Clarke and L. S. Jacyna in their *Nineteenth-Century Origins of Neuroscientific Concepts* (1987). They organize their history thematically, with chapters on the cerebrospinal axis, the nerve cell, the reflex, nerve function, brain function, and the peripheral nervous system. The authors have no particular thesis and treat these subjects as if they drifted over from some medical textbook on brain anatomy and physiology. Within each chapter the authors skip the light fantastic through numerous scientific ideas and approaches, yielding a volume rich in detail but mind-numbing when taken at a stretch. Scientists' names are attached to bundles of ideas, but they fall quickly loose from memory.

Michael Hagner's *Homo Cerebralis* (1997) at first appears an uncertain compromise between Young's book, which devoted chapters to each of the major figures, and Clarke and Jacyna's, organized around a medley of topics. In Hagner's book, Gall, Reil, and Thomas Soemmerring receive extensive consideration, but chapters on Romantic *Naturphilosophie,* aphasia, and electrophysiology (among other subjects) seem to orchestrate a cacophonous history, or at least one in which the organizing categories change from individuals to a random range of other entities. What saves the book—and, in-

---

34. Marcello Pera provides a splendid account of the theories of Galvani and Volta, and their dispute about the source of animal electricity, in his *Ambiguous Frog.*

35. See Humboldt, *Versuche über die gereizte Muskel- und Nervenfaser.*

deed, makes a powerful impression—is the strong theme that unites the disparate areas. Hagner has historically traced the dramatic changes in the way philosophers, psychologists, and anatomists have conceived of the brain from the early modern period through the mid-nineteenth century. The transition is from a theory of the brain that made it the organ or instrument of the soul to one in which the functions assigned to the soul became inscribed in the brain. The casual structure of the book works well when held together by this strong conception.

7. Genetics and Cell Theory

The nineteenth century was rife with schemes to explain heredity, but the historiography on this subject is relatively meager. Most of the work on the history of genetics has focused on the twentieth century, and especially the social application of genetics in eugenics. For the nineteenth century, Robert Olby's *Origins of Mendelism* remains indispensable. The original edition (1966) briefly surveyed the work of the Germans J. G. Koelreuter and Carl Friedrich von Gaertner at the beginning of the century; sketched the ideas of Charles Naudin, Charles Darwin, and Francis Galton in the Victorian period; focused on Gregor Mendel in the 1860s; and concluded with an account of the rediscovery of Mendel by Hugo de Vries, Carl Correns, and Erich Tschermak von Seysenegg in 1900. The second edition (1985) greatly amplified these brief considerations without spilling over the boundaries of a modestly sized book. The importance of Olby's work lies in his convincing argument that Mendel neither attempted to discover the laws of heredity nor conceived character traits as the result of two, allelelike elements, as in the later theory of the gene. Olby maintains that Mendel's real interest in his famous paper on *pisum* lay in settling the question originally raised by evolutionists. More precisely, for Mendel the question was: Did hybridization produce new, fertile forms and thus explain the origin of species? Olby's account may demote Mendel from being a man "ahead of his time," but it yet retains him as one of the great experimenters of his age.

Another area of biological ferment in the nineteenth century that yet wants the crystallizing monograph is cell theory—especially the work of Jakob Matthias Schleiden, Karl Naegeli, Theodor Schwann, and Rudolf Virchow.[36] Up to now, discussions of cell theory have been largely confined to the periodical literature. Paul Weindling has, however, sharply described the later developments of cell theory in his *Darwinism and Social Darwinism in Imperial Germany* (1991). Weindling uses Oscar Hertwig's work on cell theory to

36. Erwin Ackerknecht's *Rudolf Virchow* (1953) remains the best source for discussions of Virchow's various accomplishments.

good historiographic purpose: with it he explores forcefully a number of related confluences, especially political and social applications of evolutionary
theory.

8. Biography in the History of Biology

Weindling's book has the structure of an old-fashion life-and-works, except in
reverse—works first, then life. He follows out Hertwig's thought in developmental and cellular anatomy, and then, halfway through the book, returns to
consider the details of Hertwig's education and professorial activities, as well
as his social and political views. Though the structure of the book is odd, the
clarity of presentation, particularly of the scientific issues, makes this a compelling example of the power of the biographical approach to invade the
inner sanctum, where knowledge and imagination, cultural beliefs and personal desires, magically create new ideas and theories in science. But for the
historian, magic must be explained, or at least, ideas must be reconstructed
out of the various elements encountered in a life, so that it seems as though
those ideas must inevitably have come to birth. The artistry of the historian,
which Weindling displays, sustains the feeling of magic while laying out the
logic of development.

Biography is not in favor among social historians. This is perhaps because
the biographical approach suggests the importance of particular individuals
for understanding larger movements in history. Most social historians, if they
do not quite step back to observe the *longue durée,* yet appeal to or tacitly assume in their accounts the dominance of more proximate, yet nonpersonal
forces, such as economic structures, class and gender rankings, and mass political movements. Concentration on individuals appears retrogressively romantic. Yet important science, which has been the traditional subject of the
history of science, hardly seems to be a mass phenomenon (the suggestions
of sociologists like Bruno Latour notwithstanding).[37] The biographical approach, at its best, captures ideas as they emerge from the mind of the biologist, still carrying the signs of the causal complexes—intellectual and
passional, social and individual, scientific and philosophic—that gave them
birth. Those ideas, once loosed upon the ambient culture, then take on a
shape expressive of both their heredity and their contingent circumstances.
Several important biographies in the history of nineteenth-century biology
exemplify this genre in quite telling ways.

Frederic L. Holmes's *Claude Bernard and Animal Chemistry* (1974) established
the methods for his subsequent publications and has served as a model for
other historians. With the help of Bernard's laboratory notebooks, Holmes

37. See, e.g., Latour, *Science in Action.*

takes the reader by slow march through five years (1842–46) of the French scientist's work on digestive physiology and shows how Bernard, despite some
conceptual limitations, performed the kinds of experiments that would set
the pattern for his more illustrious later years.[38] It is biography of sorts, if we
assume Bernard lived his life exhaustively at the bench and within the confines
of his notebooks. In Holmes's study, we do not learn to whom Bernard had
been born, even that he had been born—perhaps we are to assume the latter
and ignore the former as of no consequence. Holmes mentions that Bernard
overcame a depression at the time of his marriage into a wealthy family, but
supposes the scientist's awakened spirits might have stemmed as much from
his formulating a new way to study digestion—a hypothesis that highlights
one of the two proverbial interests of a Frenchman, while throwing the other
into the shadows. Manuscript availability drove this biography and swamped
other passages that might have led to a deeper understanding of the physiologist. Nonetheless, by great sympathetic intelligence, Holmes accomplishes
something of considerable value: through his meticulous re-creation of the
day-to-day work of Bernard, we come to understand intimately the tedium
and frustration (sometimes too vividly) and magical insights attendant upon
the work of a scientist of the first rank. Holmes shows us what it was to do science in midcentury Paris, as well as what it is to create significant historical
scholarship.

Philip Pauly has crafted a rather different kind of biography in *Controlling
Life: Jacques Loeb and the Engineering Ideal in Biology* (1987). He plaits science
with the details of Loeb's life, following the physiologist from his early education and training in Germany to the various centers of learning at which he
worked in his adopted country, the United States—to which he migrated in
pursuance of a romance. Nor does Pauly finish with him at death, but trails
Loeb's influence on subsequent biologists and psychologists (like John B.
Watson and B. F. Skinner). Pauly develops the story from a definite perspective, namely Loeb as a biologist who wanted to engineer life. He portrays
Loeb, with the help of massive archival research, as one always uneasy with
more speculative and philosophical approaches to his subject, one who
sought to shape and control living organisms for practical ends. The engineer,
of course, had to be cognizant of political, social, and psychological forces
that fostered or impeded his work. Pauly likewise thinks about these matters
and captures Loeb's science with their help. Loeb, Pauly explains, suffered scientific humiliation as a young turk and thereafter restrained his polemical impulses, at least in public. And in America, Loeb, an assimilated Jew, aided
W. E. B. Dubois in opposing racism, though with some caution. Loeb was not

---

38. Holmes also makes exhaustive use of laboratory notebooks in studies of two other great
scientists; see his *Lavoisier and the Chemistry of Life* and *Hans Krebs*.

a scientist of the first rank, comparable, say, to Bernard. Under Pauly's control, however, the reader never loses interest in seeing how Loeb's ideas developed or, as was often the case, came to naught.

If Pauly finds a grandeur in the life of a good, second-rank scientist, Gerald Geison discovers shabbiness in the life of a premier scientist, a discovery he relates with comparable artistic intelligence and, for our delectation, a bit of schadenfreude. Geison's *Private Science of Louis Pasteur* (1995), like Holmes's book on Bernard, makes use of stacks of laboratory notebooks, though more selectively and for quite different ends. He investigates several of the most important episodes in Pasteur's scientific life (e.g., his theory of biologically produced fermentation, his anti–spontaneous generation debates, his discovery of the anthrax vaccine, and his use of rabies vaccine), with the aim of showing how the science became transformed, under the impact of religious, political, and psychological forces, as it moved from the private sphere of the laboratory to the public sphere of demonstration and publication. In one sense, Geison leaves us with a quite deflated picture of Pasteur's science, since he indicts the Frenchman for near-fraud in the anthrax case and suggests ethical improprieties in the handling of the rabies case. He probes Pasteur's personality and social beliefs to show exactly how science, at least Pasteur's, came aborning through a very complex parentage. Yet for all the idiosyncratic angles that Geison portrays, we recognize Pasteur's science as authentic and still capable of achieving truth. Geison's biography, richly ornamented with the similitude of life, reveals to us a genius, yet one whose scientific persona often obscured less happy personal traits.

Geison's biography of Pasteur judiciously limns the interaction of science and "extrascientific" factors—and does so in the kind of grave language the occasion and personality seem to demand. Adrian Desmond and James Moore betray a different sensibility in their biography of Darwin. *Darwin: The Life of a Tormented Evolutionist* (1991) jumps into action from the first pages, though it wobbles a bit deliriously through much of the remaining 700 pages. The book exhibits lively writing, a grand conception, and a flat outcome. Unlike Holmes, Pauly, and Geison, who give due account of their subjects's intricate science, Desmond and Moore only gesture toward Darwin's science. They are much more interested in juxtaposing the revolutionary politics of socialists and dissenters with the revolutionary scientific conceptions of the naturalist. With each chapter we hear of rioting in the streets, then glimpse Darwin as he peeks out of his London window, but only for a moment, and then turns back to filling his notebooks with the material that will have great intellectual consequence. By such juxtapositions, the authors artfully suggest, but never explicitly declare, that Darwin's science somehow sucked its energy from the revolution that never quite got started in England. The biography has enough narrative drive to make one almost forget its deficiencies, so fine is the art.

But art is displayed more deeply in scientific biography when it molds the science seamlessly to the cultural and personal context of the scientist. Pauly and Geison do this for Loeb and Pasteur, and Janet Browne does it for Darwin in her *Charles Darwin Voyaging* (1995), the first of two volumes. I might make a rounded tale by comparing her work with that of Eiseley and Himmelfarb, with whose biographies of Darwin I began this essay. Browne has studied the published works, manuscripts, and letters of Darwin, as well as the piles of material still weighing down the shelves in the Cambridge archives. Neither Eiseley nor Himmelfarb used the archival material. Eiseley knew well Darwin's theory but did not like it, at least in its implications for human beings; Himmelfarb did not really understand the theory and liked even less what she imagined it to be. Browne, by contrast, is at home with Darwin's science, and she exhibits no nostalgia for a simpler and less complicated picture of humanity. Her intimate acquaintance with the scientific and personal material, the social times and large cast of Darwin's contemporaries, the features of the theory and the rival objections—all of this means she is equipped to write a superior biography, and that she does. Like Eiseley and Himmelfarb, Browne has a lucid and near elegant style that makes the reading of her book a pleasure. She shows the power of a scientific biography artfully wrought. Like Pauly and Geison, she so masterfully stitches together the fabric of Darwin's science with his cultural surrounds that the seams cannot be detected. The biographies composed by Pauly, Geison, and Browne realize in good measure, I believe, what might be thought the ideal of a cultural history of science, to which I now turn.

9. The Structure of Cultural History of Science

The volumes I have discussed in this survey of nineteenth-century history of biology have exhibited many styles of organization and composition. Some march biological theories in tight formation past the reader's eye, which can be attractive, even smart at times. These books, which have important uses, I will refer to as intellectual histories of science. Others attempt to integrate scientific theory with the larger psychological, social, or religious concerns that might have formed the mind and habits of a scientist—that is, a cultural history of science. Of course, many of the volumes are to be found hovering between intellectual and cultural history. When done well, cultural histories, I believe, achieve the explanatory aims that good history must have. But, of course, they are not always done well, and intellectual histories like Jahn et al.'s *Geschichte der Biologie* will be preferred both for their information and for the explanations that might be constructed with their aid. Cultural history, though, has a potential that simple intellectual history lacks. I would like here briefly to indicate in what I think that potential resides and the structure of its ideal realization.

The historian, of whatever kind, begins work with some central event or series of events that he or she wishes historically to understand, that is, to explain. The central event—e.g., the origin of the American Civil War or Darwin's discovery and formulation of natural selection—serves as the beacon by which antecedent and consequent events (the causally contextual events) are illuminated, selected, and organized: some as leading up to the central event, others as being produced by it. Without a central event to serve as criterion of selection, the historian could not begin to filter out relevant antecedent events from the infinity the world offers at any moment. How that event is initially described will furnish the power and potential of the historical explanation, for it will be in light of the central event *under a certain description* that the explanatory causes (i.e., the antecedent events) and the significance (in terms of consequent events) will be isolated and organized into a coherent narrative. The description of the central event (e.g., Darwinian natural selection) will usually be modified as the historian dialectically moves from contextual events to the central event and back again. The central event may initially be understood through the artifacts with which the historian deals—in the case of both intellectual and cultural historians, the central event will initially be revealed in the documents of a given individual or community of individuals. Thus, as a second stage, the historian collects and reads the relevant books, papers, letters, notebooks, etc., assessing their relevancy in light of the central event.

The collection and assessment of artifacts will expand as the historian abstracts their meaningful contents and patterns. These patterns will then be compared with one another and ordered temporally. So, for instance, the historian might compare the logical patterns abstracted from descriptions of natural selection in the *Origin of Species* with similar patterns discovered in, say, Darwin's earlier essays and notebooks. Initially the historian will assume a developmental sequence, in which one pattern evolves over time and assumes the shape of what appears to be a descendent pattern. A good deal of history of science, especially of an older but quite valuable variety, stops at this third stage. This is intellectual history of science, but not yet cultural history. Nonetheless, intellectual history forms the backbone of responsible cultural history of science. It establishes the events, theories, and ideas to be explained, and provides a fundamental structure to guide the historian. Yet it is only at the next stage that cultural history begins.

Scientists, even the most divine, do not live in Platonic, abstract space. They live in a world streaked with social relationships, penetrating passions, and the contingencies of life. The cultural historian must thus move beyond these three preliminary stages of selecting a central event, assessing artifacts, and abstracting and ordering logical patterns. I will outline four additional stages of historical construction, stages that are, of course, only logically discriminable and not necessarily temporally distinct. As a fourth stage of historical recovery, the historian will attempt to determine the mental processes of the actors—a Darwin or a Pasteur—that led to the production of those patterns of meaning abstracted in stage three. After all, these mental processes, the thoughts and beliefs, the hopes and desires of the scientist, are the proximate causes of the hypotheses and theories that form the central concern of the historian of science. Depending on the individual or community studied, access to such processes may be limited or hardly existent. If correspondence, diaries, notebooks—even tailor bills—have not been preserved, then the means of penetrating the mental landscape of the author will be quite limited, though not necessarily completely unavailable. The surviving principal documents and circumstantial evidence may offer enough leverage to break through the interior walls of the subject. Within these walls are to be found religious beliefs, metaphysical commitments, passionate loves, consuming hates, and aesthetic needs, along with scattered scientific ideas, theories, and suspicions. From this matrix will flow the scientific accomplishments to be explained. Without some foray into the mental life of the individuals studied, no adequate causal explanation can be hoped for.

Historians, of course, must always make some assumptions about the intentional and belief states of their subjects; otherwise they would never be able to regard any sentence in a document as a proposition or assertion, as opposed to a guess, hypothesis, joke, or automatic writing. The conventions of grammar and other linguistic contextual cues allow the historian to step into the mind of the actor without being fully aware that he or she is crossing a boundary. Any short story by Borges will help sensitize historiographic practice in this regard. Since the historian must project intentions and beliefs into the mind of the scientist, or those of the scientific community, he or she might as well do a more satisfactory and self-conscious job of it.

In a fifth stage of synthetic construction, the historian will wish to recover the sources of those mental processes revealed in stage four. To do this, he or she will attempt a developmental analysis—that is, a portrayal of the series of mental developments the scientist went through to arrive at the point of producing, say, the conception of nature that invests the *Origin of Species*. Ideas have a certain inertia about them. Notions formed by an individual at an earlier period will continue to exist, perhaps in somewhat altered form, at a later period. For the historian it is essential to open for inspection, to the fullest extent possible, the course of an individual scientist's mental life. To grasp more completely the mental development undergone, the historian needs to become aware of the external stimuli for that development: newly encountered ideas, newly stimulated emotional states, new relationships with other individuals. The immediate cultural and social environments in which the scientist lives will provide these external stimuli to development, and the historian must come to terms with them.

Thus, as a sixth stage of analysis, the historian will be concerned to show

how each step in the mental development of the scientist was influenced or caused by the immediate environment in which the scientist lived and worked. Thus, in the case of Darwin's conception of nature, the historian must trace the evolution of that idea against its changing cultural, social, and physical circumstances, from Darwin's reading of Humboldt on the *Beagle*, his observations on the Galapagos Islands, and his encounters with the Indians of South America to his discussions with Huxley in the 1850s. The cultural environment provides the source of new notions, and of those that rub against and reshape already established considerations; it includes, of course, the immediate scientific terrain of established theories and practices, but also the aesthetic notions, metaphysical conceits, and theological beliefs that play upon the mind of the scientist. The social environment includes the scientist's emotional attachments, the longings, desires, aversions, and hopes that fuel the production of ideas, the formation of arguments, and the construction of theories. Ideas of an abstract Platonic sort are impotent; they lie limply in the fallow ridges of the mind. Only the emotional juices can make them spring to life and take new shape. William James had it right when he observed: "The recesses of feeling, the darker, blinder strata of character are the only places in the world in which we catch real fact in the making, and directly perceive how events happen, and how work is actually done."[39]

Finally, as a seventh stage, the historian will attempt to understand, grasp, and articulate the cultural and social patterns that shaped the mental and emotional development of the scientist. So, for example, to understand the mental and emotional development of Alexander von Humboldt, whose scientific vision guided geologists and biologists in the early nineteenth century, one must recover and re-create the intellectual, cultural, and emotional community of which he was an immediate member—namely, the circle of Jena Romantics—and then, of course, extend that out to recover the intellectual, cultural, and emotional community of German science in the last part of the eighteenth century.

A historian who moves through the first three stages I have delineated, I would call an intellectual historian simply. But the one who continues through the last four stages, that person is a cultural historian. What I have outlined is, of course, an ideal type, but an ideal, I think, worthy of the historian's aspiration.

39. James, *Varieties of Religious Experience*, 395.

Three

SCIENTIFIC MEDICINE

MICHAEL HAGNER

❧

ON 1 JANUARY 1901, the *Chicago Tribune* proclaimed: "Exit the nineteenth, enter the twentieth century. . . . With the marvelous material progress of the century has come a long train of blessings. . . . It has witnessed a vast improvement in appliances for making life more enjoyable and in prolonging it by improved sanitation and greater medical and surgical skill." At the time, it seemed as if Francis Bacon's predictions had finally come true. The material enhancement of life achieved by science, technology, hygiene, and medicine made these the heroes of the nineteenth century. In the *Tribune*'s view, the century had shown no progress "in the development of beauty, or in the progress of art," so the technological and medical achievements stood out even more. The hope for the twentieth century was that values other than utility, the materialistic mammon, and the commercial spirit might take the lead. The writer speculated that "the intellect of mankind, tiring of the material, may turn towards the higher things." Not only was aesthetics essential, but so too was "humanity and a keener realisation of the brotherhood of man."[1]

One hundred years later we know that this outlook and hope was a colossal mistake: the arts developed hand in hand with the commercial spirit; humankind has not wearied of the material; and over the course of the twentieth century, humanity was scarred and beaten in ways that were previously inconceivable. In contrast, the diagnosis that the nineteenth century was the century of material progress has continued down to our own day. Naturally, this progress involved more than the spread of material goods and sociocultural benefits. It also involved the dissolution of myths, traditional values, and images, what Max Weber famously characterized as the "rationalization" and "disenchantment" of the modern world.

Weber's understanding of modernity can also be applied to the field of medicine, advances in which have been widely regarded as an integral part of material progress. The changes within medicine that made it emblematic of modern developments have often been explained through medicine's close

---

1. "The Passing of the Century," *Chicago Tribune*, 1 January 1901, 18.

relationship with the natural sciences. Physiology and cellular pathology offered perspectives on bodily functions and malfunctions that had previously seemed impossible. Toward the end of the century, bacteriology promised causal explanations of infectious diseases as a precondition for their treatment. Around 1800 a reputable doctor had to possess individual experience and skill, trustworthiness, and an air of authority; by 1900, while these virtues were certainly not obsolete, a doctor had also to live up to the expectations of scientific medicine, which included being familiar with microscopy and the basic principles of physics, chemistry, and physiology. As William F. Bynum has written, "the medicine of 1900 was closer to us almost a century later than it was to the medicine of 1790."[2]

From the second half of the nineteenth century down to our own times historians have continued to inquire as to the achievements and results of scientific medicine. Did this kind of medicine significantly improve the living conditions of humankind? Of course, contemporaries welcomed the aforementioned "greater medical and surgical skill," for instance, the development of general and local anaesthesia, antiseptic surgery, and bacteriology. The image of scientific medicine has long profited from these innovations. But it quickly became clear that medicine could classify, distinguish, and explain the etiology of diseases much more easily than it could cure them. Roy Porter's unambigious statement applies equally well to the nineteenth and twentieth centuries: "The prominence of medicine has lain only in small measure in its ability to make the sick well. This always was true, and remains so today."[3]

How then did medicine achieve such prominence? In the last three decades, at least two answers have been offered. First, scientific medicine has profoundly shaped the ways our Western culture conceives of life and death, mind and body, health and disease, and the "philosophical status of man."[4] This reconfiguration has been characterized by an increasing concentration on patholological anatomy, on objective bodily and mental signs, on measurement, instruments, graphs, and numerical values. This process has been accompanied by an optimism that saw (and sees) the aim of medicine as the improvement of the human being and that set an associated cluster of social, cultural, and metaphysical standards as its goals.

Second, the quest to improve human beings resulted in an immense expansion of medical authority. One can see, for instance, in its treatment of physiological norms and of pathological deviances, that the expansion of scientific medicine in the nineteenth century was nurtured by its self-conception as a culture and as a politics. This growth required that certain meanings and

2. Bynum, *Science and the Practice of Medicine,* xi.
3. R. Porter, *Greatest Benefit,* 6.
4. See, e.g., Foucault, *Birth of the Clinic,* xii–xiv, 198 (quote); and R. Porter, *Greatest Benefit,* 7.

values be not simply taken from the outside, but either newly created or at least significantly transformed. Examples include the radical polarities between man and woman, black and white, and intelligence and stupidity, each of which could be distributed under the rubrics of normal and pathological.[5]

Given the remarkable impact of nineteenth-century medicine, some questions seem inevitable: What do we understand by "scientific medicine"? Can it be strictly separated from the art of medicine, as some nineteenth-century protagonists claimed? When and why did representatives of scientific medicine begin to draw such a sharp divide between the traditional practice of medicine and the sciences, and why did this divide become so meaningful? Was this a rhetorical move through which specific practical innovations were enhanced in value and through which medicine—a heterogeneous subject that also included such fields as Hippocratism, phrenology, mesmerism, and homeopathy—attempted to unite itself and to shape its own borders? And how can we describe the relationship between medicine and science? How did medicine work as part of science? Do we have an alternative to either praising medical progress or calculating the costs of its supposed progress? Is scientific medicine completely heterogeneous or can we find overarching structures?

This essay addresses these questions without pretending to give comprehensive answers to them all. First, it presents three nineteenth-century definitions and meanings of scientific medicine. Although these definitions represent Weberean "ideal types," and although we find overlap among them, they reflect the remarkable variability in theories and practices of scientific medicine (section 1). This variability contradicts the grand narrative that represented scientific medicine as a unified phenomenon (section 2). I will argue that this unified image has its roots in the nineteenth century's own self-perception, for example, in the rise of medical history and in the critical judgment of Romantic medicine (section 3). No other single development was nearly as important for creating a unified image of scientific medicine as was the rise and transformation of physiology in the middle of the nineteenth century. Since the methodological, epistemological, and historical assessments developed by physiologists like Emil du Bois-Reymond, Hermann von Helmholtz, and Claude Bernard persisted well into the twentieth century, the examination of physiology forms the centerpiece of this essay (section 4).

Section 5 deals with the decline of the grand narrative, introduced more by post-1968 social and intellectual shifts than by any original insights within the history of medicine. Following this analytical and historiographic survey, the essay then traces major trends in recent histories of medicine and explores the possibilities of ordering notions of scientific medicine. The essay does not

---

5. See, e.g., Canguilhem, *Normal and the Pathological;* Gould, *Mismeasure of Man;* Laqueur, *Making Sex;* and Moscucci, *Science of Woman.*

aim to give a descriptive overview of the secondary literature. Instead, it takes
various accounts as points of departure for exploring new perspectives for
writing histories of nineteenth-century scientific medicine. The purpose here
is not to reanimate the old dichotomy between science and art or to confirm
the newer dichotomies between knowledge and culture or instrumental practice and ideology. Instead, the purpose is to understand these antagonisms as
historically variable ensembles or clusters, which serve, complement, and
support one another in different ways, and which then can drift apart in order
to give way to new linkages and alliances. It is precisely this historical dynamic
that might enable us to pose larger and more overarching questions "from below." My examples will be twofold: I will first present spaces of scientific medicine (e.g., hospitals and laboratories) and historiographical categories for
ordering the varieties of scientific medicine; here again physiology will play a
major role (section 6). Second, I will focus on the practical dissemination of
scientific medicine, including notions of local practices and body skills, and
the role of university teaching and scientific publications (section 7). It might
come as a surprise that this section contains an analysis of the genre of biography that has so heavily shaped the grand narrative of scientific medicine
and its heroes. However, I will argue that a revisionist conceptualization of biography can deliver important insights into the function of "lives" within a
sociocultural history of scientific medicine. The penultimate section (8) deals
with patients, and thus closes the circle to the initial argument of this essay,
namely, that scientific medicine profoundly shaped our understanding of
health and the body. Finally, in a brief epilogue, I will make some remarks
about the function of the history of medicine at the beginning of the twenty-
first century.

1. Nineteenth-Century Varieties of Scientific Medicine

When historians of medicine today speak of "scientific medicine," they often
point out the different meanings and uses of "science in medicine."[6] This his-
toricization has led to numerous fruitful insights. Besides the various practical
implementations of science, the concept of scientific medicine is all but homogeneous and became pertinent long before the second half of the nineteenth century. These notions underlie the claim that scientific medicine is a
fundamental human science. Rudolf Virchow, for example, understood it as a
unified system that "must represent the sum of all knowledge of man." For
him, scientific medicine was anthropology, and thus it could not be reduced
to experiments, diagnostics, or even therapy. This broad concept of scientific
medicine had consequences for its historical classification: "The history of

---

6. Warner, "Science in Medicine."

medicine," Virchow wrote, "is thus an integrating element of cultural history in general, and it can only be understood within the context of the general history of mankind." This claim was part of an ambitious political program including material and spiritual aspects of human progress. Virchow, who was actively involved in the revolution of 1848, conceptualized scientific medicine as political.[7]

The idea of an all-around conception of medicine was by no means an invention of the reformers of 1848; rather, it had Romantic origins. In 1807, in the course of planning the foundation of a university in Berlin, Johann Christian Reil distinguished pure science, which serves only its own purpose and has no external use, from "technical science," which is determined by its applications or goals. According to Reil, medicine is a mixture of the two. It is "the study of the nature of the organisms [Naturkunde der Organismen] in their dynamic relationship to the environment applied toward the goal of healing their diseases. The study of nature [Naturkunde] is its foundation, application is its own character." Without arguing against its character as application, Reil emphasized the pure scientific aspect, since, according to him, knowing the development of organic beings in living nature is the science "which first allows man to understand himself, which needs to precede all other scientific education [wissenschaftliche Bildung], as pure science, general, without relation, systematic."[8]

One of the central goals of Romantic Naturphilosophie was that scientific medicine function both as humanistic education [Bildung] and as knowledge [Erkenntnis]. Following Johann Gottlieb Fichte's claim that Wissenschaftler need to be examples and teachers of humankind,[9] medicine was given a philosophical status as a leading science that provided orientation and aimed at more than treating patients. The goals tied to this definition went far beyond that of mere material progress. Friedrich Tiedemann, in his opening speech to the meeting of the Society of German Natural Scientists and Physicians in Heidelberg in 1829, emphasized that natural scientists helped improve middle-class society, both materially and by offering an "ennobling of the spirit, . . . the conquering and dispersion of dangerous prejudices and the awakening of sublime opinions about the final cause of the world."[10] Scientific medicine thus became a key element in cultural development, a building block against the ideology of materialism that apparently threatened the middle-class canon of norms, values, and lifestyles. Such an antagonism between materialism and spiritualism, and between pure and applied knowl-

---

7. Virchow, "Einheits-Bestrebungen," 30.
8. Reil, "Entwurf," 51–52. See Broman, Transformation, 84–90, 182–84.
9. Fichte, Bestimmung des Gelehrten, 88–89.
10. Tiedemann, "Ansichten," 492–93.

edge, also occurred in other countries. Despite their local specificities, the debates between William Whewell and Charles Babbage or Richard Owen and Thomas Henry Huxley, or the Parisian debates on materialism involving F. J. V. Broussais, Pierre Flourens, Victor Cousin, and Paul Broca, can all serve as striking examples.[11] Although these debates did not affect definitions of scientific medicine in the same way as those in the German-speaking world, it is important to understand to what extent notions of scientific medicine were part of cultural debates in different national contexts.

Although scientific medicine had been an issue already in the early nineteenth century, it developed differently in the second half of the century. Here, too, it is important to differentiate between several ideas of scientific medicine. Claude Bernard, for example, postulated three basic parts of an experimental or scientific medicine: physiology, pathology, and therapy, which in medical practice mutually supported one another. For him, standard scientific procedure entailed first seeking the cause by which life processes had shifted from a normal to a pathological state, then working to develop therapeutic measures. According to Bernard, these steps were part of a unified process indebted to one single method—experimentation. The crucial aspect was to master the "art of securing rigorous and well-defined experiments." However, the concept of "art" here stands in sharp contrast to an aestheticizing concept of medicine as art and the doctor as artist—an idea which Bernard emphatically rejected.[12] Art here is meant as the practical and theoretical skill to stabilize experiences and to generalize, that is, to deindividualize.

If Bernard spoke from the point of view of the laboratory scientist, his perspective was not identical with that of the clinicians. Many of those who stood in the tradition of the clinical Paris School favored a positivistic knowledge, that is, thinking and practicing on the basis of cases without having a specific theory. Relying upon cases was a kind of individualized thinking that was inconceivable without the predominant authority of the doctor. This tradition was especially powerful, even after the advent of Bernard. Ernst von Leyden, a professor of internal medicine at the University of Berlin, spoke for a large segment of doctors when he described medicine as "our science," which developed through the "researching spirit of man" from a "tiny plant in the beginning of the nineteenth century" into a "proud tree, whose many branches are richly filled with flowers and fruits." For Leyden, scientific medicine, or medicine *as* science, was a dynamic process with constantly opening new areas of research. The researching doctor needed to invest effort, care, hard work, and patience in order to achieve useful scientific results. Leyden

11. See Schaffer, "OK Computer"; Rupke, *Richard Owen;* and Williams, *Physical and the Moral.*
12. Bernard, *Introduction*, 2–3.

was in full agreement with Bernard regarding the role of medicine in recognizing illness. Yet he thought that this was only one side of medicine. When it came to therapy, his views became diametrically opposed to those of Bernard. Leyden saw a discontinuity between diagnostics and therapy. According to him, the hospital, "the centerpoint of scientific medicine . . . is the mediator between science and art." The art of medicine is the "tested experience" and the "individual gift" of the doctor, which becomes relevant in that "the therapy cannot simply be built up on science since it requires refined experience."[13] In other words, when medicine researches the cause of illness it is science, but when it provides therapy for illnesses it is at least in part art. In contrast, Bernard's all-encompassing claim can be compared in its totality to the goals of the Romantics and Virchow, even if he introduced an entirely different content or methods.

In presenting these three different definitions of scientific medicine, I make no claim to completeness. These examples represent a development within the nineteenth century from a more general to a more specific definition of scientific medicine, but this does not mean that the views expressed by Reil and Virchow had become obsolete by 1900. On the contrary, such views remained popular among doctors, even if they were not necessarily still termed "scientific medicine." Reil's notion of *scientific medicine as culture* and Virchow's concept of *scientific medicine as politics* were driving forces for the expansion of scientific medicine. In the second half of the nineteenth century in particular, these understandings of scientific medicine extended to wide areas of society and its values, perhaps most importantly affecting or shaping issues such as gender, morality, sexuality, criminality, definitions of social evils, and the calculation of labor productivity. In contrast to Reil's and Virchow's claims, Bernard established the predominance of one discipline—physiology. The making of physiology as a comprehensive, leading science, as I will discuss below, had crucial consequences for the acceptance of a unified picture of scientific medicine. On the other hand, Leyden was fully aware of the merits of *medicine as a science,* but as a clinician he wanted to maintain the specificity of hospital work and, correspondingly, different professional experiences and identities.

Even if these three overlapping categories—medical science as culture, medical science as politics, and medicine as a science—prove useful in approaching the phenomenon of scientific medicine in the nineteenth century, and even if fruitful research questions can be formulated for each approach, a

13. Leyden, "Die deutsche Klinik," 1–3, 11–12. Similar statements can be found by American physicians. In 1877, it was said, the "chief excellence" of medicine "is, not that it is scientific, but that it is redemptive." Quoted in Warner, *Therapeutic Perspective,* 11. See also Rosenberg, *Explaining Epidemics,* 28–30.

unified picture does not easily or necessarily emerge. Given that medicine increasingly relied upon so-called theoretical or foundational disciplines like
physiology, pathological anatomy, or chemistry, there is no unequivocal answer as to whether the term "science" applied to medical research, teaching,
the definition and construction of scientific objects, or to a way of treating
patients. Similarly, what do we understand by the concept "medicine":
surgery, psychiatry, or internal medicine? Each subject has had its own specific
history, as a science in relation to the foundational sciences and in terms of its
own scientific standards and practices. Despite these quandaries, a unified picture has dominated our outlook and deserves more detailed explanation.

2. The Long Nineteenth Century as an Epic

From the nineteenth century onward, the making of scientific medicine has
been told as an epic. It has been modified by various tellers, but no truly new
story has emerged. In the words of Guenter Risse: "To this day, the 'big epic of
medicine' remains a popular historical genre, a master narrative or 'Meister-
erzählung' legitimating our professional identity."[14] Although Risse is concerned here with "stories of individual doctors' lives and careers," and hence
includes pre–nineteenth-century physicians, the master narrative also explains the dynamics of modern science and medicine. According to this view,
the key components that shaped modern medicine—besides ingenious and
creative individuals—were industrialization, mechanization, rationalization,
the formation of the modern university, and the differentiation of the scientific disciplines. Hence the question is not whether such a grand narrative is
useful in evaluating historical processes; rather, the question is how it has
functioned and why it has been so successful. For this reason, I will now introduce some theoretical considerations that at first may seem to have little to
do with science and medicine. Through them, however, I hope to show why
nineteenth-century scientific medicine was so long perceived as a unified construction.

What is an epic?[15] The literary epic relies on oppositions: there are protagonists and antagonists, heroes and villains. The villains are associated with
darkness, confusion, sterility, and age; the heroes with sunrise, clarity, fruit-
fulness, and youth. A good epic is made up of many small episodes and narrative strategies, which nevertheless are held together by a master plot. Thus
the epic deals with wish-dreams and their fulfillment. The heroes are characterized by an aristocratic charisma. They struggle against enemies and solve
puzzles and problems, but they do not work in the sense that other mere mor-

---

14. Risse, "Reflected Experience," 201.
15. In the following I rely on W. Clark, "Narratology," 5, 20, 31–33, 48.

tals would. This means that money almost never plays a role: in good science, ideas are free. In the epic telling, science is clean and pure, idealized, intellectualized, free from profane labor, and raised to a secularized ritual. In other words, science stands on the highest level of culture, together with art and morality.

Major nineteenth- and twentieth-century writers have dramatized the nineteenth century as an epic. Their representations include the division of actors into good and bad, the heroic struggle, the telos, the purity of scientific research and its ennobling as culture. The heroes need not be individual geniuses like Virchow, Louis Pasteur, William Osler, or Jean-Martin Charcot, although many have been. Collective or nonhuman heroes can just as easily be raised in an ideal-typical fashion to the level of historical subject. Thus it has been possible to use and understand "the nineteenth century," "the Paris School," "the laboratory," "bacteriology," "the stethoscope," and "the kymograph," to name but a few, in the collective singular in order to hold the master plot together.[16] The notion of sociocultural progress created a need for this kind of master plot. Supposedly, every individual advancement contributes to the enhancement of mankind.

With the historical-philosophical charging of objects, instruments, spaces, and disciplines, plural processes were unified, histories became history, and the idea of progress received its metaphysical superstructure. At the same time, antagonisms or enemies were formed. In the first instance these were the old myths and prejudices, clerical superstitions and magic, whose new representatives were the unscientific charlatans—phrenologists, homeopaths, mesmerists, and, from a certain point on, the early–nineteenth-century Romantic *Naturphilosophen*. Authors like Cuvier, Bernard, Huxley, and Helmholtz did not tire of stressing the role of progress against such enemies.[17]

An additional class of enemies was made up of "the bacteria" and "syphilis," "degeneration" and "madness." These concepts were in turn subsumed under the overarching concept that named perhaps the most powerful enemy of all, one that appeared in various shapes but was not identical to any specific

16. On the meaning of the term *Kollektivsingular* see Koselleck, *Vergangene Zukunft*, 50–51, 264, and Jardine, "Inner History."

17. An early example (1810) of a historical review on scientific progress is Cuvier, *Rapport historique*. This review highlighted the progress of the last twenty years and thus historicized a short time period. This genre was widely extended in the course of the nineteenth century and became a powerful instrument for disseminating popular ideas about science and medicine. See, e.g., Eble, *Versuch einer pragmatischen Geschichte* (1836), and the famous inaugural speeches by Virchow, du Bois-Reymond, Helmholtz, or Haeckel on the Gesellschaft Deutscher Naturforscher und Ärzte (collected in Engelhardt, *Forschung und Fortschritt*). On popularization see Bensaude-Vincent and Rasmussen, eds., *La science populaire*; Lightman, "Voices of Nature"; and Daum, *Wissenschaftspopularisierung*.

illness or epidemic: the pathological. Nineteenth-century thinkers did not invent the pathological, but they did generalize it beyond the way that their predecessors had defined it. "The pathological" was probably the most powerful and most consequential collective singular idea that defined medicine. Without the spread of the pathological to all possible areas, the rise of scientific medicine would have been unthinkable. While the seventeenth and eighteenth centuries were occupied with ridding the world of miracles, demons, and wizardry, the nineteenth century defined the pathological and built it up, struggled against it, stemmed it, found it in hidden corners, and heightened it rhetorically in order to characterize its own deeds as heroic. Inherent in this obsession with the pathological was the casting of various phenomena, like psychic diseases, syphilis, criminality, prostitution, homosexuality, and alcoholism as the costs of modernity.

This reconception reveals the historical self-awareness of scientific medicine as culture and as politics. The confidence with which physical criteria and signs were named in order to make categorical distinctions like man and woman, black and white, intelligent and stupid, normal and pathological led to a situation in which this knowledge and the language linked to it spread to other areas and helped to shape or change their judgments, values, and symbols. In this sense it is appropriate to speak of one culture, and this was exactly the claim made by scientific medicine: that it was the "absolute organ of culture." This formula, first used by du Bois-Reymond, spelled progress in the language of technology and scientific medicine, which thus became cultivated and made attractive to an educated middle-class society. The latter's self-conception was based on the propagation of culture as a form of life. In this context it was important to reformulate the achievements of scientific medicine as an epic and to reinterpret many-faceted problems into a collective singular idea.

3. The Origins of the History of Medicine and the Stigmatization of Romantic Medicine

The history of medicine and science bear a large responsibility for the emergence of the epic that I have just described. To his catchphrase cited above, du Bois-Reymond added that the "history of the natural sciences is the real history of mankind."[18] This was not, as one might think, linked to the demand that the history of medicine and science should have a special place in the canon of academic subjects. Until the mid-nineteenth century, history had been seen as an ordering discipline in which a familiarity with the literature helped to increase knowledge and separated the important from the unim-

18. Du Bois-Reymond, "Kulturgeschichte und Naturwissenschaft," 596.

portant. History evaluated and legitimated current achievements. Thereafter, the present and the future took on ever more weight, and the past lost ever more of its relevance for scientific inquiry.[19] History now became a litmus test with a teleological focus. When du Bois-Reymond declared the history of science to be the history of mankind, he meant that the stage of cultural development of a people or culture can be read from the stage of their scientific development at a particular time and place. This differed little from the claims of the general historians; instead of raising political maturity, democracy, or cultural development to the level of the historical subject, du Bois-Reymond inserted science.[20] The present became the dominant reference point: science made itself the foundation of politics and culture and valued its own labor as mankind's highest possible achievement. It followed from this strategy that the past could not serve as a reference point for evaluating current achievements. On the contrary, the past became meaningful only in that it pointed to or radically differed from the present. Around 1900, this cult of the present put the history of medicine on the defensive. It tried to save itself by emphasizing its usefulness to doctors.[21]

Two breaks in the nineteenth century can be diagnosed that devalued the achievements of predecessors or older contemporaries and furthered self-legitimation. Around 1800, Romantic *Naturphilosophie* saw itself as overcoming the dusty collection and notation of natural objects and phenomena, a mindless empiricism and an undirected lust for knowledge that held on to descriptions and details. To be sure, the Romantics did not refer to history in order to defend themselves; rather, they wanted to formulate deep insights into the mechanics of nature and to develop a general theory that would integrate individual findings into a completed design of nature.[22] Scarcely one generation later, *Naturphilosophie* was itself thrown onto the rubbish heap of history, as the age of natural science was praised as a heroic act, as a release from the chains of slavery.

The history of medicine and science has always had a difficult time with the Romantic period. For a long time, Romantic *Naturphilosophie* and medicine were viewed as a distinct way of thinking that was dominated by speculative concepts and theories and almost devoid of observation and experiment. Such a unified negative assessment is now out of the question. In a recent survey article, Trevor Levere went so far as to claim that it is fruitless "to define and delimit the sphere of Romanticism with a clear boundary."[23]

---

19. On the transformation of history within medicine and the sciences see Heischkel, "Geschichte der Medizingeschichtsschreibung," and Engelhardt, *Historisches Bewußtsein*.

20. Gradmann, "Naturwissenschaft, Kulturgeschichte und Bildungsbegriff."

21. Kümmel, "Legitimierungsstrategien der Medizingeschichte," 83–86.

22. See, e.g., Schelling, *Von der Weltseele*, 527, 564–65, and Burdach, *Physiologie*, 17.

23. Levere, "Romanticism," 468. See also Broman, *Transformation*, 90–92.

This declaration of surrender reflects a general tendency within the history of medicine and science, according to which Romanticism is no longer defined according to conceptual criteria, but rather is explained in terms of a historical period, between 1780 and 1830.[24]

Another consequence of the revisionist view has been an internationalization of Romanticism: whereas it had previously been regarded as an almost exclusively German phenomenon, Romantic movements in other European countries have come into focus, though Germany remains the epicenter.[25] As a consequence of these shifts, scholars have traced linkages between *naturphilosophische* theories, which are often obscure and difficult to understand, and broader social events, like the formation of new disciplines and the reform of the medical profession.[26] In this context, Thomas Broman speaks of "physicians in the avant-garde."

Yet one set of questions remains unanswered: What happened within the German cultural elite that made medicine so seductive? How could it happen that medicine, according to Friedrich Schelling's programmatic statement, was a practical science and a form of knowledge wrapped into one, a forum for questions that were of equal interest to doctors and physiologists, philosophers and natururalists?[27] Was it a certain style of reasoning? Was it the matrix for a new, experimental understanding of the mind-body? Or was it a sociocultural experiment that allowed new forms of communication, including new forms of sensibility, inspiration, and creativity? And finally, if medicine played a dominant role as a catalyst for a new unified understanding of natural science, what were the practices connected to this?

Levere has justifiably complained that in Andrew Cunningham and Nick Jardine's otherwise broad survey *Romanticism and the Sciences* there is no overview of Romantic medicine. There is also no such current survey in English, and here a problem of language might be mentioned. Considering the continuing interest in the Romantic period, it may be telling that Levere, for example, does not cite any German-language work in his review article, despite the fact that in the last few years much work has been done in German. While some of this work is perhaps not pathbreaking, in many ways it has expanded our understanding of the period.[28] Still, there is no clearly identifiable

24. See Cunningham and Jardine, eds., *Romanticism and the Sciences*. In German scholarship, the epoch between 1780 and 1830 is characterized as the "Goethezeit." See, e.g., Mann and Dumont, eds., *Soemmerring und die Gelehrten der Goethezeit*.

25. R. Porter and Teich, *Romanticism in National Context*.

26. See Stichweh, *Zur Entstehung des modernen Systems;* Gregory, "Kant, Schelling and the Administration of Science"; and Broman, *Transformation*, chaps. 5–6.

27. Schelling, "Vorrede," vi; Broman, *Transformation*, 96.

28. Tsouyopoulos, *Andreas Röschlaub;* Lammel, *Nosologische und therapeutische Konzeptionen;* Lohff, *Suche nach der Wissenschaftlichkeit;* Wiesing, *Kunst oder Wissenschaft*.

trend in these works, and this is typical for the historical assessment of Romantic medicine and *Naturphilosophie*.

The long-standing rejection of the Romantic period, which over generations of scholars has been only slightly modified in its form of argument and choice of words, had a great advantage. It permitted a simplifying and compact definition of Romanticism. This definition made possible a clear divide between, on the one hand, a technologically and industrially oriented modernity and, on the other, a rapturous antimodern metaphysics of nature that briefly brought scientific progress to a standstill. This polarization influenced not only those who overcame Romanticism; it also crystallized in the following generations and was adopted without reservation by the history of medicine and science. Later apologists and latter-day imitators of Romanticism, who saw in the Romantic veneration of nature the most powerful countermodel to modernity, had exactly the same conception of Romanticism, albeit with an opposite interpretation.[29]

Thanks to the multifaceted research of the last twenty years, the negative image of Romanticism has become more or less obsolete. German historiography has largely limited itself to establishing that *naturphilosophische* ideas were not hostile to practice; instead, through a methodological confrontation with practice, they sought to negotiate how such a science might look. At the same time, it has become undisputed that Schelling's *Naturphilosophie* offered constructive additions to scientific methodology and to the theory of experimentation, that empiricism played an important role in his work, and that the theory of *Naturphilosophie* was compatible with practical medicine.[30] Yet the relationship between *Naturphilosophie* and medicine or science is difficult to reveal. I have recently argued that *Naturphilosophie* subtly transformed the new doctrine of localizing mental faculties in the brain, thus making it acceptable to a German audience. Two other examples show how differently the scientific contributions of *Naturphilosophie* can be interpreted. Initially, Timothy Lenoir's argument that the development of biology in Germany largely circumvented Romantic *Naturphilosophie* enjoyed quite a positive reception, but by now almost no one accepts this position. Conversely, it was long assumed that *Naturphilosophie* played a major role in Julius Robert Mayer's formulation of the law of the conservation of energy. Kenneth Caneva has recently challenged this assumption.[31]

The overdue rehabilitation of Romanticism, which has reinscribed it as a multifaceted, heterogeneous endeavor, has also led to several difficulties. If,

29. See Lammel, *Nosologische und therapeutische Konzeptionen*, 11–34, and Wiesing, *Kunst oder Wissenschaft*, 22–43.

30. Besides the literature cited in note 28, see Risse, "Kant, Schelling and the Early Search," and Figlio, "Metaphor of Organisation."

31. Hagner, *Homo Cerebralis*, 151–223; Lenoir, *Strategy of Life*; Caneva, *Robert Mayer*.

with a few exceptions, Romanticism can be integrated into the overall development of the sciences, then we might ask what is particular about Romanticism. If the primary interest of integrating *Naturphilosophie* into medicine and natural science was the development of a new, better-grounded methodology, then were not the rapturous tones and irrationality, the speculations and constructions, nothing but marginal phenomena? In short, if the long assumed discontinuity really has the character of continuity, what remains of the concept "Romanticism" beyond a heuristic general term for the era between 1780 and 1830? In order to answer these questions, it seems more promising to look at the specific practices and objects of each of the sciences than to examine global programmatic statements. One example of the incompatibility of approaches to a certain research practice is the role assigned to experiment.

During the Romantic period, a controversial debate occurred about the empirical value of experimental results. Comparative anatomists attacked the artificiality and limited significance of experimentation, while at the same time, galvanic experiments and self-experiments dealing with the subjective physiology of perception characterized Romantic natural research and helped develop the experimental culture of the nineteenth century.[32] It cannot be claimed that experiment played a determined role in Romantic *Naturphilosophie*. Instead, one might say that the era of Romanticism created a platform on which different styles and cultures of scientific practice were tried and sometimes established. This scenario led to important bifurcations, for example between morphology and experimental physiology.

Such an account might complement Jardine's attempt to understand the relationship among *Naturphilosophie,* medicine, and natural science as an insistent questioning and a conceptual as well as a practical exploration in which the boundaries of the sayable and the doable shifted.[33] In this process, philosophy served as a quarry for scientific activity and was used as needed. The innovative potential of such a pragmatic understanding of the variable historical role of philosophy becomes clear when seen in contrast to the traditional view. For far too long scientific innovations in the Romantic period remained outside the range of vision of the history of science because the positivism of the nineteenth and twentieth centuries and a theory-dominated history of science were convinced that philosophy's influence on science was one-dimensional, a transfer without transformation. According to this model, Ro-

32. On Romantic (self-)experimentation see Riese, "Impact of Romanticism"; Schaffer, "Genius in Romantic Natural Philosophy"; Strickland, "Galvanic Disciplines," and *Ideology of Self-knowledge;* and Hagner, "Psychophysiologie und Selbsterfahrung." On early nineteenth-century antivivisection see Figlio, "Metaphor of Organisation" and "Historiography of Scientific Medicine"; and Temkin, "Basic Science."
33. Jardine, *Scenes of Inquiry,* 11–55.

mantic natural research was seen as succeeding only "unintentionally," when a false philosophy was eliminated by good empirical method or when a speculative idea or phrase served as the spark that led to an important discovery.

The enigmatic and ambivalent character of Romantic natural research is, however, not the only reason why it continues to hold scholars' interest. I believe there are also solid historiographical reasons that have to do with our understanding of the entire nineteenth century. The rehabilitation of Romanticism made the epic—written by scholars of the nineteenth century and followed by their twentieth-century successors—significantly less attractive. A new look at Romanticism means a new look at the entire century, and this applies especially to the relationship between medicine and the sciences. Medicine was science, and not the application of science, especially of experimental physiology and pathological anatomy. This was not specific to German *Naturphilosophie,* since, as John Lesch and Russell Maulitz have shown with the example of experimental physiology and pathology, medicine held a similar status in France at the same time.[34] Instead, we need to ask to what extent those medical doctors who promoted experimental physiology, comparative and pathological anatomy, and embryology debated about the *correct* kind of scientific medicine.

4. Why Physiology?

Romanticism provides one focal point for studying nineteenth-century scientific medicine. Another is the emergence of physiology, which has garnered the attention of historians at least as much as Romanticism. Whenever twentieth-century medical history introduced a new theoretical approach, it used nineteenth-century physiology as a testing ground. In the mid-twentieth century, when historical analysis focused on the most important theories, concepts, and ideas, authors from very different intellectual traditions—e.g., Georges Canguilhem, Owsei Temkin, and Karl Rothschuh—applied the history of ideas to their studies of nineteenth-century French and German physiology.[35] In the 1980s, when social histories of science became dominant and focused on the institutionalization of scientific medicine and the formation of disciplines, nineteenth-century physiology again became a preferred subject of investigation.[36] In the late 1980s and early 1990s, when "science in action" became, for some scholars, the order of the day, physiological laboratory

34. Maulitz, *Morbid Appearances;* Lesch, *Science and Medicine.* See also Warner, "History of Science," 186–89.

35. Temkin, "Philosophical Background," "Materialism in French and German Physiology," and "Basic Science"; Rothschuh, *History of Physiology;* Canguilhem, *Etudes,* 127–71, 226–304.

36. Pertinent examples are Coleman and Holmes, *Investigative Enterprise;* Lenoir, *Instituting Science;* and Tuchman, *Science, Medicine and the State,* 113–67.

practices, experiments, and instruments inspired even more detailed and microscopic studies.[37]

How can we explain this privileging of physiology? There is broad agreement that the separation of physiology from anatomy after 1800 can be described in two complementary ways: first, as a transition from vitalism to mechanism; second, as a shift from "animated anatomy" to a concept of bodily functions independent of anatomical structures. These descriptions are not identical. The abandonment of the concept of "vital force" in favor of physical and chemical forces purportedly permitted a causal understanding of bodily events. This shift characterized German physiology (or "organic physics") around 1850, in particular the students of the Berlin physiologist Johannes Müller: Helmholtz, du Bois-Reymond, Ernst Brücke, and their ally Carl Ludwig. But this "analytical mechanics of all life processes"[38] was by no means the only option.

The replacement of "animated anatomy" by physiology first occurred in the context of Romantic *Naturphilosophie*. Here physiology conceived of organic forces in terms of their systemic functional effects, as opposed to taking organs as the point of departure. Franz von Walther, one of Müller's medical professors in Bonn, employed the term "function" as a way to emphasize the entire organism's activity. In French physiology, especially that of François Magendie, a more refined differentiation between the different aspects of bodily functions emerged, which then became the basis for Müller's physiology.[39] The introduction of function as an operational concept went hand in hand with the idea that living organisms followed special laws in addition to physical and chemical ones. Temkin has characterized this position as "vitalistic materialism." In France, this standpoint shaped physiological research from Magendie's and Pierre Flourens's work in the 1820s and 1830s down through Claude Bernard's theory of the "internal environment."[40] In Germany, vital materialism became the starting point for two different approaches to physiology, both of which arose in Müller's laboratory. For Müller and one group of his disciples, microscopy, not animal experimenta-

37. See Cunningham and Williams, *Laboratory Revolution*, and Rheinberger and Hagner, *Experimentalisierung des Lebens*.

38. Du Bois-Reymond, *Untersuchungen*, xxxv. See Cranefield, "Organic Physics."

39. Walther, *Physiologie des Menschen*, 28, 38, 122; Magendie, *Précis élémentaire*, 15–23. On Müller see Hagner and Wahrig-Schmidt, eds., *Müller*, and Fasbender, *Physiologie der Funktionen*. Fasbender has transcribed and edited an 1826 lecture by Müller on physiology and gives an extensive interpretation of the concept of function in early-nineteenth-century physiology.

40. Temkin, "Materialism in French and German Physiology," 323. There is an ongoing discussion on how to locate Bernard's position between materialism and vitalism. See Holmes, "Milieu Intérieur" and "Claude Bernard"; Mendelsohn, "Cell Theory" and "Physical Models"; Canguilhem, *Etudes*, 143–55; and Grmek, "Claude Bernard."

tion, was the main research method. In this group, the topics investigated were cell formation and cellular metabolism, not animal electricity or propagation of nerve impulses. But while Theodor Schwann's pronounced reductionism led to the formulation of the cell theory and became the point of departure for du Bois-Reymond and his colleagues, Müller's "vitalist recasting of Schwann's cell theory" led to the principle "omnis cellula e cellula," which was "established by two of Müller's disciples with teleomechanist inclination, Robert Remak and Rudolf Virchow."[41] The development of modern physiology was therefore never unified, neither theoretically nor practically. Even if the classical vitalism formulated in the late eighteenth century became meaningless, teleological thinking nonetheless persisted in cell theory.

The nineteenth-century vision of physiology was equally varied in its relationship to clinical medicine. While French experimental physiology developed within the favorable environment of the Paris Clinical School, and while Magendie, Bernard, and other French physiologists lectured on clinical issues,[42] German organic physicists intentionally worked at a distance from the clinic, following the Romantic ideal of pure science. Du Bois-Reymond would have liked to separate physiology from medicine and integrate it into "the theoretical natural sciences," adjacent to physics and chemistry. He denounced, furthermore, "the entire medical profession as obscurantist and bigoted" and claimed that physicians should aways be governed by the scientific mind as it had developed and operated in physiology. Other physiologists were more cautious in their assessments. While Helmholtz once argued that scientific medicine depended on physiology, he later showed great respect for the clinical tradition and praised its skill and experience.[43] Contradictory statements about the scientific value of clinics reflect an ongoing tension between clinical medicine and experimental physiology, on the one hand, and the physiologists' distinct self-confidence, on the other. This self-confidence, which had developed within a few decades, benefited the entire medical profession because it allowed doctors to emphasize particular demands and values in political and public spheres.

What was the concrete relationship between physiology and medicine like? On the one hand, there were fruitful collaborations between clinicians and physiologists, for example, between Ludwig and Wunderlich in Leipzig and between Henle and Carl Pfeufer in Heidelberg.[44] On the other hand,

41. Duchesneau, "Vitalism and Anti-Vitalism," 250–51. See also Lohff, "Johannes Müllers Rezeption"; Rheinberger, "Zum Organismusbild der Physiologie"; and Lenoir, *Strategy of Life,* chap. 3.
42. Lesch, *Science and Medicine.*
43. Du Bois-Reymond, *Untersuchungen,* 1, and "Der physiologische Unterricht"; Helmholtz, "Über das Ziel" and "Antwortrede."
44. Tuchman, "From the Lecture to the Laboratory;" Lenoir, *Instituting Science,* 96–130.

Wunderlich wrote retrospectively that Johannes Müller's *Handbuch der Physiologie* did not have any direct effect on medicine because it was too exclusionary and initially found its way into medicine through pathology.[45] Again, the early nineteenth-century ideal of pure science shines through. Although French physiology was much closer to pathology and to clinical demands, physiologists from Magendie to Bernard regarded experiment as the key method for building an understanding of the normal and the pathological body. The notion that the planned experiment in the laboratory was superior to the observation of clinical, individual cases and that the road of progress led from the laboratory to the hospital, and not the other way around, was a powerful nineteenth-century myth. In this myth, experience is only valid when exact, objective, and measurable.

Although physiology helped to establish the methodological boundaries of scientific medicine, it did not delimit them. Sometimes physiology even objected to the observations of clinical medicine. A striking example involves the cerebral localization of mental qualities. After Pierre Flourens had experimentally refuted Franz Joseph Gall's doctrine of cerebral organology in 1824, physiologists ceased looking for clinical evidence for localizations. Not until the 1860s did the combination of clinical case studies and pathological anatomy bring localization theory back into play.[46]

If physiology had multiple relationships with clinical medicine, then the question arises as to why it retained a key position in attempts to reach a unified, comprehensive understanding of scientific medicine. At least three reasons merit consideration. First, with its experimental approach, instruments, and measuring devices—which reflected the ideals of quantification, precision, and objectivity—physiology became a model for clinical medicine, even though it was never fully accepted or applied in all of its aspects. Entire generations of doctors acquired an idea of scientific practice through demonstrations, lectures, textbooks, and especially laboratory work. These ideals were reinforced, at least in part, through quantitative diagnostic methods, chemical analyses, and the clinical use of the microscope and other instruments. In this respect, the laboratory served as a site of medical innovation. It functioned as a kind of symbolic capital that insured the scientific character, and it had an increasing significance in medical training. It would be useful to look at the extent to which the clinicians who worked in the laboratories of Ludwig, Bernard, or Michael Foster oriented their clinical practices toward the organizational forms and social interactions of the laboratory—not so much in terms of the hospital's organization and the patient's treatment, but in terms

---

45. Wunderlich, *Geschichte der Medizin*, 350.

46. See Clarke and Jacyna, *Nineteenth-Century Origins;* Harrington, *Medicine, Mind and the Double Brain;* and Hagner, *Homo Cerebralis.*

of the epistemic relevance of the hospital, that is, the possibility of formulating new scientific questions and promoting new research projects.

Second, the methods and ideals and the instruments and practices of physiology spread far beyond medicine's boundaries. For instance, the experimental methods of physiology became the basis of Wilhelm Wundt's physiological psychology. In these techniques, Wundt saw an essential impetus for the renovation of philosophy, which for him was particularly valuable. The road to experimental psychology in turn led back to medical departments when Wundt's student, the psychiatrist Emil Kraepelin, introduced experimental testing and measuring devices for diagnosing mental illnesses.[47] Other examples of the dissemination of physiological techniques include the influence, direct or indirect, of instruments and inscription devices developed by Ludwig, Helmholtz, and Etienne-Jules Marey on widely different fields, such as modern art, aesthetics, and linguistics.[48] Finally, the physiological models of bodily processes were closely linked to economic, industrial, and military uses.[49]

Third, physiology was omnipresent in nineteenth-century discourse and culture. It became a point of departure for new theories, served as a stage for ideological debates between materialists and dualists, and provided a guiding concept for various thinkers. Auguste Comte regarded physiology as a necessary precondition for sociology and the human sciences in general. In the 1850s, physiologists played a part, and theories and practices of physiology became disputed matters, in the so-called materialism controversy in Germany.[50] The physiologies of love and hate and of beauty and pleasure, as conceived by Paolo Mantegazza, sold marvelously in many European countries. And in Friedrich Nietzsche's critical diagnoses of European culture, physiology served as a guiding concept.[51] Finally, the clarity and assertiveness with which physiologists presented their theories and practices had a significant impact on modernity's critics. In the last third of the nineteenth century, physiologists, pathologists, and bacteriologists all performed animal vivisections, but only the physiologists were targeted by antivivisectionists. This occurred not merely because of the purported cruelty of the experiments, but also thanks to physiology's general concept of the body, which was regarded as materialistic and threatening to public morals.[52] These diverse examples il-

47. Woodward and Ash, *Problematic Science*; Bakel, "Über die Dauer"; Steinberg, *Kraepelin in Leipzig*.

48. Brain, The Graphic Method; Braun, *Picturing Time*; Crary, *Techniques of the Observer*.

49. Rabinbach, *Human Motor*; Sarasin and Tanner, *Physiologie und industrielle Gesellschaft*; Gunga, *Leben und Werk*.

50. Lepenies, *Ende der Naturgeschichte*, 169–96; Gregory, *Scientific Materialism*.

51. On Nietzsche see Wahrig-Schmidt, "Irgendwie, jedenfalls, physiologisch," and Stingelin, "Moral und Physiologie."

52. Rupke, *Vivisection*.

lustrate aspects of physiology that had no direct connection to medicine. This productive branching of physiology gave it a unique power to define the body, the soul, and many facets of human life. Neither theory and practice nor desire and reality had to be in accord in order to write a meaningful chapter of the nineteenth-century epic.

5. The Fall of the Nineteenth-Century Epic

The changes that had originated in physiology—precision measurements and quantitative definitions of bodily functions, graphical inscriptions and the reduction of physiology to physical and chemical law—were a most effective driving force of a broader transformation. Various aspects of this transformation particularly affected the relationships between people and their diseases and between patients and doctors: the development of the modern hospital, physical examinations of the body, the bureaucratic establishment of clinical recordings, and a new image of the physician as a scientist who no longer dealt with sick people but rather with pathological entities. These changes were often represented as success stories; yet from the outset they also received harsh criticism. Proponents of scientific medicine like Wunderlich and Leyden were aware that they were dealing with sick human beings and not abstract diseases, and that medicine had social and human obligations. More importantly, critical voices arose (mainly in the last third of the nineteenth century) and argued that medicine was losing its human side and degrading the body to a functioning machine or causing the patient qua individual to disappear entirely. Though it is impossible to deal here in detail with the history of alternative medicine,[53] it is of interest that a deep mistrust in scientific medicine—as seen in the rise of antipsychiatry and the ecological and feminist movements in the 1970s and 1980s—became part of the critique of Western capitalist civilization in the generation of 1968. Political engagement became the starting point for a critical and social turn of history of medicine in the 1970s.[54]

This critical turn and the bad reputation of the nineteenth century can be seen as a reaction to the history of medicine as epic, which had praised the goals of scientific medicine and the promises made in its name. Keywords like "race," "gender," "normality," "experimentalization," "control," "discipline," "exclusion," and "bio-power" are shorthand for the costs of the victory that

53. On alternative medicine see Bynum and Porter, *Medical Fringe and Medical Orthodoxy;* Cooter, *Studies in the History of Alternative Medicine;* and Jütte, *Geschichte der Alternativen Medizin.* On Germany after 1870 see Dinges, *Medizinkritische Bewegungen.*

54. McKeown, *Role of Medicine;* Illich, *Limits to Medicine.* See also Jewson, "Disappearance of the Sick-man"; Fissell, "Disappearance of the Patient's Narrative"; Duden, *Woman beneath the Skin;* and Figlio, "Historiography of Scientific Medicine."

medicine and the human sciences have achieved since the nineteenth century.[55] The scientific turn of medicine, its substitution of a patient-centered discourse with objectifiable signs, its arbitrary establishment of normality and averages, and its definitional power in terms of health and illness, was given a central role in the project of a modern "disenchantment of the world," to use Weber's famous phrase. Buzzwords and phrases like "medicalization," "professionalization and differentiation of medicine," and "the birth of the clinic" have set the framework for research on nineteenth-century medicine, derived from social history on the one side and Foucault's analysis of power on the other. Although these sources are in many ways distinct, both see body and mind, health and illness, as administered and regulated in an ever-denser net of knowledge, institutions, and instruments. From this network emerge controls, norms, and regulations that serve to make body and mind available to physiological experimentation and graphic inscription, to set normative values based on statistical analyses, and to establish epidemiological and hygienic measures.[56]

These historical studies have led to invaluable insights into the mechanisms of nineteenth-century scientific medicine, and they are all, in one way or another, indebted to Foucaultian ideas like discourse, body politic, power, and control. However, Foucault himself also made other explanatory attempts, above all with his notion of the *grande mythe eschatologique* of the nineteenth century. Developments like microbiology and the hospital, and electromagnetism and the human sciences, were supposed to contribute to people's liberation from "alienation" and "determination." In other words: man made himself an object in order to become a subject of his own freedom.[57] Difficulties arose when something was found that had not been sought, namely, the nature or essence or *fameux homme,* or *propre de l'homme.* Phenomena like madness and neurosis were analyzed, and the unconscious, which was determined by nerve impulses and instincts, became a crucial category. The unconscious functioned according to mechanical rules and was located in a discursive space that no longer had anything to do with freedom, human existence, or independence. Here a negative eschatology emerged, a theological variation of the thesis of the nemesis, which Thomas McKeown and Ivan Illich have developed in more sociological terms. In this light, Foucault can be regarded as the *last* author of the nineteenth century. For he was an author who considered man's becoming a subject to be a contingent his-

55. In addition to the literature cited in note 5, see Pick, *Faces of Degeneration;* Fischer-Homberger, *Krankheit Frau;* Jordanova, *Sexual Visions;* and Showalter, *Female Malady.*

56. See Rabinbach, *Human Motor,* chaps. 4–6; Chadarevian, "Graphical Method"; Brain, *Graphic Method,* chap. 2; T. Porter, *Rise of Statistical Thinking;* Matthews, *Quantification;* Hess, *Normierung der Gesundheit;* and Schlich and Gradmann, *Strategien der Kausalität.*

57. Foucault, *Dits et écrits,* 1:663.

torical phenomenon, yet he did not shy away from making the entire century into an historical subject and agent. Here Foucault was largely in agreement with older authors like Nietzsche, Weber, Benjamin, or Canguilhem. In one point, however, Foucault differed from them. He not only criticized the nineteenth century; he wrote the epic without heroes. Where Nietzsche attacked the bourgeois subject, Weber placed bureaucracy in the center of his analysis, and Benjamin made the city an agent, Foucault employed only inert discourses, abstract dynamics, and diverse institutions. The prison, the hospital, and the psychiatric institution are real places for Foucault, but they always point to another site; they are the sites in which an incomprehensible power becomes effective.

More recent tendencies in the history of medicine have gone beyond an uncritical adoration or condemnation of Foucault by challenging the (optimistic or pessimistic) grand narrative of the nineteenth century. John Harley Warner recently noted that in current history of medicine the word "science" appears only in quotation marks. Gone are the days, Warner wrote, when it could be self-confidently claimed that science is a unified body of ideas, practices, and methodological rules that contributed to the making of medicine. Science in medicine, he continued, is instead a highly complex web that emerges at different times and places. It cannot be known in advance whether a rhetorical formula intended to enforce particular ideals or apply certain symbols and practices in a specific context may be at issue. Warner cited historical studies wherein the rhetorical and ideological meaning of science, the language of empiricism, and belief in science were favored over the scientific content that changed medical practice. Two problems thus emerged: first, how belief in science could spread so dramatically, among doctors and in society, if it often lacked practical advantages worth mentioning, and second, the dichotomy, engendered by the focus on rhetoric and ideology, between belief and knowledge, biology and culture, and the construction of illness as a social or pathophysiological process. This dualism has been relativized with the appearance of AIDS, and thus it is now the goal "to construct more complex, less reductionist historical narratives. . . . What we stand to gain from historiographic pluralism," Warner argued, "is not an integration of history of medicine with the history of science, but a more integrated understanding of the multifaceted meanings of science in medicine."[58]

This emphasis on "historiographic pluralism" and "multifaceted meanings" does more than rearticulate a mistrust in grand narratives. It also largely ignores some important problems. Warner himself named a few: "the ideas of medical elites, the technical content of medical knowledge and practice, and the dynamics of conceptual change in the biomedical sciences."[59] Whe-

---

58. Warner, "History of Science," 166–69, 173–75, 177 (quote), 180, 183.
59. Ibid., 173. See also Rosenberg, *Explaining Epidemics*, 4–5.

ther or not one mourns this loss, it is true that, along with a positivist conception of history, the tradition of the history of ideas within the history of medicine has been thrown overboard. Along with these changes, interest in the relationship between philosophy and medicine seems to have disappeared, although it is precisely in this relationship—with physiology as a mediator—that one important key to the establishment of scientific medicine might be hidden.

This leads to a further point: The concentration on case studies and on synchronic comparative studies, in itself entirely correct and valuable, has led to a situation where only a few dare to entertain larger questions and to follow these across a longer time period. As a result, the question of the relationship between medicine and the sciences during the nineteenth century has been remarkably marginalized within the history of medicine over the past few years.[60] The point may be turned into a historiographical question: How can we move from the many different examples and meanings of scientific medicine for different groups to an overarching question or perspective that is more than a historical *Wunderkammer* in which to marvel at individual artifacts? Can we illustrate certain regularities or lines of development that offer a kind of explanation different from that of traditional Whig history (which could find no more original metaphor than the development of a puny seedling to a lushly sprouting tree)? Is it not worth the trouble to write histories other than those of a single hospital, a single university, a single discipline, a single city or region, a single instrument or technique, histories that are often limited in time to about thirty years? Since the history of medicine has depended so much on general history, and above all on social history, is it still worth bringing in other perspectives, like literary studies, art history, or cultural anthropology? In the following sections of this essay, I would like to approach some of these questions by referring to current developments in the history of science and medicine.

6. The Sciences within Medicine and Medicine as Science

It is virtually a truism that the treatment of a sick person follows different rules in a hospital than in a private doctor's office or at home. In the hospital, the question of who is the guest and who the host does as much to alter the physician's gaze as does easy access to greater resources. The resources and conditions that made hospital medicine possible in the nineteenth century were sometimes simple but nonetheless effective: greater availability of instruments and devices, and thus the increased possibility of physiological, microscopic, or chemical investigation in the immediate proximity of the pa-

---

60. An exception is Bynum, *Science and the Practice of Medicine*, although he is perfectly aware of Warner's point about the multiple meanings of science in medicine.

tient; enhanced experience due to the comparative treatment of numerous patients in similar, controllable conditions; the fundamental transformation of the clientele of a hospital from older invalids in need of care to younger patients with acute illnesses who would subsequently be released; and larger medical staffs, which resulted in an intensified exchange of knowledge among doctors and, in the case of a patient's death, a shorter path from the sickbed to the dissecting table.[61]

Vienna's General Hospital (Allgemeines Krankenhaus) was the first large modern hospital. Opened by the decree of Joseph II in 1784, it contained two thousand beds. The goal of the enlightened absolutist administration in this project, primarily to incorporate medicine into the bureaucratic state apparatus and subsequently to rationalize medicine, manifested itself above all in collecting the sick into one place under a unified administration.[62] During the first half of the nineteenth century the Paris Hospital, with its medical training emphasizing dissection, its close observation and auscultation, and its introduction of physical diagnostics and pathology, established itself as a center of scientific medicine and exerted a great attraction for foreign students.[63] In the second half of the nineteenth century, the Vienna General and various German university hospitals assumed the leading role. These hospitals were the sites for the emergence of the clinical laboratory and instruments and for the professionalization of the medical doctor in training and self-conception,[64] and they provided for the constitution of new disciplines and areas of knowledge. Extensive histories of the formation of clinical disciplines are rare, with the exception of psychiatry, which has attracted many historians.

One reason for this keen interest in psychiatry is that, according to Foucault and the antipsychiatric movement in the 1960s and 1970s, psychiatric patients were the victims of institutionalization, repression, and discipline. Here scientific medicine seemed to show its violent and inhuman side. Recent historical studies have modified this repression hypothesis, but they have also shown that the more psychiatry relied on basic sciences like neuroanatomy, pathology, and experimental psychology, the more obvious became the gulf between those diagnostic efforts and therapeutic nihilism. In the late nine-

61. On hospitals as sites of scientific medicine see Rosenberg, *Care of Strangers;* Vogel, *Invention of the Modern Hospital;* Granshaw and Porter, *Hospital in History;* Murken, *Vom Armenhospital zum Grossklinikum;* and Risse, *Mending Bodies.*

62. Lesky, *Die Wiener Medizinische Schule.*

63. Ackerknecht, *Medicine at the Paris Hospital;* Foucault, *Birth of the Clinic;* Maulitz, *Morbid Appearances;* Warner, *Against the Spirit of System.*

64. On medical technology see Reiser, *Medicine and the Reign of Technology.* On professionalization see Huerkamp, *Aufstieg der Ärzte;* Ramsay, "Politics of Professional Monopoly"; Warner, *Therapeutic Perspective;* and Bynum, *Science and the Practice of Medicine,* chap. 7.

teenth century, many psychiatrists in France, Britain, and Germany made crucial contributions to nosology, psychology, anatomy, and physiology. Inspired by these developments, they were most active in shaping discussions about values, morality, and society, even though such discussions did not significantly benefit their patients.[65]

Another question to be posed is the extent to which the Parisian and Viennese hospitals actually represented scientific medicine in the nineteenth century in a paradigmatic way. According to Warner, American medical students in Paris were less interested in soaking up the medical theories taught by the most famous teachers than in the unique opportunity to train at the sickbeds. Their studies in Paris had more to do with the acquisition of practical knowledge and experience than with textbook knowledge. By midcentury, when Paris no longer presented the advantages that it had at the beginning of the century, foreign students quickly turned elsewhere, mainly to Germany.[66]

Rhetoric about the hospital changed around 1850; the hospital then began to be represented as indispensable to the practice of scientific medicine, and physical methods of investigation, case histories, microscopic and chemical investigations, as well as dissection records were established. The transformation of clinical thermometry is a typical example of this phenomenon. Although its method had long been established, clinical thermometry became significant for hospitals thanks to the usefulness of graphically representing fever curves. For Wunderlich, the Leipzig clinician, it was important that the measurements exhibit "physical precision [Exaktheit]."[67] Wunderlich wanted to demonstrate that clinical methods met physical physiology's ideals of objectivity and precision. At the same time, he and a number of other medical doctors intended to practice "the study of physiological medicine [Heilkunde]," which set itself against the "localizationism [Lokalisationismus]" of pathological anatomy. While they did not see localizationism as unnecessary, they thought it ignored the physiological dynamism of illness, the temporal course, and the patient's general condition. Since medical doctors after Wunderlich were occupied with the patient and not the illness, thermometry served multiple functions: it was scientifically reliable and better for the patient than other scientific methods.[68]

This example again illustrates the complex relationship between physiol-

65. The repression hypothesis has been taken by Foucault in many of his short texts in Foucault, *Dits et écrits*. For a more balanced view see the summary by R. Porter, *Greatest Benefit*, chap. 16. On scientific psychiatry and the professionalization of psychiatrists see, e.g., Bynum, Porter, and Shepherd, *Anatomy of Madness;* Goldstein, *Console and Classify;* and Kaufmann, *Aufklärung.*

66. Warner, *Against the Spirit of System,* 300–364.

67. Hess, "Entdeckung des Krankenhauses," 95–97, 99–101; Wunderlich, *Verhalten der Eigenwärme,* 58. For Great Britain and the United States, see the overview by Booth, "Clinical Research."

68. Wunderlich, *Geschichte der Medizin,* 311–18, 357–66.

ogy and clinical practice. Although thermometry did not emerge from physiology, its techniques, such as precision measurement and the graphical method, contributed to its reputation. Moreover, in this example physiology is understood as leading to an holistic understanding of illness rather than to localizing it in a particular organ. What does this clinical use of physiology mean for the relationship between medicine and science? A number of historians have argued for a polarization between bench and bedside, while others have suggested an integrationist model.[69] The model of polarization between anatomy and physiology, experimental pathology and physiology, and laboratory and clinic has its place within the contexts of professionalization and of disciplinary division. Perhaps even more importantly, this debate may help determine whether the physician's place was at the bedside or at the bench.

The place of the physician was a crucial issue within the context of the rise of bacteriology in the 1880s. While Robert Koch was extremely skeptical about clinical knowledge, and hence focused on laboratory research when establishing medical bacteriology, German clinicians like Leyden, Friedrich Theodor Frerichs, Bernard Naunyn, and others welcomed the innovations of bacteriology. At the same time, however, they "proposed a set of techniques of clinical investigation that promised to be every bit as rigorously 'scientific,' but which returned their practitioners to the bedside."[70] The reservations by clinicians about the dominance of bacteriology only increased when, around 1890, Koch unsuccessfully ventured into the clinics with his tuberculosis cure. More generally, this example suggests that academic physicians often made practical use of results from physiology, bacteriology, and chemistry. They appreciated the advantages of "science" without handing over their clinical authority to any single scientific field. In brain research, for example, the general acceptance of the localization of mental qualities was largely dependent upon the usefulness of localization in clinical neurology.[71] In short, as we have already seen in the case of physiology, science meant different things to different actors and in different contexts.

Taking the view that there were various types of scientific practices within medicine rather than any strict separation between science and its application, one arrives at the question of the extent to which the hospital was transformed as a space of knowledge in which physicians worked scientifically. According to Michel de Certeau, space is a dynamic network of various elements and depends upon the variability of time: "Space occurs as the effect

69. See Geison, "Divided We Stand"; La Berge, "Dichotomy or Integration?"; Jacyna, "Laboratory and the Clinic"; Moulin, "Bacteriological Research"; Lawrence, "Incommunicable Knowledge"; and Warner, "Fall and Rise of Professional Mystery."

70. Maulitz, "Physiologist versus Bacteriologist." On Koch's experimental bacteriology see Gradmann, Medizin und Mikrobiologie, chap. 5.

71. Star, Regions of the Mind; Hagner, "Aspects of Brain Localization."

produced by the operations that orient it, situate it, temporalize it, and make it function in a polyvalent unity of conflictual programs or contractual proximities."[72] This is a rather unusual understanding of space. Certeau aims to conceptualize space as an "anthropological space" as opposed to a "geometric space." Following Maurice Merleau-Ponty in this respect, Certeau understands anthropological spaces as being created by human experiences, perceptions, and practices. What can such a conceptualization mean for an understanding of spaces of knowledge like hospitals, laboratories, and the dissection hall? These spaces consist of stable structures serving the process of research, including a building's architecture, electrical facilities, instruments, apparatuses, etc. But such an account says nothing about the intrinsic variability of, for example, laboratory equipment. Laboratories are as variable as the experiments performed within their walls. Therefore, it is necessary to ask how the spaces are shaped within which the object—the patient, the corpse, the tissue, the experimental animal, and so on—is located and relocated, established and dismissed. In short, the creation of a space is linked to a history, and it remains a space of knowledge production only so long as it is variable and has open boundaries. If this is no longer the case, the space degenerates into a storehouse, if not a junk room, filled with objects that are no longer needed for a specific research enterprise.

If the laboratory is so defined, the sharp divide between hospital and laboratory is no longer particularly useful. As the example of thermometry indicates, very similar practices and methods can be used in both spaces, even if they have only a partially identical function. It is, however, crucial to historicize these spaces of knowledge, to look at strategies concerning the inclusion and exclusion of knowledge and skill as well as material cultures and epistemic objects. Both the collaboration between hospital and laboratory—as shown by Lenoir for Leipzig—and the entirely different function of pathological experiments—which Peter Schmiedebach has described for the case of the Berlin pathologists Virchow and Ludwig Traube—could serve here as test cases. Traube's laboratory was an integral part of his clinic, while Virchow represented pathology as a science independent of the hospital, though it also served the purposes of clinical medicine.[73]

The importance of experimental physiology in the formation of a unified understanding of scientific medicine should not obscure the dependence of medicine upon other technologies of knowledge production, such as pathologic anatomy, organic chemistry, and microscopy. Although chemistry as an experimental endeavor had much in common with physiology and although microscopy was practiced by physiologists like Müller, Jan Purkyně and

72. Certeau, *Practice of Everyday Life,* 117.
73. Lenoir, *Instituting Science,* 96–130; Schmiedebach, "Pathologie bei Virchow und Traube."

Brücke, it is not useful to subsume all of these practices under the heading of physiology. Considering that physiology did not become of much importance before the middle of the nineteenth century and that other tools and methods of scientific medicine developed either earlier or independently of physiology, it seems appropriate to look for a more refined classification. John Pickstone has recently attempted to differentiate various "ways of knowing." Although he focuses on problematizing the boundaries of medicine as well as of science and technology, two of those so-called types of knowledge relate to the issues discussed here. One type, the analytic-comparative, developed between 1780 and 1840 and treated its objects and phenomena as divisible into individual elements. Objects were no longer described and classified based simply on their surface but instead were linked to deeper structures and functions. Simultaneously, the dead body and its components became the preferred object of study. Another type was the experimental. Developed around 1850, its main feature was controlling life phenomena; this could be primarily achieved through animal experimentation. According to Pickstone, the strength of the experimental type lay in its promise to transform medicine: "If something can be controlled in laboratory animals, it can, in principle, be controlled in human patients."[74]

One great advantage of Pickstone's "ways of knowing" approach is that it combines larger ensembles of scientific practice that do not rigidly follow the regime of the disciplines and that cannot be fixed by their different intentions or aims. On the other hand, this typology—like all typologies—has the disadvantage of being static. It requires investigation of how the types themselves changed historically. I have already mentioned the different uses of physiology. Bacteriology, as it was practiced by Koch, would fit in both the analytic and the experimental types. New analytic instruments and techniques (like the stethoscope, ophthalmoscope, and microscope) required doctors to acquire new social and technical abilities ranging from specific training of the senses (in terms of optical and acoustical methods of diagnosis) through experimental skills (in terms of systems of inscription) to the translation "of graphic symbols into pathological meaning."[75]

The microscope was, in this context, perhaps the most important instrument. It began its triumphant march in the second third of the nineteenth century in Germany, initially in cell theory and pathology, later in bacteriology and in the hospital. As the late Roy Porter has written: "By mid-century,

---

74. Pickstone, "Objects and Objectives," 16, and *Ways of Knowing*. See also Figlio, "Historiography of Scientific Medicine," and Pauly, *Controlling Life*.

75. Frank, "Telltale Heart," 274. On the stethoscope see Lachmund, *Der abgehorchte Körper* and "Making Sense of Sound"; on thermometry see Hess, *Der wohltemperierte Mensch;* and on the early history of the ophthalmoscope see Tuchman, "Helmholtz," 29–41.

the microscope was giving medicine new eyes." But how these "new eyes" were used, what they could and could not see, and what the medical students trained in microscopy could later do with their knowledge deserves further research. What was the relationship between instrument makers, companies, advertising strategies, and doctors? When did the microscope begin to appear as an emblem of scientific medicine, for example, in photographic portraits of physicians? In which different cultural and social settings was the microscope localized? For example, Purkyně's *naturphilosophisch* praise of the microscope as an instrument for uncovering the unity of the organic world was radically different from Matthias Schleiden's pragmatic introduction of it in terms of function and use.[76] Histories of the microscope and other, less complex instruments—the fever thermometer, the ophthalmoscope, the laryngoscope, the electrocardiograph, and electrotherapeutic instruments or X rays—would be less a part of the history of scientific instruments in the classical sense and, instead, more about the history of their scientific, social, and cultural import. To that extent, it might be justified to call for biographies of instruments.[77]

The historicity of Pickstone's types becomes most clear in the spatial, instrumental, and cultural changes that determined experimental physiology. Sven Dierig, in his analysis of du Bois-Reymond's physiological work in Berlin, has argued that the space into which du Bois-Reymond's mode of doing science was inserted developed from a "parlor physiology" in the 1840s, into a "workshop physiology" in the 1850s and 1860s, and finally, from the 1870s onward, into a large-scale "industrial physiology." The division into multiple, independent departments corresponded to the division of labor among coworkers. The technical facilities and running of the institute were closely tied to the technological-urban development of Berlin.[78] Physiology no longer drew on the heroic labor of the individual but instead on a factory-like workplace, and these changes ran parallel to the development of labor in the nineteenth century. Even though physiology also developed in smaller cities, this model offers a promising point of departure for comparing scientific with other forms of labor. The question is whether similar developments

76. R. Porter, *Greatest Benefit*, 321. Purkyně, "Mikroskop"; Schleiden, "On the Use of the Microscope." On the romantic understanding of the cell theory see Jacyna, "Romantic Programme." La Berge, "History of Science," gives an overview and further references. On the importance of microscopical teaching courses in Paris and in Germany see La Berge, "Medical Microscopy," and Kremer, "Building Institutes."

77. See the overview by Rothschuh, "Bedeutung apparativer Hilfsmittel." In addition to the references cited in note 75, see on electrocardiography Lawrence, "Moderns and Ancients," and Borck, "Herzstrom"; on electrotherapy, Bryan, Wilhelm Erb's Electrotherapeutics, chaps. 9–10; and on X rays, Pasveer, Shadows of Knowledge.

78. Dierig, *Experimentierplatz in der Moderne*.

in other spaces of knowledge and forms of labor can be found in other medical disciplines and, above all, in clinical science as it was carried out in the hospital.

Of course, the division of labor and specialization are not a new development in the history of medicine. The development of cardiology in England, for example, was possible only because those who favored specialization made laboratory knowledge the pivotal point in the definition of heart disease.[79] Yet the degree to which the function of laboratory knowledge was different for clinical doctors is clear from the fact that laboratory tests in the late nineteenth century often had the function of confirming the clinician's extant knowledge rather than setting it on a new foundation. It seems clear that what was at issue was not so much the use of scientific theories, but rather the transfer of instruments, persons, and practical know-how.

Until now, scholars have used two strategies to explain why certain forms of knowledge develop at a given time and place: The first involves the opposition between individualized diagnostic and therapeutic skill and experience, on the one side, and standardized, objective, and impersonal technique, on the other. The second concerns the investigation of scientific schools in which certain theories, technologies, research objects, and thematic issues are favored.[80] It might be true that in investigating specific scientific schools, the strict opposition between individualized skills and objective scientific method makes little sense, since the two go hand in hand. In order to examine this idea more closely, the historicity of certain basic scientific values must be studied more closely. What significance did categories like "objectivity," "rationality," "evidence," "precision," or "experience" have in the formation of scientific medicine? Were these categories themselves changed in the process and variously adapted to the needs of medicine, or did they simultaneously assume different meanings in different places?[81]

7. The Dissemination of Scientific Medicine

The issue of "the doctor," a subject that has already been mentioned in the context of "individualized skill," is somewhat analogous to those of the spaces and instruments discussed in the previous section. This issue has not received much attention beyond the genre of biography. In biographical studies, however, skill has often enough been reduced to essentialistic categories

79. Lawrence, "Moderns and Ancients."

80. Lawrence, "Incommunicable Knowledge"; Sturdy and Cooter, "Science, Scientific Management." On scientific schools, see Geison and Holmes, *Research Schools*.

81. For the history of science see Daston, "Objectivity and the Escape from Perspective" and "Objectivity versus Truth"; Daston and Galison, "Image of Objectivity"; and Wise, *Values of Precision*.

like creatitivity or genius. The French Annales school made the first attempt to historicize the relationships between an author, his works, and his times. By introducing the concept of "mental equipment" *(outillage mental)*, Lucien Febvre sought for a specific way to reconstruct the tools, mentalities, and representations of a given civilization. He conceived of mental equipment as a collective entity that created the framework for individual thought and practices, as exemplified in his biographies of Martin Luther and François Rabelais.[82] How, then, can the concept be put to use in reconstructing the notion of the research process and the ways in which notions of scientific medicine were disseminated within the medical community?

How did the specific attitudes and practices that modified, transformed, or broke with traditional habits come to be mobilized, and what physical and mental processes of disciplining, what abilities and mental repertoire, participated in this? The concept of genius was long used to explain scientists' originality; it was thus seen as an individualized counterpart to technical development or supraindividual rationality and objectivity. Apart from psychoanalytically motivated attempts at explanation, this individualized and "embodied" approach to scientific practices has been largely abandoned. Instead, structuralist history, shaped mainly by Foucault's concept of bodily discipline, interpreted so-called individual skills and patterns of behavior as determined by discursive structures. Foucault's rigid claim that the mind is only the result of certain practices of bodily discipline, along with his focus on surveillance and power, underestimated the flexibility of human styles and habits.

Recent developments in the cultural history of science have, however, reinspired an interest in the notion of "embodied knowledge." Still, there are historiographical problems in writing this kind of cultural history. The social historian Jürgen Kocka has criticized cultural history by saying that one can hardly grasp the "airy" zone of values, traditions, and practices in a precise fashion.[83] In the history of science, Michael Polanyi's concept of "tacit knowledge" has provided ammunition to those voicing this objection. Kathryn Olesko has complained that the unspecifiable nature of tacit knowledge has fatal historiographical consequences: if "tacit knowledge" is not specifiable, then "the actual mechanisms for acquiring skills and values central to the formation of a school f[a]ll outside the domain of direct historical investigation."[84]

The methodological problems with direct historical investigation should not, however, lead to the erroneous conclusion that tacit knowledge is a *quan-*

---

82. See Chartier, "Intellectual History," 18–22.
83. Kocka, "Bürgertum und Bürgerlichkeit," 44.
84. Olesko, "Tacit Knowledge," 20. See Polanyi, *Tacit Dimension.*

tité négligeable. Rather, the issue is how skills, gestures, findings, preferences, perceptions, and so on can be historically analyzed, independently of whether or not they are reflected in historical agents. Alain Corbin, in his study of the history and anthropology of perception, has pointed out that historians often fall into the trap "of confusing the reality of the employment of the senses and the picture of this employment decreed by observers."[85] This kind of problem can arise, for example, when the world of the patient is reconstructed solely on the basis of medical case histories and hospital files. Of course, this danger lurks around all historical documents—textbooks, public speeches, letters, diaries, autobiographies, and so on. But the issue is not whether something like an authentic situation can be reconstructed; it is, rather, whether or not it is possible to reconstruct the relations between given mental equipment and material structures in a given historical situation. Which expectations developed around the doctor who referred to science? What communicative abilities did he have? What masculine codes of honor were surgeons taught, for example, which even today block access to the profession for women? What virtues, passions, and idiosyncrasies were supposed to characterize the scientific doctor? What was the relationship between the mental characteristics demanded by the work of scientists and physicians and the mental characteristics formed by the culture as part of the general image of the scientific doctor?[86] Some of these questions have been addressed in laboratory studies. For example, Karin Knorr-Cetina has introduced the term "local idiosyncrasies" to refer to local know-how and local interpretations of methodological rules in research laboratories.[87] Such focusing on the local production of knowledge has been part of the practical turn within the history of science for about the past twenty years, but the concept of local knowledge has only recently been brought together with categories like embodied knowledge and the persona of the scientist.[88]

Studies of the microstructures of scientific practice have little to say about the question of why scientific medicine has become so powerful and attractive. It is a remarkable fact that in medicine—as in almost no other modern profession—an ideal was successfully constructed within only a few decades, an ideal that held true far into the twentieth century and, at least in part, still holds true. Scientific medicine was (and is) a profession that for generations of young men (it remains a male genre) was (and is) worth striving toward, and its power seems to no small extent to have been determined by the image of the scientific doctor. This image can be partly but not fully explained by con-

85. Corbin, "History and Anthropology of the Senses," 187.
86. Similar questions are raised by Warner, *Therapeutic Perspective,* viii.
87. Knorr-Cetina, *Manufacture of Knowledge.*
88. Lawrence and Shapin, *Science Incarnate;* Daston and Sibum, "Scientific Personae."

ventional social categories like career motivations and the consolidation of authority or prestige. Furthermore, scientific schools created identities, narratives, and myths; formed genealogies; and made groups of doctors and scientists into culturally visible entities. Scientific schools thus served primarily to attract young students; however, they were also known and recognized beyond medical circles, far into the educated middle classes.[89]

Before the codes of scientific medicine were successfully communicated to a broader public, generations of medical students first needed to be trained in the cognitive, cultural, practical, and social aspects of scientific medicine, since it was they who would bring this new medicine into immediate contact with the patient—not only in the capitals and university hospitals, but also in the provinces and in private practice. In cultural history, scholars like Robert Darnton and Roger Chartier have studied the material requirements and the effects of book production as well as the use of printed texts by academic and nonacademic readers.[90] We still know far too little about the ways in which conceptions of scientific medicine were inscribed and how they were evaluated. What were the reading practices of medical students and doctors? Did they prefer textbooks, monographs, or journals?[91] The material and economic side of medical textbook and journal production also deserves attention, including such factors as typesetting, quality of reproduction, price, and distribution. And more generally, what is the history of the relevant publishing houses?

A second large field of study revolves around the question of how scientific medicine was anchored in university teaching.[92] The studies of Richard Kremer, Lenoir, and others have shown the importance of microscopy tutorials, chemical and bacteriological investigations, and practicing new methods of diagnosis at the sickbed.[93] But this was only a part of the university curriculum. The relation between the practical and the theoretical aspects of study remains largely unclear. What was communicated in lectures? What questions were asked of the students on examinations? What kinds of subjects

89. Zloczower, Career Opportunities; Ben-David, "Scientific Productivity"; R. Turner, "Growth of Professorial Research"; Geison and Holmes, Research Schools. From a prosopographic perspective see R. Turner, In the Eye's Mind.

90. See Darnton, Business of Enlightenment; Chartier, Cultural Uses of Print and Order of Books. See also Frasca-Spada and Jardine, Books and the Sciences.

91. On medical journals, see Bynum, Lock, and Porter, Medical Journals; Broman, "J. C. Reil and the 'Journalization' of Physiology"; and Bonah, Sciences physiologiques. On popular books and journals in Germany see Daum, Wissenschaftspopularisierung, 237–376.

92. See Numbers, Education of American Physicians.

93. On Germany see Lenoir, Instituting Science, chap. 5; Kremer, "Building Institutes"; and Tuchman, "From the Lecture to the Laboratory." On the United States see Numbers, Education of the American Physician. On France see Weisz, Medical Mandarins, and Moulin, "Bacteriological Research."

were appropriate for doctoral dissertations? To date there has scarcely been any research in this area; dissertations, lecture notes, and examinations (as far as they are still extant) have barely been analyzed. Letters from students should also come into question here.[94] Each of these sources would help to illuminate how the content, criteria, and values of scientific medicine were imparted to students.

The most relevant genre for disseminating scientific medicine is biography. Its ongoing fascination lies in its double appeal: on the one hand, it is arguably the most popular historiographical genre; on the other hand, autobiographies and biographies (including obituaries, newspaper articles, and eulogies) are important sources for historians because they played a significant role in spreading the reputation of scientific medicine and of the physician. The same is of course true for other fields, from art to politics to science. Biography has not, however, been regarded very highly in the history of science after World War II, in part due to its notorious role in shaping the myths of "great men." In 1979, Thomas Hankins noted: "The bad old history of science of the early twentieth century, which we have all been taught to abhor, was largely biographical. Books from this period usually consist of a series of illustrious names, each followed by birth and death dates, an occasional anecdote, and a description of that person's 'discoveries.' History was the assigning of priorities—every worker at the temple of science receiving credit for the bricks that he personally laid."[95] Even if more recent historiography has abandoned such a naive approach, there is an ongoing debate about the new role of biography. Its limited scope and its concern mainly "with a special type of scientist: the great scientists whose works have been of pioneering importance"[96] has deterred social historians as well as historians of ideas. Other historians do not see a wide gulf between (conservative) biography and (progressive) social history. In the words of Charles Rosenberg: "I have always thought very positively about biography and never felt that there was a necessary contradiction between the writing of an individual life and history generally . . . a life can be construed as a sampling device—as a controlled and internally coherent batch of data, a chronologically ordered set of realities and relationships perceived and understood by a particular actor."[97]

Rosenberg's interpretation raises the question of what methodological and historiographical demands current biographies must fullfil. Even if the hagiographic biography shows no signs of dying out as a popular genre, it

94. Warner, *Against the Spirit of System.*

95. Hankins, "In Defence of Biography," 2–3.

96. Kragh, *Introduction,* 173.

97. Rosenberg, *Explaining Epidemics,* 215. For recent overviews of biography see Shortland and Yeo, *Telling Lives in Science,* and Gradmann, "Leben in der Medizin." For theoretical remarks on writing biographies see Raulff, *Der unsichtbare Augenblick,* 118–42.

cannot set the standard for reputable scholarly biographies. More promising would be a biographical practice that does take the individual's life seriously, while at the same time investigating the "chronologically ordered set of realities and relationships" as a mixture of life and legend, events and (self-)constructions. How did the image of Helmholtz, Osler, or Bernard develop, and in turn form a whole generation of scientists? How did Pasteur and Koch shape their scientific egos? What scientific practices were tied to this shaping, and how did work and personality fuse? The embedding of an individual life in collective ideas, values, and practices of a specific time demonstrates that biographical greatness is not the result of free-floating intelligence, but of multiple social processes that reveal something about the composition of the world in which the subject lived. From this perspective, Michael Shortland and Richard Yeo have argued that "Steven Shapin's recent work shows how a biographical focus on Robert Boyle can move to an historical analysis of social norms: Boyle's life can be appreciated as an exemplar of what it meant to be a natural philosopher in the early modern period."[98] The alliance between sociocultural history and biography, then, might uncover what it meant to be a scientific physician in Victorian England, Wilhelminian Germany, or fin-de-siècle France.[99] Finally, we should not forget that the narrative reevocation of these worlds includes descriptions, images, metaphors and anecdotes. Biography is thus also a work of literature that should be judged according to its literary merit.

8. Patients as Consumers

Neither publishing practices nor university teaching has, as yet, told us very much about practical work at the bedside. With the historiographic turn toward the "therapeutic" perspective in the 1980s, the meaning of healing became an object of study in the history of medicine. Rosenberg has noted that therapies in the first half of the nineteenth century worked and that countless patients accepted bloodletting or purging as a matter of course; at the least, this shows that some historians have asked the questions why, in what contexts, and with what arguments new therapies emerged and older ones were replaced.[100] At stake are chemical and pharmacological developments as well as changes in the mentality of doctors and patients, along with new forms of knowing and experiencing one's body. The question is how these new perceptions were established. If, for example, it could be shown that middle-class

---

98. Shortland and Yeo, "Introduction," 37; and see Shapin, *Social History of Truth.*

99. Though not a biography in a traditional sense, another example is Geison, *Private Science.*

100. Rosenberg, *Explaining Epidemics*, 9–31. See also Warner, *Therapeutic Perspective;* Vogel and Rosenberg, *Therapeutic Revolution;* and Weisz, *Medical Mandarins,* chap. 7.

patients associated scientific medicine with certain methods of diagnosis, certain therapeutic interventions, an exact record of the course of illness, a sharpness of observational ability, and the doctor's healing effect, and that all these had effects on their modes of perception or experiences of their own bodies, then it would be possible to draw conclusions about the practical effects and the cultural dominance of scientific medicine.

As argued above, in the social history of medicine the patient was often characterized as the victim of a paternalistic medical system. An alternative view regarding the patient as a consumer would mean emphasizing the patient as an active participant who tries again and again to find a balance between the organized world of normalizing and disciplining the body, of doctors and hospitals, and of instruments, diagnostic techniques, and therapies. Following the pathbreaking work of Roy and Dorothy Porter and others, we have learned a lot about the world of patients, but most of the studies to date deal either with the early modern period or with economic aspects. Only a few studies explicitly emphasize the relationship between scientific innovations in medicine and patients' experiences and wishes.[101] Edward Shorter, focusing on the nineteenth century, has pointed out the irony of the doctor-patient relationship in the fact "that the prestige of the doctor rested not upon his improved ability to cure, but rather to understand disease and to establish an accurate prognosis."[102] Is it sufficient to assume that patients were satisfied with certainty? Or were more complex experiences involved?

The dissemination of the stethoscope, which represented the symbolic power of medicine as no other scientific instrument did, offers a striking example of a new adjustment among doctor, instrument, and patient. René Théophile Hyacinthe Laennec published his important work in 1819; by 1825, stethoscopes were being displayed in the windows of Paris medical shops, and very rapidly translations of Laennec's account appeared in English, German, and Italian. But these facts still do not explain the overwhelming symbolic power attributed to the stethoscope. As Bynum wrote: "By 1832 . . . the stethoscope symbolized the progressive forces in medicine. . . . By mid-century the stethoscope had come near to its present symbolization of the profession itself."[103] This triumph requires explanation: If one considers that auscultation went through two entirely different, basically incompatible courses of development in France and Germany, then how could the new identity of the competent medical doctor build itself on this instrument?

---

101. R. Porter, "Patient's View"; D. Porter and Porter, *Patient's Progress;* Duden, *Woman beneath the Skin.* On the economic aspect see Digby, *Making a Medical Living.* On patients' autobiographies in the nineteenth century, see Lachmund and Stollberg, *Patientenwelten.* For a useful overview see Loetz, "Medikalisierung."

102. Shorter, "History of the Doctor-Patient Relationship," 791.

103. Laennec, *Treatise;* Bynum, *Science and the Practice of Medicine,* 37–41, quote on 40–41.

What changes in the physical relationship between doctor and patient were at work here? Why was this technology accepted without opposition? What role did advertising play? Did the stethoscope find a place on shelves at home? When was it given as a gift? One key to the instrument's success seems to lie in the fact that it created a new bond between doctor and patient, as Jens Lachmund has shown.[104] For the patients, "noises" became part of their reality and thus their knowledge of illness. The sounds heard through stethoscopes were at first considered strange, but they were rapidly individualized. This kind of permanent transformation in the patient's self-perception was of great importance to the power of scientific medicine.

Finally, it is often overlooked that consumers of scientific medicine were not necessarily sick or bound to the social role of patient. Consumers were also readers of popular treatises and magazines, visitors to exhibitions, and listeners at public speeches. In his analysis of hygienic literature and its reception in France and Germany, Philip Sarasin has convincingly argued that the normalizing power of medicine has been overrated in its importance for the construction of the modern body. Instead, discourses about hygiene shaped an individualized understanding and concerned the self-regulation of health. Because therapy was often regarded as ineffective, such discourses sometimes stood in opposition to medical treatment. The main characteristics of hygiene, in contrast, were the cunning combination of diathetic ideas and values that reached back to antiquity and modern physiological theories of bodily functions that were located at the center of scientific medicine. For the hygienists, bodies were "excitable machines" with needs and desires, and the task was to find a well-tempered way to take care of the body. Not coincidentally, Sarasin has relied upon the late work of Foucault and transferred the concept of the *soucis de soi* from antiquity to the nineteenth century.[105] It remains an open question as to whether and to what extent scientific medicine has served as a tool for the "technologies of the self."

9. Epilogue

We have now returned to our point of departure: namely, how did the concept of "scientific medicine," despite its multiplicity of meanings in the nineteenth century, achieve stability and permanence? What recognizable structures made it possible, over a certain period of time, for a powerful understanding of scientific medicine to become dominant? I have argued that physiology served as the glue that held scientific medicine together; its prac-

104. Lachmund, *Der abgehorchte Körper.*

105. Sarasin, *Reizbare Maschinen;* Martin, Gutman, and Hutton, *Technologies of the Self;* Foucault, *Souci de soi.* See also Tomes, "Private Side of Public Health."

tices and methodology contributed to this cohesion, as did its role in creating the grand narrative or epic of scientific medicine. This epic worked remarkably well in overshadowing the various, sometimes contradictory, meanings of scientific medicine and the sharp conflicts between the bench and the bedside. In the latter part of the twentieth century, a revised epic of scientific medicine, dominated by discipline, bio-power, and "silencing the patient," has been the engine for a more critical social history of medicine. Now that the twentieth century has ended and the twenty-first has begun, we have a greater perspective on the nineteenth, allowing us further to historicize the latter. To be sure, we shall certainly not write an entirely different history of the nineteenth century, but it is perhaps time to adopt a more balanced view of it, and so not immediately assume that the course of all our joys and sorrows was definitively set in that period. Certainly a number of historians of medicine have already begun to adopt this viewpoint. But if both the optimistic history of modernization and the pessimistic history of discipline and bio-power have now lost their attraction and historical plausibility, a new question arises: Is an intensive occupation with the nineteenth century even worth the effort? Apart from professional historians of the nineteenth century, to whom should such research be addressed?

More than twenty years ago, Bynum welcomed the opening of the history of medicine to social history, anthropology, and literary history. But at the same time he presciently warned against losing contact with the audience of medical doctors.[106] Since then, this gap has grown still larger. The far-reaching consequences for the academic existence of the history of medicine are obvious. History of medicine will find itself ever more marginalized within the medical faculties and schools, and could even disappear from them entirely, becoming another specialization within general history or within the history of science. To be sure, there are many good reasons for establishing history of medicine within history departments.[107] On the other hand, confronted with the rise of medical ethics, a subject that many clinicians seem to find useful in their clinical work, historians of medicine have recently pondered the question of what they can offer physicians. The various answers emphasize a cultural perspective, accepting the postmodern pluralism of viewpoints and diversity of interests, or arguing for an expansion of the humanities within medical schools: Guenter Risse, for instance, writes that "history's expanded universe of voices and agents, power brokers and consumers, provides frames for orientation and identity, teaches perspective and skepticism, reveals the contingency and clash of meanings, the power of knowl-

---

106. Bynum, "Health, Disease and Medical Care," 252.
107. For an overview on history of medicine in Britain see Pickstone, "Development and Present State."

edge." This pluralistic account might help to "shape new and realistic boundaries and objectives for the medicine of the future." More specifically, Alfons Labisch argues that "history within medicine" might help to solve medical problems because "the topics derive from internal problems of the current development of medicine."[108]

Either approach can become fruitful only if a refined historiographical reflexivity, derived from the historical disciplines, is united with an interest in scientific and medical content. This kind of historiography would truly be a complex and difficult undertaking. It would mean relativizing the dichotomies between science and art, knowledge and culture, instrumental practice and ideology, and instead conceiving of these antagonisms as historically variable ensembles that need to be constantly reexamined. How will the wider audience for this kind of study look? In the course of the nineteenth and twentieth centuries, the history of medicine has undergone several attempts to improve medicine and to give it a more human face. These attempts have not been very successful, so there is not much reason for optimism. But it may also turn out that medical doctors, bioengineers, health managers, and lifestyle advocates may see the advantages of a culturally enriched history of their professions. Until the outcome becomes clearer, we shall have to live with this tension between the high-quality standards of historiography and professional interests.

Acknowledgments

For their helpful comments, criticisms, and suggestions I thank David Cahan, Christoph Gradmann, Jens Lachmund, and Andreas Mayer. I am particularly indebted to Laura Otis, who helped me turn this essay into readable English.

108. Risse, "Reflected Experience," 212; Labisch, "Von Sprengels 'pragmatischer Medizingeschichte,'" 249.

Four

THE EARTH SCIENCES

DAVID R. OLDROYD

❧

TO APPRECIATE the extensive but scattered literature on the history of the earth sciences[1] one must first consider the meaning of the term "earth science," its origins, and its principal authors. The term is recent, coming into widespread use in the 1960s and 1970s, accompanying the establishment of the plate-tectonics paradigm, and today refers to all the sciences concerned with the earth, including, for example, oceanography and paleoclimatology. The term is thus broader than "geology," traditionally concerned with rocks, minerals, and fossils and their historical processes of formation and change. In the seventeenth and eighteenth centuries, the field of "earth sciences" was divided into the study of fossils, crystals, minerals, and rocks, and "theories of the earth," which were often concerned with its origin. Though the term was proposed in the eighteenth century, "geology" was established in its modern form in the nineteenth century and was soon extended, developing various subdisciplines, as we shall see presently.

There are in the world today perhaps thirty or forty scholars whose principal professional or intellectual efforts are devoted to the history of geology or the earth sciences, and much of their attention is devoted to the study of the nineteenth century. Some are trained historians, but many are older or retired scientists. Almost all have geological as well as historical experience. These writers divide more or less clearly into two groups: the scientist-historians and the historians of geology. More work has been done by the scientist-historians. The emphasis is different for the two groups. The scientist-historians are devoted to the accurate "recording of the history of geology";[2] their work is often internalist in character, with attention to museum holdings, geological sites, and so on. The historians of science, by contrast, give greater attention to social issues, institutional histories, the intellectual context of geology, patronage, scientific controversies, philosophical problems arising from the

---

1. Sarjeant, *Geologists and the History of Geology*. For ideas on the historiography of geology, see particularly Greene, "History of Geology," and Good, "Toward a History."

2. The words used to describe the work of those receiving the Sue Tyler Friedman Medal of the Geological Society, London, for contributions to the study of the history of geology.

geological literature, and the actual practice of geology, which is sometimes taken for granted by geologist-historians. The latter have the great advantage of knowing where the important problems lie, or have lain, within geology. However, the internalist / externalist divide is weakening, and does not co-incide exactly with that between scientist-historians and historians of science. As we shall see more clearly in what follows, the field of the study of nineteenth-century earth sciences lacks unity.

1. Divisions of the Field of Earth Sciences in the Nineteenth Century and Its Historiography

By the end of the nineteenth century, various subdivisions had emerged in writings about the earth sciences or geology. First, the field had been divided into quite a large number of subject matters: notably mineralogy, crystallography, petrography, igneous petrology, metamorphic petrology, sedimentary petrology and sedimentary processes, paleontology, stratigraphy and mapping, tectonics, geomorphology and glaciology, volcanoes and volcanism, geodesy, gravimetry, seismology, geomagnetics, the earth in relation to cosmology (for example, the origin and age of the earth), geochemistry, theories of the earth's interior, and economic geology. Second, studies had focused on particular regions: for example, Britain, Australia, the Alps, the several American or German states, the "Old Red Sandstone country," an employer's estate, a coal field. Third, practitioners had been categorized as members of different "schools": Huttonians, Wernerians, catastrophists, Lyellians, geosyncline theorists, supporters of notable controversialists such as Adam Sedgwick or Léonce Elie de Beaumont. Fourth, the field could be characterized in terms of techniques or methods of study: for example, surveying–fieldwork–mapping, "museum collection," paleontological reconstruction, mathematical models of the earth's behavior, microscopic examination of rocks, or the experimental modeling of geological processes, especially the foldings of strata and the formation of crystalline rocks. The field could also be studied according to various conceptions of the stratigraphic column or boundaries therein, for example, the Jurassic System or the Precambrian–Cambrian boundary. Historians might also wish to subdivide the field sociologically, for example, according to the class or occupation of those working in it.

Though the foregoing divisions present "natural" categories for the study of the field in the nineteenth century, historians of geology have not in practice always followed them. First, not all of the divisions were in place at the beginning of the century. Second, much history of geology has focused on "great men," particular controversies, or topics thought to be of more general scientific or philosophical interest. Moreover, some approaches are simply easier to follow through than others. While it can be a major undertaking to produce a detailed intellectual biography of a figure such as Georges Cuvier, the topic is relatively well defined in space and time. To study, say, the history of experimental modeling in geology could involve research over many scattered sources in many languages and might not always lead to particularly interesting results. Besides, such a study would not attach in a natural way to the existing literature in the history of geology (though it could connect with writings on experimentation). In practice, then, some topics, such as Charles Lyell, have received a great deal of attention, while large areas of the earth sciences—perhaps most notably petrology—have hitherto received only cursory treatment.

These divisions can be apprehended by looking at the broad features of the historiography of geology for the nineteenth century.

2. Early Historical Studies

The first major account of the history of geology in English was by Lyell, who argued in his *Principles of Geology* (1830–33) that geology should be distinct from geogony. Past and present conditions should be presumed similar; the laws of nature were constant, though the intensity of action of geological forces might change somewhat over time. The appearances of violent events could and should be attributed to present-day agencies acting over immense periods of time. The geologist's method should be to examine present processes and use them to account for the earth's past. Lyell was, using William Whewell's term, a "uniformitarian."[3]

Though Lyell's history set out a framework for understanding early geology and was *prima facie* a "neutral" piece of writing, it has been shown to have been deeply polemical, being used as a rhetorical tool to support his ideas about the nature of geology as a science.[4] Lyell wrote sympathetically of past philosophers and geologists who held views akin to his own and somewhat slightingly of those with views similar to those of his opponents—geologists whom Whewell was shortly to name "catastrophists." Further, Lyell conflated catastrophists with "Mosaic" geologists, who sought to relate geology and Scripture. Yet the two were logically distinct. Geologists such as William Buckland or Cuvier could be "catastrophists" on empirical, not biblical, grounds (though Buckland and many others imagined the Noachian Deluge could have been a geological agent). Lyell's history subtly suggested that only uniformitarian methodology was scientific. However, Jan Klaver has claimed, contra some commentators on Lyell's historiography (Martin Rudwick and

---

3. [Whewell,] review of Lyell.
4. Rudwick, "Strategy of Lyell's *Principles of Geology*." See also Rudwick's introduction to the 1990 reprint of *Principles* and R. Porter, "Charles Lyell."

Roy Porter), that in discussing earlier geologists—even "proto-uniformitari-ans"—Lyell played down the importance of his uniformitarian predecessors, thereby enhancing an impression of novelty for his own ideas; and he attacked Genesis indirectly by making adverse comments on the cosmological opin-ions of the Koran.[5]

The polymath Whewell, Master of Trinity College, Cambridge, was, among other things, a historian of geology—of what he termed the "palaeti-ological sciences," which dealt with ancient causes. He thought that such sci-ences might "properly be called *historical*," but that this would be confusing, as the term "history" (as in "natural history") did not always connote the idea of time. Hence Whewell's neologism.[6] Whewell wanted to consider the de-velopment of the sciences through time, since he thought that the history of science revealed a gradual clarification of "fundamental" ideas (for example, force). Thus his informative sketch of the history of geology "attempt[ed] to discover general laws in geology"[7] as successively revealed by geologists. In part, such laws were, up to Whewell's day, chiefly phenomenological in char-acter: Elie de Beaumont's supposed geometrical patterns in mountain ranges, which might have arisen as the earth cooled and contracted. Or, Whewell noted, there seemed to be a general pattern in the appearance of fossils in the stratigraphic column. Further, the distribution of volcanic eruptions and earthquakes seemed not to be random. Whewell was perhaps the most "philosophical" of all historians of geology, in that his historiography was linked to a sophisticated and original philosophy of science. However, Klaver has shown that Whewell took some short cuts and drew on Lyell's historiog-raphy significantly.[8]

Probably the most interesting nineteenth-century history of geology was Archibald Geikie's *Founders of Geology* (1897). Geikie, a Scotsman, director-general of the Geological Survey of Great Britain, and a splendid writer, gave a nationalist twist to his writings. He spoke of the "Scottish School of Geol-ogy," and his account was more sympathetic toward the followers of his eighteenth-century countryman James Hutton—who espoused a cyclic view of the earth's past, with heat as a prime agency ("Vulcanist" theory)—than to disciples of Abraham Werner, teacher at the Freiberg Mining Academy in Sax-ony in the eighteenth century, who advocated water as the source of crys-talline rocks, which were supposedly deposited on the earth's core in a linear order from a universal ocean ("Neptunist" theory).

Geikie thought Hutton's views were more "correct" than Werner's (con-

5. Klaver, *Geology and Religious Sentiment*, 32–33.
6. Whewell, *History of the Inductive Sciences*, 3:400–401.
7. Ibid., 3:444.
8. Klaver, *Geology and Religious Sentiment*, 147–48.

sidering, for example, the low solubility of silica in water), and he scorned the Wernerian emphasis on the study of the external characters of minerals. But in so writing Geikie had insufficient regard for the "style" of eighteenth-century mineralogy, which, as in other areas such as botany, emphasized external characteristics for the purposes of classification. Moreover, he gave insufficient credit to Werner's teaching. His success as a teacher was almost held against him—as if Werner beguiled students into accepting implausible theories. Further, Geikie did not emphasize sufficiently the importance of Werner's recognition that strata were deposited in a determinable order (even if the basis of his order was unsatisfactory) and the way in which in the early nineteenth century Werner's students endeavored to use this order, established in Saxony, as a "paradigm" for stratigraphic work elsewhere. Finally, Geikie gave Werner's influence insufficient credit in relation to the development of "neo-Wernerianism" in the second half of the nineteenth-century. For example, in the 1850s Gabriel Auguste Daubrée performed experiments in which he subjected substances to high temperatures and pressures, with or without water present. He concluded that crystalline rocks could, in the presence of steam under high pressure, be converted to other crystalline types without wholesale melting. To take a second example, the Canadian surveyor Thomas Sterry Hunt displayed neo-Wernerian ideas in his attempts to develop sequences of chemical reactions that could have occurred as the earth cooled, thereby giving rise to the observed sequence of rocks.[9] In Geikie, then, we see leanings towards nationalism, whiggery, and historiographical anachronism.

Yet Geikie's book was interesting because it took a personal view and was written in a lively fashion. By contrast, the writings of his contemporaries such as Karl von Zittel were dull (though accurate) compilations.[10] If one wished to know the several branches of geological work undertaken in the nineteenth century, then one would turn to Zittel rather than Geikie. But for pleasurable reading and an engaging story, Geikie, regardless of his whiggish proclivities, is the appropriate choice. Surprisingly, Geikie gave negligible attention to Lyell, whereas Zittel provided a succinct, favorable account. The omission is puzzling, given that Geikie was, like Lyell, a Scot and a uniformitarian, at least methodologically, if not substantively (see section 4). Geikie coined the maxim "The present is the key to the past" (though he did not initiate the idea).[11] Zittel's book, for its part, is dull *because* it was written by a scientist, subject to scientists' "norms." Maintaining a veneer of objectivity,

9. Brock, "Chemical Geology or Geological Chemistry?"

10. Zittel, *History of Geology*.

11. Hutton had earlier enunciated the idea: "In examining things present, we have data from which to reason with regard to what has been; and from what has actually been, we have data for concluding with regard to that which is to happen hereafter." (Hutton, "Theory of the Earth," 217).

many scientists and scientist-historians have tended to shelter behind the sup-position that "facts" are value-free, unsullied by human passions or frailties. Besides being diligent collectors of data, scientists are represented as hard-working, dispassionate, kindly, dedicated, sometimes heroic (rather than as self-serving, money-grubbing, metaphysically biased). For the early geolo-gist-historians, any social patina that facts acquired was "polish" (rather than "rust"), and arose from the positive attributes of scientists. But for the most part, their accounts were so "internalist" in character that neither contextual polishing nor rusting of facts was evident, geology apparently being con-ducted in a social vacuum.

To illustrate: George Merrill's *First One Hundred Years of American Geology* (1924) acknowledged hardly any controversies during that long period, with the exception of the so-called Taconic Controversy. But the account of this episode hardly conveys the depth and intensity of the dispute concerning the older rocks of northwest America. Reading between the lines, one can see that Merrill did not like Sterry Hunt or Jules Marcou, two of the protagonists, both of whom had "Continental" tendencies in their thinking. Merrill as-serted that Sterry Hunt propounded "fanciful" opinions, in a manner that was "characteristically emphatic and apparently decisive."[12] Marcou's deploy-ment of Joachim Barrande's Bohemian ideas[13] "swelled the literature and confused the question until for a time the correct solution seemed hope-less."[14] The fact that Marcou spent (on and off) twenty-one years in the field on the question went unmentioned. Years later, Cecil Schneer recorded that the controversy involved a libel suit.[15]

Merrill's work is also interesting for the way in which he discussed geolo-gists' moral characters. Concerning Ferdinand Hayden, one of the great explorer-geologists of the American Midwest, Merrill wrote that his "honesty and integrity were undoubted and his work for the Government and for sci-ence was a labor of love."[16] Medically trained, Hayden volunteered for the Union Army at the outbreak of the Civil War, rising to colonel for his "meri-torious conduct."[17] He might eventually have become director of the Federal Survey but was prevented by illness and advancing age. How different is Mike

12. Merrill, *First One Hundred Years,* 601–2.

13. Barrande's "Theory of Colonies," developed in Bohemia, allowed deviations from the stan-dard stratigraphic succession of fossils. It was sometimes used to "explain away" stratigraphic anomalies caused by earth movements.

14. Merrill, *First One Hundred Years,* 604.

15. Schneer, "Great Taconic Controversy." See also Yochelson, "Question of Primordial and Cambrian / Taconic." On Marcou, see also Durand-Delga, "Jules Marcou," and Durand-Delga and Moreau, "Un savant dérangeant."

16. Merrill, *First One Hundred Years,* 526.

17. Ibid., 525.

Foster's recent account of Hayden.[18] Here we learn that Hayden plagiarized an anthropologist, entered the army for career reasons, and was an inveterate womanizer—the disease that prevented his being considered for the directorship was, Foster concludes, probably venereal. I mention this not to pillory either Merrill or Hayden (who had many positive qualities) but to illustrate how sanitized were the early geological histories. The early geologist-historians, working mostly from published sources, were looking for different kinds of information than that used, for example, by Foster.

3. The Postwar Period: The Intellectual Background of Geology

Professional historians (even professional historians of science) took little interest in geology until after World War II. However, in the postwar period some geologists, such as Victor and Joan Eyles in England and George White in America, began taking a serious interest in the history of geology and amassed huge collections of early geological texts. They published some important papers using historical techniques, such as the approximate dating of issues of the maps of William Smith (the engineer / surveyor who published the first geological map of England and Wales in 1815) according to the information depicted thereon.[19] Nonetheless, the first postwar writer to take a historian's view of geology was Charles Gillispie, in his lively *Genesis and Geology* (1951), which focused on the relationship between geology and religion in nineteenth-century Britain. This was no hagiography. It described with gusto (and some partisanship) such issues as the debates between Vulcanists and Neptunists, uniformitarians and catastrophists. Gillispie sought to show that biblical beliefs, natural theology, Neptunism, and catastrophism were interconnected. Persons involved in forging such a synthesis, such as the Irish chemist and mineralogist Richard Kirwan and the Swiss traveler Jean André de Luc, were almost ridiculed.

Gillispie's book was a classic example of whig historiography, and to an extent it was misleading, with its dash of anti-Irish and anti-German sentiment. Yet *Genesis and Geology* successfully demonstrated that geology was a science with significant metaphysical implications. Texts such as Elie Halévy's *Growth of Philosophic Radicalism*[20] or Georgii Plekhanov's *Essays in the History of Materialism* were as important to Gillispie as scientific texts, and he raised issues not considered by scientist-historians—such as the role of journals and learned societies, and of controversies in constructing scientific knowledge.

18. Foster, *Strange Genius.*
19. Eyles and Eyles, "On the Different Issues."
20. This has much to do with the history of utilitarianism, economic theory, and philosophy of law.

As Nicolaas Rupke has written: "Geology was removed from an abstract level of cognitive isolation and placed in the hustle and bustle of the society in which the practitioners lived and worked."[21] Victor Eyles's review of *Genesis and Geology*, by contrast, objected that Gillispie was not a geologist, or even a scientist.[22] Here was a demarcation dispute: in Eyles's view, only geologists should write history of geology. Yet Gillispie had shown that natural theology and design were in some respects as important for nineteenth-century geologists as field exposures. However, in so doing he had pulled geological historiography unduly toward ideas rather than practices, and in some cases he may, for a considerable time, have thrown an unwarranted cloud over the work of geologists such as Kirwan or Buckland, his emphasis on their metaphysical predilections having obscured their geological achievements. Later work by Klaver has taken a closer look at the issues broached by Gillispie, with attention to the views of a number of major nineteenth-century geologists, such as Adam Sedgwick, and poets, such as Tennyson. Klaver suggests that Gillispie "described with inappropriate gusto an intellectual battle that never took place."[23] Be that as it may, *Genesis and Geology* was an early manifestation of one of the preoccupations of the Darwin industry: setting the intellectual and social scene out of which Darwin's ideas emerged. In particular, this involved investigation of Lyell's ideas, so signally underplayed by Geikie. This led Gillispie to consider uniformitarianism and its promulgation, reception, and associated controversies in England. As Gillispie neatly put it: "If Buckland feared that without cataclysms there was no God, Lyell was as fundamentally apprehensive lest, without uniformity, there be no science." Assuredly *some* battle took place. The uniformitarianism issue was further pursued in the 1960s and 1970s, both historically and historiographically.

4. Meanings of Uniformitarianism

The issue was treated in detail by the Dutch Calvinist scholar Reijer Hooykaas in his *Natural Law and Divine Miracle: The Principle of Uniformity in Geology, Biology and Theology* (1963).[24] For Hooykaas, the issue was alive, since a physical principle of uniformity conflicted with Christian beliefs about miracles and geological uniformitarianism was at odds with Genesis read as history. Hooykaas sieved the nineteenth-century geological literature and showed that uniformitarianism was espoused in many writings besides Lyell's. Hooykaas also pointed out that "uniformitarian" doctrine had several connotations. There

21. Rupke, "Gillispie's *Genesis and Geology*," 263.
22. Eyles, "Scientific Thought."
23. Klaver, *Geology and Religious Sentiment*, xii.
24. Hooykaas (1952) was one of those who reviewed Gillispie positively.

was the question of whether laws of nature were constant. There was the belief that present and past causes were the same in kind and energy, and produced the same effects. There was the belief that past and present geological circumstances were essentially the same. (Not many took this Lyellian view.) Alternatively, one might suppose that circumstances were changing but in a regular or uniform fashion. Finally, uniformitarianism could be regarded as a methodological rule—to the effect that when explaining the geological past one should choose hypotheses with maximum analogy to present circumstances. Hooykaas favored the last alternative. He also emphasized that catastrophists were not necessarily biblical literalists. They could be exemplary empiricists.[25]

Despite Hooykaas's close analysis, not everyone was satisfied. Stephen Jay Gould, in his first scientific paper, "Is Uniformitarianism Necessary?," distinguished between "substantive" and "methodological" uniformitarianism.[26] The first held that past conditions were much the same as the present—and few people believed *that* for the whole of the earth's history. Uniformitarianism as a methodological maxim, as has been indicated, instead enjoins one to adopt those hypotheses that offer maximum similarity to or analogy with present-day ("actual") events or processes. However, if one pressed "methodological uniformitarians," it seemed that their method led them no further than to the unexciting position that the laws of nature were constant. So Gould was inclined to think that uniformitarianism was *not* necessary as a special principle for geology.

The issue of multiple meanings for the term "uniformitarianism" was also addressed by Rudwick's "semantic analysis."[27] Rudwick made a number of distinctions regarding conceptions of past causes: such causes might be seen as "naturalistic" or "supernaturalistic"; as "actualistic" or "non-actualistic" (i.e., analogous or not analogous to actual causes now in operation);[28] as similar or dissimilar in "degree" or in "kind" to those now operating; and as having produced changes in a "gradualistic" or a "saltatory" manner. Further, geological circumstances might be conceived as approximately "steady-state" or "directional" (or progressive versus developmental). This last distinction was especially important, since it related to one of the main controversies of Lyell's day. Most of his contemporaries were "directionalists" (that was the "paradigm").[29] Lyell was not. Darwin was (though the direction of change was not predetermined in his theory).

25. Hooykaas, *Catastrophism in Geology.*
26. Gould, "Is Uniformitarianism Necessary?"
27. Rudwick, "Uniformity and Progression."
28. In this usage, which derives from Continental terms like *actualisme,* "actual" means "present." (See Gould, *Time's Arrow,* 120.) So an actualist approach would be one that assumes, in Geikie's words, that "the present is the key to the past." (Geikie, *Founders of Geology,* 299.)
29. The "directionalists" could be either "gradualists" or "saltationists."

Rudwick's categories have subsequently been somewhat modified by Gould.[30] Uniformitarianism could refer to assumptions of: the uniformity of the laws of nature; uniformity of process ("actualism" or "the present is the key to the past"); uniformity of rate, or "gradualism"; or uniformity of state, or "non-progressionism." The first two assumptions are primarily methodological, the latter two substantive (though they could, I suppose, also be treated methodologically). Thus Gould reaffirmed his dichotomy and the issues were further clarified. However, it remains a question how one could be a "serious" methodological uniformitarian without being committed to some kind of substantive uniformitarianism.

The issue has been revisited recently by the Turkish geologist and historian of geology A. M. Celâl Şengör. In an important historical and philosophical discussion of the work of Hutton and Werner, he compares the basic elements of their theories and ways of thinking with those of Adam Smith and Karl Marx, respectively.[31] He uses the term "actualism," as have some Continental writers, in reference to the hypothesis that "all past geological action was like all present geological action." And "uniformitarianism" is taken to be the hypothesis that "does not allow a significant deviation in nature and rate of geological progress throughout geological time." This distinguishes processes from states of affairs. Şengör rejects Gould's distinction between substantive and methodological uniformitarianism, saying that no one has in fact ever been a substantive uniformitarian and that Gould's methodological uniformitarianism would be tantamount to accepting divine revelation, which would have to serve as the basis of inductive inference. In some respects, the arch-uniformitarians, Hutton and Lyell, would not be substantive uniformitarians according to Gould's dichotomy, as initially proposed. Personally, I think that the four distinctions made by Rudwick serve us better than trying to make a distinction between "actualism" and "uniformitarianism."

5. Controversies about Lyell

Because geological historiography is thinly spread and conducted chiefly by scientist-historians who prefer to avoid polemics in print, few geologists have been researched in detail by historians from different historiographical perspectives. Lyell is an exception, his method having formerly had paradigmatic status.[32] His first biographer, Thomas Bonney, regarded him as an empiricist and associated this with uniformitarianism.[33] Lyell's principal biographer,

---

30. Gould, *Time's Arrow*.

31. Şengör, *Is the Present the Key?*

32. Longwell and Flint, *Introduction to Physical Geology* (1955), 385, represented it as the "cornerstone of geologic philosophy, probably the greatest single contribution geologists have made to scientific thought." The statement was, however, withdrawn from the 1962 edition.

33. Bonney, *Charles Lyell and Modern Geology*.

Leonard Wilson, studying manuscript records of Lyell's travels and fieldwork, has presented his geological theory and uniformitarianism (in a broad sense, as understood by Whewell) as a product of his observations.[34] For example, Lyell's early work showed the present-day accumulation of calcareous matter in lochs in Forfarshire. From this, one could explain the occurrence of similar deposits in neighboring lochs where limestone was no longer forming; thereby one could understand the formation of the calcareous sediments of the Paris Basin, studied by Cuvier and Alexandre Brongniart. For Lyell, this was sound geological reasoning, based, as it were, on "Newtonian" ratiocination. Wilson used such considerations to argue for the empirical basis of Lyell's uniformitarianism.

Rudwick, by contrast, taking his lead from Hooykaas, emphasized the a priori aspects of Lyell's geology. In particular, he described how Lyell reasoned about Etna, using evidence from historically recorded eruptions and the overall size of the mountain to infer that it was of very considerable age; and as it overlay geologically young sediments, these too must have been extremely old in human terms. So the earth as a whole must be really ancient. Rudwick suggested, however, that Lyell's ideas and approach were established in part *before* he visited Sicily, so they could hardly have been direct inductions from Sicilian experience.[35]

In several later publications, Rudwick has analyzed Lyell's thinking, examining the strategies used in composing his major work, including his rhetorical use of history.[36] He has looked at the intellectual resources (or conceptual analogies with human historiography, linguistics, demography, and political economy) upon which Lyell drew in developing his uniformitarianism.[37] In this respect, Rudwick has extended the historiographic approach pioneered by Gillispie. Wilson, however, in a strongly worded paper that attacks the work of Hooykaas, Rudwick, and Porter, has vigorously reaffirmed his view that Lyell's uniformitarianism was empirically based, not an a priori methodological rule.[38]

Regarding Lyell's debt to philosophy of science, Rachel Laudan has shown how Lyell wanted to model his geological philosophy on that of Newton and Scottish Enlightenment philosophers.[39] Newton sought "verae causae" to account for observable phenomena. Lyell wanted to do likewise, and his

---

34. Wilson, *Charles Lyell.*

35. Rudwick, "Lyell on Etna."

36. Rudwick, "Strategy of Lyell's *Principles of Geology,*" "Historical Analogies," and Introduction.

37. Rudwick, "Transposed Concepts."

38. Wilson, "Geology on the Eve." Wilson hinted that Hooykaas's writing on Lyell was influenced by religious considerations.

39. Laudan, "Role of Methodology" and *From Mineralogy to Geology.*

methodology (actualistic, in the sense of the second of Rudwick's dichotomies) made this possible. For if Lyell could show that, say, lava flows like those of the present could form a mountain such as Etna, then one could understand the formation (and destruction) of volcanic regions such as the Auvergne.

A Spanish voice has been raised against the Anglophone discussions of Lyell.[40] Logician Alberto Elena has pointed to problems in looking at Lyell through the lens of Thomas Kuhn's philosophy. Did Lyell's work give geology its first paradigm, as is sometimes supposed? Did he effect a scientific revolution, changing the community from one paradigm to another? Or was there, contra Wilson, no Lyellian paradigm at all? Elena has argued that those who think Lyell gave geology its first paradigm have been misled by his historical rhetoric. There was geoscientific work before Lyell. But did it have the form of a coherent paradigm—"directionalism"—as Rudwick has suggested? Elena thinks not, since "directionalism" was formulated after Lyell (for example, by Elie de Beaumont) as much as before him; and in any case Lyell's "uniformitarianism," or nondirectionalism, never made much progress, especially beyond the English-speaking world.[41] For Elena, the "Lyellian revolution" is a historical myth first propagated by Gillispie. Be this as it may, Elena at least shows that one should be cautious about representing Lyell as the "founder" of scientific geology or his *Principles* as effecting a (Kuhnian) scientific revolution. So perhaps Geikie, writing well before the advent of the Darwin industry, showed good judgment in not giving too much attention to Lyell.

In all this, we see, starting in the 1960s, the historiography of geology being given new foci by historians (and philosophers) of science, moving away from hagiography and the straightforward compilation of "facts." Also, there has been a sort of replay among twentieth-century geohistorians of the ideological battles that occurred in Lyell's time. They have had their own philosophical and historiographical agendas, which have, to some extent, been reflected in their writings on Lyell.

6. The Origins of Geology and Its Historiographic Periodization

Few attempts have been made to periodize geology's history generally for the nineteenth century, for geological studies began at different times and rates in different parts of the world. One could talk about pre- and post-Lyellian or perhaps pre- or post-Darwinian geology. But such divisions would have little general significance. Geology was in a reconnaissance phase in the nineteenth century, as the world was explored and mapped. The field had many subdivi-

---

40. Elena, "Imaginary Lyellian Revolution."
41. Bartholomew, "Non-Progress of Non-Progression."

sions, as indicated above. It was small enough for the subject's leaders to have general command of most of it. Even so, there was no agreed unifying theory. Geological research slowed significantly during the First World War, and then accelerated, but there was no marked theoretical shift at that time. During the nineteenth century there was much increase in empirical knowledge, and theories kept changing in the various branches of geology. But the development was quite gradual, and mostly cumulative—though there were keen debates about the elaboration of the stratigraphic column, with disagreements between opposed controversialists about the placement of boundaries, and certainly new ideas of great importance, such as isostasy, were proposed. Only in the late 1960s, with the advent of the plate-tectonic "paradigm," did geology undergo such a major change that historiographic periodization would there be natural and warranted. However, within the nineteenth century, Şengör has seen the work of Eduard Suess (around 1875) as marking a radical change in the development of theories of mountain building.[42]

There have been significant differences of opinion concerning the "origin" of geology. The term itself was originated by de Luc in 1778, but as the terrestrial analogue of cosmology, not in its modern sense. Dennis Dean sees Hutton as the "founder of modern geology."[43] Laudan sees Werner's lithostratigraphy as providing a paradigm that set geology moving on the Continent, enabling geologists to deploy an established approach and pattern.[44] In a recent study, Rudwick has seen de Luc, the man pilloried by Gillispie, as a central figure in the emergence of geology as an historical science, even though his contribution was linked to biblical chronology.[45] By Rudwick's analysis, de Luc observed present processes, such as the formation of peat in north Germany or the extension of the Rhone delta into Lake Geneva, and estimated how long they might have been proceeding. He found the times to be in approximate agreement with the supposedly known time that had elapsed (on scriptural evidence) since the Noachian Flood. In fact, we would say, de Luc was providing an approximate empirical estimate of the time since the last ice age. De Luc, however, was dealing with what Rudwick suggests was a "binary geotheoretic model": earth history was divided into *histoire moderne* and *histoire ancienne,* the latter meaning either "ancient" or "former history." (Such a distinction had been made earlier, as in lectures in Paris by the chemist G. F. Rouelle.) De Luc had "chronometers" for his modern history, such as his accumulation of deltaic sediments. So although he might properly be called a scriptural geologist he was also, as Rudwick puts it, providing "at least the

42. Şengör, "Classical Theories of Orogenesis."
43. Dean, *James Hutton and the History of Geology.*
44. Laudan, *From Mineralogy to Geology.*
45. Rudwick, "Jean-André de Luc."

rudiments of a quantified chronology for the Earth itself, based on the observable rates of actual causes or ordinary natural processes."

Most of de Luc's ideas were published in the eighteenth century, but he exerted a significant influence on what followed in the nineteenth century in
that his "binary history" was taken up by Cuvier, who likewise linked earth
history to human history (as in his comments on the organisms found buried
in the pyramids). The point of interest here is that recent scholarship offers a
view of the scriptural geologist de Luc at odds with the negative picture of
him presented in Gillispie's *Genesis and Geology* (which had, however, correctly
remarked de Luc's "binary" geohistory).

Werner, of course, studied and lectured with a background of mining and
technology. In Britain, Hugh Torrens has shown how miners, surveyors, and
engineers provided the technological basis for survey work, establishing the
order of strata so that predictions could be made as to where particular rock
types might or might not be located underground. Such knowledge fed into
theoretical geology.[46] Torrens sees this industrial input as an essential ingredi-
ent in the emergence of modern geology, though he does not argue according
to a Marxist "base-and-superstructure" model. Such a view might commend
itself to Martin Guntau who, writing in the former East Germany, emphasized the industrial origins of geology. He held that new ideas were made possible by new knowledge produced by mining and that, reciprocally, mining
generated a need for geological knowledge.[47]

In my own work, I have seen the emergence of geology as a historical enterprise as a manifestation of a general change of worldview around 1800,
when so many phenomena came to be seen from a historical perspective.[48]
Thus Werner's "genetic" account of the earth's past (which saw it developing
irreversibly in one direction in an almost predetermined way from some supposed starting point, according to the laws of nature and without relevant
contingencies) was superseded by the "historical" geology of, say, Cuvier,
who sought to build up an account of the earth's past according to the nature
and disposition of the "monuments" of its past that researchers could find in
its rocks, using analogies based on modern organisms for his paleontological
reconstructions.

Rudwick, then, has argued that Cuvier was the key figure in the emergence
of geology as a historical enterprise.[49] Werner was a cabinet geologist who
sorted his specimens and arranged them in his museum according to taxo-

46. See, for example, Torrens, "Some Thoughts" and "Timeless Order."

47. Guntau, *Die Genesis der Geologie* and "Emergence of Geology."

48. Oldroyd, "Historicism and the Rise of Historical Geology." I have also attempted to relate
the emergence of geology to Michel Foucault's periodization of Western culture; see Albury and
Oldroyd, "From Renaissance Mineral Studies to Historical Geology."

49. Rudwick, "Cuvier and Brongniart" and "Smith, Cuvier et Brongniart."

nomic criteria. His strata were classified temporally on lithological grounds: organic fossils were not related to the rocks in which they occurred. William Smith sorted his fossils and arranged them in a table according to their positions in the earth's strata, and in temporal sequence, the strata being tabulated according to the fossils they contained. So fossils and rocks were linked. But Smith's (catastrophist) theory of the earth was not particularly important to him, and though it might be regarded as a "genetic" conception (Rudwick calls it "geognostic") the genetic / historical distinction would not have been likely to have interested Smith, whose concerns were primarily technical and practical (or economic). His interest in organic fossils chiefly had to do with their significance as stratigraphic indicators, important for such enterprises as agriculture and coal prospecting. As for the eighteenth-century "theories of the earth," whether linear, like that proposed by Comte de Buffon, keeper of the Jardin du Roi in Paris, in his *Histoire naturelle,* or cyclic, like Hutton's, Rudwick regards them as "systems," for which he prefers the term "epigenetic" to "genetic."

But for Rudwick Cuvier offered something different. In collaboration with Alexandre Brongniart, Cuvier studied the rocks of the Paris Basin, considering fossils, lithologies, and the conditions of existence that obtained when the various strata were deposited, the latter being deducible (on "actualistic" grounds) from the sediments and fossil types. Piecing bits of evidence together—as if they were items in an archive studied by a historian—Cuvier and Brongniart sought to provide a "geohistory" of the Paris region.[50] It was this fusion of four traditions (specimen collection and classification; geognosy, or mineral or rock identification, classification, and ordering; system building; and historical inquiry) that yielded modern geology, and, according to Rudwick, marked the emergence of geohistorical research.

While Western historians have not been greatly interested in the periodization of geology in the nineteenth century, it has concerned Russian historians. The blind historian of geology Vladimir Tikhomirov related the development of metal industries to the successive socioeconomic policies of Peter the Great, Catherine the Great, and Nicholas I. He characterized the first third of the nineteenth century in Russia as one of industrial stagnation and hence of limited interest in mineral prospecting and geological survey, whereas during the second third, with the rise of capitalism and the bourgeoisie, much survey work was accomplished, based on paleontological stratigraphy. But geology was regarded as a subversive science and was attacked by the church.[51] A somewhat later volume by A. I. Ravikovich attempted a periodization of European geology according to the hegemony of

---

50. Cuvier and Brongniart, "Essai sur la géographie minéralogique."
51. Tikhomirov, *Geology in Russia,* chap. 4.

catastrophism, then uniformitarianism, and then evolutionism. But her divisions were imprecise, as she acknowledged that there were uniformitarian ideas around already during the catastrophist epoch (first quarter of the nineteenth century).[52]

7. The "Visual Language" of Geology

Rudwick has also pioneered the study of geological illustrations. As a former paleontologist, accustomed to geologists' making their ideas intelligible with diagrams, he was surprised by the lack of attention given them.[53] In his publications on this topic, Rudwick first scrutinized the caricatures of Lyell made by then British Survey director Henry De la Beche and showed how they enabled one to understand the differences of opinion between the two. De la Beche thought Lyell's uniformitarianism dogmatic and speculative, and his parodical caricatures reflected that assessment.[54]

Rudwick's caricature study was followed by a more general paper on geological illustration.[55] He regarded illustrations as a "language" that should be examined as closely as verbal texts. One could see how the development of illustration techniques influenced what was printed and how.[56] Scrutiny of early maps and sections revealed much about the ideas of those who produced them.[57] One could, for example, compare the "block-diagram" sections of John Farey (1811) with the sections of Thomas Webster (1816), the former being analogous to "ahistorical" surveyor's diagrams, the latter manifesting a geologist's concern with the historical development of a region's structure. Rudwick reduced the complexities of his topic to a single diagram. He followed up his study with a volume showing the evolution of representations of geological time.[58]

Other authors have also sought to read the language of visual representation. Rupke, for example, in his study of pictures illustrating the idea of continental drift, claims that they reveal a hitherto unsuspected "Eurocentrism." This, he suggests, may have been a subsidiary cause of the long American antipathy towards mobilist theory.[59] We also have studies pointing in the other

---

52. Ravikovich, *Development of the Main Theoretical Tendencies*.
53. Rudwick and Oldroyd, "Martin Rudwick."
54. Rudwick, "Caricature as a Source."
55. Rudwick, "Emergence of a Visual Language."
56. For another good study of this kind, see Blum, " 'A Better Style of Art.' "
57. For further information on the history of geological maps, see, for example, Dudich, ed., *Contributions to the History;* Cook, "From False Starts to Firm Beginnings"; and Butcher, "Advent of Colour-Printed Geological Maps."
58. Rudwick, *Scenes from Deep Time*.
59. Rupke, "Eurocentric Ideology." See also his " 'End of History.' "

direction, so to speak: there is work that examines the influence of geological thought on the history of art.[60] Perhaps one can learn something about the history of ways of thinking about the earth by examining changes in art over time?[61]

On a different tack, a beginning has been made fairly recently by Richard Howarth to examine the history of the kinds of diagrams used as technical tools for the analysis, presentation, and understanding of numerical data in geology. He began with a study of the history of ternary diagrams,[62] considering several physical sciences (e.g., color theory) as well as matters specifically to do with petrology, such as the eutectics of three-component systems. This work has been followed up by three further investigations, both of which span the nineteenth and twentieth centuries.[63] Howarth reports the French petrologist Auguste Michel Lévy as being the first, in 1897, to represent the composition of mixtures of rock-forming compounds with the help of triangular grids. Around the end of the nineteenth century, a considerable amount of statistical work (as, for example, in the study of sediments) began to be recorded graphically. It is evident that geology could not have proceeded or developed without such aids to the analysis and presentation of data.

Of course, throughout the nineteenth century information about strata and rock types was routinely presented in the form of geological maps. Black-and-white maps were usual in journals and monographs; governmental survey maps were often in color, which was initially added by hand. Color printing of geological maps was first attempted in the 1820s, increased from the 1840s onwards, and became quite widespread by the 1860s.[64] From the early days of the nineteenth century, field geologists had used hand-colored maps when presenting papers before learned societies, even if their published maps were usually uncolored.

But there was other information to be represented graphically, on maps or accompanying diagrams, and Howarth has also studied the history of these figures. Dips and strikes were routinely measured with a clinometer and recorded on maps by a variety of notations (typically a T, oriented such that a longer crossbar denoted the direction of strike and a shorter upright the direction of maximum dip, perhaps with a number indicating the angle of dip).

60. For example: Pointon, "Geology and Landscape Painting," and Klonk, *Science and the Perception of Nature.*

61. For my commentary on Klonk's book, and her forthright reply, see Oldroyd and Klonk, "Picturing the Phenomena." For a study of information about ideas about the earth elicited from paintings, but from an earlier period, see Montgomery, "Eye and the Rock."

62. Howarth, "Sources for a History of the Ternary Diagram."

63. Howarth, "Graphical Methods," "Measurement, Portrayal and Analysis," and "From Graphical Display to Dynamic Model."

64. Cook, "From False Starts to Firm Beginnings."

Cleavage or foliation orientations, or joints, could likewise be represented once the distinction between cleavage and bedding was published by Sedgwick in 1835.[65] (The distinction had long been recognized by quarrymen.) Starting in the Lake District in the early 1820s, Sedgwick had recorded endless details of dips and strikes in his notebooks, but the pattern in these could not be recognized easily without graphical representation. And it was a pattern that Sedgwick was looking for, hoping that his observations would eventually show that mountain ranges lay on preferred lines of orientation, according to the theories of Elie de Beaumont. Sedgwick, like other physical scientists of his day at Cambridge, wanted to reveal the *laws* that governed the distribution of structures on the earth's surface.[66] (The laws, the geologists thought, might have to do with the earth's cooling and contraction since its formation, or perhaps some electrical cause.) For such a goal, suitable representation of data was needful, as were "visual" techniques for averaging data. Besides dips and strikes, recorded data might refer to joints or various forms of lineation, such as that produced by stretching (though Sedgwick did not himself deal with stretching phenomena).

Howarth has described various forms of visual representation of orientation data, among which the "circular frequency-polygon" or "rose diagram," developed in the second half of the nineteenth century (and named after the thirty-two-point figure traditionally shown on the faces of mariners' compasses), proved notably popular and effective as a means of representing statistical information about joint or fault orientations or other formations of geological interest. Then in the early twentieth century a new branch of study known as petrofabric analysis developed from the work of Bruno Sander, Walter Schmidt, and others: data for fabric orientations and attitudes were plotted on stereograms with the help of stereographic nets and could then be averaged.[67] Such work led to many developments in structural geology, which became a new department of the earth sciences in the twentieth century. There can be little doubt that structural geology could never have become established as a major field of study without the advances made in the visual representation of textural and structural information and its diagrammatic averaging.

8. The Geological "Underworld"

The preceding sections (3–7) have emphasized the intellectualist aspects of geology. But geology is specifically concerned with the earth, and people have

65. Sedgwick, "Remarks on the Structure of Large Mineral Masses," 469–75.
66. C. Smith, "Geologists and Mathematicians."
67. Howarth, "Measurement, Portrayal and Analysis."

taken an interest in it to a large extent because of the materials that may be extracted from it. It is surprising, then, that historians have so neglected industrial geology, often ignoring the activities of surveyors, miners, chemical analysts, and the like.

To be sure, a few historians, notably Torrens, have investigated the "substratum" of geologists in the first half of the nineteenth century. He has shown how men such as Farey and Smith did consulting work for landowners, advising on the mineral potentials of their estates. Such work involved significant geological mapping. Torrens's most striking example is a colored geological map of part of Derbyshire, prepared by Farey in 1812 for Sir Joseph Banks.[68] This extraordinary map, with twenty-four separately colored strata,[69] looks to have been produced in the late nineteenth century. It shows the extent to which the "underground" geologists were technically ahead of their social betters in the Geological Society—who took the view that Farey's observations were "too circumstantially minute for common use." "General principles stripped of all useless details . . . [were] much more attractive."[70] Torrens located Farey's map in California, which illustrates some of the difficulties involved in uncovering the geological underground. Torrens has published further important essays on the "technical" origins of geology around 1800.[71] But work of this kind, involving knowledge of the history of technology as much as, say, the history of ideas, has hardly begun for the second half of the nineteenth century.

For Torrens's period, a synthetic essay by Roy Porter covers the earlier years of the nineteenth century and examines the relationship between underground knowledge and "elite" geology during the British Industrial Revolution.[72] He maintained that the fragmentation of the provincial knowledge grievously limited its effectiveness. Laissez-faire government attitudes led to mining accidents and technological backwardness, relative to Continental developments. The issue had to do with the different traditions of education: Whereas in England the Oxbridge system dominated, producing a geological elite with limited interest in technology, on the Continent there was a tradition of high-level instruction in mining and engineering, the details of which are not well known in the Anglophone world.[73] Thus there was a closer integration between theoretical and practical knowledge on the Continent, where metallurgical practices, particularly, influenced ideas about volcanic

68. Torrens, "Patronage and Problems."

69. Reproduced in color in ibid., 66.

70. Ibid., 67.

71. Torrens, *Practice of British Geology*.

72. R. Porter, "Industrial Revolution." This follows on from Porter's *Making of Geology*.

73. But see Brianta, "Education and Training," or Guntau, "Geologische Institutionen" and "Zur Geologie und Mineralogie."

action.[74] Yet despite the British divide between industrial practice and geological theory, Britain was the leading industrial power in the nineteenth century and also had many of the leading geologists. Torrens justifiably takes issue with Porter's view, sanctioned by Rupke,[75] that the rise of geology—at least in Britain—had little to do with mining, quarrying, and agriculture and their practices.[76]

Relatedly, there has recently been considerable interest in the work of amateur geologists, particularly in Scotland. For example, the work of Hugh Miller, a Scottish stonemason, bank clerk, geologist, and journalist, has attracted attention. Recent studies have revealed much about his psychic condition, his social philosophy, and how he sought to present himself to his public in staged photographic images.[77] Such work provides valuable models for the detailed biographical study of geologists more generally.

From a different perspective, Michael Collie and John Diemer have studied the Morayshire clergyman George Gordon, showing the important part played by amateurs in the tapestry of nineteenth-century geology. Through his correspondence, Gordon functioned as a clearing house for knowledge of the natural history of his region; by studying him one thus gains insight into the social system of geology in the nineteenth century, and the social and cognitive relationships between the metropolitan experts and peripheral amateurs are displayed.[78]

There are some older studies of geological amateur institutions, such as Woodward's study of the Geological Society of London,[79] Macnair and Mort's volume on the Geological Society of Glasgow,[80] and Sweeting's history of the Geologists' Association.[81] There is, however, scope for renewed institutional work on such bodies, from a more critical perspective.[82] The recent study by Simon Knell on the early history of museums and collections in nineteenth-century England provides an excellent picture of the important work of amateur and professional collectors and their relationships to the leading geologists of the time, such as Sedgwick.[83] Besides giving much useful information on the social situation of the "little" men and women of

---

74. Fritscher, "Vulkane und Hochöfen."

75. Rupke, "'End of History,'" 85.

76. Torrens, "Some Thoughts on the Forgotten History of Mineral Exploration."

77. Shortland, *Hugh Miller.* See also idem, *Hugh Miller's Memoir.*

78. Collie, *Huxley at Work;* Collie and Diemer, *Murchison in Moray;* Keillar and Smith, *George Gordon;* Collie and Bennett, *George Gordon;* Collie, "George Gordon (1801–1893)."

79. Woodward, *History of the Geological Society of London.*

80. Macnair and Mort, *History of the Geological Society of Glasgow.*

81. Sweeting, *Geologists' Association.*

82. A more recent study of the Geological Society of America has some of the characteristics of the older English works: Eckel, *Geological Society of America.*

83. Knell, *Culture of English Geology.*

British geology at that time, Knell considers ideas about the relation of paleontological collections to the actual occurrence of fossils in the rocks. Institutional investigations, where they deal with the nineteenth century, could involve prosopographical investigations of networks of amateur geologists, an area little covered for the second half of the century. However, the study of the geological underworld is a daunting task because of the scarcity of documentary evidence concerning the smaller practitioners.

9. The Stratigraphic Column

The development of the stratigraphic column has attracted considerable attention. Theoretical questions involved in dividing the column are entertainingly described by Derek Ager, and William Berry has outlined the history of the establishment of its subdivisions.[84] The establishment of some of the subdivisions involved much controversy, and unraveling the details of the arguments can involve great labor, historians' fieldwork being important. I have myself studied the establishment of the Precambrian–Cambrian boundary in Britain.[85] (However, there can be snags in historical fieldwork.)[86]

James Secord has studied in detail the great controversy between Sedgwick and Roderick Murchison about the Cambrian and Silurian systems and the placement of the boundary between the two.[87] The debate's broad outline had long been known, but until Secord's work the details were obscure. As might be expected, there was right and wrong on both sides. Neither contestant was above rewriting history to further his case. The British contest was paralleled in the Taconic Controversy, an American controversy that would merit a monographic study on the scale of Secord's investigation. Another instance is that of the Ordovician System, proposed in 1879 by the amateur geologist Charles Lapworth to resolve the ongoing Cambrian–Silurian debate. The Ordovician's history remains to be written, though a brief essay by Michael Bassett provides a starting point.[88]

The establishment of the Devonian System has been the object of a major project by Rudwick.[89] This study is notable for the way in which it incorporates ideas from the sociology of knowledge, and of all the books on the history of geology (Hooykaas's possibly excepted) it has attracted the greatest attention from those interested in the philosophy and sociology of science. The "Great Devonian Controversy" was concerned with the interpretation of

84. Ager, *Nature of the Stratigraphical Record;* Berry, *Growth of a Prehistoric Time Scale.*
85. Oldroyd, "Archaean Controversy in Britain."
86. Oldroyd, "Sir Archibald Geikie" and "Use of Non-Written Sources."
87. Secord, *Controversy in Victorian Geology.*
88. Bassett, "100 Years of Ordovician Geology."
89. Rudwick, *Great Devonian Controversy.*

the geological horizon of certain beds in Devon, and their correlation with the better-known strata of the Old Red Sandstone. Rudwick utilized Harry Collins's concept of a "core set": the scientists whose *ideas count* in the settlement of a scientific controversy.[90] Rudwick gave pictorial representations of the relative social status of the participants in the controversy by means of concentric circles, with members of the core set occupying the central position, amateur collectors at the periphery, and persons of intermediate status figured at intermediate positions. He also explicated the changing fortunes of different strata by means of diagrams, which showed the particular (theoretical) units rising or falling in the stratigraphic column (or appearing and disappearing) as the controversy ran its course. The project demonstrated that stratigraphic subdivisions are *not* transcendentally (or empirically) given but are constructed in the agonistic field of the geological community. The subdivisions are arrived at when there is resolution of the controversy and consensus in the community.[91] Yet empirical information is involved in the achievement of the consensus. Rudwick eschews the (absurd) idea that geological knowledge is "nothing but" the result of social forces.

The Jurassic strata were those among which Smith had the greatest success in the early years. They were also important for the establishment of the concept of stages and fossil zones, with the work of Alcide d'Orbigny, Friedrich August von Quenstedt, and his pupil Albert Oppel. However, there is no detailed account of the work of such men, fundamental though it is to the understanding of the history of stratigraphy (and paleontology).[92] For d'Orbigny, the idea of stages made sense since he was both an actualist and a "creationist / catastrophist," rejecting progressive development. A study of the history of ideas about, and work within, the Jurassic System would be an excellent topic for a major investigation.

The subdivision of the Tertiary by Lyell has been expounded by Rudwick and given further exposition by Gould.[93] Lyell's geology included the appearance of new species, produced by some process that, while unknown, resulted in their being adapted to local environmental conditions. Species also

90. Collins, "Place of the 'Core-Set.'"

91. In the nineteenth century there were no formally established forums for the achievement of consensus. The International Geological Congresses were established (1878) in part to assist the establishment of agreed subdivisions of the stratigraphic column, and in the twentieth century "boundary commissions" were appointed to undertake such work. Though their work has been generally successful, it has sometimes been bedeviled by national rivalries; and the fact that commissioners *vote* on the question of stratotypes and stratotype boundaries establishes beyond doubt that the compilation of the stratigraphic column is an essentially social process.

92. But see Hölder, *Geologie und Paläontologie*, 439–46 and 477–78, for Oppel and his concept of zones; see also Martin, "Albert Oppel." For d'Orbigny, see Gaudant, "Actualisme, antiprogessionisme, catastrophisme et créationisme"; Tintant, "Alcide d'Orbigny"; and Taquet, *Alcide d'Orbigny*.

93. Rudwick, "Charles Lyell's Dream"; Gould, *Time's Arrow*.

died out when the environment became unfavorable. Thus there was a slow cycle of forms, with the Tertiary corresponding approximately to one complete cycle of species. The Tertiary could thus be subdivided according to the percentage of extant species present. Gould likens the processes of species creation and extinction to events in a bean bag, with old beans being taken out and new ones put in their place, the total being kept constant. But there was a snag. Lyell also believed that if, by chance, environmental conditions returned to those of some previous era, then newly produced species would be like the old. Thus he imagined that under suitable circumstances the iguanodon (or some similar creature) might reappear. But this would be akin to putting old beans back in the bag, which would confuse the method for subdividing the Tertiary (or any other part of the stratigraphic column where Lyell's method might be applied). To my knowledge, the point has been overlooked previously.

There is, furthermore, scope for historical work on the Carboniferous, Permian, Triassic, Jurassic, and Cretaceous systems. The immensely long Precambrian has also received rather little historiographic attention, though here the essays in a volume edited by Walter Kupsch and William Sarjeant, and also Stephen Brush's work, warrant mention. Much has been written on the Quaternary by those interested in the history of geomorphology. Collaborative research is needed on the history of the stratigraphic column, for example between Russian- and English-speaking geologists, to ascertain how Murchison et al. arrived at the concept of the Permian.

10. Tectonics and Related Matters

Mott Greene has written a well-received book, entitled *Geology in the Nineteenth Century* (1982), that provides a comprehensive account of nineteenth-century ideas about mountain building, with discussions of Elie de Beaumont, British theorists such as John Herschel and De la Beche, Americans such as Henry and William Rogers, the grand synthesis of the Austrian Suess, the Swiss nappe theorists (for example, Arnold Escher von der Linth, Bernhard Studer, and Albert Heim), the work of geophysical theorists such as John Pratt, Osmond Fisher, and George Airy, arising from topographic survey work in India and associated gravimetric determinations, and the tectonic ideas of Thomas Chamberlin, which were linked to his "planetesimal" hypothesis about the earth's origin.

These intricate discussions were integrated with ideas about the earth's age and its possible inner constitution—gas, liquid, or solid? The earth's age was a subject of much discussion, as physicists, headed by William Thomson, argued that it could not be as old as Lyellians and Darwinians desired. Such topics have been explored by, for example, Joe Burchfield and Stephen

Brush.[94] Brian Shipley has, however, shown how the physicist John Perry took the side of the biologists and geologists.[95] Questions of mountain building were closely connected also to ideas about cratons, geosynclines, and isostasy—topics studied particularly by Celâl Şengör, Robert Dott, and Naomi Oreskes, respectively.[96]

Robert Muir Wood, in a volume chiefly concerned with the history of plate tectonics but also addressing, in early chapters, nineteenth-century theories of mountain building, offers a very different approach from that of Greene.[97] Wood situates his account within the framework of the literature on the sociology of knowledge, considering ways in which theoretical ideas are influenced by—or at least reflect—the social situation or circumstances of the persons who formulate the ideas. The approach is stimulating, if not always convincing. For example, Wood refers to Elie de Beaumont's model of crustal folding, according to a regular geometrical arrangement of mountain fold-belts, as the earth supposedly cooled. This hypothesis, Wood suggests, was inspired by crystallographic studies at the Ecole des Mines and was analogous to Haussmann's scheme of city planning for Paris. Haussmann, Wood shows, was interested in geography and geology, and while Elie de Beaumont's theory is now a museum piece, "the geometry lives on in the radiations and perspectives of Paris, when Haussmann turned the crystalline state into a city."[98] Wood's book is full of similar, perhaps strained, similes, and the causal connections are hardly established by reference to the notion of a *Zeitgeist*. Nevertheless, *The Dark Side of the Earth* is one of the most stimulating books in the literature of the history of geology.

Şengör provides quite a different analysis of theories of mountain building. Initially, he divided tectonic theorists into two twentieth-century camps—the Wegener–Argand ("mobilist") and the Kober–Stille ("fixist")—and considered their relationship to the work of the nineteenth-century geological giant Suess.[99] Later, Şengör proposed that the roots of his dichotomy lay in antiquity and the physicotheological ideas of seventeenth- and eighteenth-century "theories of the earth."

Şengör's model is briefly as follows. According to one view (which reached

94. Burchfield, *Lord Kelvin and the Age of the Earth;* Brush, "Nineteenth-Century Debates" and *Transmuted Past.*

95. Shipley, "'Had Lord Kelvin a Right?'"

96. Şengör, "Continental Interiors and Cratons"; Dott, "Geosynclinal Concept," "Geosyncline," "James Hall's Discovery," and "James Dwight Dana's Old Tectonics"; Oreskes, *Rejection of Continental Drift.*

97. Wood, *Dark Side of the Earth.*

98. Ibid., 18.

99. Şengör, "Eduard Suess' Relations." The ideas in this article were a development of those adumbrated in one published in Japanese in 1979, which subsequently appeared as "Classical Theories of Orogenesis."

its apotheosis in the work of the Kober–Stille school, but persisted among
some postwar plate-tectonics theorists and twentieth-century Russian "fix-
ists" such as Vladimir Beloussov), episodes of mountain building occur
worldwide simultaneously, in an episodic, deterministic manner. On the alternative view (exemplified by the Wegener–Argand school), orogenies are
random, local, irregular episodes. Şengör associates the former with the
traditions of Aristotelianism, Werner's mining tradition, notions of divine
"planning," and catastrophism (understood broadly). The latter view, he argues, had its roots in Greek atomism, indeterminism, absence of divine "planning," and uniformitarianism (again in a broad sense). In the "Aristotelian"
tradition we have such figures as Werner, Leopold von Buch, Cuvier, Elie de
Beaumont, James Dana, Chamberlin, Leopold Kober, Hans Stille, and Beloussov. Amongst the "atomists" we find Hutton, Lyell, Suess, Alfred Wegener, and Emile Argand. Şengör refers to the alternative standpoints as
Leitbilder.[100] One might also call them "themata."[101] His thinking develops
that of Eugene Wegmann, who traced the inheritance of the ideas of Hutton
and Werner, likewise deploying the notion of *Leitbilder*.[102] For discussion of
nineteenth-century tectonic theorists, and especially Suess, Şengör's paper
"Classical Theories of Orogenesis" is particularly valuable. Regrettably, his
ideas, written by a scientist-historian and not published in history of science
journals, have hitherto had rather little impact on historians of science,
though his reputation amongst geologists is immense. It should be noted,
however, that in contrast with Wegmann and Şengör, François Ellenberger,
the French geologist and historian of geology, saw the work of the French tectonic theorist Marcel Bertrand as providing a synthesis between the traditions
of Hutton and Werner, even in the nineteenth century.[103]

While the specifics of Şengör's claims deserve scrutiny, he has offered a
program for understanding the history of tectonic theories and fascinating insights into nineteenth-century geology as a whole. For example, instead of
Wood's handwaving argument, which evokes Parisian street planning to "explain" the form of Elie de Beaumont's tectonic theory, Şengör quotes passages from the Frenchman's texts and shows how particular sentences were
related to the ideas of Werner, the Wernerian von Buch, and Cuvier. He does
the same for Lyell's commentary on Elie de Beaumont, and thereby demonstrates how the two schools (catastrophist and uniformitarian) were opposed
to one another at that time.

---

100. Şengör, "Timing of Orogenic Events."
101. Holton, *Scientific Imagination*.
102. Wegmann, "Das Erbe Werners und Huttons."
103. Ellenberger, "Marcel Bertrand et 'l'orogenèse programmée.'"

Related to tectonics is my study of the controversies about the structures and stratigraphic sequence of the mountains of northwest Scotland.[104] I traveled through the contested areas, trying to examine the exposures from the point of view of the disputants. I deployed ideas from Rudwick (and Collins) on "core sets," from Bruno Latour's *Science in Action,* and from the work of Ernan McMullin on the closure of controversies.[105] In a sense I did work analogous to the recommendations of Robin Collingwood,[106] in that I sought to reenact past experience, or put myself in the boots of earlier geologists. But subsequently discovered manuscripts showed both the extent to which historians are prisoners of the evidence available to them and the risks in "thought reading" or Collingwoodian reenactments.[107]

11. Volcanism and Earthquakes

Rudwick has examined the classic work of George Scrope in the Auvergne, which so influenced Lyell, as well as the latter's use of Mount Etna as a chronometer to demonstrate the great age of the earth, as previously discussed.[108] With Beryl Hamilton, I have reviewed several conflicting theories about sources of magma that were developed in the mid-nineteenth century, and have discussed a nineteenth-century controversy about the former volcanoes of the Hebrides.[109] Given that some igneous rocks were acidic, others basic, and others of intermediate composition, it was a significant question whether there are chemically different magmas within the earth, or whether there is essentially one kind, which somehow differentiates within the earth before or during crystallization. Amongst German geologists, there was some lingering of the (neo-Neptunist) idea that the nature (either texture or composition) of crystalline rocks was related to their age. Some rather complicated and speculative ideas about the processes involved in the rise of magma were proposed, for example, by Ferdinand von Richthofen, who thought that acidic magma generally preceded basic.[110]

The literature on Italian volcanism and seismology is vast, much of it historical in character, in that those studying Italian volcanoes and earthquakes have often included historical records in their accounts or scientific analyses, as did Lyell. For example, in their volume on Etna, D. K. Chester et al. begin

104. Oldroyd, *Highlands Controversy.*
105. Latour, *Science in Action;* McMullin, "Scientific Controversy."
106. Collingwood, *Autobiography* and *Idea of History.*
107. Oldroyd, "Sir Archibald Geikie." See also Oldroyd, "Use of Non-Written Sources."
108. Rudwick, "Poulett Scrope" and "Lyell on Etna."
109. Oldroyd and Hamilton, "Geikie and Judd."
110. Richthofen, "Principles of the Natural System."

with a literature survey of the researches done on Etna, and describe, with illustrations, recent volcanic activity.[111] A comprehensive bibliographic survey of the writings on Italian volcanoes has been published by Ilaria Cerbai and Claudia Principe, and Antonio Nazzaro has written a volume on the history of Vesuvius's volcanism and the geological work carried out there, illustrating the history of volcanic theory.[112] Nazzaro seeks to gain a scientific understanding of the volcano's action by studying the history of its eruptions, so his historical work serves to assist modern scientific enquiries.

I am unaware of publications that deal generally with the history of ideas about volcanoes in the nineteenth century, but a recent large collection of papers contains much information on the topic.[113] Another recent book, by the Icelandic geologist Haraldur Sigurdsson, addresses theories about the sources of the earth's internal heat and the formation of molten magma, with consideration of nineteenth-century arguments based on thermodynamics and empirical investigations of the relationships between melting points and pressure. However, even the often-mentioned debate between the supporters of the theory of "craters of elevation," such as Alexander von Humboldt and von Buch, and those who believed that volcanic cones were built up by successive lava flows, has not, to my knowledge, been studied in detail. Ideas on volcanoes were also linked to ideas about the inner constitution of the earth, studied in particular by Brush and Sigurdsson.[114]

While there is information on the occurrence of earthquakes, to my knowledge there is no recent comprehensive study on the history of earthquake theories in the nineteenth century. There is Charles Davison's old book on the history of seismology,[115] and an authoritative study by Dean of the work of Robert Mallet,[116] who investigated the great Neapolitan earthquake of 1857 on behalf of the Royal Society.[117] Mallet produced a remarkable map showing the global distribution of earthquakes. It matches similar maps produced in the twentieth century, which provide important evidence in favor of plate-tectonic theory. Jan Kozák and Marie-Claude Thompson have collected illustrations of many historically important earthquakes, some of which refer to the nineteenth century.[118] Robert Dott has examined the history of ideas about the relationship between volcanism, island arcs, and mountain building, his study containing a significant amount of information about

111. Chester et al., *Mount Etna*.
112. Cerbai and Principe, *Bibliography of Historic Activity*; Nazzaro, *Il Vesuvio*.
113. Morello, *Volcanoes and History*.
114. Brush, "Nineteenth-Century Debates" and *Transmuted Past*; Sigurdsson, *Melting the Earth*.
115. Davison, *Founders of Seismology*.
116. Dean, "Robert Mallett and the Founding of Seismology."
117. See also Cox, *Robert Mallett*.
118. Kozák and Thompson, *Historical Earthquakes in Europe*.

nineteenth-century theories. He notes that Alpine geologists interpreted ophiolites as material from an ancient oceanic floor.[119]

There have been detailed studies of the effect of earthquakes on people's thinking in the eighteenth century, notably the great Lisbon earthquake, which is said to have had a strongly adverse effect on the supposed cogency of the argument from design and deism,[120] and artistic representations of that famous earthquake have also been studied,[121] but I know of nothing similar by way of studies of the influence of earthquakes and volcanoes on nineteenth-century natural theology.

12. Petrography / Petrology

The secondary literature on this vast topic is meager, and there is an urgent need for monographs in the area. A useful starting point is Hatten Yoder's list of major names, dates, and references.[122] Also useful are a synthetic article by W. Nieuwenkamp and a dictionary by Sergei Tomkeieff, which gives all the major petrographic terms and their various meanings, with references to the publications where the rock names were first used.[123] Such works provide, however, only a groundwork for a general history of nineteenth-century petrology. A detailed history would require discussion of problems of classification, chemical and crystallographic analysis, origins, crystallization, differentiation, succession, and hybridization of magmas, the experimental study of minerals and rocks, eutectics, ideas on cleavage, metamorphism, metasomatism, the use of the petrographic microscope, and much else besides. Even biographical information on major petrographers such as Johann Friedrich Breithaupt, Harry Rosenbusch, Ferdinand Zirkel, Auguste Michel-Lévy, Alfred Harker, Jethro Teall, Joseph Iddings, and so on, is scattered and scant, though there is a biography of Henry Clifton Sorby.[124] Nieuwenkamp helpfully sets out in diagrammatic form the range of petrogenic theories developed in the nineteenth century.

At the beginning of the nineteenth century, the issue of the relative roles of heat and water in the formation of crystalline rocks was debated in the Vulcanist / Neptunist dispute, which dwelt on the origin of basalt and its relationship to the lavas that poured out of volcanoes. The problem, as shown by Bernhard Fritscher's detailed study, involved questions of practical and theo-

119. Dott, "Recognition of the Tectonic Significance of Volcanism."
120. See Kendrick, *Lisbon Earthquake;* Willey, *Eighteenth-Century Background.*
121. Kozák and Thompson, *Historical Earthquakes in Europe.*
122. Yoder, "Timetable of Petrology."
123. Nieuwenkamp, "Trends in Nineteenth Century Petrology"; Tomkeieff, *Dictionary of Petrology.*
124. Higham, *Very Scientific Gentleman.*

retical chemistry, the issue of actualism, and much empirical work in the field and laboratory.[125] Contrary to what might be supposed from the old histories such as Geikie's, the Neptunists were in many ways more empirical or phenomenological than the Vulcanists and Plutonists. Emile den Tex has described the empirical grounds by which the debates were eventually resolved.[126]

Darwin's work on igneous rocks and his efforts to account for their diversity should not be overlooked. He was interested in the occurrence of silica-rich and silica-poor lavas, and wondered whether they might have separated from the same source according to differences in melting point. If crystals of higher melting point (principally quartz) crystallize first, the chemical composition of the remaining liquid will be altered, though the whole might be expected to retain the original bulk composition. Darwin, however, suggested that chemical partition may occur as a result of "gravity settling" of the initial crop of crystals. In consequence, two (or more) rock types might be produced from the same initial melt. This topic is of fundamental importance in petrology, and the idea has taken many forms through to the present. To follow through this question would be a major contribution to the understanding of the history of petrology, but it is chiefly a twentieth-century topic.

In studying early petrological writings it is often difficult to know what the observers were observing and why they described them as they did. This applies to Darwin's work no less than to that of others. However, at the end of the nineteenth century the Cambridge petrologist Alfred Harker sectioned some of Darwin's igneous specimens from the *Beagle* voyage, and Paul Pearson has examined Harker's slides as well as Darwin's corresponding macrospecimens.[127] This approach allows a greatly improved understanding of Darwin's thinking on igneous petrology, and Pearson's procedures could be generalized.

The origin of granite was a persistent problem through the nineteenth century. Though this coarse crystalline rock "looks" igneous, a possible aqueous origin was encouraged by the fact that it could not be fused and then cooled to form a mass like the starting material. A thorough historical study of ideas about plutonic rocks like granite is needed, though den Tex has made a start for the nineteenth century.[128] Among the Plutonists, he suggested, two schools developed: the "transformists" and the "magmatists." The transformists followed Hutton in supposing that granites were recycled sediments, but some of them (quasi-Wernerians) supposed that water was necessary for

125. Fritscher, *Vulkanismusstreit und Geochemie* and "Vulkane und Hochöfen."
126. Den Tex, "Clinchers in the Basalt Controversy."
127. Pearson, "Charles Darwin on the Origin and Diversity of Igneous Rocks."
128. Den Tex, "Punctuated Equilibria" and "Helicoidal and Punctuated Cyclicity."

the conversion. Sorby's discovery of water in cavities in quartz grains in granite encouraged such neo-Neptunist views,[129] as did experiments on the action of superheated water in sealed vessels on various mineral mixtures.[130] The magmatists were more directly descendants of Hutton, and saw igneous rocks, including granite, as originating from subterranean reservoirs of molten rock. There were also "wet" and "dry" magmatists. The latter were centered on Heidelberg and Paris and achieved success in simulating rock formation by slow cooling of melts. The "wet" magmatists, by contrast, thought that water was necessary for the formation of granite but that the magma source was juvenile, not recycled rock. In the twentieth century, as Herbert Read has shown, battle became joined between "migmatists" and "magmatists."[131]

The history of igneous petrology in the nineteenth century is intimately connected with that of the twentieth century, where it has been one of the most contentious fields of geology, or even of science (though this is hardly realized by the public at large). A convenient end point for the study of nineteenth-century igneous petrology might be 1910, when experimental investigations of rock melts were initiated by Norman Bowen at the Carnegie Institution.

If the history of igneous petrology is sparsely represented, the situation is even less satisfactory for sedimentary and metamorphic petrology. For the former, there are, to my knowledge, no monographic studies and little other material. For the history of metamorphism, however, a recent internalist survey of the history of metamorphic petrology by Jacques Touret and Timo Nijland well shows the nineteenth-century origins of the twentieth-century schools of migmatism and magmatism.[132] Gabriel Gohau has described the work of French nineteenth-century experimentalists such as Auguste Daubrée, Joseph Durocher, Achille Delesse, and Auguste Michel-Lévy.[133] He has seen a divide among nineteenth-century theories of metamorphism that can be traced back to separate and distinct Huttonian and Wernerian roots. Justus Roth has published a synopsis of ideas up to 1871.[134] There are no comprehensive, up-to-date histories of mineralogy known to me.

---

129. Sorby, "On the Microscopical Structure of Crystals."

130. Den Tex calls the men who espoused these ideas about granite "wet transformists." There were later "dry transformists," who hypothesized the existence of "basic fronts" around granite masses, these being surrounding regions in which acidic granitic materials had been "scavenged" from the country rock.

131. Read, *Granite Controversy*. Migmatists envisaged the formation of gneisses and granites by the percolation of hot solutions through sediments, under pressure.

132. Touret and Nijland, "Metamorphism Today."

133. Gohau, "Evolution des idées sur le métamorphisme."

134. Roth, "Über die Lehre vom Metamorphismus."

13. Paleontology

While the literature on petrology is undeveloped, that on paleontology, thanks to its connection with evolutionary biology, is relatively abundant and developed. For general texts, there are, for example, Yvette Gayrard-Valy's short general history of paleontology and Eric Buffetaut's book on the history of vertebrate paleontology.[135] On the other hand, should one wish to find a book that deals with the history of, say, ideas about graptolites, there seems to be nowhere to turn.[136]

One of the first studies to shape the field in the modern period of science historiography was Rudwick's *Meaning of Fossils* (1972). It dealt with the history of *ideas* about fossils—their scientific, religious, and metaphysical / philosophical significances—from the Renaissance to the time of Darwin.[137] He sought to rescue Cuvier from the Anglophone emphasis on the virtues of Huttonian and Lyellian geology, showing the relationship between Cuvier's comparative anatomy and directionalist stratigraphy. As we have seen, Rudwick regards Cuvier as the founder of "geohistorical" investigation. In a recent book, he has provided translations of some of the less known of Cuvier's writings, interspersed with explanatory commentary and historical background.[138] This offers almost a new genre for the study of the history of geology, though one might also think of it as an extreme version of the classic historiography of Alexandre Koyré, with its exceptionally long quotations from primary sources and accompanying exegesis.[139] The philosopher Michel Foucault has also emphasized the importance of Cuvier's comparative anatomy and has seen him as effecting a fundamental transition between eighteenth-century natural history and Darwinian evolutionary biology, even though Cuvier was anything but an evolutionist.[140]

Another history of paleontology from the 1970s, Peter Bowler's *Fossils and Progress*, focused on the relationship between fossils and the "idea of progress." Bowler has subsequently explored ideas about evolution in the post-Darwinian period and has found that in many cases biologists in the closing decades of the nineteenth-century favored non-Darwinian models, such as neo-Lamarckism and orthogenesis.[141] In his important book *Life's Splendid*

135. Gayrard-Valy, *Story of Fossils;* Buffetaut, *Short History of Vertebrate Palaeontology.*

136. There are comprehensive "annals" of graptolite research in Britain to 1913, which would make an essential source for a history of the topic. See Elles and Wood, *Monograph of British Graptolites.*

137. Rudwick was by training a paleontologist and an authority on brachiopods.

138. Rudwick, *Georges Cuvier.*

139. See, as a sample, Koyré, *Etudes Galiléennes.*

140. Foucault, "La situation de Cuvier."

141. Bowler, *Eclipse of Darwinism.*

Drama (1996), Bowler examines the complex ways in which ideas about evolutionary morphology, paleontology, and biogeography were intertwined. Nineteenth-century biologists and paleontologists had tried to integrate their work with shifting ideas about former distributions of land masses, tectonic theories, convergent or parallel evolution, notions of climatic change, the age of the earth, and so on. The field was a tangle; it provides ample work for historians of science. Bowler eased the situation by focusing on specific debates, such as the transitions from invertebrates to vertebrates, fish to amphibia, reptiles to mammals, and the origin of flight. He also examined the problem of "patterns in the past." Patterns of classification and phylogenetic evolution have likewise been studied by Robert O'Hara, who, in another paper, has pointed out how texts and illustrations in evolutionary works have manifested a "whiggish" view of evolution, always culminating in *Homo sapiens*.[142] Investigations such as O'Hara's are not specifically concerned with the history of paleontology, but they thread together with paleontology and stratigraphy in complex and interesting ways.

Ronald Rainger has noted a tendency toward philosophical idealism in late nineteenth-century paleontology.[143] This was especially so for studies of vertebrates, where a number of disassembled bones might need to be reassembled, or where missing parts might need to be "imagined" and constructed to make up a total skeleton. In such cases, some general notion of what the organism might have been like, if preserved whole, was required. (This, of course, had been a problem treated earlier by Cuvier.) The "idealistic" emphasis was thus linked to the prior empirical and Linnean interest in type specimens, or even archetypes. This linkage has been studied from a biological perspective for the so-called Cuvier–Geoffroy debate.[144] It reappeared with Richard Owen, founder of the British Museum of Natural History and the subject of an exemplary intellectual biography by Rupke.[145]

Owen was not strictly a geologist or paleontologist, and certainly not a fieldman, relying on others to send him materials for study. But his paleontological contributions were immense, his work on the extinct megafauna of

142. O'Hara, "Representations of the Natural System" and "Telling the Tree."

143. Rainger, "Paleontology and Philosophy."

144. See Appel, *Cuvier-Geoffroy Debate*, and Laurent, *Paléontologie et évolution en France*.

145. Rupke, *Richard Owen*. Further detail on Owen, and information on the history of vertebrate paleontology, appears in Dennis Dean's important biography of Mantell: *Gideon Mantell and the Discovery of Dinosaurs* (1999). A highly informative account of the personal characters of Mantell and Owen is given by the journalist Deborah Cadbury in her *Dinosaur Hunters* (2000). This book, incidentally, shows that science writers can sometimes produce more vivid or revealing insights into scientists' personalities and motivations than do professional historians of science. There has recently been quite a spate of interesting semipopular books on the history of paleontology and stratigraphy. On the question of books of this type and their relationship to the work of professional historians of science, see Miller, "Sobel Effect."

the southern hemisphere being particularly remarkable—as, for example, in his recognition of the main features of the New Zealand moa from a single bone. He carried out a major survey of the anatomies and classification of British fossil reptiles, and introduced the term "dinosaur."[146] Rupke shows that Owen was neither a Platonist nor a biblical fundamentalist; and, though opposed to Darwinian theory, he had a kind of evolutionary theory of his own (a "combined orthogenetic-mutational mechanism" with resort to "Lamarckian atrophy," as Rupke puts it).[147] Owen did, however, hold to a kind of "transcendental anatomy." This was convenient for his museum displays, as he did not need to show reconstructed organisms in their supposed ecological habitats in order to be true to his philosophy. His articulated specimens could stand isolated in the British Museum's halls.

Rupke sees a kind of faultline running between the London transcendentalists, with Owen at their head, and the Oxbridge functionalists, such as Buckland, a hyperactualist (albeit a catastrophist).[148] Rupke differs from Adrian Desmond, who sees a divide between the Oxbridge elite and the radical medical fraternity in London, headed by the likes of Robert Grant.[149] Here one can discern two alternative views of history, reflecting in part the philosophical, social, and political proclivities of their authors.[150] It is probably in such excellent studies as these—on the borderline with the history of biology—that we see history of geology in its most politically oriented mode. This is also a rare case where attention has been focused sufficiently for alternative views as to the social context of geological ideas to be pitted against one another.

Goulven Laurent has undertaken a detailed study of nineteenth-century French paleontology and paleontologists in relation to evolutionary theory in several publications and in particular has traced, year by year, the debates

146. Owen, "Report on British Fossil Reptiles." This paper was delivered to the British Association at its meeting in Plymouth in 1841 but not published until the following year. It is commonly thought that Owen introduced the term "dinosaur" in 1841, but the detective work of Torrens, making use of his study of the surviving fossil specimens that Owen examined in the period between the oral presentation of his paper and its subsequent publication, reveals that Owen thought up the name in the time between oral presentation and publication. See Torrens, "When Did the Dinosaur Get Its Name?" Here we see, in a small way, the value of museum work to the historian of geology.

147. Rupke, *Richard Owen*, 250.

148. Buckland compared marks on bones excavated from Yorkshire caves with marks made on bones by modern hyenas. The experiment convinced him, and others, that hyenas had formerly occupied the Yorkshire caverns.

149. Desmond, *Archetypes and Ancestors*.

150. Desmond's political radicalism is seen also in his *Politics of Evolution* as well as in *Darwin*, a biography he coauthored with James Moore.

between Lamarck and Cuvier.[151] He has also addressed the question of continuity or discontinuity in the stratigraphic record, suggesting that interpretation according to the latter doctrine gradually gave way to the former after the death of Cuvier. Laurent has proposed an obscure figure, Frédéric Gérard, as having been the first to present a clear exposition of a scientific theory of "evolution" (1844–45), using that term rather than "transformism"; he further maintains that an anticatastrophist Lamarckian Belgian geologist, Jean-Baptiste d'Omalius d'Halloy, provided fossil evidence sufficient to support the theory.[152] It should be remembered that the establishment by Alcide d'Orbigny and Albert Oppel of the fundamental stratigraphic concept of the "(bio)zone" rested on "catastrophist" suppositions (which were denied by Gérard). Laurent's historiography is also reminiscent of the tradition of Alexandre Koyré, in that he lays out long extracts from primary sources and develops his argument accordingly.[153] Laurent consistently emphasizes the importance of Lamarck in the history of ideas about organic remains, and sees Gérard as belonging to the Lamarckian heritage.

14. Geomorphology

Studies of the history of landforms in the nineteenth century have been dominated by consideration of diluvialism, glaciation, and landscape evolution. The field has been well covered in its internalist aspects, with studies by such historians as Gordon [Herries] Davies, Keith Tinkler, and Richard Chorley, Antony Dunn, and Robert Beckinsale.[154] Reaching back into the eighteenth century, Hutton and Playfair's geomorphological ideas have been closely analyzed by Davies and Dean.[155] Davies's book describes geologists' efforts to come to terms with Louis Agassiz's glacial theory. Lyell, for example, vacillated, worrying whether glacial theory could or could not be satisfactorily accommodated to his multifaceted uniformitarianism. Rupke has sorted out many of the details of controversies concerning the Noachian flood (earlier studied by Gillispie and Hooykaas),[156] and the topic is also treated by the neodiluvialist geographer Richard Huggett.[157]

---

151. Laurent, "Paléontologie(s) et évolution au début du XIXe siècle." See also Laurent, *Paléontologie et évolution en France* and *Naissance du transformisme*.

152. Omalius d'Halloy, "Note sur la succession."

153. Laurent, *Naissance du transformisme*.

154. Davies, *Earth in Decay;* Tinkler, *Short History of Geomorphology;* Chorley, Dunn, and Beckinsale, *History of the Study of Landforms*.

155. Davies, *Earth in Decay;* Dean, "James Hutton's Place."

156. Rupke, *Great Chain of History*.

157. Huggett, *Cataclysms and Earth History*.

Davies has particularly emphasized the importance for geomorphology of Joseph Beete Jukes's studies of Irish river patterns (1862). Jukes realized that rivers can cut through hills, if the rate of cutting approximates the rate at which land is elevated.[158] So we have "antecedent drainage patterns," later so called by John Wesley Powell[159] and admirably described by Chorley, Dunn, and Beckinsale in their study of geomorphology in relation to explorations in the American West by the likes of Powell, William Morris Davis, and Clarence Dutton. These explorers laid the foundations of scientific studies of landforms, introducing ideas on the "life histories" of landscapes and elaborating differences between erosive processes in different climates. Such "organic" and evolutionary metaphors were popular until well into the twentieth century.

A major issue in geomorphology in the first half of the nineteenth century was the relative erosive power of rivers and seas. Darwin and Lyell tended to emphasize marine processes, and this led to Darwin's "great failure" in geology, in his study of the "parallel roads" of Glen Roy, Scotland—markings on the sides of the glen, which Darwin believed were produced by the sea having stood at different levels relative to the land, but which were later regarded as shorelines of glacially dammed lakes.[160] Rudwick's study of the Glen Roy debate displays the intertwining of observations, social relations, and ideas about method.[161] Dorothy Sack's analysis of investigations at Lake Bonneville, Utah, though less methodologically oriented, presents an analogous study.[162]

15. The Earth as a Planet

Agassiz's glacial hypothesis was "overdetermined,"[163] but a few decades later a member of the Scottish Survey, James Croll, proposed a theory that seemed quite satisfactory: changes in the ellipticity of the earth's orbit, together with the effect of the precessional motion of its axis of rotation—due to secular changes in planetary orbits—could cause climatic changes. The hypothesis implied the occurrence of widely spaced glacial epochs, as in the Permian and the Pleistocene, and also interglacials within the larger glacial phases, arising from precession of the equinoxes. The theory depended on complex calcula-

---

158. Herries Davies, *North from the Hook.*

159. Powell, *Exploration of the Colorado River,* 163.

160. Darwin, "Observations on the Parallel Roads of Glen Roy." This embarrassing "failure" may have been part of the reason why Darwin moved from geology to biology in the second part of his career.

161. Rudwick, "Darwin and Glen Roy."

162. Sack, "Reconstructing the Chronology of Lake Bonneville."

163. Lugg, "Overdetermined Problems in Science."

tions but was deducible from Newtonian mechanics and astronomy. It required that the climates differed in the two hemispheres in glacial epochs.

The full story of the arguments about Croll's theory has yet to be told, but it is outlined by John and Katherine Imbrie,[164] nineteenth-century history being harnessed to support modern scientific theory. Croll's ideas were developed by the Serbian mathematician Milutin Milankovich in the 1920s and 1930s, and arguments continue to the present as to causes of the occurrence of glacial epochs. A thorough study of the gradual development of glacial theory in the nineteenth century, culminating in the theory of four glacial phases in the Pleistocene (in the work of Penck and Brückner) is needed.[165]

For about thirty years after Lyell's *Principles*, geologists were wary of discussing cosmological questions, and when they did so, as in the case of Sterry Hunt, they were sometimes regarded with disfavor. So it was mostly physicists and astronomers, or outsiders like Croll,[166] rather than field geologists, who sought to develop grander geological theories, thinking of the earth as a planetary object. Still, when physicists intruded into the geological domain they were not always welcome. For example, William Thomson (Lord Kelvin) argued—on the basis of ideas about the earth as a cooling body and considerations of its shape and the effect of tidal friction—that the planet was about twenty million years old.[167] This was unsatisfactory to both geologists and evolutionary biologists, and to an extent the geologists just went their own way. Nevertheless, the issue was important. Burchfield has sorted out the details well.

There were also geophysical discussions, and the nineteenth-century models of the earth have been analyzed by Brush, who illustrated them diagrammatically and showed how and why each was proposed and received.[168] The Cambridge mathematician and geologist William Hopkins argued that the rotating earth could not have a liquid interior as it would be unstable to tidal forces. Geologists again deferred to mathematical and physical expertise, but in the early twentieth century seismological evidence suggested that there was indeed a liquid core within a large solid "mantle" below the crust. In the early nineteenth century, there were debates about the relation between Kant's and Laplace's nebular hypothesis and geology, which generated tec-

164. Imbrie and Imbrie, *Ice Ages*.

165. Penck and Brückner, *Alpen im Eiszeitalter*. See also Schultz, "Debate over Multiple Glaciation," and Eagan, "Multiple Glaciation Debate." Eagan shows that preference for single or multiple glaciation could be dependent on the social and political circumstances in different countries, as well as the amount of fieldwork performed.

166. Croll was employed in routine administrative work in the Scottish branch of the Geological Survey but did little fieldwork.

167. Burchfield, *Lord Kelvin and the Age of the Earth*.

168. Brush, "Nineteenth-Century Debates."

tonic theories such as that of Elie de Beaumont. According to Philip Lawrence, nineteenth-century directionalism originated in the nebular hypothesis as well as in stratigraphic evidence.[169] The borderland between geology
and astronomy has been closely surveyed by Brush.[170]

As a planetary body, the earth is subject to meteoric impacts, and many
now believe that bolides have had a profound (though occasional) effect on
terrestrial history and the course of biological evolution.[171] The topic of meteoric or cometary impacts was not of special importance in the nineteenth
century (though the term "astrogeology" was coined then by V. V. Lesevich).
However, detailed historical study on meteorites has been undertaken by Ursula Marvin,[172] and Ron Westrum has provided a social and philosophical
analysis of early meteorite reports and the way they were received by scientists. His study provides an entrée to the general problem of how scientists
judge the a priori probability of reports of anomalous occurrences.[173]

16. Geophysics

In geodesy, much work (empirical and theoretical) on the figure of the earth
was accomplished in the eighteenth century. In the middle years of the nineteenth century, geodetic determinations were a major factor leading to new
ideas in tectonic theory;[174] and Oreskes has shown how gravimetry and
geodesics were essential to the nineteenth- and early-twentieth-century development of ideas about isostasy, which dominated the negative American
attitude toward continental drift in the twentieth century.[175]

I have suggested, using ideas from Don Ihde, that a useful way to think
about seismological investigation is to consider its instruments as extensions
of the human sense organs, and that the reading of seismograms is a complex
exercise in hermeneutics.[176] The "world" (here the earth's interior) is never
inspected directly. Instead, one views and interprets an "instrument–world"
couple. Thus the question of whether or not the theoretical entities of
the seismologist "refer" (in a philosophical sense) requires consideration.
The problem applies more particularly to twentieth-century seismology,
but it originated in the nineteenth century; and one can hardly understand

---

169. Lawrence, "Heaven and Earth."
170. Brush, *Nebulous Earth* and *Transmuted Past*.
171. Glen, *Mass-Extinction Debates*.
172. See, for example, Marvin, "Chladni" and "Meteorites, the Moon, and the History of Geology."
173. Westrum, "Science and Social Intelligence."
174. Greene, *Geology in the Nineteenth Century*.
175. Oreskes, *Rejection of Continental Drift*.
176. Oldroyd, *Thinking about the Earth*; Ihde, *Technology and the Life World*.

the twentieth-century history without considering the instruments and techniques of the preceding century. There is information available about nineteenth-century seismological instruments, particularly in Italy, Germany, and Japan, where nineteenth-century seismology was chiefly developed.[177] Likewise, there is information available on geomagnetic and gravimetric instruments.[178]

Some connecting schema is required to make sense of the plethora of instruments and associated theories, and here the work of Gaston Bachelard, with his notion of "phenomenotechnics," may be useful.[179] Such an approach has been deployed by John Schuster and Graeme Wachirs in their study of eighteenth-century electrical machines.[180] Their idea was to see how instruments evolved in concert with electrical theory. One can see the instruments as having family relationships, as various types were successively refined and modified. Bachelard held that instruments and theories developed in tandem, or were "coupled." One could not develop without the other. Thus, for a Bachelardian, understanding the history of theory would be impossible without historical study of the associated instruments. The Bachelardian methodology could usefully be applied to all branches of geophysics, but to date this has not been seriously attempted. In general, geophysicists are rather interested in historical records for the reason that they can provide empirical nutriment for the development of ideas in, say, geomagnetism or seismology. But such interest is not history for history's sake.

17. Geochemistry

Geochemistry is an extremely complex field, concerned with, among other things, the circulation and distribution of chemical elements in the earth, waters, and atmosphere. It involves chemical analyses of rocks and minerals, and the use of the information thus obtained for the purposes of petrography and petrology. Moreover, it involves quantitative study of physicochemical systems, especially silicates, in the crystallization of magmas, such as was carried out at the Carnegie Institution from the early years of the twentieth century. Geochemistry is also concerned with ore bodies and mineralization, and the processes involved as elements move from one reservoir to another, in which cyclic movements of living organisms may play an essential role. There were nineteenth-century attempts to work out a "genetic" history of the globe, ac-

177. See, e.g., Ferrari, *Two Hundred Years of Seismic Instruments;* Dewey and Byerly, "Early History of Seismometry"; and Wartnaby, "John Milne."

178. Multhauf and Good, *Brief History of Geomagnetism;* McConnell, *Geophysics and Geomagnetism;* Swick, *Modern Methods.*

179. Bachelard, *Rationalisme appliqué.*

180. Schuster and Wachirs, "Natural Philosophy."

cording to supposed natural sequences of chemical reactions. This has been called "chemical geology"[181] but can be regarded as a kind of geochemistry.

Geochemistry is largely a science of the twentieth century, but its earlier roots are significant. For its history (as yet unwritten),[182] one would consider, for example, the cyclic theories of Hutton and Lamarck; the hoped-for universal (or cosmic) science of Alexander von Humboldt; and the work of the Swiss chemist Christian Schönbein (who coined the term "geochemistry" in 1838),[183] of Dana (on coral reefs particularly), of Gustav Bischof (who regarded the earth as a vast chemical laboratory),[184] of Suess (who coined the term "biosphere"),[185] of the Norwegian Johan Vogt, of the Russian school of Vladimir Vernadsky and his coworkers, and of Sterry Hunt. To my knowledge there has been no such systematic study of the field for the nineteenth century. A recent translation of Vernadsky's *Biosphere*, however, gives essential information about his ideas.[186] His work has attracted much interest and admiration in Russia.

18. Some Generalizations

Attempts at synthesis in the historiography of the earth sciences are not the norm. Much of the literature consists of minor biographical studies or accounts of the geology of small regions. Studies of the interrelations of the different branches of the earth sciences are scarce, and the history of geological research in many parts of the world, especially Africa, is almost unknown, even to specialists.[187] Large areas, notably Asia, present problems for scholars not fluent in the relevant languages. Though the philosophical assumptions of geologists have interested historians, there are few philosophical studies of the history of geology, and those that have been written have chiefly deployed well-known philosophies of science such as those of Karl Popper or Thomas Kuhn. More sophisticated philosophies, such as those currently emanating from cognitive science or sociology of knowledge, are little used. There is, however, a valuable German analysis of the way geological science is practiced, with some historical information.[188]

---

181. Brock, "Chemical Geology or Geological Chemistry?"
182. See, however, Manten, "Historical Foundations"; Boyle, "Geochemistry in the Geological Survey of Canada"; Lapo, *Traces of Bygone Biospheres;* Vinogradov, "Development of V. I. Vernadsky's Ideas"; and Westbroek, *Life as a Geological Force.*
183. Schönbein, "On the Causes of the Change of Colour."
184. Bischof, *Elements of Chemical and Physical Geology.*
185. Suess, *Die Entstehung der Alpen,* 159.
186. Vernadsky, *Biosphere.*
187. But see Mohr, *Bibliography of the Discovery.*
188. Engelhardt and Zimmerman, *Theory of Earth Science.*

There is much scope for detailed studies of the history of geology from a social or sociological perspective. Rudwick's *Great Devonian Controversy* shows what can be done, but his book is so complex that it is unlikely to have imitators in the near future. Perhaps the most pressing current needs are for general studies in the histories of mineralogy and petrology, metamorphism, and invertebrate paleontology in the nineteenth century. We also need monographic studies of the history of geological work in particular regions, such as the Himalayas, and particular countries, such as Japan. One can also envisage detailed studies of the history of research in fairly small regions, such as the Auvergne. These could serve as lenses through which the development of geological theory and practice could be viewed. There are already detailed studies relating to the history of mapping and fieldwork, but comparisons would be useful between work in such countries as Britain or France and in "remoter" areas, such as the American West or China.[189] Biographical studies of major geologists such as Bertrand, Dana, Dutton, Elie de Beaumont, Lapworth, Suess, and Zirkel are also needed.[190] More work could be done on the history of the relationship between geology and industry, though it has been given considerable emphasis in the work of Hugh Torrens on William Smith and others. For geology and military affairs, there are now two books.[191] There are studies linking geology to such cultural movements as Romanticism[192] and historicism,[193] but other connections—with, for example, positivism, Marxism, colonialism, or Hegelianism—could presumably be explored with profit. Whether such "history of ideas" approaches tell us things of essential import for understanding the history of the earth sciences may be open to question (especially for scientist-historians), as compared with getting a firmer grasp of the technical, social, and institutional histories of the earth sciences. Be that as it may, regardless of the perspective or topic, the range of possible historical exploration of nineteenth-century earth science is vast and fascinating; and the search to unite understanding of the field's various elements and approaches needs to continue.

189. For China, see Oldroyd and Yang, "On Being the First Western Geologist in China."

190. But on Bertrand, see Trümpy and Lemoine, "Marcel Bertrand." On Dana, see Pirsson, "Biographical Memoir." On Suess, see Österreichische Geologische Gesellschaft, *Eduard Suess.*

191. Rose and Nathanail, *Geology and Warfare;* Underwood and Guth, *Military Geology.*

192. See Rupke, "Study of Fossils." Certainly, one can see a connection between the earth sciences, Romanticism, and interest in travel in the nineteenth century, perhaps most notably in the work of Alexander von Humboldt. Rupke also, rightly I think, draws attention to the Romantic interest in caves, where many important paleontological collections were made in the nineteenth century. Interestingly, the popular scriptural geologist Granville Penn wrote: "Geology, indeed . . . may be called a romantic science." (Penn, *Conversations on Geology,* 5.)

193. See Oldroyd, "Historicism and the Rise of Historical Geology."

ACKNOWLEDGMENTS

I am indebted to Gabriel Gohau, Martin Guntau, Jan Kozák, Goulven Laurent, Antonio Nazzaro, and Celâl Şengör for graciously sending me some of their publications; to Bernhard Fritscher for information about German-language biographies of nineteenth-century geologists; to Ludmila Stern for information about Russian geological histories; and to Hugh Torrens for numerous matters. Professor Şengör furnished a critical reading of a draft and made several helpful suggestions. David Cahan has been wonderfully patient and, along with Joel Score, has improved my text immeasurably by his judicious editing.

Five
MATHEMATICS

JOSEPH DAUBEN

SINCE THE MID-1970s, the literature on the history of nineteenth-century mathematics has grown more rapidly than that for any other period of the history of mathematics.[1] This remarkable increase is reflected in the number of new journals devoted to the history of mathematics (six) that have appeared since *Historia Mathematica* was launched in 1974. One of these, *Studies in the History of Modern Mathematics,* is dedicated specifically to history of nineteenth- and early twentieth-century mathematics.[2] The amount of literature (itself in a variety of languages) and the inherent technical difficulties involved in understanding nineteenth-century mathematics are formidable. It has been estimated that "the nineteenth century alone contributed about five times as much to mathematics as had all preceding history."[3]

To explore the extent to which our understanding of the history of nineteenth-century mathematics has changed over the course of the past century, including both dramatic and fundamental changes in the character of mathematics and the nature of historical scholarship itself, is one of two goals of this essay. The other is to identify and evaluate the most important contributions to that history and to suggest potential directions for future research.

1. May, "Growth and Quality."

2. Dauben, *"Historia Mathematica"* and *"Historia Mathematicae."*

3. Bell, *Development of Mathematics,* 17, quoted in Boyer, *History of Mathematics,* 224. For a survey of the nearly exponential increase in mathematical publications, see May, "Quantitative Growth"; on the nature and expansion of studies related to history of mathematics in particular, see Dauben, *History of Mathematics,* introduction. On the difficulties of doing history of mathematics in the modern period, especially due to the technical demands of the material, see Grabiner, "Mathematician," and Weil, "History of Mathematics." May's voluminous *Bibliography and Research Manual* (1973), still a young researcher's best introduction to the field, contains separate sections devoted to the nineteenth century. *Historia Mathematica* includes a special section for abstracts; each issue provides a useful guide to the rapidly growing literature. *Mathematical Reviews* also reserves a special category for history of nineteenth-century mathematics. The annual bibliographies published by *Isis* are likewise extremely helpful in orienting readers to the current literature. Dauben, *History of Mathematics,* a comprehensive critical bibliography, appeared in 1985; it contains a special section for the history of nineteenth-century mathematics. Albert C. Lewis has recently coordinated production of a revised edition, which was released on CD-ROM by the American Mathematical Society in 2000.

The essay first discusses the uniqueness of the history of mathematics among the histories of the sciences, in particular taking up the transformation of mathematics from its applied to its pure form during the nineteenth century (section 1), and then addresses the subject of mathematicians writing the history of mathematics (section 2). It then turns specifically to the history of nineteenth-century mathematics, beginning with nineteenth-century histories (section 3), John Theodore Merz on the history of nineteenth-century mathematics (section 4), histories written prior to and during World War I (section 5), and the works of Eric Temple Bell and Carl Boyer (section 6). Thereafter it takes up a series of more specialized historiographical issues: the influence of Thomas Kuhn's analysis of science on mathematical revolutions and communities (section 7), Marxism and the history of mathematics (section 8), the emergence of the social history of mathematics (section 9), the social construction of mathematics (section 10), and postmodernist interpretations of nineteenth-century mathematics (section 11). Section 12 provides a lengthy discussion of three traditional approaches to the history of nineteenth-century mathematics: biographies, surveys, and special topics. Finally, section 13 draws attention to the constitutional conservatism and plight of historians of mathematics within the history of science community.

Unlike the histories of the other sciences in the nineteenth century, the history of mathematics has been written largely for an audience of its own specialists, i.e., mathematicians, who constitute its largest and most appreciative followers but whose standards and expectations tend to be severely technical and much narrower than would satisfy most historians of science, or intellectual historians generally. Initially at least, and almost exclusively until after World War II, those who have written this history have mostly been mathematicians; given the technical sophistication of nineteenth-century mathematics, this is perhaps understandable. Without a basic understanding of the mathematics in question, it is difficult to imagine how one might expect to write historically about the subject in an insightful or acceptable way.

Moreover, textbooks on the history of mathematics, a type of literature that is for the most part absent from the other natural and social sciences, have had a wide following among mathematicians, and they have been employed with increasing regularity in undergraduate teaching in particular. This stems no doubt in part from a characteristic unique to mathematics among the sciences, the seemingly timeless validity of its results, a factor that creates unique problems and determines the special character of the historiography of the field. This matter is discussed again in the last section of this essay.

It should also be remembered that although the history of modern mathematics reaches back well into the nineteenth century, the study of its practitioners, specialties, institutions, and journals—and of social and other professional matters, including gender issues, relations with society, and postmodernist concerns—has been largely a product of the past twenty-five years. Indeed, the history of mathematics has now established itself as an active and productive specialization within both mathematics and history of science, and, among other accomplishments, has developed important links with the philosophy and sociology of science and with mathematics education.

In large measure, the complex history of the history of mathematics is due to the special nature of mathematics itself. Its peculiar epistemological character has been expressed in countless ways, but perhaps nowhere more succinctly than as follows:

> In theoretical physics it is but seldom necessary to master in detail a work published over thirty years ago, or even to remember such a work was ever written. But in mathematics the man who is ignorant of what Pythagoras said in Croton in 500 B.C. about the square on the longest side of a right-angled triangle, or who forgets what someone in China proved last week about inequalities, is likely to be lost. The whole terrific mass of well-established mathematics, from the ancient Babylonians to the modern Chinese and Japanese, is as good today as it ever was.[4]

Moreover, there has always been good reason to advocate serious history of mathematics over textbook accounts of mathematics, which usually obscure the origins and thus tend to minimize if not misrepresent the difficulty by which new results were won. As the historian of mathematics Morris Kline puts it, the final published presentations found in journals and textbooks

> fail to show the struggles of the creative process, the frustrations, and the long arduous road mathematicians must travel to attain a sizable structure. Once aware of this, the student will not only gain insight but derive courage to pursue tenaciously his own problems and not be dismayed by the incompleteness or deficiencies in his own work. Indeed the account of how mathematicians stumbled, groped their way through obscurities, and arrived piecemeal at their results should give heart to any tyro in research.[5]

This was exactly the advice Karl Weierstrass gave his students in the nineteenth century—"read the masters"—and the same idea has continued to serve both mathematicians and historians well. There is perhaps no better example of this than the careful reading Harold Edwards has given of the great nineteenth-century mathematician Bernhard Riemann. As Edwards acknowledges, "No secondary source can duplicate Riemann's insight."[6]

Likewise, George Polya emphasizes the importance of understanding how

4. Bell, *Mathematics, Queen and Servant*, 7–8.
5. Kline, *Mathematical Thought*, ix.
6. Edwards, *Riemann's Zeta Function*, x.

mathematicians have come to their great discoveries, not just learning how they presented new discoveries in their final published versions. Polya draws a distinction between demonstrative reasoning and plausible reasoning:

> The difference between the two kinds of reasoning is great and manifold. Demonstrative reasoning is safe, beyond controversy, and final. Plausible reasoning is hazardous, controversial, and provisional. . . . Finished mathematics presented in a finished form appears as purely demonstrative, consisting of proofs only. Yet mathematics in the making resembles any other human knowledge in the making. You have to guess a mathematical theorem before you can prove it.[7]

History of mathematics can help the mathematician identify and study the crucial moments in the development of the subject in the nineteenth century, just as for the historian of science the history of mathematics creates a foundation upon which a full understanding of the other sciences may depend. As George Sarton firmly believed: "Take the mathematical developments out of the history of science, and you suppress the skeleton which supported and kept together all the rest."[8]

1. From Applied to Pure Mathematics

At the beginning of the nineteenth century, mathematics served as an instrument of great utility in applications primarily to astronomy and various branches of physics; by the end, it had become an increasingly abstract, independent discipline, more indispensable than ever to the sciences when applied, but also devoted at its most advanced levels to pure mathematics. If the general movement was from natural philosophy to individual disciplines of science, mathematics was moving from "handmaiden" and "servant" of the sciences, as Bell put it, to autonomous subject, with its own journals, institutions, professional identity, and international orientation. Not only were mathematicians interested in their subject per se as an object of study, by the end of the century they were increasingly preoccupied with its foundations and philosophical questions about its nature. They sought to understand what set mathematics apart from the other sciences and aimed to show that it was the most perfect and certain of the sciences. By the end of the century, metamathematics, like mathematics itself, had emerged as a subject of considerable interest.

Nineteenth-century mathematics has often been characterized as ever more concerned with generality and ever sharper in its self-criticism. At the start of the century there arose an interest in rigor, as displayed by Bernhard

---

7. Polya, *Induction and Analysis*, v–vi.
8. Sarton, *Study of the History of Mathematics*, 4.

Bolzano and Augustin-Louis Cauchy; by the end, mathematicians like David Hilbert, Ernst Zermelo, Bertrand Russell, and Alfred North Whitehead sought to secure the foundations of mathematics on purely axiomatic grounds. Similarly, the rise of abstract algebra and set theory provided new, increasingly solid foundations for more general and comprehensive mathematics. But with greater generality came concerns about the logical consistency of mathematics, prompted in part by the antinomies of set theory and the paradoxes of modern symbolic logic. This view of the course of nineteenth-century mathematics is one that has not changed much over the years, at least in primarily internalist histories of nineteenth-century mathematics.

What constituted mathematics had, however, changed dramatically in the course of the century. Early views that mathematics was the science of quantity, numbers, or measurement gave way to views reflecting greater interest in pure rather than applied mathematics and to definitions of the subject that were increasingly operational or abstract. William Rowan Hamilton, for example, characterized mathematics as the study of pure time, whereas Benjamin Peirce subsequently defined it as "the science which draws necessary conclusions." Peirce's son, Charles Sanders Peirce, viewed mathematics pragmatically: he believed it to be the science of relations. Felix Klein took a psychological approach, defining it as the science of "self-evident things."[9] By the end of the nineteenth century, at least, much of what was most revolutionary in mathematics—nondifferentiable yet everywhere-continuous functions, non-Euclidean geometries, and transfinite set theory, for example—was far from self-evident.

By the end of the nineteenth century, and for virtually all of the twentieth, what mathematics is has been interpreted in increasingly general terms. In 1898, Whitehead took mathematics in its widest signification to be the development "of all types of formal, necessary, deductive reasoning." His colleague, Russell, held that "pure mathematics is the class of all propositions of the form 'p implies q,' where p and q are propositions."[10] Alternatively, the French group of mathematicians known collectively as Nicholas Bourbaki reflects a twentieth-century view that considers mathematics in terms of structures:

> From the axiomatic point of view, mathematics appears thus as a storehouse of abstract forms—the mathematical structures; and so it happens—without our knowing why—that certain aspects of empirical reality fit themselves into these forms, as if through a kind of preadaptation.[11]

9. Klein, *Vorlesungen*, 19–20.
10. Whitehead and Russell, as quoted in Wilder, *Evolution of Mathematical Concepts*, 105.
11. Bourbaki, "Architecture of Mathematics," 231.

2. Mathematicians Writing the History of Mathematics

By virtue of the special character of mathematical knowledge, which (from a Platonist perspective at least) appears to enjoy a timeless validity, mathematicians have always had a special interest in the history of their subject. They usually approach this history from a very specific, not to say narrow, perspective. As Judith Grabiner has expressed it: "The history of mathematics, as written by mathematicians, tends to be technical, to focus on the content of specific papers. It is written on a high mathematical level, and deals with significant mathematics."[12] William Aspray and Philip Kitcher have, however, noted the inherent limitation of Platonist interpretations for the history of mathematics: The difficulty with a traditionally "Platonist" interpretation for the history of mathematics is that it views the subject as independent of the individuals who make mathematics happen, and of the cultures and traditions within which they are working. The timeless nature of Platonist mathematics may serve to account for why the same mathematical discoveries often seem to have been made quite independently in very different parts of the world, or at very different times, or even simultaneously by different individuals, but this does not resolve a much deeper problem. Acknowledging that different but equally valid forms of mathematics may emerge in different cultures and times leaves open the question of whether those forms are truly equivalent.[13] This may serve more as a warning than an indictment, for the very acknowledgment that different but equally valid forms of mathematics may emerge in different cultures and times leaves open the question of whether those forms are truly the same.

The famous school of French mathematicians, Bourbaki, was highly influenced by philosophical and foundational concerns, and this has carried over to the group's interest in the history of mathematics.[14] Bourbaki's efforts have resulted in a highly influential series of books presenting the most important branches of modern mathematics from a rigorous, axiomatic perspective that draws upon the abstract approaches of set theory and modern algebra. The Bourbaki viewpoint, however, represents the internalist, severely technical interpretation of the history of mathematics, a result of the group's overall program: Bourbaki was determined to unearth the key structures that reveal the underlying unity of pure mathematics, in course of which history served certain didactic ends.

Bourbaki devoted a volume to history of mathematics, the *Eléments de l'histoire des mathématiques* (1984). This is little more than a collection of his-

---

12. Grabiner, "Mathematician," 439.

13. A cogent discussion of these problems may be found in Kitcher, "Mathematical Naturalism," esp. 310–16.

14. On Bourbaki, see Beaulieu, "Paris Café."

torical notes culled from volumes of the Bourbaki series otherwise devoted to different branches of pure mathematics. *Eléments* is intended to be accessible (according to its editors) to anyone with "a sound classical mathematical background, of undergraduate standard," although often the level of discussion surpasses that of most undergraduates. Not surprisingly, areas of peripheral interest to Bourbaki—differential geometry, algebraic geometry, the calculus of variations, the theory of analytic functions, differential equations, and partial differential equations—are covered only marginally.[15]

Bourbaki reads history in terms of concepts, and thus their volume represents the strictest sort of technical history of mathematics, with virtually no bibliographic or "anecdotal" information about the mathematicians or works under discussion. Throughout the focus is meant to highlight "as clearly as possible what were the guiding ideas, and how these ideas developed and reacted the ones on the others."[16] The treatment of the history of the calculus is typical; emphasizing the rigor introduced by Cauchy, Bourbaki regarded the history of the calculus in the nineteenth century as primarily concerned with restoring the importance of the definite integral (rather than regarding the integral as antiderivative, as Isaac Newton and Gottfried Wilhelm Leibniz had done): "Cauchy claimed to prove the existence of the integral, that is to say the convergence of the 'Riemann sums,' for an arbitrary continuous function; and his proof, which would have become correct if it were based on the theorem on uniform continuity of functions continuous in a closed interval, is stripped of all convincing value for lack of this notion." Bourbaki is anachronistic here not only in referring to Riemann sums, which came considerably later than Cauchy's efforts in the 1820s to bring new rigor to the calculus, but in judging Cauchy not by the standards of his own day but wholly in the light of hindsight, i.e., applying what we now know to be the role that uniform continuity plays in determining questions about convergence.[17]

In sum, Bourbaki's approach to the history of mathematics is excessively conservative and rather dated, yet it typifies the approach to the history of mathematics shared by most mathematicians writing history of mathematics today. Working entirely within the confines of their own academic community, they remain largely unaffected by the broader interests and methods developed by sociologists, philosophers, historians, and historians of science in the latter part of the twentieth century.

15. David Rowe has described Bourbaki's view of the history of mathematics as follows: "Modern subjects of algebra and topology garnered considerable attention, less streamlined fields rooted in classical analysis, universal methods, and statistics were either pushed to the periphery of Bourbaki's unified structural picture or simply never appeared in it at all." (Rowe, "New Trends and Old Images," 4.)

16. Bourbaki, *History of Mathematics*, preface.

17. Ibid., 198.

Such works by twentieth-century mathematicians are basically textbooks that use history as window dressing, as a source of entertaining interludes included to give students some relief from the real point of the books, which is the presentation of mathematics. Representative volumes, in increasing order of seriousness (and consequently, increasing value historically), include W. S. Anglin, *Mathematics: A Concise History and Philosophy* (1994), Jay R. Goldman, *The Queen of Mathematics: A Historically Motivated Guide to Number Theory* (1998), and Charles Henry Edwards, *The Historical Development of the Calculus* (1979). Each will be described succinctly here, if only to explain why such works do not satisfy certain basic criteria to qualify as serious contributions to the history of mathematics in the nineteenth century.

Anglin devotes six (of forty) chapters to the nineteenth century; that on algebra, to take one example, is less than two pages and claims basically that the abstract nature of algebra today arose from one "key event," namely the discovery of noncommutative multiplication in the case of quaternions for which $xy \neq yx$. Goldman, for his part, writes using anachronistic symbolism and attributes to Georg Cantor a proof that the set of complex numbers is uncountable, based on the countability of the set of all polynomials in one variable with integer coefficients. Yet Cantor neither used such notation nor demonstrated any conclusions about complex numbers, though he did establish the uncountability of the real numbers.[18] Rather than cite any historical accounts of this material (see section 12, below), Goldman cites Garrett Birkhoff and Saunders Mac Lane's *Survey of Modern Algebra.* As for Edwards's *Historical Development of the Calculus,* here "history" provides a chronological framework within which mathematical results are described but with little attention to historical context. Edwards explains this by saying that he is writing primarily for "the wider mathematical community"—those who study, teach, and use calculus—rather than for historians of mathematics. Nevertheless, Edwards captures nicely the major reason why mathematicians are so interested in the history of their subject: "Although the study of the history of mathematics has an intrinsic appeal of its own, its chief *raison d'être* is surely the illumination of mathematics itself. For example, the gradual unfolding of the integral concept—from the volume computations of Archimedes to the intuitive integrals of Newton and Leibniz and finally the definitions of Cauchy, Riemann, and Lebesgue—cannot fail to promote a more mature appreciation of modern theories of integration."[19]

From the above three examples it should be clear that mathematicians appreciate the value of history to enhance their presentations of technical subject matter. In the best cases, as Edwards suggests, history can help to pro-

18. Dauben, *Cantor*; Goldman, *Queen of Mathematics,* 426.
19. Edwards, *Historical Development,* vii.

mote better understanding of the most recent mathematical results. But such works do not represent serious attempts to write history of mathematics. They are, at their best, books about mathematics that include history, but not the sorts of works to which the rest of this essay will be devoted. In what follows, attention will be focused on works whose primary concern is clearly to advance the history of nineteenth-century mathematics.

3. Nineteenth-Century Histories of the History of Nineteenth-Century Mathematics

The history of nineteenth-century mathematics was already a subject of study before the century was out. Most such accounts, inevitably written by mathematicians, may be charitably described as chronological reconstructions of a basically internalist nature. Typical of these is Karl Fink's *Geschichte der Elementar Mathematik* (1890).[20] Although primarily devoted to material prior to the nineteenth century, the book nevertheless brought arithmetic into the nineteenth century with discussion of insurance and mortality tables, the algebra of Niels Henrik Abel and Evariste Galois, elliptical and Abelian functions, and "the more rigorous tendency of mathematics." Moreover, Fink discussed developments in geometry and trigonometry and gave separate coverage to non-Euclidean geometry and topology. As Fink's English translators noted in 1900, despite the approach to the "scientific history [of mathematics]" inaugurated by Jean Etienne Montucla, and notwithstanding the four-volume work of Moritz Cantor, few books covered the nineteenth century. Of those that did, Fink's was "the most systematic attempt in this direction." Most other histories of mathematics, they complained, offered nothing more than "a store of anecdotes, . . . tales of no historic value," and in general missed "the real history of science." Even so, Fink's history is hardly an exemplar; Fink boasted, for example, that he "omits biography entirely," deeming individual biographies irrelevant to the timeless discoveries that represent the body of mathematics. And so Fink virtually ignored context in favor of content. Still, he was among the few to treat the history of mathematics in the nineteenth century, albeit from a strictly internalist perspective, through which he attempted to consider systematically "the growth of arithmetic, algebra, geometry, and trigonometry."[21]

There was one collective effort early in this century that is germane to the history of nineteenth-century mathematics in a very specific way. Undoubtedly the most ambitious project undertaken by mathematicians concerning the history of mathematics was the *Encyclopädie der mathematischen Wis-*

---

20. Fink, *Brief History.*
21. Fink, *Brief History,* iii–iv.

senschaften, plans for which were laid as early as 1880. The many volumes of this extraordinary work were published between 1898 and 1935. In accordance with the guidelines of the project, which aimed to report on the development of all aspects of pure and applied mathematics up to contemporary work, the articles provide only raw information. Historiographical questions are ignored almost entirely, and mathematical results are usually just listed, with no discussion of their proofs. Anachronistic notations occur, and only rarely is attention given to motivations or to details concerning historical figures. Nevertheless, for an overview of what mathematicians accomplished in the nineteenth century, in terms of theoretical results and applications, nothing matches the expertise and detail that this series provides.[22]

The first American to write seriously about the history of nineteenth-century mathematics was Florian Cajori, who produced several such histories before the end of the century. The first, undertaken with a typically American sense of practicality, understood the value of the history of mathematics for instructional purposes: *The Teaching and History of Mathematics in the United States* (1890) was written for the U.S. Office of Education. Three years later Cajori published the first edition of *A History of Mathematics,* and in 1919 a revised and enlarged second edition. Here Cajori gave an even more detailed survey of nineteenth-century mathematics (nearly half of the second edition was devoted to it). Although a rather standard history of mathematics, Cajori's *History* covered all of the technical developments one would expect, and included some biographical details of major contributors. Considering the fact that mathematics itself was only just emerging as a serious discipline in the United States at the end of the nineteenth century, it is remarkable that there was any historical interest in the subject at all.[23]

4. Merz and History of Mathematics as Intellectual History

The most comprehensive intellectual history of the nineteenth century was Merz's *History of European Thought in the Nineteenth Century* (four volumes, 1904–12). It is a classic example of the history of ideas, one with an explicitly avowed philosophical purpose, "to contribute something towards a unification of thought."[24] Merz viewed mathematics as the foundation upon which all of the sciences are based. Insofar as the natural sciences represented hallmark achievements of the nineteenth century, mathematics was necessarily

---

22. Lorey, *Das Studium der Mathematik,* 223–36; Grattan-Guinness, "Does History of Science Treat of the History of Science?" 61.

23. Parshall and Rowe, "American Mathematics Comes of Age" and *Emergence of the American Mathematical Research Community.*

24. Merz, *History of European Thought,* 1:v.

an important focus of Merz's analysis. Nor was his approach superficial or un-informed. Merz appreciated many of the details of mathematics as it had de-veloped and changed in the course of the century and presented them in considerable detail at the end of his second volume.

The first part of Merz's history ended with an account of "the develop-ment of mathematical thought in this century." Merz sought to provide a "record" to capture the "inner life of an age."[25] Yet his book was primarily lim-ited to subjects and themes prominent in French, German, and English thought in the nineteenth century; he largely ignores the important contribu-tions made by mathematicians in other countries (Italy, Russia, and the Scan-dinavian countries). This represents a major lacuna in his account. Still, the main and most important developments are basically covered, albeit in a rather pedestrian fashion.[26]

More importantly, however, the general overview Merz provided has with-stood the test of time. As Merz characterized it, science in the course of the nineteenth century became "international,"[27] a view certainly in keeping with his belief that it was possible to perceive the unity of knowledge as it was then understood. Nor did Merz shrink from considering the larger social pic-ture of how science and mathematics were to be understood as communities and institutions. He described, for example, the significance for science of the Ecole Polytechnique, the importance of journals to emerging professions, and ways in which commerce and industry made it possible to communicate more rapidly by telegraph, railway, and steamship.[28]

Throughout his study, Merz was convinced that "mathematical thought will play an increasingly important part in the progress of science and cul-ture." "No science," he wrote, "has advanced and changed during the 19th century more than that of mathematics." Indeed, thanks to mathematics and "the increasing application of mathematical methods of measuring and cal-culating, our thought has become truly scientific."[29] Merz saw that as the cen-tury drew to an end, both scientific concerns and philosophical interests were powerful stimuli to mathematics. And he was aware of Klein's *Evanston Collo-quium* (1893), where Klein had stressed the important connections between pure mathematics and applied science. Without going into any technical de-tails, Merz managed to identify many of the significant new areas of mathe-matics that had taken center stage at the close of the century and to indicate

25. Ibid., 1:8.

26. For details of the histories of the history of mathematics in Italy, Scandinavia, and Russia /
USSR, see the contributions by Bottazzini, Andersen, and Demidov, respectively, in Dauben and
Scriba, *Writing the History of Mathematics*.

27. Merz, *History of European Thought*, 1:19.

28. Ibid., 2:305.

29. Ibid., 2:626, 736.

the extent to which they were to be important to the future development of science and, consequently, to society.

Merz also appreciated the value of modern, analytical methods in mathematics as compared to older, synthetic ones. He was aware of the newest mathematical sciences, crystallography and probability, and he appreciated Napoleon's interest in the exact sciences and the positive atmosphere this created for French mathematics early in the century. He also understood the importance of Pierre Simon (Marquis de) Laplace and Carl Friedrich Gauss, of Carl Gustav Jacob Jacobi's school, of what he called the "mathematical spirit," of how mathematical formulae had what he called a "focalising effect," of geometric axioms, of infinitesimal methods, and, finally, of the "old formula": "*simplex sigillum veri.*"[30] But if simplicity were to be taken as a mark of truth, it was not always so transparent to nonmathematically minded historians, who did not make much of the sciences in their histories of the nineteenth century. In this regard, Merz and his *History* remain unusual.

5. Historiography of Mathematics prior to and during World War I

W. W. Rouse Ball devoted a chapter of his book, *A Short Account of the History of Mathematics* (1908), to the nineteenth century. He focused on "the lives and discoveries of those to whom the progress of the science is mainly due," offering brief biographical sketches of numerous mathematicians. Ball admitted to using modern notation, rendering proofs that are "sometimes translated into a more convenient and familiar language."[31] Ball also included specific topics, noting, for example, how in the nineteenth century the concept of number was "discussed at considerable length."[32] Transcendental, irrational, and transfinite numbers were described as comprising "one of the most flourishing branches of modern mathematics." Non-Euclidean geometry made in Ball's view an important "splash," but he devoted only one line to set theory.[33] Ball regarded the century as "commencing a new period" in the history of mathematics. He noted that the literature was vast, and that most histories of mathematics did not cover the subject at all, although he cited nearly a dozen works that did treat the nineteenth century in one form or another, including Klein's *Lectures on Mathematics* (his Evanston Colloquium lectures), and the chapter David Eugene Smith contributed on mathematics in the nineteenth century to Mansfield Merriman and Robert S. Woodward's *Higher Mathematics* (1896).

30. Ibid., 1:401.
31. Ball, *Short Account*, v. The first edition of Ball's book appeared in 1888. By the fourth edition (1908), he had added considerable material covering the nineteenth century.
32. Ibid., 460.
33. Ibid.

Richard Courant and Otto Neugebauer, the editors of Klein's *Vorlesungen über die Entwicklung der Mathematik im 19. Jahrhundert,* reported that Klein's lectures exerted a "potent magic" as they circulated in typewritten copies among his colleagues.[34] Klein actually delivered the lectures—part of a project to address *"Kultur der Gegenwart"* (contemporary culture)—before a small circle of listeners at his home during the First World War. He hoped to complete the lectures as a sort of final overview of his life's work, but he never did. The published lectures are incomplete and often read more like an outline than a finished work. The editors noted that Klein did not pay as much attention to number theory, algebra, or set theory as these areas perhaps deserved, and that he had planned chapters on Henri Poincaré and Sophus Lie that were never realized. Nevertheless, Dirk Struik, who also helped edit Klein's lectures for publication, believed that "they remain, with all their personal recollections, the most vivid account of the mathematics of this period."[35]

In addition to technical details, Klein appreciated the broader institutional and cultural background that shaped nineteenth-century science. After a survey of Gauss and his contributions to both pure and applied mathematics, Klein, like Merz, went on to consider France and the significance of the Ecole Polytechnique for mathematics in the early decades of the century, noting the contributions of its great teachers to mechanics and mathematical physics, to geometry, and to analysis and algebra. He then considered the founding of August Leopold Crelle's *Journal für die reine und angewandte Mathematik* (1826) and the rise of pure mathematics in Germany. Here he emphasized the development of algebraic geometry, mechanics, and mathematical physics in Germany and England; the general theory of functions of a complex variable; algebraic geometry; algebraic numbers; the theory of algebraic functions; group theory; and function theory.

While Klein covered the major and obvious figures of the century, history for him consisted of little more than brief biographical and chronological sketches. Nevertheless, the attention he paid as a mathematician to institutional histories—including local and national organizations, international congresses, and journals for the history of mathematics—was unusual. Klein was especially aware of the important influence that the French Revolution had exerted upon the sciences generally, including mathematics. The Revolution fostered not only what he called the "democratization" of the sciences; it also promoted their greater diffusion and increased specialization. As a result, teaching became much more important, and the disappearance of class restrictions on the professions helped to encourage study of the sciences. In turn, those who taught the sciences were no longer primarily academicians,

34. Klein, *Vorlesungen,* v.
35. Rowe, "Interview with Dirk Jan Struik," 19.

but rather instructors trained especially for teaching in secondary schools and specialized institutions for the sciences. The latter included a growing number of polytechnic schools founded on the model of the Ecole Polytechnique in various parts of Europe.[36]

As Klein considered the value of history in organizing mathematics, he explained how modern mathematics was increasingly concerned with rigor and foundations: "Gradually, however, a more critical spirit asserted itself and demanded a logical justification for the innovations made with such assurance, the establishment, as it were, of law and order after the long and victorious campaign. This was the time of Gauss and Abel, of Cauchy and Dirichlet. . . . Hence arose the demand for exclusively arithmetical methods of proof."[37] This reflects a view of the nineteenth century expressed more directly by Claude Henri de Saint-Simon, who remarked, "The philosophy of the 18th century was revolutionary; the philosophy of the 19th century must organize."[38]

6. The Cases of Bell and Boyer

Eric Temple Bell, who disclaimed being a historian of mathematics, took a popular approach to history. In a series of books—*The Queen of the Sciences* (1931), *The Handmaiden of the Sciences* (1937), *Men of Mathematics* (1937), *The Development of Mathematics* (1940), and *Mathematics, Queen and Servant of Science* (1951)—Bell presented lively accounts of mathematics along with gossipy, anecdotal sketches of the lives of mathematicians.[39] He did so without pretense to erudite scholarship, attention to primary sources, or any attempt to recount the evolution of mathematics in technical detail. In many respects Bell was a pioneer, and he offered many readers their first encounter with history of mathematics. Nevertheless, his popularizations did not reflect the best academic scholarship on the history of nineteenth-century mathematics that the postwar era was to produce.

Among the new breed of historians to appear at the time of the Second World War was Carl Boyer. Where Bell took a popular approach to the history of mathematics, Boyer aimed at teachers and mathematicians interested in better understanding the conceptual development of mathematics. Trained in mathematics at Columbia, Boyer also studied intellectual history there. His dissertation, published as *The Concepts of the Calculus* (1939), appeared at about the same time as Guido Castelnuovo's *Le origini del calcolo in-*

36. Klein, *Vorlesungen*, 3.
37. Klein, "Arithmetizing of Mathematics," 241.
38. Jahnke and Otte, "On 'Science as a Language,' " 78.
39. For a biography of Bell, see Reid, *Search for E. T. Bell*.

finitesimale nell'era moderna (1938).[40] Boyer was especially concerned to address students. He noted that his was "not a history of the calculus in all its aspects, but a suggestive outline of the development of the basic concepts, and as such it should be of use to students of mathematics and to scholars in the field of the history of thought."[41] When his book was reprinted ten years later, he changed its title to *The History of the Calculus and Its Conceptual Development*, in part to avoid being mistaken for a textbook and in part to clarify that the book was about the history of the calculus. He hoped that it might "serve effectively to alleviate the lack of mutual understanding too often existing between the humanities and the sciences" and that it would have a beneficial effect upon teachers, for he firmly believed in the value of "the role that the history of mathematics and science can play in the cultivation among professional workers in the fields of a sense of proportion with respect to their subjects."[42]

In assessing the history of nineteenth-century mathematics, Boyer characterized it as a period of rigor, contrasting it to the eighteenth-century "period of indecision." He claimed specifically that the year 1816 "marks a new epoch" in the history of mathematics. Not only was Lacroix's *Traité élémentaire* translated into English that year, wherein the calculus was approached in terms of limits rather than Newton's fluxions; but also Bolzano's famous *Rein analytischer Beweis des Lehrsatzes*, which emphasized the value of careful reasoning in mathematics and logic generally, appeared. To Boyer, this last was a hallmark of the rigorous formulations (from Cauchy to Karl Weierstrass) for the calculus that were pursued throughout the century. Boyer closed his study of the calculus with a description of the precise definitions of the continuum of real numbers pioneered by Richard Dedekind and Cantor. What is surprising about his history, however, is that Riemann is mentioned only once, and Lebesgue and others not at all.

Boyer typified the small group of historians of mathematics who found themselves, after the war, in departments of mathematics, teaching courses in the history of mathematics and writing for an increasingly educated and articulate readership that included mathematicians, historians of mathematics, and historians of science. As reflected in his *History of Analytic Geometry* (1956), Boyer wrote the history of mathematics from a mathematical rather than a historical perspective, omitting biographical details "because often they have little bearing upon the growth of concepts."[43] Rather than focus on individu-

40. Boyer's dissertation, *Concepts of the Calculus* (1939), was later reprinted with the more accurate title *History of the Calculus and Its Conceptual Development* (1949).

41. Boyer, *Concepts of the Calculus*, iii.

42. Boyer, *History of the Calculus*, v.

43. Boyer, *History of Analytic Geometry*, viii.

als, Boyer emphasized the "broad general picture" and concentrated almost exclusively on internalist developments. Moreover, Boyer was concerned with priorities, not so much in the sense of how a particular discovery was made as of who had made it or first introduced a particular notation or formula. As he emphasized: "In view of the frequency with which the normal forms appear in L'Hulier's work, it is surprising to find that they often are ascribed to later mathematicians, notably to Cauchy in 1826 or to Magnus in 1833 or to Hesse in 1861!"[44]

For Boyer, the nineteenth century represented the "Golden Age" of analytic geometry. Surprisingly, he attributed the subject's success at least in part to the esprit de corps of its proponents.[45] He noted above all the transforming role Julius Plücker had played, especially through his students (e.g., Klein, Lie, and Rudolf Friedrich August Clebsch). He argued that the works of Gaspard Monge and Plücker emancipated the subject from "the constructions of pure geometry—it was arithmetized."[46] It was this, he claimed, that so greatly impressed Auguste Comte, who drew a direct connection between the tendency to arithmetize mathematics and its bearing upon positivistic philosophy.

7. Kuhn's Influence on Mathematical Revolutions and Communities

Kuhn's widely read and influential *The Structure of Scientific Revolutions* (1962; revised 1970) has been influential on several levels for the history of nineteenth-century mathematics, including most directly the question of whether or not revolutions ever occur in mathematics.[47] This question has been largely answered in the negative by mathematicians who hold that mathematics is a cumulative science that grows by the addition of new theorems to form a timeless body of knowledge. Revolutions in the Kuhnian sense, the displacement of older, falsified paradigms by revolutionary, incompatible or "incommensurable" new ones, are prima facie impossible, virtually by definition, due to the very nature of mathematics. This view has been most forcefully advanced by Michael Crowe in a series of "laws" for the history of mathematics, of which the tenth is: "Revolutions never occur in the history of mathematics."[48] Yet consider, for example, Cantor's self-proclaimed "revolution" in mathematics with the advent of transfinite set theory toward the end of the

44. Ibid., 223.
45. Ibid., 225.
46. Ibid., 226.
47. Kuhn, *Structure of Scientific Revolutions*. See the discussions of Kuhn's work by Ben-David, *Scientist's Role in Society*, 6–13, and by Friedman, "On the Sociology of Scientific Knowledge," esp. 250–51.
48. Crowe, "Ten 'Laws,'" 165.

nineteenth century[49] or the simultaneous discovery by several mathematicians of non-Euclidean geometries, also in the nineteenth century. Indeed, a variety of "revolutions" in mathematics are considered as individual case studies in *Revolutions in Mathematics,* a collection of papers edited by Donald Gillies.[50]

Kuhn's work has also been influential in serving to sharpen questions of how theories grow and change in different paradigmatic contexts. One result of his emphasis on the significance of scientific communities is the increased attention paid to science networks in understanding how new theories emerge and eventually gain currency among scientists. Among recent investigations devoted to mathematical communities of the nineteenth century are Della Fenster and Karen Parshall's studies of the American mathematical community toward the end of the century,[51] as well as Parshall and David Rowe's *Emergence of the American Mathematical Research Community, 1876–1900.*[52]

Bruce Chandler and Wilhelm Magnus, for their part, consider the mathematical community within the history of combinatorial group theory.[53] Not only have they produced an impressive study in terms of the technical details provided; they are especially innovative in their consideration of the "modes of communication and in tracing the various avenues and routes whereby distribution of research of group theory occurred." They devote several chapters specifically to "modes of communication," "geographical distribution of research and effects of migration," and the "organization of knowledge."[54]

8. Marxism and the History of Mathematics

Like Kuhn's *Structure,* Marxist approaches to the history of mathematics have emphasized noncognitive dimensions that cause scientific change. Among

---

49. See Dauben, "Conceptual Revolutions" and "Revolutions Revisited," and Gillies, *Revolutions in Mathematics.*

50. In addition to the collected papers edited by Gillies, see Trudeau, *Non-Euclidean Revolution,* and Krüger, Daston, and Heidelberger, eds., *Probabilistic Revolution.* For an alternative view on the question of revolutions in mathematics, see Grattan-Guinness, "Scientific Revolutions as Convolutions?"

51. Fenster and Parshall, "Profile of the American Mathematical Research Community." The same volume also includes a study of the specific subcommunity of women (Fenster and Parshall, "Women in the American Mathematical Research Community").

52. For a briefer overview, see Parshall and Rowe, "American Mathematics Comes of Age."

53. See, e.g., Chandler and Magnus, *History of Combinatorial Group Theory,* pt. 1, chap. 9, which offers limited historical help to the reader. This chapter of "biographical notes" is barely more than two pages long and lists only obituaries drawn from three general reference sources. (Ibid., 68–70.)

54. Ibid., 58–67, 187–206.

the first to use Marxism as a framework within which to approach the history of science were John Desmond Bernal, John Burdon Sanderson Haldane, and Boris Hessen. The usefulness of Marxism for the history of mathematics, especially for the nineteenth century, was demonstrated most eloquently and persuasively in the writings of Struik, whose *Concise History of Mathematics* has had an enormous impact on the history of the subject since the book first appeared in 1948. (It has subsequently been translated into sixteen languages.)[55] A founding editor of *Science and Society*, Struik pioneered the Marxist approach to studying the history of mathematics. By his own account, it was in the wake of the Russian Revolution, and while assisting in editing Klein's lectures on nineteenth-century mathematics, that he was first attracted to the ideas of Marxist history. His "confidence [in historical materialism] was strengthened a few years later when I read Boris Hessen's landmark paper on seventeenth-century English science," as well as the writings of Bernal, Haldane, Joseph Needham, Lancelot Hogben, and Hyman Levy.[56]

There have been basically two distinct approaches to applying dialectical materialism and fundamental Marxist principles to the history of science. One is that taken largely by (former) Soviet and other Eastern-bloc (especially East German) historians of science; the other is that of academic historians in the West, who tend to see Marxism less tendentiously and more philosophically.[57] In the hands of Russian authors, like Konstantin Alexeyevich Rybnikov and Sof'ya Aleksandrovna Yanovskaya, the former approach directed considerable effort to the study and compilation of a critical edition of the *Mathematical Manuscripts of Karl Marx*.[58] A curious example of a Marxist approach to history of mathematics is the biography of Cantor by Walter Purkert and Hans Joachim Ilgauds, published first in an East German edition (Teubner, 1985) and soon after in a "revised" version (Birkhäuser, 1987). The earlier edition included ideological discussions of Cantor's work from a Marxist perspective, most of which were removed for the Birkhäuser edition. For

55. Rowe, "Interview with Dirk Jan Struik," 14.

56. Dirk Struik, quoted in ibid., 22–23. Of the Marxist contributors to the history of science, the only ones other than Struik (several of whose papers are cited in the bibliography) to have made substantial contributions to Marxist readings of history of mathematics are Needham and Hogben, and only the latter offers anything relevant to history of mathematics in the nineteenth century.

57. On the East-bloc approach to history of mathematics, see Wussing, "Zur gesellschaftlichen Stellung der Mathematik"; the Marxist approach most typical in the West is reflected in the works of Struik.

58. Yanovskaya, "On Marx's mathematical manuscripts"; see also Lombardo-Radice, "Dai 'Manoscritti Matematici' di K. Marx," as well as Marx, *Mathematical Manuscripts*. For a general evaluation of scholarship on Marx and his mathematics, see Dauben, "Marx, Mao, and Mathematics" and "Mathematics and Ideology."

example, in commenting on the dialectics of mathematical knowledge, Purkert and Ilgauds wrote in the Teubner edition that "only the materialistic point of view can provide successful explanations."[59] In the Birkhäuser version, however, they substituted a new paragraph in which Cantor's Platonism is discussed instead![60] Virtually all references to Marx, Engels, and Lenin have been deleted from the revised edition.[61]

Such changes of heart have not been typical of Western authors, for whom Marxism has been influential in ways quite different from the overtly ideological role it typically played in work by Soviet and Eastern European scholars during the Cold War. In the West, ideology has served very different ends for writers like Struik and Needham. The latter's magisterial series, *Science and Civilisation in China*, has focused on the productive values of technology and has investigated the history of mathematics in social contexts guided by the precepts of Marxism and a concern for the constant interaction among science, technology, and society. Yet little of this material concerns the history of nineteenth-century mathematics. On the other hand, Marx and the subject of mathematics in the eighteenth and nineteenth centuries interpreted in sharply ideological terms is clearly evident in the writings of Chinese scholars working in the wake of the Cultural Revolution. Although restricted in their access to historical documents and original sources, their views nevertheless show the kinds of analyses Marxism can provide for the history of nineteenth-century mathematics.[62]

9. The Emergence of the Social History of Mathematics

Ever since Robert K. Merton's *Science, Technology, and Society in Seventeenth-Century England* (1938) showed the significance of scientific roles and communities, historians and sociologists have appreciated the necessity of studying social contexts in order to understand more fully the advance of modern science. Merton did so, it should be noted, without insisting that all scientific

---

59. Purkert and Ilgauds, *Cantor* (1985), 65.

60. Purkert and Ilgauds, *Cantor* (1987), 107.

61. See the review by Dauben of both versions of Purkert and Ilgauds's biography of Cantor. Both versions emphasize (wrongly) that "Cantor never said he was Jewish." The revised version of Meschkowski's biography of Cantor (1983), however, makes it clear that Cantor was descended from Portuguese Jews in Denmark (see Meschkowski, *Cantor*, 234–35). Meschkowski's book is not listed in Purkert and Ilgauds's bibliography, although it had already appeared when they published the first version of their biography.

62. Detailed discussion of publications devoted to Marx and history of mathematics, including material pertinent to the nineteenth century, may be found in Dauben, "Marx, Mao, and Mathematics" and "Mathematics and Ideology."

knowledge is socially constructed or that scientific truth can only be understood as a relative concept determined contextually, as later claimed by Barry Barnes and David Bloor, among others.[63]

By the early 1970s, historians of mathematics had begun to take the social approach more seriously, and before the decade was out Henk J. M. Bos and Herbert Mehrtens's important article on the interaction of mathematics and society had appeared. Bos and Mehrtens identified three areas of special interest for the social history of mathematics: traditional historiography, biography, and institutional history of mathematics.[64] Several years later, they (and Ivo Schneider) edited a forward-looking book, *Social History of Nineteenth-Century Mathematics* (1981).[65] As this collection of essays makes clear, biographies, institutional histories, and histories of national developments have long been familiar elements in the history of mathematics, joined more recently by interest in the roles played by journals and an awareness that economic and productive forces, as well as questions of race, sex, gender, and the like, may also be relevant. Since the 1970s, historians of mathematics have increasingly acknowledged that ignoring such factors can significantly diminish the value of histories of mathematics.

Bos, Mehrtens, and Schneider make clear their awareness of trends among historians toward a broader view of history through their use of Michel Foucault's influential *The Order of Things* (1970) and *The Archaeology of Knowledge* (1972). But they are also aware that the history of mathematics is a unique sort of intellectual history that may be more resistant than others to sociological interpretations. As they wrote: "Mathematics typifies the most objective, most coercive type of knowledge, and therefore seems to be least affected by social influences."[66] Nevertheless, as more recent biographies and institutional as well as national histories of mathematics demonstrate, even the most conservative and traditional scholars now seem to acknowledge that mathematics too is a social activity.

Three illustrative examples from *Social History of Nineteenth-Century Mathematics* should suffice here: Luke Hodgkin, in "Mathematics and Revolution from Lacroix to Cauchy," used Foucault's idea of "discursive formation" to examine the role of Lacroix's textbooks, along with autobiographical fragments of Stendhal to demonstrate the value of "discursive formation" in analyzing the history of mathematics. Gert Schubring examined a specific mathematical phenomenon in the context of wider social issues in "The Con-

---

63. Barnes and Bloor, "Relativism, Rationalism, and the Sociology of Knowledge." See also the extensive critique of Barnes and Bloor's views in Friedman, "On the Sociology of Scientific Knowledge."

64. Bos and Mehrtens, "Interaction of Mathematics and Society."

65. Bos, Mehrtens, and Schneider, *Social History*, ix.

66. Ibid.

ception of Pure Mathematics as an Instrument in the Professionalization of Mathematics." And Struik, writing about mathematics in the early nineteenth century, showed how sociopolitical struggles and revolutions constituted significant influences on the history of mathematics in that century.[67]

As if to confirm that the social history of mathematics had indeed come into its own, there appeared, also in 1981, *Epistemological and Social Problems of the Sciences in the Early Nineteenth Century*, a volume edited by Niels Heinrich Jahnke and Michael Otte. Although the book was concerned with the sciences in general, an entire section was devoted to mathematics. Winfried Scharlau's contribution may be taken as representative. Regarding the French Revolution as a signal and familiar turning point, he argued, like Struik before him, that the French and Industrial revolutions set the conditions for the emergence of pure mathematics during the nineteenth century.[68]

Among more recent works devoted to history of nineteenth-century mathematics with a strong emphasis on social and contextual questions is Joan Richards' study of the reception of non-Euclidean geometry in Victorian England, which the philosopher Michael Friedman has described as a model "where themes from the philosophy of geometry—culminating in the early philosophy of geometry of Bertrand Russell—are illuminatingly integrated with social and intellectual history of mathematics."[69] More comprehensive insights into the social aspects of American mathematics have recently been provided in publications by Fenster and Parshall, as well as Parshall and Rowe's study of the emergence of the American mathematical research community (discussed in section 7).[70]

Institutional histories are now playing an important role in understanding the social context within which mathematicians work. An excellent example for the nineteenth century, drawing upon a deep and extensive knowledge of archival sources, is Kurt-R. Biermann's *Die Mathematik und ihre Dozenten an der Berliner Universität 1810–1933*. Other noteworthy examples, often inspired by celebratory occasions, include Helmut Gericke's review of the history of the Mathematical Research Institute in Oberwolfach, Germany, and Thomas

67. Hodgkin, "Mathematics and Revolution"; Schubring, "Conception of Pure Mathematics"; Struik, "Mathematics in the Early Part of the Nineteenth Century."

68. Scharlau, "Origins of Pure Mathematics," 334.

69. Richards, *Mathematical Visions*. See Friedman, "On the Sociology of Scientific Knowledge," 270. Historians of mathematics have been more critical than Friedman of Richards's book; see, for example, the essay review by Albert C. Lewis, who writes that "English mathematics, and the Cambridge scene within it, are not meaningfully related [in Richards's book] to the broader context that helped to shape them" (273).

70. Fenster and Parshall, "Profile of the American Mathematical Research Community"; and "Women in the American Mathematical Research Community." See also Parshall and Rowe, "American Mathematics Comes of Age" and *Emergence of the American Mathematical Research Community*.

Scott Fiske's account, "The Beginnings of the American Mathematical Society." (The society has issued a three-volume collection of papers consisting almost entirely of reflections by mathematicians on topics related to the history of recent mathematics, some relevant to the nineteenth century.)[71] Finally, the history of journals has also been the subject of recent studies; this too is an important part of institutional history that deserves more detailed study in the future. A collection of papers devoted to this subject, and dealing primarily with the nineteenth century, has been edited by Elena Ausejo and Mariano Hormigón.[72] Other early examples of useful studies of journals include several publications by Wolfgang Eccarius on Crelle's *Journal.*[73] Nonetheless, the history of periodicals is an area that awaits much further exploration.

Few compelling psychological studies of nineteenth-century mathematicians have as yet been provided, although Joseph Brent's of Peirce takes into consideration important psychological questions about his subject and Nathalie Charraud applies Lacanian psychology to Cantor. Ivor Grattan-Guinness has emphasized that Cantor's mental illness was almost certainly endogenous, meaning that he would have suffered from manic depression whatever his profession had been, and there seems to be no direct correlation between his work on the infinite and the nervous breakdowns he suffered. But Joseph Dauben has pointed out that Cantor's depression may nevertheless have contributed to his defense of transfinite set theory, especially when Cantor was confronted with formidable opposition to his work.[74] On the question of psychology and mathematicians in general, mention should be made of the important psychosocial study undertaken by the French mathematician Jacques Hadamard, who sent questionnaires to hundreds of mathematicians in an attempt to illuminate the psychology of mathematical invention.[75]

10. The Social Construction of Mathematics

Emergence of the social history of nineteenth-century mathematics has gradually led, for a few scholars, into the social construction of mathematics.

---

71. Gericke, "Das Mathematische Forschungsinstitut"; Fiske, "Beginnings."

72. Ausejo and Hormigón, *Messengers of Mathematics.*

73. Eccarius, "August Leopold Crelle als Herausgeber," "August Leopold Crelle als Förderer," and "Der Gegensatz."

74. Among recent, detailed portraits of Cantor, all of which draw heavily upon previously unpublished or only recently available sources, see Charraud, *Infini et inconscient;* Dauben, *Cantor;* Grattan-Guinness, "Towards a Biography"; Meschkowski, *Probleme des Unendlichen;* and the 1985 and 1987 versions of Purkert and Ilgauds, *Cantor.*

75. Hadamard, *Psychology of Invention.* Another aspect of the psychology of mathematics is considered in Cajori, *Teaching and History,* 391–94.

The history of mathematics, like the history of science as a whole, has recently been subjected to the view that all mathematical knowledge is relative and that social contexts virtually determine and "construct" mathematics. For example, Bloor, a scholar who advocates the "strong programme" in the sociology of science, has written on nineteenth-century British algebra. His approach to the history of nineteenth-century mathematics and his strong programme seek to show that between provable, necessary results and individual choices of topics, methods, and concepts, it is possible to find "social interests embedded in the style and content of mathematics."[76] He explored differences that he identified between Hamilton, on the one hand, and the Cambridge group (George Peacock, Charles Babbage, William Whewell, and John Frederick William Herschel), on the other, over the nature of algebra. Bloor raised the question as to why those who were leaders in their field disagreed about "the fundamental nature of their science." His answer was a sociological one related to Hamilton's metaphysics and his development of quaternions (a noncommutative algebra).[77]

Mathematicians have not always seen it this way. Bell had earlier dismissed metaphysics as irrelevant to Hamilton's development of quaternions.[78] E. T. Whittaker had argued that quaternions arose naturally from the problem of how to generalize the two-dimensional geometric representation of complex numbers to three dimensions. More recently, however, Crowe has suggested that metaphysics did indeed play a decisive role in Hamilton's discovery of quaternions. And Thomas Hankins, who has also studied Hamilton's mathematics extensively, shares this view: "Hamilton's metaphysical theories helped direct his mathematical researches."[79] Whether it was metaphysics or something else (e.g., luck) that prompted the insights that led to the discovery of quaternions, once discovered, their mathematical properties were a fait accompli, which could then be investigated by mathematicians of all persuasions. Metaphysics may have "directed" Hamilton's research, but this is not to say that it had anything to do with the noncommutative nature of quaternion multiplication. Although social context may reveal something about how or why Hamilton's theory developed when and as it did, there seems no convincing evidence that social context, specifically his metaphysics, determined the ultimate theoretical character of quaternions, and indeed the mathemati-

76. Bloor, "Hamilton and Peacock."

77. Ibid. In light of discussion following Bloor's oral presentation of his paper at a meeting in Berlin (1979), he thanked several colleagues "for gently but firmly making clear to me how far the present paper is from being an acceptable piece of historical scholarship. The point is well taken: the most that I dare claim is that it may have some interest as an illustration of what a certain type of explanation may look like when applied to mathematicians." (Ibid., 232)

78. Bell, *Men of Mathematics*, 358.

79. Hankins, *William Rowan Hamilton*.

cal facts of Hamilton's quaternions, once defined, may be applied on equal terms in diverse social contexts. The danger in the strong programme is that it overlooks the imperatives of scientific content, reducing the history of mathematics to little more than a study of social context. As Joseph Ben-David has maintained, social context may be relevant to evaluating the circumstances under which scientific theories emerge, but it should not be expected to have any bearing on the internal content of theories themselves.[80]

11. Postmodernist Interpretations of Nineteenth-Century Mathematics

Not unrelated to the social constructionist approach are attempts at postmodernist interpretations of nineteenth-century mathematics. For the most part, these have been far from successful or convincing. Two examples may suffice to give readers an idea of the spectrum of work written in response to the ideas of Jacques Derrida, Jean Baudrillard, and Michel Foucault, among others, as captured in a recent book edited by Timothy Lenoir, *Inscribing Science*.[81] In one essay, Brian Rotman cites Derrida as a "resource for criticizing disembodied Platonism in mathematics," and in attempting to do so has recourse to a "semiotics of mathematics." Lenoir describes Rotman's approach as assuming that "mathematics is an immanently embodied form of semiotic practice" and that, in such mathematical demands as "Take the function f(x)," "the subject is an idealized, truncated simulacrum, the Agent."[82] Yet what, for example, is one to make of Rotman's assertions that "$2 + 3 = 3 + 2$" is a "prediction"? In trying to deconstruct mathematics or to provide postmodernist accounts of something so basic as arithmetic, it is important to have a solid grasp of what the mathematics in question is really about if such approaches are to attract the interest rather than the skepticism of mathematicians. If postmodernist attempts to write about the history of mathematics have so far been unconvincing, this may be in part because mathematicians have little patience with new-fangled approaches by theorists who seem not to understand fully the material about which they are writing.

The second, and more successful example, is Mehrtens's book *Moderne,*

80. Ben-David, *Scientist's Role in Society,* 7–13.

81. Specifically, Lenoir says that *Inscribing Science* was conceived as an answer to Sokal, "Transgressing the Boundaries," which was a hoax article designed to show the dangers of the "relativist malaise" Sokal takes postmodern philosophy to represent, a point that he explains in his "A Physicist Experiments with Cultural Studies." Lenoir's book is also directed against Paul Gross and Norman Levitt's *Higher Superstition.* See Lenoir, *Inscribing Science,* 13–14.

82. Lenoir, *Inscribing Science,* 13–14; Rotman, "Technology of Mathematical Persuasion."

Sprache, Mathematik (1990), which presents a penetrating analysis of foundations debates, especially in Germany after World War I, drawing specifically on Foucault and adopting in part a semiotic view of mathematics. Basically, Mehrtens uses language and the techniques of semiotics to evaluate modern mathematics.[83] Because the foundational issues Mehrtens considers originated in the set theory and mathematical logic pioneered by figures like Cantor, Dedekind, Gottlob Frege, and Giuseppe Peano in the nineteenth century, it merits notice here.

12. Biographies, Surveys, and Special Topics

All historical research depends upon the availability of primary sources—the correspondence, notes, unpublished papers, and so on that are the necessary basis for biographical research as well as for the technical or institutional histories increasingly gaining favor. Comprehensive, scholarly efforts are currently underway to publish the papers of several nineteenth-century mathematicians. Of these, the Peirce edition project, part of which is devoted to Peirce's mathematics and mathematical logic, has demonstrated the highest standards of editorial professionalism.[84] Equally meticulous is the ongoing series of volumes devoted to the collected works of Bolzano.

Of special value are previously unpublished treatises and collections of correspondence, and historians of mathematics have diligently produced such works, often with extensive critical commentaries that constitute substantial contributions of their own to the history of nineteenth-century mathematics. Representative editions here are Grattan-Guinness's *Joseph Fourier, 1768–1830: A Survey of His Life and Work, Based on a Critical Edition of His Monograph on the Propagation of Heat, Presented to the Institut de France in 1807* (1972), and Erwin Neuenschwander's *Riemann's Einführung in die Funktionentheorie: Eine quellenkritische Edition seiner Vorlesungen mit einer Bibliographie zur Wirkungsgeschichte der Riemannschen Funktionentheorie* (1996). Among collections of correspondence, Biermann has published the letters of Gauss, Peter Gustav Lejeune Dirichlet, and Alexander von Humboldt, among many others.[85] Parshall has published the correspondence of James Joseph Sylvester.[86] Con-

---

83. Sigurdsson, review of Mehrtens, *Moderne, Sprache, Mathematik,* 157. Rowe offers a detailed analysis and some pointed reservations of Mehrtens's work in Rowe, "Perspective on Hilbert."

84. Even prior to the Peirce Edition Project, which has issued five volumes in its chronological, critical edition of Peirce's works, Carolyn Eisele had published four volumes (in five books) of the mathematical papers of Charles S. Peirce. See Eisele, *Studies,* as well as her edited volume, *New Elements.*

85. Biermann, *Dirichlet, Gauss,* and *von Humboldt.*

86. Parshall, *Sylvester.*

versely, Herbert Meschkowski's studies of Cantor depended heavily on archival sources, and led him later to produce, with Winfried Nilson, an edited collection of letters from Cantor's *Briefbücher*.[87]

For the nineteenth century, many excellent biographical studies have appeared that make detailed use of previously unpublished letters, lecture materials, and archival resources. These have been based largely on collections available in Europe and North America; only recently have such resources elsewhere begun to be used systematically. The biographies of important contributors to mathematics in other parts of the world will doubtless claim the attention of writers in the years ahead, and histories of mathematics and mathematicians in Latin America, Africa, and Asia will doubtless become subjects of increasing interest. Considerable historical work dealing with China and Japan in particular is already underway.[88]

It is impossible to do justice here to even a small number of the many recent and excellent individual biographies of nineteenth-century mathematicians, although the best new biographies merit brief attention. In part due to the inherent fascination with the infinite and the importance, philosophically and technically, that it has assumed for mathematics, Cantor has been the subject of no fewer than seven recent studies, each of a very different character. The earliest book-length biography to draw on extensive primary sources is Meschkowski's *Probleme des Unendlichen* (1967; revised 1983). There followed Grattan-Guinness's monograph, "Towards a Biography of Georg Cantor" (1971), which is based on detailed archival research, and Dauben's *Georg Cantor: His Mathematics and Philosophy of the Infinite* (1979), which portrays Cantor's life against the background of his personal and philosophical concerns. Less historical and much more heavily philosophical in their aims are the recent studies of Cantor's work by Michael Hallett (1984) and Shaughan Lavine (1994), while Charraud's study of Cantor, *Infini et inconscient* (1994), offers a psychoanalytic portrait.[89] Charraud emphasizes the personal relations between Cantor and his father and gives a detailed account of Cantor's last years in the Halle *Nervenklinik* and the psychological disorders he suffered while there. The two versions of the Cantor biography by Purkert and Ilgauds mentioned in section 8 offer useful additional material.

The best biographies of nineteenth-century mathematicians succeed by exploring the many facets of an individual's life and times in such a way as to also illuminate the larger community of mathematicians, the institutions,

87. Meschkowski and Nilson, *Cantor*.

88. For limited discussion of mathematics in nineteenth-century China, see Li and Du, *Chinese Mathematics*, 232–66, and the more recent, revised English version of Martzloff, *History of Chinese Mathematics*, 341–89.

89. Meschkowski, *Cantor*; Grattan-Guinness, "Towards a Biography"; Dauben, *Cantor*; Hallett, *Cantorian Set Theory*; Lavine, *Understanding the Infinite*; Charraud, *Infini et inconscient*.

and the entire social network of meetings, journals, and activities that supported them. This is apparent from such comprehensive biographies as that
of Cauchy by Bruno Belhoste (1991), which draws upon extensive archival
studies to produce a well-developed picture of Cauchy in the context of the
political and social milieu of his times.[90] Also important for setting the background for Cauchy's (and Lagrange's) work is Grabiner's *Origins of Cauchy's
Rigorous Calculus* (1981). Equally impressive for its breadth of scholarship and
attention to voluminous amounts of personal archival material is Jesper
Lützen's biography of Liouville.[91] Brent's *Charles Sanders Peirce* also makes
much use of archival material and clarifies many details about Peirce and his
controversial life that had previously remained unknown to most Peirce
scholars. Other notable biographies of nineteenth-century figures include
Jean Dhombres and Jean-Bernard Robert on Fourier, Charles Gillispie (with
Grattan-Guiness and Robert Fox) on Laplace, Vladimir Maz'ia and Tatyana
Shaposhnikova on Hadamard, and Øystein Ore on Abel.[92]

Three biographical accounts of nineteenth-century women mathematicians stand out. Louis Bucciarelli and Nancy Dworsky have given a detailed
characterization of the life and times of Sophie Germain.[93] And two biographies of the well-known Russian-born mathematician Sofia Kovalevskaia have appeared: Roger Cooke has written the more technical account;
Anne Hibner Koblitz's explores more details relating to Kovalevskaia's controversial personal and political life.[94] These and other works share an appreciation of the importance of using primary sources to illuminate the life
and work of an individual within the larger social and political contexts of
the times in which they lived. Finally, one might note the few attempts at
mingling fictional creations with historical fact to re-create a subject's life
and works.[95]

There is still no single survey of the history of nineteenth-century mathematics. In 1968, Boyer published *A History of Mathematics,* intended in part as a
textbook, and included carefully chosen problems and exercises at the end of
each chapter to sharpen and test the students' understanding of the history
they were studying. His approach to the history of mathematics remained as
conventional as his earlier works. For the nineteenth century, he covered ma-

---

90. Belhoste, *Cauchy.*

91. Lützen, *Liouville.*

92. Dhombres and Robert, *Fourier;* Gillispie, Grattan-Guinness, and Fox, *Laplace;* Maz'ia and
Shaposhnikova, *Hadamard;* Ore, *Abel.*

93. Bucciarelli and Dworsky, *Germain.*

94. Cooke, *Mathematics of Sonya Kovalevskaya;* Koblitz, *Kovalevskaia.*

95. Infeld, *Whom the God's Love.* Galois has been the subject not only of novels but also of three
documentary films. The best recent account of his life and career is Toti Rigatelli, *Galois.* See also
Rothman, "Genius and Biographers."

terial familiar from Klein's *Vorlesungen* and Struik's *Concise History,* offering somewhat more detail but far less technical subject matter.

In 1972, Morris Kline published a monumental work, *Mathematical Thought from Ancient to Modern Times.* When it was reissued in 1990, he explained that this was part of his "long-time efforts to humanize the subject of mathematics." Opposed in particular to the overly abstract approach of Bourbaki, Kline wanted to "set the record straight." Above all, he sought to give calculus, differential equations, and classical mechanics the attention he thought they deserved. This was in large measure due to his belief that an appreciation of applied mathematics and of the important interactions between pure and applied mathematics were needed if one was to understand properly the history of nineteenth-century mathematics. Like Boyer, Kline approached the history of mathematics predominantly in an internalist mode; as Rowe has suggested, "[Kline's] image of mathematics could hardly be described as radically new. . . . Indeed, it may well mark the end of an era dominated by holistic images in the history of mathematics."[96]

Following Boyer and Kline, the standards for rigor and historical veracity in textbook writing have been maintained by several more recent histories of mathematics, notably Victor Katz's *History of Mathematics* (1993) and Grattan-Guinness's *Rainbow of Mathematics.*[97] These works, written with mathematical detail but with substantial concern for establishing the historical context of the mathematics they discuss, represent the cutting edge of their genre and are especially valuable in regard to the recent period. Katz's work is based on primary and secondary sources, taking into account the cultural context in which mathematics has developed and avoiding ethnocentrism.[98] Like Boyer's, it includes problems to sharpen students' understanding of the roles of problems, proofs, and techniques in the historical progress of mathematics. Katz pays substantial, if conventional, attention to nineteenth-century mathematics and shows a greater awareness than any of his predecessors of both the technical mathematical issues and the most recent historical scholarship. Likewise, Grattan-Guinness encourages a fundamentally historical and cultural approach to the history of mathematics. He also stresses applied mathematics and appreciates the cultural contexts both within particular mathematical communities and, as an international community has emerged in the modern period, across local and national boundaries.

Various others have also offered accounts of the history of nineteenth-century mathematics, with differing degrees of success. In addition to *Mathemat-*

---

96. Rowe, "New Trends and Old Images," 4.

97. The latter appeared first as *The Fontana History of Mathematics* (1997), then as *The Norton History of Mathematics* (1998).

98. Rowe, review of Katz, *History of Mathematics.*

ics of the Nineteenth Century, a series of books edited by Andrey Nikolaevich Kolmogorov and Adolf Pavlovich Yushkevich, there is Jean Dieudonné's *Abrégé de l'histoire mathématiques,* which places particular emphasis on the nineteenth century. Both of these works provide very technical, almost exclusively internalist accounts of mathematical details. In sum, virtually nothing in the way of a comprehensive textbook for the nineteenth century is available for student use, one giving sound knowledge of the history, social context, and major figures of the century as well as mastery of the technical details of the mathematics, both pure and applied. This may well be too great a task for any one person, but such a study would certainly be a substantial contribution to the history of nineteenth-century science.

Increasing attention has been directed to members of minority groups in science, including women. When the *Dictionary of Scientific Biography* was published in the 1970s, for example, it included biographies of only six women mathematicians: Maria Gaetana Agnesi, Gabrielle-Emilie Le Tonnelier de Breteuil (Marquise du Châtelet), Germain, Hypatia, Sofia Vasilyevna Kovalevskaia, and Emmy Noether. By contrast, Louise S. Grinstein and Paul J. Campbell's *Women of Mathematics: A Biobibliographic Sourcebook* (1987), the most comprehensive and useful of the many works now in print devoted to this subject, presents forty-three essays on the contributions made to mathematics since 1860 by women who were notably "exceptions," especially as "pioneers" in graduate education.[99] What is remarkable, these essays make clear, is the extent to which women were able to pursue mathematics at all, given the tremendous opposition and prejudice against their abilities. Mary Somerville, for example, was not allowed to buy books at a bookstore in Scotland because she was a woman, and Christine Ladd-Franklin, despite her graduate work at the highest level at the Johns Hopkins University in physics, mathematics, and psychology, only received her degree forty-two years after having finished her doctoral work. Princeton did not even admit women as graduate students in mathematics until the 1960s.

Although women have attracted by far the greatest amount of interest in special studies devoted to the history of mathematics, more limited efforts have been directed to issues of race, sexuality, and sexual preference as they relate to history of mathematics in the nineteenth century. The subject of black mathematicians has been of special interest in the United States, and several studies have undertaken to identify important examples.[100] The sample of studies related to sexuality is very limited, although some attention

---

99. LaDuke, Introduction, xvii. Another useful bibliographic compilation is Høyrup, *Women and Mathematics.* See also L. Fox, Brody, and Tobin, *Women and the Mathematical Mystique;* Henrion, *Women in Mathematics;* and Fennema and Leder, *Mathematics and Gender.*

100. For example, Newell et al., *Black Mathematicians and Their Works.*

has been paid to homosexuality, rumored or otherwise, in the lives of Charles S. Peirce and James Mills Peirce.[101]

Approaching the history of mathematics by way of special topics or sub-disciplines has long been a favorite genre of mathematicians and historians of mathematics. Early examples that have stood the test of time are Julian Lowell Coolidge's *History of Geometric Methods* and Roberto Bonola's *Non-Euclidean Geometry*. Excellent contemporary examples include studies by Crowe on vector analysis; Jeremy Gray on non-Euclidean geometries, on linear differential equations, and on group theory from Riemann to Poincaré; Thomas Hawkins on Lebesgue's theory of integration and on Lie groups and algebras; Dale Johnson on topology; Mehrtens on lattice theory; Erhard Scholz on manifold theory; and Hans Wussing on group theory.[102] Mention should also be made of Umberto Bottazzini's study of real and complex analysis; Harold Edwards's books on Riemann's zeta function, Fermat's last theorem, and Galois theory; Grattan-Guinness's survey of the history of the foundations of analysis from Euler to Riemann; and Luboš Nový's *Origins of Modern Algebra*.[103] Each of these studies makes excellent use of archival material or provides an especially rich context for its subject. In addition, numerous collections of papers have focused on particular mathematical subjects; for example, the recent volumes on algebra and geometry in Italy edited by Aldo Brigaglia, Ciro Cileberto, and E. Semesi, or the *Introduction to the History of Algebra* edited by Scholz and intended especially for teachers, exemplifying how closely history of mathematics may be linked with mathematics education. Another approach, exemplified by Esther Phillips's collection of papers, *Studies in the History of Mathematics,* is intended to meet "the greater need and demand for historical information related to the teaching of advanced topics and to the early development of the major fields of contemporary research."[104]

But in terms of breadth, scope, and detail of material presented and evaluated, it would be hard to match Grattan-Guinness's monumental, three-volume study of French applied mathematics in the early nineteenth century, *Convolutions in French Mathematics, 1800–1840: From the Calculus and Mechanics*

101. E.g., Brent, *Peirce,* and Kennedy, "Case for James Mills Peirce."

102. Crowe, *History of Vector Analysis;* Gray, *Ideas of Space* and "Non-Euclidean Geometry"; Hawkins, *Lebesgue's Theory of Integration* and *Emergence of the Theory of Lie Groups;* Johnson, "Prelude to Dimension Theory" and "Problem of Invariance"; Mehrtens, *Entstehung der Verbandstheorie;* Scholz, *Geschichte der Algebra;* Wussing, *Genesis of the Abstract Group Concept.*

103. Bottazzini, *Higher Calculus;* Edwards, *Riemann's Zeta Function, Fermat's Last Theorem,* and *Galois Theory;* Grattan-Guinness, *Foundations of Mathematical Analysis;* Nový, *Origins of Modern Algebra.*

104. Phillips, *Studies in the History of Mathematics,* i.

to *Mathematical Analysis and Mathematical Physics*. Grattan-Guinness focuses on the period 1800–30, when France dominated mathematical analysis and mathematical physics. His story includes institutions and individuals, the calculus, mechanics, engineering, theories of heat, new areas of mathematical physics, physical optics, connections between electricity and magnetism, elasticity theory, and energy mechanics. He also discusses the influence of new serials like Crelle's *Journal*. While he regards the period 1800–15 as building upon late-eighteenth-century traditions, he sees that from 1815 to 1830 as one "of major changes in pure and applied mathematics and then in engineering, which were joined to the old thought in the ensuing years (1830–1840)."[105] What distinguishes this work from most and represents the maturity of contemporary historiography of mathematics is the attention Grattan-Guinness pays to contextual issues, including institutional and philosophical concerns, addressing such matters as "how it means: similarity of structure between mathematical and physical theories," "physical guidance of mathematical investigations," "the place of 'experientiability' in formulating physical theories," and "science as one or sciences as many." He also considers mathematical notation, and asks such questions as "Are physical theories true or hypothetical?" On the subject of mathematical *Denkweisen*, he writes: "To sum up, *Denkweisen* involve both epistemological and methodological aspects of the work of a savant, and in instances such as Lagrange, maybe psychological ones also."

Throughout, Grattan-Guinness emphasizes that his concern is with how mathematics was carried out rather than the specific problems to which it was applied.[106] Still, he aims to fill some of the "huge gaps" that have existed between these extremes by providing both mathematical details and consideration of the historiographical issues surrounding them. The nature of the relationship between pure and applied mathematics in early-nineteenth-century France constitutes what he terms a "vast, neglected area of the history of science," adding that what he has been able to provide is only a "selective, suggestive history."[107] Even so, it may certainly be taken as a model of its kind, and it may be hoped that future studies of nineteenth-century mathematics will adopt his approach to writing comprehensive studies of particular aspects of nineteenth-century mathematics.[108]

---

105. Grattan-Guinness, *Convolutions in French Mathematics*, 1:6.

106. Ibid., 1:57.

107. Ibid., 1:63.

108. For another, more limited appreciation of the need to study the connections between pure and applied mathematics, see Scholz, *Symmetrie, Gruppe, Dualität*.

13. Constitutional Conservatism
and the Plight of Historians of Mathematics

In closing, it is worth considering why historians of mathematics generally, and those who devote themselves to the nineteenth century in particular, seem so conservative in their approaches to a subject that can clearly be treated with profit from different perspectives with a variety of emphases. Working mathematicians will doubtless continue to have a penchant for internal, technical histories; it is unlikely that their concerns will broaden significantly to include considerations of a more sociological, philosophical, or even historical nature. Worse still, a general parochialism among mathematicians often precludes sympathy for works by authors who are not fellow mathematicians. As Aspray and Kitcher observed following a meeting designed to promote dialogue among mathematicians, philosophers, and historians of mathematics, they were "less sanguine about the prospects for dialogue," especially given "the tendency for some of the mathematicians present to discuss the work of historians and philosophers as ignorant invasions of the mathematicians' professional turf."[109] Unless such attitudes undergo some unforeseeable and radical transformation, it is doubtful that more than a handful of mathematicians will undertake broad historical research devoted to nineteenth-century mathematics. The majority will continue to prefer the highly technical, predominantly internalist accounts that typify the approach of most mathematicians to the subject.

Unfortunately, historians of science are unlikely to take more informed interest in the history of mathematics either. The first issue of the new series of *Osiris* offers a contentious case in point. When the History of Science Society reinstituted the series in 1985, its premiere issue, "Historical Writing on American Science," was devoted to presenting "state of the art" surveys of the histories of the various sciences in America but, remarkably, did not include the history of mathematics. Recently one of the issue's editors went so far as to interject a reply into a book review that appeared in *Isis* reiterating a complaint made by Parshall and Rowe in their book about this glaring omission. The explanation was simply that, despite considerable effort, no suitable author could be found to write an article.[110] And yet, an informal canvass of four prominent historians of mathematics, some of whom had just published important studies of nineteenth-century mathematics in America, found that they were not among those approached about writing on mathematics for the

---

109. Aspray and Kitcher, "Opinionated Introduction," 24.

110. As the editor of *Isis* explained, "On the contrary, the coeditors of *Osiris* 1 sought far and wide for an author of an article on the history of American mathematics, but none of the suitable persons was willing to take on the task." See Grattan-Guinness, review of Parshall and Rowe, *Emergence*.

Osiris volume. The point is less to criticize the editors than to suggest that the failure to include mathematics in the first issue of *Osiris* is symptomatic of a much deeper problem facing historians of mathematics within the larger community of historians of science. Although history of mathematics was once regarded as the essential foundation for the history of science, deserving a place at its very center, as George Sarton maintained, historians of science have recently relegated it to the periphery or ignored it altogether. As Stephen Brush observed: "Mathematical reasoning is generally admitted to be an essential feature of the development of science during the past four centuries. Yet with the recent emergence of history of science as an independent discipline, there has been a marked tendency to omit any serious critical analysis of such reasoning."[111] Perhaps Grattan-Guinness put it best in describing the plight of the history of mathematics: "too mathematical for historians and too historical for mathematicians."[112] To present the history of mathematics in all its historical richness requires technical, historical, linguistic, and specific research skills. The amount of technical material one needs to master to discuss convincingly the history of nineteenth-century mathematics is considerable; at the same time, to write the history of mathematics without the mathematics, to approach it only in terms of institutions, roles, or semiotic significance, is to remove from consideration the essential matter that constitutes the subject itself. In contrast to mathematicians and scientists, when historians, historians of science, sociologists, or social historians write about the history of modern science it is as if the history of mathematics had no part to play at all. And this fact is reflected most clearly in the first volume of the new series of *Osiris*.

Among major American Ph.D. programs in the history of science, hardly any include a historian of mathematics. It is a virtual certainty that as the next generation of historians of science is trained, at least in the U.S., few will have been exposed to the history of mathematics, and most will never have given it any thought. Fortunately, however, the ranks of serious historians of nineteenth-century mathematics are slowly expanding, gaining more from mathematics than history of science. It may well be that such historians will increasingly consider questions of social context, gender, institutional histories, and any number of alternative studies of mathematics prompted by postmodernist interests and concerns, but no doubt even those with broader concerns will still be interested primarily in technical, internal histories. That,

---

111. Brush, "Wave Theory of Heat," esp. 145, and *Kind of Motion*, cited in Grattan-Guinness, *Convolutions in French Mathematics*, 1:61.

112. Grattan-Guinness, "Does History of Science Treat of the History of Science?" 158. And still more bluntly: "Especially conspicuous to me is the almost entire absence of the history of mathematics from the history of science as it is practiced." Idem, "History of Science Journals," 196.

after all, is what the history of *mathematics* is clearly about. Nonetheless, *history* of mathematics should certainly be about more than technical mathematics alone. As this essay has sought to show, it must come to terms with issues of context, roles, and not only the importance of mathematics for nineteenth-century science but, as Merz saw so acutely at the beginning of the twentieth century, its pervasive cultural significance as well.

Six

PHYSICS

JED Z. BUCHWALD AND SUNGOOK HONG

※

A HISTORY PURPORTING to deal with "physics" throughout the nineteenth century must decide what the term encompasses. No definition, however adroitly chosen, can represent what people throughout the period took the term to mean, because its significance changed during the course of the century as discoveries were made and as new concepts arose, and because the term had different local meanings at any given time. The *Index* to the Royal Society Catalogue of Scientific Papers, printed in 1914, nicely exemplifies the complexity of definition that the historian faces. In it we find that "physics" is concerned with special topics such as Brownian motion, as well as the broader categories of heat, light, sound, electricity, and magnetism. Mechanics has a separate listing. Under these headings, we find many subjects that are nowadays usually thought to be embraced by industrial or applied, as distinct from fundamental, physics (e.g., microscopy). This overlap of categories poses a problem for the historian who seeks to write a survey of the field, since one cannot include everything. We will nevertheless follow in spirit the notion, common circa 1900, that "physics" is the general science of matter and energy, exclusive of the properties by which specific substances may combine to form new substances (i.e., chemistry)—recognizing that "energy" becomes problematic as a category before midcentury. This definition works reasonably well, encompassing much nineteenth-century activity that many then and now would recognize as "physics," and it is also wide enough to include such subjects as metrology, with its tight links to industry.

By the turn of the twentieth century higher education and research in physics in our sense of the term were carried out by professionals who formed a self-conscious, international community. Education in the subject had itself become highly systematic: textbooks, with attached problem sets, as well as laboratory training, were common, and in many cases there was comparatively little difference between texts used in different places. Research was dominated by physicists from four countries: Britain, Germany, France, and the United States. Each country had nearly the same per-capita number of physicists and spent about the same amount on physics education and research. Several prominent universities, including those in or at Cambridge, Berlin, Göttingen, Leyden, and Baltimore (Johns Hopkins), dominated research, and national laboratories for research into units and standards had been established in Germany, England, and the United States.[1]

Although the content and character of research did vary considerably from place to place, a common set of problems and laboratory techniques can be identified. Physicists throughout Europe and America, as well as in Japan, were particularly concerned with six areas, some of which had become prominent only in the 1890s: the nature of X rays; the character and behavior of electrons; the properties of the ether; the statistical description of gases, liquids, and solids; the phenomenon of radioactivity; and the long-wavelength regime of electromagnetic waves. Investigators generally attacked problems in these areas with a common set of mathematical techniques and experimental methods, and confirmed or refuted one another's results. Experiments remained closely coupled to much of this work, and common sets of techniques and tools facilitated communication across national boundaries and even theoretical and laboratory regimes. Increasing mathematical sophistication in many areas of research created a need for greater precision in measurement, while more exact measurement itself prompted advances in theory. By 1900, a physical hypothesis had little chance of attracting serious attention unless it was both quantitative and capable of being tightly bound to existing bodies of knowledge.

The magnitude of the change can be best appreciated by considering the situation in the mid-eighteenth century. Physics, or, more precisely, natural philosophy, was then neither confined to specialized institutions nor dominated by self-conscious professionals who spoke primarily to one another. There were of course scientific societies, such as the Royal Society of London and the Académie des Sciences in Paris, but discussions were also conducted in salons and coffeehouses. Moreover, boundaries between the professional and the amateur were neither well-defined nor strongly policed. Public demonstrations of Newtonian effects, of electrical wonders, of optical illusions, among many other things, formed the natural-philosophical world. Rational mechanics was considered part of mathematics, and astronomy retained its own goals, methods, and social systems. The quantitative spirit that had long existed in astronomy, and that had become predominant in mechanics, had as yet had little effect on natural philosophy. Also, there were substantial differences between the kinds of natural philosophy done at different places, for theories and experimental practice were often highly localized. The physics carried out by Cambridge Newtonians, for example, had little in common with the Newtonianism of Joseph Priestley, let alone that of French Newtonians in the Académie. And only a small portion of natural philosophy was

---

1. Forman, Heilbron, and Weart, "Physics circa 1900."

quantitative or tied to exacting experiment. Nor did systematic education exist.[2] As late as 1816, the French *physicien* Jean-Baptiste Biot could write:

> Everyone who has had occasion to make extensive researches has seen with regret the scattered state of the materials of this fine science, and the uncertainty under which it still labors. One result is admitted in one country, and another in another. Here one numerical value is constantly employed, while in another place it is regarded as doubtful or inaccurate. Even the general principles are far from being universally adopted. . . . What it wants is union. It is the function of the parts that makes a single body of it; it is a fixing of the data and the principles which gives the same direction to all efforts.[3]

During the nineteenth century physics was transformed into a professional, unified, quantitative, and exact discipline with methods that markedly distinguished it from astronomy, chemistry, and mathematics. We begin our discussion with two related issues. The first concerns the transformation of eighteenth-century "natural philosophy" into nineteenth-century "physics" (section 1); the second, closely tied to the first, involves the mathematization of physics (section 2). We then consider questions that have been much discussed in recent years concerning the methodology of nineteenth-century physics, turning our attention to a series of cases that illuminate the relation between theory and experiment (sections 3–6). Finally, we turn to the interaction between physics and technology (section 7), a topic that has recently been the subject of extensive commentary. In our conclusion we briefly outline the striking changes that physical practice underwent at the end of the nineteenth century.

Before turning to these topics, we need briefly to discuss issues of periodization, which raise many of the same problems—and for many of the same reasons—that the term "physics" does. Since there is no universally accepted scheme, we will use our own. We suggest that physics as a separate discipline with distinctive methods—exact, quantitative, and experimental—can be reasonably well discerned by the end of the first third of the nineteenth century. During the second third, unifying themes such as the principle of energy, as well as dynamical conceptions, served to integrate subjects that had hitherto been only loosely connected. During this period physics was firmly institutionalized in universities, and academic physicists extended their influence by demonstrating the utility of physics through interactions with

---

2. For natural philosophy in the eighteenth century, see Heilbron, *Electricity;* Stewart, *Rise of Public Science;* and Golinski, *Science as Public Culture.* Varieties of local practice in eighteenth-century natural philosophy are discussed in Schaffer, "Machine Philosophy." For its audience, see R. Porter, "Science, Provincial Culture, and Public Opinion." For rational mechanics, see Truesdell, "Program."

3. Biot, *Traité de physique,* 1: ii–vii, quoted in Crosland and Smith, "Transmission of Physics," 8.

telegraphists, optical instrument makers, civil engineers, and others in indus-try. During the last third of the century, continuum physics—grounded on energy principles and linked to dynamics—was elaborated, notably in the area of electromagnetism. Physical laboratories and institutes became com-mon during this time. Starting in the 1890s, substantially as a result of such dazzling experimental endeavors as those that led to the discovery of electro-magnetic waves, X rays, the electron, and radioactivity, physicists began to grapple on paper and in the laboratory directly with the microworld. A new form of physics—microphysics—emerged, and it eventually displaced much of the ether-based, dynamical, and continuum work that had previously been thought of as lying at the discipline's frontiers.

1. From Natural Philosophy to Physics

The contrast between eighteenth-century natural philosophers and nineteenth-century physicists can be succinctly illustrated by the clash between Domi-nique François Arago and Biot over the application of Etienne Louis Malus's scheme of ray statistics to chromatic polarization, a new optical phenomenon discovered by Arago. Both Arago and Biot had attended the Ecole Polytech-nique (founded in 1794), whose instructors placed new emphasis on quantifi-cation and experiment in all areas of what had been natural philosophy. But where Biot absorbed and championed the new spirit, Arago did not. Arago's optical memoirs, in which he sought to produce a general theory that would unify disparate phenomena, remained qualitative and discursive. He never tried to generate formulas, and his work did not contain numerical, much less tabular, data of any kind. In this last respect, his work exemplified that of eighteenth-century natural philosophers. Biot, by contrast, produced lengthy papers filled with formulas and extensive tables. He stood firmly with the new spirit epitomized in general by Pierre Simon Laplace and in optics by Malus. This emergent, and highly influential, tradition insisted upon the careful tab-ular presentation of numerical data and the generation of formulas capable of encompassing the material at hand.[4]

These changes were hardly the only ones that occurred in the first half of the nineteenth century. According to P. M. Harman, the areas that were even-tually grouped under the term "physics" were first joined through the overar-ching concepts of energy and mechanical explanation.[5] The germ of these unifying concepts, which eventually resulted in specific forms of practice, can be found, argues Crosbie Smith, in several eighteenth-century figures, espe-cially the Scottish natural philosopher John Robison. He had divided physics

---

4. Buchwald, *Wave Theory*, 67–107.
5. Harman, *Energy, Force, and Matter*. See also Purrington, *Physics*.

into natural history and natural philosophy, as well as chemistry and mechanical philosophy. Mechanical philosophy led to "mechanical history," which itself consisted of four subdisciplines: astronomy, studies on the force of cohesion (involving the theory of machines, hydrostatics, hydraulics, and pneumatics), electricity and magnetism, and optics. Robison saw two links between these four subdisciplines. First, all concentrated on "forces" existing in nature: gravitation, cohesion, electrical and magnetic forces, and forces involving light and material particles. Second, all were concerned with motions of bodies in one fashion or another. Smith argues that Robison's emphasis on the unity provided by this underlying dynamical emphasis was shared by John Playfair and eventually by James David Forbes, William Thomson (later Lord Kelvin), and William J. M. Rankine. Smith contrasts the views of these Scotsmen with those of Cambridge mathematicians such as William Whewell, who divided the physical sciences into "mixed mathematics," or "mixed sciences" (mechanics, hydrostatics, pneumatics, optics, and astronomy), and (as yet) nonmathematized "experimental philosophy," or "applied sciences" (magnetism, electricity, heat, light, chemistry, sound, and metrology). Unlike the Scottish scheme, Whewell's did not envision a central unity. Physics as a unified system thus seems to have a Scottish origin.[6]

Susan Cannon asserted in her *Science in Culture* that "physics itself was invented" in the nineteenth century, but disagreed with Smith over where it happened. Cannon argued that under the leadership of André-Marie Ampère, Sadi Carnot, Jean Baptiste Joseph Fourier, and Augustin Jean Fresnel physics was invented in France in the years 1810–30. She even pointed to a specific date: 1816, the year Biot published his *Traité de physique expérimentale et mathématique*. She further argued that the *Traité*'s publication showed that Biot possessed "something like our concept of physics." Physics was then transferred to Britain, and in particular to Scotland, where the first generation of physicists included such men as Forbes and Thomson. According to Cannon, James Clerk Maxwell can be thought of as a "fully-formed" physicist not because of his training at Cambridge but because he had previously studied under the physicist Forbes. In Cannon's view, the new spirit of unified physics consisted essentially in the idea of a general unity among disparate areas, which she saw embodied in John Herschel's *Preliminary Discourse on the Study of Natural Philosophy* (1832).[7]

Buchwald, for his part, emphasizes the change in France in standards of reporting, calculation, and experiment; Harman, Smith, and Cannon instead concentrate on the new emphasis on unity in physics, differing, however, over

6. C. Smith, "Mechanical Philosophy."
7. Cannon, "The Invention of Physics," chap. 4 in her *Science in Culture*, 111–36. For the transmission of French physics to Britain, see Crosland and Smith, "Transmission of Physics."

where the emphasis first emerged. Once created out of natural philosophy, though, wherever that may have occurred, physics rapidly became a professional discipline, something that natural philosophy had never been. Indeed, Simon Schaffer asserts a consequential link between the professionalization of physics and the "end of natural philosophy," which he argues for on the basis of contemporary discovery narratives. The end of democratic natural philosophy and the beginning of hierarchical modern science can be discerned in the emergence of a particular kind of historical narrative concerning earlier discoveries. Schaffer examines four of these—concerning the discoveries of Uranus, oxygen, the inverse-square law for electricity, and photosynthesis— and argues that they share two significant features. First, all emphasize what Schaffer terms "heroic authorization," or the tight coupling of the discovery with a unique heroic genius (respectively, William Herschel, Antoine Laurent Lavoisier, Charles Augustin de Coulomb, and Joseph Priestley). Second, the specific techniques of these heroic authors, Schaffer argues, become canonical for the corresponding scientific discipline. This standardized technique then becomes the link in a social hierarchy characterized by a dichotomy between the heroic discoverer and ordinary practitioners, the latter having to be trained in the techniques invented by the hero. Schaffer argues that we find here the first signs of a separation between the contexts of discovery (the heroic producer) and of justification (the yeoman cultivator).[8]

What of France and Germany? We have already seen that Cannon located the origins of unified physics in France, and that Buchwald located the emergence of experimental canons and insistent quantification there. Throughout the eighteenth century in France, *physique* had consisted of two separate disciplines: *physique générale* and *physique particulière*. After midcentury the former meant Newtonian mechanics, while the latter connoted experimental science in general, but sometimes meant specific studies in heat, light, sound, electricity, and magnetism.[9] During the first quarter of the nineteenth century, due largely to the work of Coulomb, Laplace, Malus, Biot, Siméon-Denis Poisson, Fourier, and Fresnel, these disciplines were bound together, although there were many powerful differences in specific beliefs among these men. We shall return to this point in more detail in section 2.[10]

8. Schaffer, "Scientific Discoveries." For a critique, see Alborn, "'End of Natural Philosophy.'" The historiography of natural philosophy in the seventeenth and eighteenth centuries has been intimately linked to criticism of the allegedly whiggish historical practice of using the term "science" for these periods, on which see Cunningham, "Getting the Game Right."

9. This classificatory scheme originated with Denis Diderot and was brought to maturity by Auguste Comte. For a physicist's exposition, see Haüy, *Traité élémentaire de physique*.

10. Crosland, *Society of Arcueil;* R. Fox, "Laplacian Physics"; Grattan-Guinness, "Mathematical Physics in France" and *Convolutions in French Mathematics;* Frankel, "J. B. Biot"; Silliman, "Fresnel and the Emergence of Physics"; Friedman, "Creation of a New Science"; Grattan-Guinness and Ravetz, *Fourier.*

In Germany, as Kenneth Caneva has shown, the study of electricity and magnetism changed radically in the first half of the nineteenth century. Before the 1820s, theories and experiments involving electric and magnetic phenomena (by scientists such as Thomas Seebeck and Johann S. C. Schweigger) were notable examples of what Caneva terms "concretizing science"—science characterized by belief in experience as the epistemologically primary source for scientific knowledge, and by the assumption of an inherent distinction between physical and mathematical modes of explanation. By the late 1840s, a completely different approach, which Caneva terms "abstracting science," had emerged. This new approach was marked by adherence to precision in experiment and by the relatively free use of physicomathematical hypotheses. Georg Simon Ohm in the 1820s pioneered this new spirit, Karl Gustav Jacob Jacobi and Carl Friedrich Gauss continued it in the 1830s, and Wilhelm Weber, Gustav Fechner, and Franz Neumann fully developed it in the 1840s.[11]

The transformation from concretizing to abstracting physics in Germany accompanied, and was continued by, the development of "theoretical physics" as a subdiscipline separate from experimental physics. This may appear confusing, since we have thus far stressed historians' insistence on the unification of mathematical and experimental physics that had taken place since the late eighteenth century. Two points require clarification. As Christa Jungnickel and Russell McCormmach have exhaustively demonstrated, German universities, which had scarce resources, created distinct chairs in theoretical physics in order to supplement the teaching and research of the existing experimental physicist without provoking conflicts over resources.[12] Second, nineteenth-century theoretical physics was neither a substitute for, nor a linear continuation of, the procedures and principles that had guided eighteenth-century rational mechanics. German theoretical physics was born along with such new fields as electrodynamics, physical optics, thermodynamics, and statistical physics. These fields had not been mathematized in the eighteenth century; some scarcely existed at all.

2. Mathematization and Laplacian Physics

Thomas Kuhn long ago divided the physical sciences historically into what he termed the "classical" and the "Baconian." The classical sciences included astronomy, statics, optics, harmonics, and mathematics. These had been mathematical since antiquity, and, except for harmonics, all underwent major changes during the seventeenth century—a transformation that, Kuhn be-

---

11. Caneva, "From Galvanism to Electrodynamics."

12. Jungnickel and McCormmach, *Intellectual Mastery of Nature*. For Franz Neumann's seminar, see Olesko, *Physics as a Calling*.

lieved, owed more to the development of "new ways of looking at old phenomena than to a series of unanticipated experimental discoveries." Observations and experiments played, he argued, minor roles in this great change. Kuhn's Baconian sciences included electricity, magnetism, chemistry, sound, heat, and light. Practitioners in these fields constructed and used entirely new instruments, such as the barometer, the air pump, and the microscope, with which they performed experiments in highly artificial circumstances. The gap between the classical and Baconian sciences, which had social as well as intellectual dimensions, remained largely intact until well into the eighteenth century. Mathematization of the Baconian sciences narrowed the gap, and here Laplace and his collaborators played crucial roles.[13]

In 1974, Robert Fox powerfully influenced subsequent historical work through his analysis of what he termed "Laplacian physics."[14] According to Fox, the fourth volume of Laplace's *Traité de mécanique céleste* (1805) provided a basis for a "truly Laplacian style of science." Laplace there examined in great mathematical detail two mundane phenomena—capillary action and the refraction of light in air—by employing the assumption of short-range forces that act between material particles proper and those that constitute light. Moreover, Laplace, and those who worked under his inspiration and influence, compared these theoretical results with experimental data. His work set standards for *physiciens* such as Biot, Malus, and Poisson. Biot's extensive studies of voltaic electricity, the propagation of sound, and chromatic polarization; Malus's work on double refraction and polarization; and Poisson's on electricity and magnetism were all directly imbued with the spirit of Laplacian physics.[15]

Just as significant, Laplacian practice incorporated a striking change in the way that scientific work was done and presented. During the last quarter of the eighteenth century, and especially after the founding of the Ecole Polytechnique, French *physiciens* had become profoundly concerned with precision, with the invention of new instruments, and with putting existing instruments to new uses. They had also begun to insist on presenting their results in the form of numerical tables, which had been uncommon previously.[16] Eugene Frankel suggested that these and similar features constituted

13. Kuhn, "Mathematical versus Experimental Traditions"; see also Kuhn, "Function of Measurement," 220.

14. R. Fox, "Laplacian Physics." Sutton, in his study of the debate concerning the voltaic pile in 1800–1801, argues that the program was already influential in 1800. See Sutton, "Politics of Science."

15. For Biot, see Frankel, "J. B. Biot." For Malus, see Buchwald, *Wave Theory*, 23–66, and Frankel, "Search for a Corpuscular Theory." For Poisson, see Arnold, "Mécanique Physique," and Home, "Poisson's Memoirs."

16. For such changes in the style of practices in physics, see Buchwald, *Wave Theory*, and Heilbron, "Weighing Imponderables."

a core part of the Laplacian program, thereby extending Fox's original emphasis on theoretical structure and enabling Frankel to extend the boundaries of what one might call the Laplacian way of doing science. Frankel distinguished four major features: first, the use of new instruments or techniques to measure physical entities, such as heat; second, a new-found concern with accuracy and precision; third, insistence on expressing experimental results numerically and as fully as possible; and fourth, the assimilation of quantitative data or algebraic relations into the paradigm of forces acting at a distance between particles.[17] More recently, Buchwald has argued that a central component of this change was the substantial replacement, in France, of geometric by algebraic modes of representation. He has further asserted that certain features of the form of practice that resulted, as well as the particular social and personal character of French physics in the early 1800s, had striking effects on the development of optics.[18] Many of these elements—including algebraization and the first three features in Frankel's list—were already present in the late eighteenth century and should not, properly speaking, be linked specifically to Laplace. However, they did not come together until after 1805, and the overall Laplacian program, which certainly incorporated these elements, did not reach its peak of influence until circa 1815.

A central factor in these developments, mentioned above and receiving increased attention in recent histories, was specific to France, namely the establishment of new educational institutions during the French Revolution. The most influential of these was the Parisian Ecole Polytechnique. It trained students not only in military and civil engineering, but also in mathematics and the physical sciences, including heat, optics, electricity, and magnetism. The mathematician Gaspard Monge early aimed to produce a practical education for engineers; theoretical studies in mathematics and physics soon became a central part of the curriculum, largely due to Laplace's influence. Although he himself never taught at the Ecole, Laplace exerted powerful influence on it as one of the examiners and as a member of the Conseil de Perfectionnement, established in 1800 to supervise examinations. Fox has remarked that the syllabus for the physical sciences in the 1800s and 1810s strongly reflected the influence of Laplacian beliefs. Biot and Malus, two full-fledged Laplacians, were among the first Polytechniciens. Poisson and Arago studied physics and mathematics there, as did Fresnel. Not all of them assimilated every aspect of the Laplacian way, and this, as we shall see presently, had important results.[19]

As a teaching institution, the Ecole spread wide the influence of the new French approach to physical investigation. Fundamental research occurred

17. Frankel, "J. B. Biot."
18. Buchwald, *Wave Theory*, 12–19.
19. Williams, "Science, Education, and Napoleon I"; Bradley, "Scientific Education versus Military Training."

there, as elsewhere, though it is perhaps unwise to draw these sorts of distinctions at a point when organized laboratories and institutes of the kind that became common in Germany half a century later did not yet exist. However, beginning in 1807 Laplace and his friend, the chemist Claude Louis Berthollet, together inaugurated informal meetings with their young protégés, thereby forming a group—including Biot, Malus, Arago, Poisson, Jean Antoine Chaptal, and Pierre Louis Dulong—that came to be known as the Société d'Arcueil. The Société's members aimed fundamentally to produce exact data and, where feasible, to link that data mathematically with short-range forces in physical science and in chemistry.[20]

There were other venues, in addition to the Ecole and the Société, through which Laplace spread his influence, for he exerted substantial control over the prize competitions proposed periodically by the Institut. Two among these demonstrated the power of the Laplacian way: one, proposed in 1807, concerned the phenomenon of double refraction and was won in 1810 by Laplace's favored protégé, Malus; the other concerned the heats of gases and was awarded to two young members of the Société who explained latent heat in terms of the properties of "caloric," or substance of heat, a concept cultivated by Laplace.[21] These and other competitions exhibit the novel emphases and techniques of late-eighteenth- and early-nineteenth-century French physics, many of which, as we noted above, are not specific to the emphasis on short-range forces between hidden material structures insisted upon by Laplace and others.

Laplacian commitment to microphysical explanation grounded on short-range forces was never altogether dominant. Two examples are especially pertinent. Consider first an 1816 competition for the best mathematical examination of the elasticity of solids. The prize was awarded to the sole essay submitted—by Sophie Germain; the committee consisted of Laplace, Poisson, Biot, Poinsot, and Germain's main supporter, the mathematician Legendre.[22] Laplace, Biot, and Poisson did not approve Germain's physical assumptions, though they criticized only her method for generating the differential equation. They nevertheless awarded her the prize (albeit with reservations). A decade earlier (1807), Joseph Fourier had read a paper on the distribution of heat in bodies before the First Class of the Institut, a paper that (like Germain's on elasticity) did not employ Laplacian microphysics to generate its central physical novelty, the diffusion equation. Fourier's work raised mathematical as well as physical questions—for he offered trigonometric series as

20. Crosland, *Society of Arcueil*. Crosland based his narrative on the concept of a scientific society; after all, the term "Society" was used by group members themselves. As Owen Hannaway has argued, it is perhaps more informative to describe Arcueil as an early research school. See Hannaway's review of *Society of Arcueil;* see also Geison, "Scientific Change," chart 1.

21. R. Fox, "Laplacian Physics," 102–7.

22. Bucciarelli and Dworsky, *Germain*, chap. 7.

general solutions, which occasioned much discussion[23]—but a revised version of the paper won an Institut prize in 1812.[24]

Laplacian orthodoxy also received a striking blow from Fresnel's wave theory of light, whose basic principles ran directly contrary to Laplacian belief and practice in optics. When Fresnel's account of diffraction was awarded a prize by the Paris Académie in 1818, only one member of the judging panel, Arago, was willing to accept more than that Fresnel had produced a successful mathematics for diffraction. In that regard, Fresnel's success was stunning in the new French way, for by adapting Huygens's principle and elaborate integral methods he had produced formulas so accurate that the discrepancy between theory and experiment was reduced to 1.5 percent.[25]

Arago, who championed Fresnel, had an ambiguous relationship with the Laplacian community, though he was himself a graduate of the Ecole. He had been humiliated and frustrated when Biot had published work on chromatic polarization without acknowledging Arago's priority in the field, which had been opened by Malus's pioneering work on polarization. What Arago lacked, and Biot possessed, was precisely the kind of synthesis between mathematics and precise experiment so prized by the new French physics. Fresnel's wave theory of light, characterized above all by marvelous agreement between theory and data, became in Arago's hands a powerful weapon for attacking Biot—and, incidentally, for dethroning Laplacian power. As Arago's power grew, Laplacian orthodoxy abated. Indeed, by the mid-1820s partisans of wave optics had assumed positions of intellectual and institutional dominance. Arago controlled the *Annales de chimie et de physique*. He, Ampère, Fourier, Dulong, and Fresnel himself were members of the Académie, and Fourier had become its permanent secretary in 1822, defeating Biot. Biot, the most enthusiastic proponent of the old optics, was in nearly total eclipse.[26] Nor was optics the only source of trouble: In 1819, Alexis-Thérèse Petit and Dulong attacked another early success of Laplacianism, the caloric theory of heat. As Fox has noted, "By the mid-1820s the style of science that had appeared so right and unassailable in the Napoleonic period had been abandoned by the leading figures in a new generation."[27]

Historians generally agree that French physics underwent a marked de-

23. Herivel, *Fourier*, passim, discusses Laplace's, Biot's, and Poisson's attitudes toward Fourier's work. While rejecting both Fourier's nonmolecular generation of the diffusion equation and his trigonometric series solutions, all three sought to accommodate the equation itself as well as surface conditions that Fourier had developed.

24. R. Fox, "Laplacian Physics," 109–27. For Fourier, see Grattan-Guinness and Ravetz, *Fourier;* for Fourier's influence, see Garber, "Reading Mathematics." For Thomson's mathematization of electrostatics under the influence of Fourier, see Buchwald, "William Thomson."

25. For Fresnel, see Buchwald, *Wave Theory*, 113–233.

26. For the second-round debate between Arago and Biot, see ibid., 237–54.

27. R. Fox, "Laplacian Physics," 134.

cline in the production of mathematical and physical novelties by the 1830s, although a considerable amount of creative measurement took place in laboratories like that of Victor Regnault. Petit, Fresnel, and Fourier were dead by 1830; Ampère died in 1836. Arago, never among the most creative of physicists, became more involved in politics than in physics. Dulong grew ill, and the mathematician Augustin-Louis Cauchy went into exile after the fall of the Bourbons. Regnault concentrated on metrology, aiming to replace crude laws with more sophisticated ones through elaborate measuring experiments. Novel physics after the 1830s was produced preeminently in Britain and Germany, to which we now turn.

3. Theory and Experiment I: The Case of William Thomson

Many historians of physics during the 1960s and 1970s were influenced by Alexandre Koyré, and so were concerned primarily with theory. This emphasis, as historians and sociologists pointed out repeatedly in the 1980s, was coupled to a comparative neglect of experiment and of instrumentation—an imbalance that could hardly be justified on the basis of what scientists actually did, since, for example, more than 90 percent of physicists living circa 1900 were experimentalists.[28] Furthermore, during the 1960s and 1970s historians of physics focused largely on "high" theory, that is, on the deepest and most fundamental speculations and mathematical work. Theory was treated as providing a thoroughgoing, deep conceptual structure from which everything else about scientific practice followed. Embedded in instruments and overlaid onto experiments, it was seen as dictating scientific activity.

This hierarchy was largely shattered in the 1980s.[29] Two new historical conceptions of scientific practice were essential to this change. First, it was realized that Norwood Russell Hanson's concept of the "theory-ladenness" of observation,[30] which was often taken to show the epistemological priority of theory, should not be construed to mean that theory always determines or guides experimental results in a significant manner. Many results are not guided by or dependent on *relevant* theory at all. For example, experiments were frequently undertaken in Germany in the 1880s to measure the optical constants of metals. Such experiments did depend on basic elements of wave optics (viz., phase and amplitude), but they did not depend at all on belief that light constituted a mechanical or an electromagnetic disturbance. Quite the contrary—the results of such experiments could be, and were, used by scien-

---

    28. Forman, Heilbron, and Weart, "Physics circa 1900."
    29. Studies that emphasize practices in science include Hacking, *Representing and Intervening,* and Shapin and Schaffer, *Leviathan and the Air-Pump.*
    30. Hanson, *Patterns of Discovery.*

tists with considerably different views on such matters, and used to very different ends. *Relevant* theory was not active in either the construction of these experiments or their use in calculating values for metallic constants. Naturally, this need not always be the case, and experiments were constructed that depended in meaningful ways on specific theoretical points; this, however, did not entail that others with different theoretical views could not interpret the results according to their own lights, which is precisely what occurred with the Hall and Kerr effects, for example.

A second factor that destabilized the old theory-experiment hierarchy was the drawing of distinctions between experiments designed primarily to measure stable variables—thus explicitly aimed at confirming theoretical speculation—and experiments designed to reveal novel effects, that is, experiments whose theoretical foundations might be vague or undeveloped, as in the cases of the Volta, Peltier, Seebeck, Faraday, Thomson, Hall, and Kerr effects. The interconnections and distinctions between theory and experiment accordingly became much more complex, forming, as it were, a web woven of two heterogeneous, but often largely autonomous, strands, rather than a hierarchical tree of dependence.

Less noted than this change is a novel understanding of theory itself that appeared in the work of historians of physics. From the late 1970s on, historians began to view theory as one of several strands in a complex web of practice. We may assign the term "theoretical practice" to describe what historians had (and have) in mind, although the term was not used at that time. Historians of theoretical practice concentrated on the ways in which scientists used, and even invented, what are best thought of as theoretical resources—including mathematical techniques, theories in adjacent fields, and analogies, as well as conceptual and mechanical models—to mold novel structures. One might say that historians of theoretical practice were more interested in "theory in the making" than in "theory made." Some historians began to pay close attention to how resources were mobilized and used among the members of a group of physicists, and how this sharing helped produce community identity.[31]

Norton Wise's early work on Michael Faraday and Thomson provides an example of the new focus on theoretical practice.[32] Historians had long connected Thomson's discussion of Faraday's work on fields to an early paper of Maxwell's entitled "On Faraday's Line of Force" (1855), where central elements in Maxwell's subsequent field theory first appeared. Wise concentrated

31. For Hunt's and Buchwald's studies of the British Maxwellians, see below (section 4). See also Olesko, *Physics as a Calling*, on Neumann's Könisberg physics seminar, where the ethos of exactitude and the mathematics of error analysis formed the common identity for his students.
32. Wise, "Flow Analogy."

carefully on the mathematical and conceptual apparatus that underlay field practice, leading him to ask not what connected Thomson and Maxwell, but rather what distinguished their deployment of apparently similar techniques. Wise located the gap between them—and the structure of field theory as subsequently practiced—in a duality of force specific to Faraday.

Historians had long noted the duality of vectors in later field theory and had traced it (at least in a formal sense, but for magnetic effects only) to Thomson. However, Wise recognized in Faraday's work the production of an entirely novel way of thinking grounded at a primitive level on this very duality. He found in Faraday's discussions of quantity and intensity both a visual representation and an underlying, deeply quantitative structure with broad applicability. Faraday had not developed an explicit, articulated set of propositions that explained how to work with intensities and quantities; rather, he exemplified his novel understanding, a duality that distinguished between force as quantity and force as intensity, in particular problems (initially in electrostatics, later in magnetostatics).

Thomson, Wise then showed, never accepted or worked with the full panoply of resources that this distinction provided, whereas Maxwell did.[33] Although Thomson and Maxwell had much in common—the tradition of British dynamical theory, the use of differential equations (rather than integrals) to represent physical processes at a visualizable, geometrical level, and the emphasis on mechanical models—they nevertheless disagreed, with immense consequences, on this fundamental point. Such differences as these, Smith and Wise together argued, move the historian to look beyond the confines of laboratory and paper worlds, to cultural resources and local convictions.[34] These authors sought, for example, to forge a direct, consequential link between such cultural tropes as "power and efficiency" and Thomson's physics.[35] Indeed, if Smith and Wise are correct, then Thomson deployed cultural resources in his efforts to unite geometric physics with mechanical effect, because the latter "attained its potency in Thomson's natural philosophy through the steam-engine, deployed as a metaphor for work, wealth, and

33. Wise, "Mutual Embrace."

34. C. Smith and Wise, *Energy and Empire.* Their understanding is consonant with the constructivist sociology of science that first became significant in the history of science in the early 1980s, as exemplified by Pickering, *Constructing Quarks,* and Collins, *Changing Order.*

35. "'Progressive' and 'anti-metaphysical' nearly define the 'science' of Thomson and his friends, for whom opposition to metaphysics encompassed latitudinarian religion, politics above party (ostensibly), and non-hypothetical, practical knowledge. Science in this sense entailed concomitant emphases in the mathematical style of the new professionals [at Cambridge in the 1830s and 1840s]. They sought generality of expression, simplicity of technique, and utility of results, or *power* and *efficiency."* (C. Smith and Wise, *Energy and Empire,* 178–79.)

progress."[36] This particular claim (and others like it), we think, suffers from a lack of direct evidence.[37] The question of the extent to which "cultural resources" are in fact crucial to physicists' practice has in recent years been endlessly debated by historians and sociologists of science, and we do not think that the answer is altogether clear-cut, or that the ways in which such connections occur and their existence is demonstrated have been thoroughly examined. We will return in the conclusion of this essay to issues raised by models like Smith and Wise's, in which cultural resources are brought to bear on scientific activity.

4. Theory and Experiment II: The Case of the Maxwellians

Although Thomson's work pervaded British science and industry (on which, see section 7), and although he had many students, one cannot identify a group of physicists that based their work in electromagnetics on Thomsonian precepts. Maxwell's work, by contrast, did become the property of a group, the so-called Maxwellians. Maxwell's field theory, loosely assembled in his *Treatise on Electricity and Magnetism* (1873), produced a conceptual cluster around which a group of young British physicists assembled in the late 1870s and early 1880s. Although the group was not at all homogeneous, it found in Maxwell's *Treatise* novel, if difficult, concepts of charge and current, a "dynamical" emphasis on the storage and transmission of energy by electromagnetic fields, and an identity between light and electromagnetic radiation. At Cambridge, where Maxwell's *Treatise* was conveyed primarily by coaches for the intensely competitive, and highly mathematical, tripos, emphasis was placed on the Lagrangian technique of great generality that Maxwell had deployed in parts of the *Treatise*, which was used as well by Thomson and Tait in

---

36. Ibid., 256.

37. So, for example, at one important juncture the authors remark that "this incredibly cryptic paper of a mere three pages makes no reference to work, *vis viva*, mechanical effect or least action, and supplies no aids to understanding other than the concluding remark" (ibid., 271). But from it they conclude that for Thomson "the equilibrium state of the system is thus to be understood in essentially temporal terms, the terms of genesis and further development in a natural succession of states" (ibid., 274), thereby providing a hook to progression and through it to theology. It seems to us that this chain of reasoning requires more direct evidence than the cryptic three pages, though one might argue that the evidence of a common pattern in Thomson's work is compelling (on the fractal model). If we accept the argument, then Faraday's influence must be rethought, because according to Smith and Wise, "In the process [of Thomson's working through his novel understanding] Faraday's theory had acquired a power far beyond that of the descriptive lines of force, but the mathematical theory too had become a much more powerful, and quite different tool: it had become field theory" (ibid., 275). Thomson, it seems, did not "mathematize Faraday" in any meaningful sense; instead, he *created* mathematical field theory.

their work on mechanics, the two-volume *Treatise on Natural Philosophy*. Their *Treatise* contained in addition—indeed consisted of for the most part—many examples of worked problems in electric and magnetic potential theory that did not require or use concepts unique to field theory but did employ advanced mathematical structures and techniques.

Buchwald and Bruce Hunt, who have written in recent years on British Maxwellians and field theory, provide different definitions of "Maxwellian." Hunt's focus is narrower; his Maxwellians are limited to essentially three people: Oliver Lodge, George Francis FitzGerald, and Oliver Heaviside, with Heinrich Hertz as an external member. Hunt argues that this small group gains definitional legitimacy because these three identified themselves as Maxwellians; moreover, he writes, it "was mainly the[se] Maxwellians who gave Maxwell's theory the form it has since retained, and it was largely through their work that it first acquired its great reputation and breadth of application."[38] Hunt's definition has several shortcomings,[39] but he is right to imply that the form of practice that began in the 1890s to constitute electrical communications engineering is connected in Britain to these three individuals, and his structure gains plausibility from the close interpersonal network among them, a network that gradually took shape starting in the early 1880s during public debates. These debates included exchanges between Lodge and FitzGerald on the very possibility of artificially generating Maxwell's electromagnetic waves; between Heaviside, Lodge, and others, on the one hand, and the "practical" electrician William H. Preece, on the other, concerning "practice versus theory"; and, finally, among Lodge, FitzGerald, Heaviside, and J. J. Thomson over the propagation of potential and field.[40]

The Lodge-FitzGerald-Heaviside trio must be distinguished from others who tilled Maxwellian fields, in particular those who were educated at Cambridge, where they went through the rigorous, competitive training of the mathematical tripos and in some cases worked at the Cavendish Laboratory, of which Maxwell had been the first director. Buchwald's Maxwellians consist of three groups: Cavendish students; Cambridge men who worked in Maxwellian electrodynamics; and non-Cambridge Maxwellians, including

---

38. Hunt, *Maxwellians*, 2.

39. Problems with his extended discussion of the point include these: Hunt's trio never produced any confirmation of Maxwell's theory that did not also confirm theories that had nothing at all to do with fields; they were certainly not the first to explore the possibility of wire waves or to generate and manipulate such things; Maxwellian methods have little specifically to do with wire waves; the trio was not the first to (nor did they ever) demonstrate the existence of electromagnetic waves in air; Poynting's energy-flow theorem nicely fit the sort of problems that attracted interest at Cambridge; and, finally, of Hunt's three only Heaviside "recast" Maxwell's equations into the form now used.

40. Hunt, *Maxwellians*, passim.

Hunt's three. Many in this third group were trained as physicists and became members of the first generation of practicing scientist-engineers. Buchwald links his Maxwellians into a coherent network primarily through a core set of beliefs and practices concerning the utility, or lack thereof, for research purposes of energy, field energy, charge, and the effect of matter upon ether; electric conductivity; boundary conditions; mechanical models; microscopic entities; and macroscopic properties. In addition, each subgroup had its own set of beliefs and practices. Cambridge-trained Maxwellians were, for example, devoted to Lagrangian structures and to the principle of least action (or to Hamilton's principle), and they often generated new physics by modifying electromagnetic energy functions that appear in mathematical representations of this kind. Hunt's three Maxwellians—with the overwhelming exception of FitzGerald in the late 1870s (since he was the first to introduce Hamilton's principle into electromagnetics)—never attacked the kinds of problems that were eminently suited to the Cambridge tools. Further, the vector potential was suspect to Hunt's Maxwellians for several reasons, but it was widely used as a convenient analytical device by the Cambridge group, although they were not committed to its physical significance.[41]

Although Buchwald and Hunt clearly agree on a number of essential points—including the centrality for the group, persisting into the twentieth century, of what eventually crystallized as "Maxwell's equations"; the role of Hertz and Heaviside in producing the canonical form of the equations; Heaviside's role in the foundation of wave guidance; and Hertz's role as the producer of electric waves in air and the associated theoretical apparatus— Buchwald prefers an expanded Maxwellian group with a common set of practices that did *not* persist into the twentieth century. He has focused on several examples of how Maxwellian beliefs and practice differed markedly from the structures that began to be formed in the 1890s on the basis of the electron, including the Hall, Kerr, and Faraday effects.[42] Hunt's Maxwellians have living descendants in contemporary departments of electrical engineering; many of Buchwald's do not. Nevertheless, British journals of the day were filled with articles that strikingly exemplified the beliefs and practices of Buchwald's Maxwellians.

Buchwald did not discuss how Maxwellian research emerged at Cambridge during the 1870s; he concentrated instead on how problems were attacked given a consensual set of techniques and ways of posing and choosing problems. Research work did not of course emerge full-blown from Maxwell's *Treatise,* like Athena from the head of Zeus. On the contrary, the *Treatise* had to be assimilated; work was necessary to turn its riches into useful

41. Buchwald, *From Maxwell to Microphysics*, passim.
42. Ibid.; Buchwald, "Design for Experimenting."

resources. Its physical novelties were obscure, its mathematics difficult; only over the course of a decade—at Cambridge via the coaching system—were comparatively coherent structures forged from its disparate elements. Despite these difficulties, and despite the significant differences between the local contexts in which the *Treatise* was read, there were many commonalities among those who worked and researched problems in a Maxwellian vein. Indeed, it was entirely possible for a significant novel deduction—in particular, a theorem concerning energy flow in the field—to be produced in the same form, and with many of the same conclusions drawn from it, by people with such utterly different backgrounds and training as John Henry Poynting and Oliver Heaviside.

5. Theory and Experiment III: "Unarticulated Knowledge" and the Case of Helmholtzianism

The diversity and flexibility of theoretical practice within a group of scientists who otherwise had much in common has implications for understanding their experimental work. As noted in section 3, during the 1980s historians and philosophers of science began to consider experiment without treating it as a subset of theory.[43] This reconsideration was partly inspired by the insights of social constructivism, which emphasized investigations of local practices, and partly by the work of historians like Derek de Solla Price, who viewed theory and experiment as largely independent regimes.[44]

Though we substantially agree with the sentiment underlying this change in historical emphasis, we think caution is necessary when divorcing experiment from theory. Some experiments clearly are guided heavily by theory;[45] others are not. Independence occurs particularly when a laboratory tradition has become firmly established, for then experimenters can ply their trade with hardly any resort to troubling theoretical issues. Of course, theories are, in a limited sense, generally involved in the construction and manipulation of instruments, and instruments themselves lead not infrequently to new ideas and theories.[46] Yet the manner in which theories are embedded in instruments does not always, perhaps even usually, affect in a relevant manner the outcomes of experiments done with them; in this regard theories may not matter at all. Historians need to come explicitly to grips with the complexity

---

43. Historians influenced by constructivist sociology of science tend to separate experiment from theory and to associate it with such factors as authority, the construction of consensus, or difficulty in replication. In this regard, see especially Schaffer, "Glass Works." For criticisms of Schaffer's argument see Shapiro, "Gradual Acceptance."

44. Price, "Of Sealing Wax."

45. See, e.g., Hong, "Controversy."

46. Dörries, "Balances."

of theory-experiment interactions, recognizing that interactions take place at many different levels and in a variety of ways, making any simple declaration of divorce or marriage impossibly simplistic.

In this regard, it may be useful to consider issues raised by the connections between theory and experiment to illuminate certain developments in late-nineteenth-century physics. In particular, we shall use the concept of "unarticulated knowledge," which counters the notion that whatever relationship between theory and experiment exists in a particular case must be contained in a set of explicit, codified arguments. We want instead to emphasize that many core notions—and these are often the ones that bear upon the experiment-theory nexus—are not explicitly articulated at all. This unarticulated knowledge may not only shape explicit theory, it may also influence laboratory practice. Specifically, by "unarticulated knowledge" we intend knowledge that is generally unexpressed but that guides research. This is not at all the same thing as *unexpressible* knowledge, such as the kind of skill that is needed to form a beautiful piano leg on a lathe. Not at all—it is rather knowledge that is *unexpressed,* that exists below the surface of explicit discourse. Such knowledge is accordingly tacit, in the sense of unspoken, but it can be—and often eventually is—heard, particularly when a science settles into a reasonably stable form.[47]

This point is well illustrated by a highly influential development in nineteenth-century German electrodynamics. Physicists who worked in Hermann von Helmholtz's Berlin laboratory in the 1870s and early 1880s differed markedly from those who were trained by Weber. Weber and his students sharply divided theory from experiment; they considered the objects of theory (such as electric particles) to be prior to those of experiment, so that measurement in Weber's laboratory necessarily entailed significant theoretical meaning. In Helmholtz's laboratory, on the other hand, objects of theory and objects of experiments were treated as effectively equivalent, as existing on the same epistemological level, although the point was not explicitly articulated. Helmholtzian physicists considered laboratory objects not (in Weberean fashion) as composed of hidden entities known from prior theory, but rather as entities that possess states, with interactions between objects determined entirely, at any given instant, by their states and by the distance between them. The interaction was embodied mathematically in a "potential" function from which specific effects could be deduced (and which could itself be divined from knowledge of effects).[48]

This way of working affected both mathematical technique and laboratory practice. For example, in order to discover the forces between objects a

---

47. Further discussion can be found in Buchwald, "Design for Experimenting."
48. Buchwald, *Creation of Scientific Effects,* 7–42.

mathematical technique (specifically, variation of the potential function) was deployed. Variational techniques lay at the heart of Helmholtzianism, and this fact had important implications for laboratory life in that environment. Translated into laboratory technique, variational technique suggested that its instrumental analogue had to have either characteristics that changed over time or parts that changed their mutual positions. In either case the experimental device could not be static, although the changes involved could be extremely small. During the 1870s Helmholtz's laboratory concentrated on investigating novel effects that depended on changes in an object's state of charge, and the apparatus used (in particular, oscillating currents and electrostatic generators) were not formed or used in quite the same way anywhere else. Hertz's early experiments on dielectric polarization, which eventually led to the discovery of electromagnetic waves, were molded by just this context.[49] This emphasis on novelty and object states extended to other areas as well, such as investigations of gaseous discharge by Eugen Goldstein.

The unarticulated knowledge that underlay Helmholtzian practice suggested ways to tailor devices for discovery. Though it was not a blueprint for calculating, it was a design for experimenting.[50] The connections between theory and experiment accordingly appear in a different light given the notion of unarticulated knowledge. Consider an example in which theory of any sort—unarticulated or otherwise—seems not to have been operative, namely, in Hertz's construction and stabilization of the device that he used to produce electric waves in wires (and, eventually, in air).[51] This device was novel in that no one before Hertz had produced or manipulated ultrahigh frequency waves in wires with comparable control. But the device itself did not raise any relevant theoretical issues—nobody questioned its ability to produce and detect wire waves, precisely because the existence of such things was not questioned and because Hertz's device, though new, was built out of conceptually and experimentally stable components (induction coils, copper rods, capacitors, and spark gaps). Theory—in the sense of something that would be both relevant to the operation and use of the device, and that might be contested by a contemporary—was not at all built into the Hertzian apparatus. One might say that theory was already built into, for example, the induction coils, which would no doubt be true, but the relevant aspects of induction-coil theory were utterly unproblematic precisely because it had been long stabilized through the building and working of such coils. No one had to create induction coils *ex nihilo* and to figure out how they worked.

Here, then, we have a clear case in which the production of a novel appa-

---

49. Ibid., pp. 75–92, 217–62.

50. Buchwald, "Design for Experimenting," 184.

51. For details and analysis of the following discussion see Buchwald, *Creation of Scientific Effects.*

ratus seemed not, in any practical sense, to be directly related to theory of any kind. Nevertheless, Hertz's motivation for constructing and using such a device was very much connected to Helmholtzian practice, if not to any articulated theory. First, ultrahigh frequency waves in wires were desirable laboratory objects precisely because they were significant for investigating pressing theoretical questions of the day—a point that was entirely explicit. But second, the character of Hertz's device coheres with the Helmholtzian emphasis on constructing apparatus that embodied object-object interactions in a directly manipulable fashion. Hertz's device forced current-bearing objects to work against and with one another in producing an observable effect (namely, a spark). One can conceive of other ways of building and using ultrahigh-frequency apparatus that do not depend so markedly on direct interaction of this sort, an interaction that embodied the stimulating object in one part of the device and the detecting object in a distinct piece, with the two interacting directly.

Other experiments do connect tightly to articulated theory, among them the ones that Hertz performed in the spring of 1888 when he used his device to produce electric waves in air for reflection, refraction, and polarization. Here the properties of objects whose very existence was at issue were invoked directly in the experiment's design. In these experiments and their immediate predecessors, questions did arise concerning the properties of the apparatus itself in respect to the objects (electric waves) that it was supposed to produce. As a result, controversy could arise easily enough (particularly in France) concerning just what Hertz's device had produced and detected, arguments that made the device temporarily problematic as a generator of air waves. Here, then, we have a case in which a device that was stable as a producer and detector of waves in wires became unstable when applied to waves in air. Explicit, articulated theory in connection with the apparatus was not in any significant sense a factor in the wire experiments, but it was a serious factor in the air experiments. Unarticulated theory, on the other hand, was at work in both cases.

6. Theory and Experiment IV: Other Studies

The above discussion (sections 3–5) of theory and experiment has been limited to electromagnetism and, within that subject, to studies by Smith and Wise, Buchwald, and Hunt. These historians are, of course, far from the only ones who have probed issues concerning the relation of theory with experiment, even in the area of electromagnetism. Several other works particularly distinguish themselves in these respects, by virtue of both content and historiographical stance.

Manuel Doncel and José Romo as well as Friedrich Steinle have directed careful attention, based on manuscript evidence, to a topic that has long in-

trigued historians of physics: namely, how and why Faraday introduced magnetic curves into his understanding of electromagnetic induction. Doncel and
Romo have argued that magnetic curves played essentially no discovery role
at all, but that Faraday eventually used them as a resource in order to capture
salient effects that had escaped his earlier formulations. They thus see Faraday
as driven primarily by the exigencies of experiment to abandon unworkable
laws when he realized that he could reformulate his system in an entirely satisfactory manner by using magnetic curves.[52] Steinle, for his part, takes a
somewhat broader perspective and concentrates as well on the priority fears
that drove Faraday to publication. In Steinle's view Faraday's work shows a
powerful bifurcation between general considerations on causes, which he believed gripped the working Faraday only at certain stages, and what Steinle
calls Faraday's "systematizing approach," which was essentially independent
of particular causal hypotheses. This latter method, according to Steinle, led
Faraday to an understanding of induction based on magnetic curves.[53]

Daniel Siegel, in a book-length essay, has attacked one of the oldest and
most intriguing questions in the history of electrodynamics: How and why
did Maxwell invent the displacement current? The physicist's (conventional)
story has it that Maxwell must have introduced the displacement current so
that all currents could be closed as required by the differential form of Ampère's law. Historians have long known that much more was involved than
that, but there have been as many opinions as accounts of just what did happen. Siegel has taken convention seriously but inverted it. He showed that
Maxwell introduced the displacement current not for reasons of mathematical consistency (to close currents that would otherwise be open) but for reasons of mechanical consistency, to open currents that would otherwise be
closed. To make his point Siegel relied in a novel way on the significance that
the position of a physical quantity within an equation has for considering it to
be a cause or an effect: things that appear on the left-hand side of an equation
were (and are) conventionally thought of as effects produced by the things on
the right-hand side. In his attempt to build a model for the electromagnetic
ether, Siegel noted, Maxwell put the current of conduction, as effect, on the
left and the magnetic field, as cause, on the right. When he introduced what
was later seen as the displacement current, he also put it on the right. This addition was not originally intended as a current but rather as a field-cause
whose purpose was to obliterate the current that the other (magnetic) field-
cause would otherwise produce in regions where currents cannot exist.[54]

52. Romo and Doncel, "Faraday's Initial Mistake."
53. Steinle, "Work, Finish, Publish?"
54. Siegel, "Origin of Displacement Current" and *Innovation in Maxwell's Electromagnetic Theory.*

Siegel concentrated directly on Maxwell, while Doncel and Romo and Steinle attended entirely to Faraday. Though all of these historians placed their subjects in the relevant contexts, each concentrated on an individual scientist rather than on the characteristics of any group to which he may have belonged. There are clear and compelling reasons for doing so in these cases, since both Faraday and Maxwell produced work that itself constituted, as it were, nuclei around which subsequent groups condensed. The same cannot be said, at least not to the same extent, of Thomson in his work on electrodynamics. Nevertheless, one can ask whether there might not be aspects of Faraday's and Maxwell's work that connect in substantive ways to broader characteristics of their milieux—as, for example, Smith and Wise have argued at length for Thomson.

In this respect, Andrew Warwick's work on the Cambridge tripos system casts a revealing light on how pedagogy and its associated culture can affect the kind of science done. Warwick shows, among other things, that the competitive, quasi-athletic training of the aspiring wrangler was coupled tightly to the skills necessary for success in the examination. Generations of rapid-thinking mathematical sportsmen knew just what to do in order to succeed in the Cambridge system, and those among them who became physicists and mathematicians brought their nicely honed skills to bear on contemporary science.[55] Many of Buchwald's Maxwellians, though not Hunt's, were trained in this way, and their love for variational techniques, continuum analysis, boundary-value problems—indeed, for the mathematical underpinnings of Buchwald's Maxwellianism—has its roots here, although, as Warwick explicitly notes, "to what extent the detailed technical content of this training shaped the specific research problems formulated and tackled by subsequent generations of Cambridge physicists remains to be investigated."[56] The very fact that Warwick's Cambridge-trained, mathematical sportsmen did tackle problems in a distinctive fashion speaks to the powerful legacy of their training, and, we think, it can indeed be shown that the specific tools they learned structured their practice at a deep level, having an influence on, among other things, the kinds of objects that they were prepared to countenance in nature. Warwick has in effect provided the first detailed investigation into the manner in which a Maxwellian tradition was produced in a particular locale, namely Cambridge, by coaches and students who drew on mathematical and other resources that were available only there.

Finally, Olivier Darrigol has written extensively in recent years on electrodynamics, concentrating especially on issues raised by Helmholtz's version of it and by problems connected with moving bodies. His work is distinguished

55. Warwick, "Mathematical World."
56. Ibid., 318.

from that of Doncel and Romo, Steinle, and Warwick by its primary concentration on mathematical structure. In addition, Darrigol does not attempt to constitute groups, though he does discuss many different works. His history builds on previous analyses, often providing novel mathematical detail, correcting imprecise or overstated arguments, and probing as far as possible the underpinnings of electrodynamic theory. In this way difficult questions that were previously hard to formulate precisely—such as the sense in which Helmholtz's system can be reduced to Maxwell's in free ether—have been sharpened.[57] There is, however, the ever-present danger of traducing original meaning and structure, which is an inevitable concomitant of the sort of rereading in which Darrigol engages.[58]

7. "Practice versus Theory" Reconsidered

Physicists draw on material as well as conceptual resources in their practice, and these material objects include technological artifacts and processes as well as measuring devices designed explicitly for the laboratory. The nineteenth century is preeminently the age of the steam engine, the iron ship, the telegraph, the dynamo, the electric lamp, and the complex technological systems associated with them. As a result of the increasingly frequent interaction among physicists, emerging engineers, and industrialists, the character of physicists' work during this period broadened enormously, becoming much more diverse than it had been early in the century. For not only was physics applied to industry, but industry was applied to physics.

In this regard, Smith and Wise's *Energy and Empire* argues for a complex, multilayered interaction between Thomson's theories and his industrial practice, the latter taken in a broad sense to include beliefs as well as specific technological concerns. They argue that Thomson's deep involvement with the telegraphic industry led him to consolidate his laboratory at Glasgow University, which he organized around metrological principles and techniques that were pertinent to practical telegraphy (e.g., determining the capacity, current, and resistance of telegraphic cables).

Maxwell, by contrast, has been seen as nearly the prototype of the academic physicist. Yet neither Maxwell nor the Cavendish Laboratory were in-

---

57. Darrigol, "Electromagnetic Revolution."

58. This danger raises the question of what it means to assert that a scientist was mistaken. Neither Warwick, Romo and Doncel, Siegel, nor Steinle writes of their respective subjects that they were in error about a particular claim or calculation. Darrigol and Buchwald, however, do. To justify doing so, the historian's critique should illuminate the point at issue in a historically significant way, should not bring to bear knowledge that the subject could not possibly have possessed at the time, and should argue that the subject could reasonably have been convinced by a contemporary that he was in error.

sulated from the era's material culture, for metrology also linked Maxwell's electromagnetism to the British telegraph industry. Before his appointment as Cavendish director, Maxwell had been closely involved in the work of the Electrical Standards Committee of the British Association for the Advancement of Science (BAAS), and he collaborated with telegraphic engineers such as Fleeming Jenkin. Indeed, his experience with practical telegraphy led him to rethink in field-theoretic terms a puzzling phenomenon known as "residual discharge" that occurs in condensers used in long-distance submarine telegraphy.[59] The committee's work certainly did further British imperial interests by setting a standard resistance, something essential to maintaining and improving telegraphic communication among the British colonies. Schaffer argues that Maxwell was able to continue this "imperial" research in the newly established Cavendish Laboratory at Cambridge University by transferring to it the instrumental resources and know-how of the BAAS Committee. This work flourished even more under the laboratory's second director, Rayleigh, and his student Glazebrook. During Rayleigh's tenure the Cavendish Laboratory became a mecca for exact metrological research into electric standards.[60]

Throughout the nineteenth century, exact measurement occupied a central place in the endeavours of many physicists; as F. K. Richtmyer remarked, "Look after the next decimal place and physical theories will take care of themselves." In 1900, Friedrich Kohlrausch, director of the German Physikalisch-Technische Reichsanstalt (PTR), declared that "without [the measuring of nature], the progress made during the last century in the natural sciences and technology would not have been possible."[61] By the end of the nineteenth century specific metrological research was localized in specialized laboratories such as the National Physical Laboratory (NPL) in Britain, the PTR in Germany, and the National Bureau of Standard in the United States—a displacement that can be seen in the following two examples.

First, Maxwell and Rayleigh's emphasis on metrology was strikingly diminished after J. J. Thomson became the third director of the Cavendish Laboratory in 1885. Measurements involved in the maintenance of the British electrical units and standards were supervised there by Richard Glazebrook, not by Thomson, and this work was eventually transferred to a new institution, the NPL, founded in 1899.[62] Second, when the PTR was established in Germany in 1887, its first director, Helmholtz, aimed to create a genuine interaction between pure physical research and techno-industrial concerns.

---

59. Hong, "Controversy," 247–48.
60. Schaffer, "Late Victorian Metrology."
61. Quoted, respectively, from Badash, "Completeness," 57, and Cahan, *Institute for an Empire*, 129.
62. Kim, "J. J. Thomson."

The PTR's initial division of its structure into "scientific" and "technical" sections reflected this goal. However, as time passed, scientific research was increasingly organized around metrology and related issues, such as measuring instruments and standards, whereas testing for industrial needs became a central concern for the technical section. In 1914, Helmholtz's scheme of parallel scientific and technical sections was given up, and three departments—concerned with optics, heat, and electricity—were created, each with its own scientific and technical sections. Work beyond metrology found its home elsewhere, in physics departments in German universities.[63]

Theory and mathematics provide another link between physics and technology. Stuart Feffer's study of Ernst Abbe nicely illustrates how the physicist Abbe deployed his understanding of spherical aberration and of diffraction to improve the Zeiss company's microscopes and to argue for their uniquely scientific character.[64] Heaviside's career provides another illustration of the rich connections between theory and practice, since many British Maxwellians in the 1880s and 1890s saw Maxwell's field theory as strongly connected to the rapidly changing world of electrical technology.

Ido Yavetz has argued that changes in Heaviside's concerns between 1872 and 1889 show above all a powerful attempt to connect Maxwell's field theory to issues in telegraphic and telephonic engineering.[65] Heaviside began his career as a telegraph engineer, and between 1872 and 1881 he was concerned mostly with linear circuit theory: he attempted to incorporate inductance and leakage resistance, which he was certain profoundly affected signal transmission. Heaviside thereby came to grips with the dynamic nature of electric current, which led him to Maxwell's theory of electromagnetic fields. In 1887, he argued that, contrary to contemporary belief among telegraphic engineers, inductance could benefit transmission. Hunt has suggested that Maxwell's field-theoretic insights were essential for Heaviside's route to this conclusion: on Hunt's account, Heaviside reasoned on Maxwellian grounds that only the simultaneous propagation of electric and magnetic forces could prevent signal distortion, and on this basis sought a condition guaranteeing the simultaneous transmission of electric and magnetic fields.[66] Yavetz, by contrast, argues that "Heaviside was able to reach the distortionless condition precisely because he reduced the discussion to an approximation that enabled him to avoid the complexities of field theory."[67] Yavetz has demonstrated that Heaviside developed a linear circuit theory that enabled him to manipulate electri-

63. Cahan, *Institute for an Empire*.
64. Feffer, "Ernst Abbe."
65. Yavetz, *From Obscurity to Enigma* and "Oliver Heaviside."
66. Hunt, *Maxwellians*, 129–36.
67. Yavetz, *From Obscurity to Enigma*, 213.

cal parameters in a way that had little directly to do with fields. As he worked this golden vein, however, he began to forge, in Yavetz's words, an "interconnected presentation of the theories of circuits and fields."[68] It was precisely because Heaviside understood what was entirely independent of Maxwell's theory that he was able to forge illuminating links with it. In Yavetz's view, Heaviside used his knowledge of circuits—a knowledge that was bound to his own experience of telegraphic issues and problems—to move between the highly abstract realm of field theory and the complex one of practical technology. Heaviside's results spoke to few of his contemporaries in either realm, but those who did listen and understood were not the telegraphic engineers. A fitting audience for his work did not exist until a different kind of electrical engineer began to appear early in the twentieth century.

Both William Thomson and Heaviside encountered substantial opposition, though Thomson, unlike Heaviside, never adopted field concepts. Many working engineers who had accumulated technical knowledge on the basis of skills gleaned from field experience reacted negatively to their claims. Years before Heaviside, Thomson had begun forging a new kind of activity, that of the scientist-engineer, by incorporating mathematical theory and laboratory apparatus into telegraph engineering. His arguments concerning the attenuation of submarine telegraphic signals were strongly resisted by Edward O. W. Whitehouse, a nonmathematical telegrapher who was, however, largely discredited as the result of a series of public disputes with Thomson.[69] Heaviside's arguments concerning the beneficial effect of self-induction were challenged by William Preece, a practical and highly influential telegraphist, who held the position of Electrician of the Post Office before becoming, in 1894, its Engineer-in-Chief. The dispute between Heaviside and Preece grew ever more bitter and complicated, as others, such as the Maxwellian physicists Silvanus Thompson and Lodge, joined Heaviside in the so-called "theory versus practice" debate. Still, by 1890, scientists and engineers had reached consensus that theoretical understanding of Heaviside's sort was essential for the transmission of relatively high-frequency signals such as human speech.[70]

Similar debates occurred in power engineering. Sungook Hong has analyzed a series of disputes in this field between scientist-engineers such as John Ambrose Fleming, who was trained in experimental physics under Maxwell, and mechanically trained practicing engineers such as James Swinburne, who had years of experience working with large machines in the field. The debates between Fleming and Swinburne resembled the earlier ones in telegraphy

68. Ibid., 235.

69. C. Smith and Wise, *Energy and Empire*, 649–83, and Hunt, "Scientists, Engineers, and Wildman Whitehouse."

70. Hunt, "Practice vs. Theory"; Jordan, "Adoption of Self-Induction"; Yavetz, "Oliver Heaviside."

and telephony. Power engineering was largely controlled by practicing engineers who worked with and designed machinery without deploying much of the theory and mathematics that had been developed by physicists. Scientist-engineers, on the other hand, were trained in physics and sought to expand into the world of large electrical machinery. A strange effect that was difficult to explain on the basis of existing theoretical ideas, as well as related technological puzzles (such as the Ferranti effect), provoked a controversy between the groups that erupted into angry dissension. Mechanically trained power engineers, masters of their craft and of machine design, clashed with scientist-engineers, disciples of abstract, mathematical theory. Fleming was able ingeniously to manipulate his scientific resources—including calibration techniques, precision laboratory measurements, mathematical theory, controlled field experiments, and even the authority of Thomson (by then Lord Kelvin)—to end the controversy in his favor. During these debates, experimental physics entered into the practical regime of machinery. When the controversy closed, the balance of power had also shifted, creating thereby new social and professional roles for scientist-engineers like Fleming.[71]

8. Concluding Remarks

We have not sought to cover most, or even a large proportion, of the considerable terrain of the history and historiography of nineteenth-century physics. The most significant attempt to provide a comprehensive account of physics as practiced in a given region is Jungnickel and McCormmach's *Intellectual Mastery of Nature: Theoretical Physics from Ohm to Einstein*. In their two volumes the authors survey physics as practiced in German-speaking lands throughout the nineteenth century. They concentrate on the work of the physicist and build their account around its integration into an institutional setting, specifically, the German physics institute. They follow for a while a physicist or a group of physicists in a particular institution and then move elsewhere, only to return to those individuals years later in what are often transformed settings. They show in often intimate detail how specific institutional frameworks molded the concerns of physicists, developing as well the distinction between theoretical and experimental physics, one that strikingly affected the course of physics in Germany. Here, too, we are introduced to the peculiarly German "physics institute," which evolved into loci for experimental or theoretical research, each dominated by a single, commanding professor. The authors thrice provide overviews of what can be found in the *Annalen der Physik,* the major German physics journal of the day, thereby identifying

---

71. Hong, "Efficiency and Authority" and "Forging Scientific Electrical Engineering."

contemporary research issues considered to be of interest to the discipline. The scope of this history, whose title was drawn from a remark late in life by Helmholtz, and its intricate melding of theory, experiment, and working environment combine to produce something very like a biography of the discipline, an account of the birth and development of physics in Germany as an intellectual and practical activity carried out in a specific setting. Nothing similar to this work exists for the physics of any other region.[72]

Given their comprehensive approach, Jungnickel and McCormmach did not delve very deeply into the details of physical theory and experiment, and consequently did not provide intimate accounts of the many subjects that have, in part, been treated by historians of nineteenth-century physics. Nor have we touched on all, or even a majority, of these here. Among the subjects that we have not examined, perhaps the most glaring are thermodynamics and statistical physics (though Smith and Wise do discuss at some length Thomson's involvement in the creation of thermodynamics, as well as the emergence of energy as a unifying category). For heat science before thermodynamics, Fox's *Caloric Theory of Gases* remains unsurpassed in its thorough, detailed analysis of both theory and experiment; any future work must certainly begin with it. Though he has not published a book on the topic, Philip Lervig has provided deep insight into the mathematical and physical character of Sadi Carnot's work, demonstrating the power it derived from the assumption that a body's heat content is determined by its state. Clifford Truesdell's *Tragicomical History of Thermodynamics* provides analyses of many subjects, from caloric theory through thermodynamics, and offers as well critiques of foundations (in particular of Carnot's introduction of reversible processes). Donald Cardwell's *From Watt to Clausius* remains the only comprehensive treatment of thermodynamics in all countries. The energy concept itself has been discussed in numerous articles over the years, but few books have been devoted to the subject. Four notable exceptions—two older, two recent—are Erwin Hiebert's *Historical Roots of the Principle of Conservation of Energy*, Yehuda Elkana's *Discovery of the Conservation of Energy*, Kenneth Caneva's biography of Robert Mayer, and Crosbie Smith's *Science of Energy*. Hiebert reaches back to roots in several areas, while Elkana concentrates on Helmholtz; Caneva seeks a specific, and highly local, meaning of force conservation for Mayer, while Smith is concerned solely with Britain, where he sees energy as inflecting every aspect of physical practice. We lack a comprehensive treatment that takes into account these and other works over the last forty years. Diana Barkan's *Walther Nernst and the Transition to Modern Physical Science* provides a careful, detailed account of the working of thermodynam-

72. Jungnickel and McCommarch, *Intellectual Mastery of Nature*.

ics in chemistry; we lack something similar for physics (though there is of course substantial practical overlap between the two disciplines here).[73]

Statistical physics, like thermodynamics, has not received a comprehensive treatment in recent years. The fullest account of many aspects of the subject remains Stephen Brush's collection, *The Kind of Motion We Call Heat*, to which one must add *Maxwell on Molecules and Gases* by Elizabeth Garber, Brush, and Francis Everitt.[74] These authors distinguish the various aspects of kinetic from statistical physics, emphasizing in the case of Maxwell his signal creation of an entirely novel scheme—now called transport theory—in which such effects as viscosity are linked to the carriage of (in this case) momentum and (in other cases) energy by molecules from one region to another. For the most part kinetic theory did not produce much in the way of experimental novelty throughout the nineteenth century, and statistical mechanics in Boltzmann's form remained comparatively marginal to most practicing physicists until the early twentieth century. Kinetic and statistical physics were occasionally used when it seemed there was no alternative (such as in Maxwell's counterintuitive deduction that a gas's viscosity is substantially independent of its density), but for the most part these subjects did not (during the nineteenth century) generate an active regime of problem-solving practice comparable to electromagnetics, optics, or thermodynamics itself. Only after J. Willard Gibbs and Albert Einstein produced statistical mechanics in essentially the modern form—with Einstein (unlike Gibbs) having been directly concerned with achieving a general unity among the various subjects that by 1905 employed microphysical reasoning—did statistics become widely used.[75]

For most of the nineteenth century, and throughout western and central Europe, detailed concern with the microworld lay at the margins of most mathematical, theoretical, and laboratory practice.[76] Although few physicists doubted the existence of atoms, fewer still were willing to ground their work on the putative properties of the microworld. Microentities, that is, were usually thought of not as things that should be used in the first instance to construct theories but, at best, as things that are useful for reaching results known to be correct on other grounds. Certainly atoms and molecules did figure substantially in much thought about the underpinnings of matter, but the worlds of the laboratory and of calculation had little to do with such things until quite late in the century. Furthermore, nineteenth-century conceptions of atoms and molecules differed significantly from those held later, even among

73. R. Fox, *Caloric Theory;* Lervig, "On the Structure"; Truesdell, *Tragicomical History;* Cardwell, *From Watt to Clausius;* Hiebert, *Historical Roots;* Elkana, *Discovery;* Caneva, *Robert Mayer;* C. Smith, *Science of Energy;* Barkan, *Walther Nernst. See* also Kragh, "Between Physics and Chemistry."

74. Brush, *Kind of Motion;* Garber, Brush, and Everitt, *Maxwell on Molecules.*

75. Klein, *Ehrenfest,* esp. chap. 6; Renn, "Einstein's Controversy"; Navarro, "Gibbs, Einstein."

76. Buchwald, "How the Ether."

those figures who were prepared to engage in speculation. For example, many found compelling the image of an atom as a vortex ring in the ether, which made the atom a structure in and of that medium and therefore parasitic upon its properties. Atoms were not often taken to be things that the world was built out of in a fundamental sense; rather, they were themselves built out of the world (ether). This inversion, which remained common across national boundaries until the 1890s, directed attention away from microentities as basic building blocks of physical systems.

Little historical attention has been directed to the profound transition in practice that the emergence of the experimental microworld had early in the twentieth century. Many informative and well-researched books and articles have been written on developments such as black-body theory, investigations of radioactivity and of X rays, and the explosive growth in interest in the structure and behavior of the electron.[77] Yet no one to date has taken the full measure of the new physics of the early twentieth century, a physics that based practice on the properties of a world that many in the nineteenth century thought to lie, if anywhere, forever at the outermost limits of legitimate speculation.[78] Not many years before J. J. Thomson measured the charge-to-mass ratio of his new corpuscle in 1897, this microworld had figured hardly at all in his daily practice. Between the late 1890s and early 1910s experimental work that was explicitly thought to connect directly to the microworld, and theoretical work that indubitably did so, exploded: cathode rays, X rays, and radioactivity all became areas of feverish research during these years; the latter two subjects did not exist at all before the mid-1890s. Recent work by Isobel Falconer and others carefully discusses the history of cathode-ray investigations, while Bruce Wheaton's *Tiger and the Shark* remains the most careful and informative discussion of attempts to understand X rays and their implications.[79] Radioactivity has of course an extensive literature, but we feel that it lacks a new account concentrating on the several locales, with their specific resources, goals, and strictures, in which the subject emerged and rapidly became part of the increasingly-practical microworld.

These are hardly the only areas that have not been thoroughly examined by historians. One can cite, as we mentioned above, the development of equilibrium thermodynamics in the hands of Maxwell, Helmholtz, and Gibbs; the production of continuum mechanics by Stokes, Helmholtz, and Gustav Kirchhoff, among others[80] (including the novel concept of characteristic di-

---

77. Trenn, *Self-Splitting Atom*; Kuhn, *Black-Body Theory*; and Wheaton, *The Tiger and the Shark*.
78. See, however, Buchwald, "How the Ether."
79. Falconer, "Corpuscles to Electrons"; Wheaton, *Tiger and the Shark*.
80. Darrigol, "Between Hydrodynamics and Elasticity Theory" and "Turbulence," are the first recent discussions of nineteenth-century hydrodynamics that goes considerably beyond the older

mensionless numbers by Ernst Mach and Osborne Reynolds); Rayleigh's and Helmholtz's work on mathematical acoustics, including related problems (such as scalar diffraction theory) in optics; the efflorescence of novel optical devices connected to wave techniques, which led to considerable changes in optical design and theoretical practice; and many other subjects. Few have received careful attention from historians; all deserve it.

In addition to these and other specific subjects, the broader, historiographical issues that have emerged in recent writing also merit more careful thought than they have perhaps received. Two in particular strike us as particularly important. The first concerns the ways in which historians interpret their subjects' use of metaphor, which seems to us often based primarily on homologies and not on consequential relationships in the historical material proper. We remain skeptical that metaphorical transfer does function in the manner that is often asserted,[81] but, if it does, then we think historians should specify criteria of adequacy for supporting data and should also develop a clear conception of how transfer works. It is not sufficient to point out analogies, no matter how compelling they may seem to be, in the absence of clear criteria.

The second issue concerns methods for identifying the character of scientific practice—or even for deciding whether a practice exists at all. We discussed above several of the issues raised by the related problem of group identity, but here we want to point out that it is not in general possible to extract an essence of, say, Maxwellianism or ray optics that exists independently of physicists' living practice. It may be, and we think it generally is, possible to identify specific tenets, procedures, principles, devices, and so on that a particular group of physicists hold in common, and it may be further possible to show that they think of themselves as engaged in a specific activity that differentiates them from others. Moreover, we may, as Warwick has for Cambridge Maxwellians, be able to trace the construction of the research community in its specific context. But it would be otiose to extract, as it were, minimal re-

---

literature, providing a thorough discussion of the several derivations of the Navier-Stokes equation, including its relation to eighteenth-century experimental work. Darrigol also argues that the opposition perceived today between, e.g., Poisson's thoroughgoing molecularism and Fourier's diffusion equation, does not properly represent the tenor of the times, since effectively everyone, in France at least, subscribed to molecular foundationalism. This is no doubt correct in the limited sense that no one thought the world to be continuous, but the mathematicians and physicists of the day certainly did perceive fundamental differences between analyses that thoroughly employed molecules (e.g., Poisson) and those that barely used them beyond rudimentary justification (e.g., Fourier). As Darrigol shows, there were hybrids (such as Navier), and eventually people (like Cauchy) who worked both veins. (For molecular work in optics by Cauchy and others see Buchwald, "Optics and the Theory.")

81. C. Smith and Wise, *Energy and Empire,* provides several examples of metaphorical argument of the sort we have in mind; see, e.g., p. 357.

quirements for group membership that transcend what the members themselves believed. Though it might be the case that member x's use of principle y was incompatible with other aspects of the group practice that x engaged in (either logically or in some other way), unless group members became explicitly aware of the incompatibility it would be historically useless to argue that x had thereby ceased to engage in group practice. One can certainly point out that continued work in this vein would have been likely to lead to strains within the group, and even within x's own practice, and it might be possible to show that this actually did occur. But unless something of the sort happened, prevailing practice would have been substantially unaffected. For example, many optical practitioners in the 1820s were perfectly able to assimilate the principle of interference to practices that remained tied to the reality of light rays and not wave surfaces.[82]

Finally, a related issue concerns transformations in the contents of a given practice. It is quite possible for entire regimes to acquire utterly different forms without there being, initially at least, any discernible impact on empirical consequences. Thomson, for example, early on demonstrated that Faraday's space-based field imagery for electrostatics can be expressed in the mathematical language familiar to French analysts of the day. Nevertheless, we note that when such a thing happens the very different calculational practices that each group deploys are quite likely to lead over time (possibly as a result of novel experimental work) to elaborations of the original systems that breach the compatibility. This occurred in early nineteenth-century optics, and in electromagnetics as well. The historian must pay careful attention to the details of practice: to how problems are set up, what kinds of calculations are permissible, what mathematical tools are used, and what canonical instruments and devices are deployed in the laboratory—in short, to the specific set of behaviors and objects that constitutes a group's practice. Only in this way is it possible to understand when, and in what manner, a change has taken place. As Newton said of natural philosophy, and as one of the foremost historians of physics has always told his students, there is "no truth except in the details."[83]

---

82. Kipnis, *Interference of Light*.
83. Kox and Siegel, *No Truth*, referring to Martin Klein.

Seven

CHEMISTRY

BERNADETTE BENSAUDE-VINCENT

❦

1. The Presentist Perspective and Its Problems

NINETEENTH-CENTURY CHEMISTRY has been aptly described as a modern experimental science whose theoretical structures and disciplinary organization were gradually established in the academic and the industrial worlds. Most historians have aimed—and some still aim—at reporting the results and achievements of nineteenth-century chemists and the concomitant emergence and maturation of institutions that shaped and, in some cases, continue to shape the discipline's landscape. Aaron Ihde's classic and still useful volume, *The Development of Modern Chemistry,* well illustrates this presentist and standard approach to the history of chemistry.[1] The presentist perspective has prevailed to such an extent that the standard periodization of nineteenth-century chemistry often reflects the contemporary divisions of the discipline into inorganic, organic, and physical chemistry. It is generally assumed, in accord with this perspective, that Antoine Lavoisier's chemical revolution at the end of the eighteenth century and John Dalton's atomic hypothesis of the early nineteenth century laid the foundations of modern inorganic chemistry. Together, these events have been taken to constitute the first period of modern chemistry. The second period, according to this account, is characterized by the emergence of structural organic chemistry in the mid-nineteenth century. There then followed a third period, starting from the 1870s, devoted to ionic theories and the emergence of physical chemistry. Not surprisingly, this tripartite historiographical periodization fits remarkably well with the pedagogical order of modern courses of chemistry.

From the presentist perspective, moreover, the logical result of the advancement of academic chemistry is industrial chemistry. Lutz F. Haber's fundamental book, *The Chemical Industry during the Nineteenth Century: A Study of the Economic Aspect of Applied Chemistry in Europe and North America,* illustrates this point in its very title: it presupposes a connection between the advancement of science and technological innovation. From this assumption there

---

1. Ihde, *Development of Modern Chemistry.* See also Partington, *History of Chemistry,* vol. 4, and Hudson, *History of Chemistry.*

follows a "natural" periodization: a first section is devoted to the development of heavy chemistry (e.g., sulfuric acid and soda) in the late eighteenth and early nineteenth centuries; the second half of the century is characterized by the emergence of the fine chemical industry (synthetic dyes and pharmaceuticals), based on the structural formulae of organic chemistry; and a third and final section is devoted to describing Fritz Haber's invention of his eponymous process for the direct production of ammonia.

This traditional picture of the smooth and unproblematic development of a positive science was originally shaped not by historians but by chemists themselves. From Thomas Thomson in the early nineteenth century to Ihde in the twentieth, a number of chemists have acted as historians of their discipline: Hermann Kopp, Adolphe Wurtz, Albert Ladenburg, Marcellin Berthelot, Edward Thorpe, Pierre Duhem, Ida Freund, Wilhelm Ostwald, and J. R. Partington—to name only the most prominent—wrote narratives that themselves became sources for later historians. These chemist-historians were active scientists, confident in the success of their discipline, sometimes involved in scientific controversies, and often enough deeply influenced by powerful national feelings.[2]

The key feature of the presentist approach to nineteenth-century chemistry is to organize developments around a few epochal discoveries: Lavoisier's chemical revolution (1789), Dalton's atomic hypothesis (1808), Amedeo Avogadro's law (1811), Friedrich Wöhler's synthesis of urea (1828), August Kekulé's discovery of the benzene ring (1865), Dmitri Mendeleev's discovery of the periodic system (1869), Achille Le Bel and Jacobus van't Hoff's publications on the tetrahedron of carbon (1874), Svante Arrhenius's laws of electrolytic dissociation (1884), and the like. Having thus selected a number of landmarks and path-breaking discoveries, the standard historical narratives inevitably turn to a central issue: the delay in adopting innovations that appear obvious to those familiar with modern chemistry. As John Brooke has pointed out, it is a striking feature of nineteenth-century chemistry that many of these innovations first met with indifference, if not hostility, when they were initially promulgated.[3] Modern textbooks have invariably portrayed Lavoisier, Dalton, and Avogadro, for example, as heroes of modern chemistry; it is hard to believe, then, that nineteenth-century chemists, having already at hand the basic notions of element, atom, and molecule, somehow failed to recognize the atomic and molecular structure of chemical compounds. How, for instance, are we to understand that Avogadro's law, discov-

2. On the nineteenth-century chemist-historians, see Russell, "Rude and Disgraceful Beginnings"; Laudan, "Histories of the Sciences"; and Rocke, "Between Two Stools" and "History of Science."

3. Brooke, "Chemists in Their Contexts," 14.

ered as early as 1811, was rejected until midcentury, and in some cases until the
end of the century, by a number of prominent chemists? How could able
chemists, who contributed so much to the advancement of their science
through their experimental work and publications, have been so narrow-
minded as to ignore the developments of molecular physics, in particular
Rudolph Clausius's work on heat and James Clerk Maxwell's kinetic theory of
gases? Answer: strong obstacles prevented scientists from recognizing the
truth. The historiography of nineteenth-century chemistry has thus focused
on one major target: identifying the prejudices or obstacles that delayed the
acceptance of novelties.

However distorting this presentist approach may be, it has offered some
manifest advantages that may—should such an approach now be abandoned—turn into major difficulties for the future history of chemistry. The
focus on key events has encouraged the useful tradition of publishing sourcebooks or excerpts from past chemistry texts in general histories.[4] Moreover,
the clear periodization provides invaluable pedagogical resources, encourag-
ing the development of historical perspectives in scientific training. The periodization has also been useful to working chemists who, like all scientists,
fondly celebrate the founding heroes of their discipline. And finally, the presentist approach teaches (or reminds) professional historians of nineteenth-
century chemistry that the present is not itself a fixed line, but rather a
ever-moving frontier.

On the other hand, the traditional, presentist approach to nineteenth-century chemistry is more than a typical whiggish attitude of working scientists.
Professional historians are sometimes all too ready to adopt uncritically the
historiographical categories forged by chemists. Because the nineteenth century opened "in the shadow of Lavoisier," with the specter of a fundamental
paradigm shift from phlogiston theory to oxygen theory lurking everywhere,
professional historians have extensively used the notion of *conversion* introduced by Lavoisier's disciples in the midst of the controversy raised by the
new language of chemistry. This notion, which was adopted as if it were a
neutral historiographical category, has shaped the description of later developments. The whole historiography of nineteenth-century chemistry became centered on conversions to scientific change. Far from being eliminated
by the emergence of professional history of science in the 1960s, the issue of
delay became a prime concern among historians and philosophers of science
seeking to identify the intellectual or social factors at play in the reception
of scientific innovation. In accord with the Bachelardian epistemology that
prevailed in France, the main task of historians was to identify the "epistemo-

4. E.g., Jagniaux, *Histoire de la chimie;* Freund, *Study of Chemical Combination;* Benfey, *Classics in
the Theory of Chemical Combination* and *From Vital Force to Structural Formulas.*

logical obstacles" that prevented the recognition of scientific truths.[5] Further-more, antipositivistic historiography did not, strangely enough, deeply affect the search for obstacles. When the Kuhnian framework became fashionable among historians, they continued to concentrate their attention on the issue of the reception of revolutionary breakthroughs, seeking to describe the re-sponses of various scientific communities to innovations and to identify the factors that prompted or delayed their adoption. Whether the resistance to novelties was due to individual psychological prejudices (related perhaps to the training or to the age of a given chemist), to institutional strategies occa-sioned by chemists' position in the hierarchy of the chemical establishment, or to more or less loose connections with the center of power due to their ge-ographical location at the so-called periphery—all this was a matter of debate that fueled rival models of scientific change. By contrast, there was no debate about why Lavoisier should have proved victorious over Priestley. As long as historians of chemistry assumed that the phlogiston theory had to be dis-missed and that only resistance to the new paradigm had to be explained, their main task was to identify obstacles to the adoption of what is presently re-garded as (self-evident) scientific truth.

As this essay seeks to show, one major result of the historiographical stud-ies of nineteenth-century chemistry that have developed during the past two decades is the revision of the entire issue of the reception and diffusion of sci-entific innovations.

William Brock's review essay of 1983 on the historiography of chemistry directed attention to the topics that attracted much scholarship during the 1960s and 1970s (viz., the history of electrochemistry, the atomic-molecular theory, and the development of organic chemistry) and those that were ne-glected (viz., the history of physical chemistry). He also provided a bibliogra-phy, which he extended and updated ten years later in his *Norton History of Chemistry*.[6] His work, as well as Colin Russell's review *Recent Developments in the History of Chemistry* (1985), provides rich and useful bibliographical data. Hence, the present essay restricts itself to identifying the current historio-graphical trends in the study of nineteenth-century chemistry. One major aim is to emphasize the revisions of the alleged founding events of modern chemistry. More particularly, this essay is concerned with the historiographi-cal functions of biography (section 2), the status of the chemical revolution (section 3), chemical atomism (section 4), and the legends of chemistry (sec-tion 5). Following this review of several critical aspects of recent historiogra-phy, the essay addresses a number of topics and interests that have emerged

5. Bachelard, *Nouvel esprit scientifique* and *Formation de l'esprit scientifique.*

6. Brock, "History of Chemistry" and *Fontana History of Chemistry*; see also Russell, "Rude and Disgraceful Beginnings."

during the past decade. It draws attention to the concern with disciplinary identity and boundaries (section 6). It emphasizes the studies of research schools (section 7), the formation of professional chemists (including the teaching of chemistry and the role of textbooks) and of learned societies (section 8), and debates about the relations between industrial and academic chemistry (section 9). Finally, it tries to identify neglected aspects of nineteenth-century chemistry that warrant more scholarly attention in the future. Among the many potential fields to be explored, three desiderata stand out: the study of instrumentation and laboratories (section 10), the philosophy of chemistry (section 11), and the future audiences of the history of chemistry (section 12).

2. The Biographical Approach: A Safe Value

Perhaps the easiest way to reconcile the chemists' inclination for founding heroes and the historians' concerns for a contextualized and critical account is through the biographical approach to the history of chemistry. While the biographical genre is anything but novel, there are abundant and highly useful biographies of nineteenth-century chemists that contextualize their life and works. We now have readable biographies of such central figures as Claude-Louis Berthollet, thanks to Michelle Sadoun-Goupil, and of Humphry Davy, thanks to David Knight's illuminating and entertaining volume. Similarly, we have an intellectual biography of Jans Jacob Berzelius by Evan Melhado, perspectives on Michael Faraday by L. Pearce Williams and Geoffrey Cantor, and descriptive biographies of Jean-Baptiste Dumas by Marcel Chaigneau and of Jean-Baptiste Boussingault by F. W. J. McCosh. Moreover, there are recent biographies of Justus von Liebig by William Brock and of Eilhard Mitscherlich by Hans-Werner Schütt, a brief but rich essay on August Wilhelm Hofmann by Christoph Meinel, a polemical look at the French mandarin Marcellin Berthelot by Jean Jacques, and detailed but accessible biographies of Edward Frankland by Colin Russell and of Svante Arrhenius by Elisabeth Crawford. Alan Rocke opens a window on French nineteenth-century chemistry through a portrayal of Charles Adolphe Wurtz. Pierre Duhem has recently been the subject of two biographies: one in English by Stanley Jaki and one in French by Paul Brouzeng. There are now also studies of the fascinating lives of the Curies: of Pierre by Anna Hurwic and of Marie by Robert Reid, Susan Quinn, and Helena Pycior. And finally, there is a voluminous biography of Fritz Haber by Dietrich Stoltzenberg.[7] However impressive this long (and ob-

7. Sadoun-Goupil, *Le chimiste C. L. Berthollet;* Knight, *Davy;* Melhado, *Berzelius;* Williams, *Faraday;* Cantor, *Faraday;* Chaigneau, *Dumas;* McCosh, *Boussingault;* Brock, *Liebig;* Schütt, *Mitscherlich;* Meinel, "Hofmann"; Jacques, *Berthelot;* Russell, *Frankland;* Crawford, *Arrhenius;* Rocke, *Nationaliz-*

viously incomplete) list may be, much biographical work remains to be done if we are to understand nineteenth-century chemistry as a human and social activity. We await the courageous scholar who will provide, for example, a full biography of Dmitri Mendeleev or of Wilhelm Ostwald, one that convincingly unites the multifarious aspects of their variegated works and careers.

When it is not hagiographic, the biographical genre can offer invaluable resources to historians. Crawford's biography of Arrhenius, for instance, describes the life of a scientist immersed in the context of northern Europe during the late nineteenth and early twentieth centuries. Thanks to a fine-structure analysis of Arrhenius's work on electrolytic dissociation and immunochemistry, Crawford has reconstructed in some detail his investigative pathway while at the same time (through careful definition of terms) making his work accessible to the lay public. Her volume also provides an original view of the emergence of physical chemistry and the group of "Ionists," showing especially the close interaction between van't Hoff's memoirs on the law of chemical equilibrium and Arrhenius's theory. The founding myth of physical chemistry—that the theory of the Ionists first met with hostility in the scientific community and generated a great controversy—is countered by evidence that, instead, a handful of opponents faced a great mobilization by Ostwald and an "army of Ionists" who, through an active campaign, worked for the wide circulation and rapid acceptance of the notion of permanent ions.

The emphasis on local context and its influence on scientific styles is one of the most powerful features of biographies. Arrhenius, for example, decided to make a career in his native country, Sweden, even though he had an international reputation. This choice enabled him to define a style of his own and to create his own niche. He continuously changed his research field, moving from physical chemistry to cosmic physics and then to the study of climate changes. In the 1890s, he sought to apply physical chemistry to the study of toxin-antitoxin reactions and eventually positioned himself as the founder of "immunochemistry." Arrhenius thus acted as a sort of hybridizer of academic disciplines, educating himself about new fields of research through informal communication: conversations with new friends and travels to learn experimental skills at laboratories specializing in areas he wanted to investigate became regular features of his career. Arrhenius became involved in many controversies concerning such a wide range of disciplines; Crawford's biography thus covers a significant portion of the spectrum of the history of science at the turn of the twentieth century.

As much as we continue to need biographies of the leading figures of

---

ing Science; Jaki, *Uneasy Genius;* Brouzeng, *Duhem;* Hurwic, *Pierre Curie;* Reid, *Marie Curie;* Quinn, *Marie Curie;* Pycior, "Marie Curie's Anti-Natural Path" and "Pierre Curie and 'His Eminent Collaborator' "; Stoltzenberg, *Haber.*

nineteenth-century chemistry, scholars (and their publishers) also need to give more attention to lesser-known (if not obscure) figures. A promising start in this direction has been made by several younger scholars who have devoted their doctoral research to the study of "ordinary chemists." For instance, Agusti Nieto-Galàn's dissertation on the little known entrepreneurial chemist Francesc Carbonell i Bravo has opened a window on the development of chemistry in early nineteenth-century Catalonia.[8] Similarly, Brigitte van Tiggelen's thesis on Karel van Bochaute, a chemist from Louvain who published an unsuccessful system of nomenclature the same year as Lavoisier, provides an original perspective on the cultures of chemistry in the late eighteenth century.[9] Biographies of "obscure" figures should not, of course, be confined to doctoral dissertations. Moreover, prosopographical work needs to be done, since only by collecting biographical and related data on lesser-known chemists, many of whom contributed much to shaping chemistry through their teaching and research activities, can we expect to understand the processes that constituted the professionalization of chemistry.

3. Reappraisals of the Chemical Revolution

Lavoisier's image as the founder of modern chemistry was a powerful stimulus for celebrating the bicentennial of his death (1994). As might be expected, several biographies were published for this occasion and several conferences held.[10] In fact, this commemoration demonstrated both how much scholarly attention Lavoisier has attracted over the past decade and the lack of consensus among scholars. Controversies between Lavoisier scholars ruled over several fundamental issues, not the least of which was: Just what was the chemical revolution about? A number of scholars challenged the traditional view that the overthrow of phlogiston was the hard core of Lavoisier's revolution. While alternatives views have been proposed—positing, for example, the theory of composition or the theory of gases as the core issue—the question gradually became: Why should there be a hard core instead of a gradual process, a "conceptual passage" involving a wide range of chemical problems?[11]

A second issue concerned the role of experimental physics in Lavoisier's achievements. A number of scholars, following Henry Guerlac's suggestion,

---

8. Nieto-Galàn, "Ciència a Catalunya."

9. Van Tiggelen, "Un chimiste des Pays-Bas."

10. Poirier, *Lavoisier;* Donovan, *Lavoisier;* Bensaude-Vincent, *Lavoisier;* Demeulenaere-Douyère, *Il y a deux cents ans.* See also the review essay by Melhado, "Scientific Biography and Scientific Revolution."

11. See the papers by Gough, Siegfried, Perrin, and Holmes, in Donovan, *Chemical Revolution;* and see Holmes, "Boundaries of Lavoisier's Chemical Revolution."

argued that experimental physics inspired Lavoisier's quantitative method; others objected that disciplinary boundaries are not an appropriate reference for understanding the process of the chemical revolution.[12] Was the chemical revolution a revolution in chemistry or into chemistry? This controversial question in turn led to an examination of whether chemistry was an autonomous discipline before the chemical revolution. Important new approaches to eighteenth-century chemistry have come into play here. Whereas Enlightenment chemistry has traditionally been described in backward fashion, that is, as if chemists were awaiting Lavoisier, the savior who would discard erroneous doctrines and obscure practices, Frederic L. Holmes has portrayed prerevolutionary chemistry as a "maturing scientific discipline" within a thriving investigative tradition.[13] This reinterpretation has even led to debate over the extent of Lavoisier's originality.[14]

The Lavoisier-oriented historiography of the chemical revolution has also been revised in light of more thorough investigations of the works of a number of contemporary European chemists, including Lavoisier's French collaborators.[15] Once historians began paying more attention to the variety of local cultures of chemistry, for example in Edinburgh, Uppsala, Berlin, Paris, Bologna, and Coimbra, the common view of conversion to the oxygen theory began to be called into question.[16] Thanks to an active and decentralized network of chemists exchanging views, texts, and techniques, the process of diffusion of the reformed chemical language went fast. Still, the adoption of the new language of chemistry cannot accurately be described in terms of "conversions" or "resistance" to the new paradigm because adopting the new language did not necessarily imply a full adoption of Lavoisier's theory. Chemistry's language was remarkably standardized through a wide variety of specific responses to the French nomenclature. At this point it has become manifest that, because an asymmetry was implicitly assumed between the French antiphlogistonist chemistry and the phlogistonist side, the traditional perspective of the reception of Lavoisier's theory has distorted our view of this crucial episode. When the symmetry between the winners and the losers in this controversy is restored, scholars begin to realize that instead of passively adopting or rejecting the French innovation, contemporary chemists

12. Guerlac, "Chemistry as a Branch of Physics"; Melhado "Chemistry, Physics"; Perrin, "Chemistry as a Peer of Physics"; Donovan, "Lavoisier as Chemist"; Melhado, "On the Historiography of Science."

13. Holmes, *Eighteenth-Century Chemistry.*

14. Donovan, *Chemical Revolution;* Beretta, *Enlightenment of Matter.*

15. Goupil, *Lavoisier et la révolution chimique;* Grison, Goupil, and Bret, *Scientific Correspondence;* Hahn, "Lavoiser et ses collaborateurs."

16. See, for instance, Donovan, *Philosophical Chemistry;* Abbri, "Tradizione chimiche"; and Allchin, "Phlogiston after Oxygen."

across Europe discussed the new language, tried to improve it, or adapted it to local contexts. Within this process of negotiation there existed political, religious, and national interests that interfered with scientific and local traditions.[17] There is much exciting work to be done here.

4. Chemical Atomism Revisited

The rejection of Avogadro's law for about fifty years has become a classical topos for illustrating the opposition between presentist and historicist approaches. The alleged blindness of nineteenth-century chemists appears as a perfectly intelligible attitude once one considers that Avogadro's hypothesis of molecules formed of two atoms of the same element stood in contradiction to the theoretical framework of chemical atomism. In Dalton's theory, such diatomic molecules were physically impossible because of the repulsion between the atmospheres of heat of two identical atoms; in Berzelius's electrochemical paradigm, which soon became dominant in Europe, diatomic molecules were equally impossible because of the repulsion between two identical electrical charges.[18]

Increased attention to the theoretical and experimental contexts of atomic conceptions has led to a clear distinction between chemical atomism and structural theories of matter. Since Leonard Nash's and Arnold Thackray's classic studies of the origins of Dalton's atomic theory, it has been understood that Dalton's main concern was to derive weights from chemical data.[19] Moreover, Seymour Mauskopf's emphasis on the crystallographic origin of the early concept of molecule has helped identify the various sources of chemical atomism.[20] More recently still, investigations of the origins of Avogadro's and Ampère's laws have shown that a remarkable variety of backgrounds led to early molecular theories: while Avogadro's conceptions emerged from his interest in Berthollet's affinity theory, Ampère was mainly concerned with geometrical crystallographic considerations.[21]

During the past decade the reception, as opposed to the origins, of atomic and molecular theories has also attracted the attention of several scholars. Thanks to Alan J. Rocke's two book-length studies, the entire issue of the "conversion" to atomism is now seen in a different light.[22] First, the famous controversy between equivalentism and atomism can no longer be described

17. Bensaude-Vincent and Abbri, *Lavoisier in European Context.*

18. Brooke, "Avogadro's Hypothesis"; Fischer, "Avogadro."

19. Nash, "Origin of Dalton's Atomic Theory"; Thackray, "Origins of Dalton's Chemical Atomic Theory" and *Dalton.*

20. Mauskopf, "Crystals and Compounds."

21. Ciardi, *L'atomo fantasma;* Scheidecker and Locqueneux, "La théorie mathématique de la combinaison chimique"; Scheidecker, "L'hypothèse d'Avogadro et d'Ampère" and "Baudrimont."

22. Rocke, *Chemical Atomism* and *Quiet Revolution.*

simply as a fight between Baconians and speculative chemists. Despite the opinion spread by equivalentists—that equivalent weights, in contrast to atomic weights, were purely matters of fact or empirical data free of theory—we now know that the tables of equivalents were as loaded with hypotheses and conventions as were the tables of atomic weights. The very title of Rocke's second study, *The Quiet Revolution: Hermann Kolbe and the Science of Organic Chemistry*, symbolizes the new historiographical direction. Instead of focusing on paradigm shifts and solemn moments of conversion, instead of listing obstacles to the adoption of novelties, Rocke identifies precisely which positive reasons brought about changed attitudes toward atomism. He points to diverse but converging data and circumstances that gradually led to the adoption of Avogadro's law during the 1850s. By thus emphasizing the quiet process of theory change, Rocke is able to reinterpret the historical meaning of the Karlsruhe Conference (1860) as more a symbolic event than as an effective instrument for conversion.

Such revisions of our understanding of nineteenth-century chemical atomism are, interestingly enough, due neither to the discovery of new texts—as was, for instance, the case for our understanding of the role of alchemy in the scientific revolution—nor to new perspectives developed by the social studies of sciences. The reinterpretation of chemical atomism has emerged mainly from efforts to contextualize the discoveries within the framework of contemporary chemistry, to restore the original understanding of discoveries that were subsequently celebrated as founding events. Although the resources of intellectual history à la Hélène Metzger's work—namely, the view that historians should try to think as contemporaries of the subject under examination thought[23]—have proven sufficient for revising the dominant presentist viewpoint about chemical atomism, we now need more refined analyses of the debates about atomism. Only by paying attention to the local settings that helped shape arguments in the controversies between atomism and equivalentism, for example, can we expect to understand the theoretical orientations of nineteenth-century chemistry. Mi Gyung Kim's recent studies of Dumas's linguistic practices and the changing attitudes within the institutional context of the Paris Academy of Science exemplify the promises of such a multifaceted cultural approach.[24]

5. Legends in Chemistry

In a relentless quest to grasp the meaning of discoveries within their original context, John Brooke has dismantled one of the most famous legends of nineteenth-century chemistry: that concerning how Wöhler synthesized

---

23. Metzger, *La méthode philosophique;* Freudenthal, *Studies on Hélène Metzger.*
24. Kim, "Layers of Chemical Language" and "Constructing Symbolic Spaces."

urea in 1828 and thereby ended belief in the existence of vital force. Brooke has shown that, since the synthesis was not directly from the four elements involved (carbon, hydrogen, oxygen, and nitrogen), it could not in any way have dispelled the notion of vital force. Seen within its context, however, this synthesis does appear as a great event since it provided a striking example of two different compounds (ammonium cyanate and urea) composed of the same constituents in the same proportion. It thus prompted the creation of the basic notion of isomerism.[25] Peter Ramberg describes the emergence of the myth of the synthesis of urea in nineteenth-century textbooks and emphasizes "its paradoxical dual role as both the founding moment of organic chemistry and as the agent that unified organic and inorganic chemistry under the same principles."[26]

Similarly, historians of chemistry working jointly with psychologists have subjected Kekulé's dream, from which the benzene ring supposedly originated, to critical analysis. The famous vision of a snake chewing on its own tail, which purportedly came to Kekulé as he sat before a fireplace in Ghent, turns out to be a story invented by him in 1890, some twenty-five years later. He opportunely hid the work of predecessors, like Auguste Laurent, and, more directly, the rival claims for discovery by Archibald Scott Couper and the Russian chemist Aleksander Boutlerov.[27]

Mendeleev's construction of the periodic system, to cite a third example of legends in chemistry, was (and is) presented in chemistry textbooks as the first step toward an understanding of the electronic structure of the atom. Its meaning, like those of the synthesis of urea and Kekulé's dream, has also been reassessed within the context of its discovery. J. W. van Spronsen conducted an exhaustive study of the various periodic systems invented in the nineteenth century, and he found that Mendeleev was but one among a host of chemists who attempted to classify the elements. Van Spronsen concluded that the discovery of the periodic system should be credited to at least six chemists: Alexandre Emile Béguyer de Chancourtois, William Odling, John Alexander R. Newlands, Gustav Hinrichs, Julius Lothar Meyer, and Mendeleev.[28] Moreover, recent studies now provide a fuller understanding of Mendeleev's creative process within the context of contemporary debates. They emphasize three main points: First, Mendeleev's motivation, like that of many chemistry professors of the day, was to find a pedagogical technique for organizing basic chemical knowledge about an ever-increasing number of elements. Second, Mendeleev initially sought to arrange the elements according

25. Brooke, "Wöhler's Urea."
26. Ramberg, "Death of Vitalism."
27. Wotiz and Rudotsy, "Unknown Kekulé"; Wotiz, *Kekulé Riddle*.
28. Van Spronsen, *Periodic System*.

to chemical analogies and through comparisons of their combining powers. Contrary to what he later declared when he pointed to the Karlsruhe Conference and Stanislao Cannizzaro's atomic weights as the crucial step toward the discovery of the periodic law, he did not initially assume that increasing atomic weights provided the clue for a system of the elements. Third, unlike most rival classifiers of the elements, Mendeleev did not try to confirm William Prout's hypothesis of a primary matter; instead, he firmly believed that the periodic law supported the view that elements were indivisible and not transmutable entities. He strongly protested against the use of his discovery to support Prout's hypothesis and, after the discovery of the electron and of radioactivity, he desperately employed mechanical interpretations in order to argue that the elements were not transmutable.[29] Hence, far from anticipating future developments in atomic theory, Mendeleev sought to systematize and reorganize contemporary knowledge. Much work remains to be done, however, before we can understand the full significance of Mendeleev's achievement. In this respect, Stephen Brush's case study of the reception of the periodic system in America and Britain has usefully sought to identify the role of successful predictions in theory changes.[30] More such studies are needed, however, in order to illuminate the epistemological status of nineteenth-century chemistry and, especially, the specific and close connection that existed between theorizing and pressures or requirements of teaching.

6. Disciplinary Identity and Boundaries

Most scholarly efforts in the history of chemistry during the 1980s were devoted to minute case studies. Then suddenly in the early 1990s, several publications attempting to reconstruct the development of chemistry from ancient times to the present appeared simultaneously.[31] This boom has revived the nineteenth-century tradition of providing a broad historical picture or panorama and has raised some interesting questions. For example: What exactly should the writing of the history of a science such as chemistry be like?[32] Should it seek to present a chronological and objective account of the

29. Graham, *Science in Russia*, 45–55; Bensaude-Vincent, "L'éther, élément chimique" and "Mendeleev's Periodic System."

30. Brush, "Reception of Mendeleev's Periodic Law."

31. Aftalion, *Histoire de la chimie*; Mierzecki, *Historical Development*; Hudson, *History of Chemistry*; Knight, *Ideas in Chemistry*; Brock, *Fontana History of Chemistry*; Bensaude-Vincent and Stengers, *Histoire de la chimie*; Levere, *Transforming Matter*.

32. An international colloquium on "Writing the History of Chemistry," held at the University of Paris X–Nanterre in May 1993, offered an opportunity for comparing the various narratives and authors' choices.

development of a discipline? Should it aim to construct a full narrative with actors, dramatic events, and perhaps fierce battles? Just who are the relevant actors? Are they simply the chemists, *tout court?* And what roles are there to play for chemical instruments and instrument makers, chemical institutions, chemical concepts, and the like?

The question of the disciplinary identity of chemistry has emerged as a major focus from these recent general histories. How can we define chemistry from antiquity to the present without projecting our current definitions or understandings onto the past? Significantly, this major question has provided the organizing principle for a number of general histories of chemistry that try to identify chemistry's changing identity over time.

In this perspective, the nineteenth century appears as a golden age, a period when chemistry was perceived to have become a fundamental science. Historians often use the phrase "mature science" when speaking of nineteenth-century chemistry, even though they are reluctant to convey the image of a natural process. This biological metaphor and example show the difficulty in eliminating an essentialist viewpoint derived from the present state of the discipline. In order to avoid this pitfall, it is convenient to define nineteenth-century chemistry from three standpoints. First, as regards its cognitive territory, chemistry considerably extended (and diminished) its domain. In 1800, chemistry included light, heat, and electricity within its territory, while the subject of respiration, now studied as absorption of oxygen, brought chemistry and the life processes close together. Second, as regard its epistemology, chemistry appeared as a model positivist science. It is remarkable that in the controversy between atomists and equivalentists, both sides legitimated their claims with reference to positivism. On the one hand, equivalentists like Berthelot or Henri Sainte-Claire Deville illustrated positivist prudence because they rejected hypotheses and wanted to confine the science of chemistry to observational and experimental data. On the other, committed atomists like Laurent, Charles Gerhardt, or Kekulé could also be considered positivists because they promoted their structural formulae without assuming that they had any ontological reality. Finally, in a broader cultural perspective, chemistry enjoyed a high social prestige during the nineteenth century, thanks to an increasing number of respectable professors. A number of chemists (Liebig and Kekulé, for instance) were knighted and several of them—like the Frenchmen Jean-Antoine Chaptal, Dumas, and Berthelot, who became ministers of state—played important roles in the public sphere or embraced political careers.

Chemistry's changing boundaries with physics and biology have also shaped its identity. The borders with physics have been renegotiated numerous times: from the attempts made by Dumas in the 1830s to distinguish the chemical atom from the physical to the promotion of physical chemistry later in the century by the Ionists—above all, Ostwald, Arrhenius, and van't Hoff. A number of recent publications on physical chemistry have opportunely filled the lacuna deplored by Brock in 1983.[33] For example, Mary Jo Nye, in describing how the discipline evolved its own, autonomous theoretical framework as it was shaped within the larger context of natural philosophy, suggests that nothing like a smooth, linear process occurred. Instead, a new perspective on molecules emerged, one that differed from the views of both organic chemists and physicists. Ironically, the development of physical chemistry as a discipline from the 1870s on did not bridge physics and chemistry; rather, it brought inorganic and organic chemistry closer through their common adoption of concepts describing reaction dynamics.

The emergence of atomic science at the turn of the twentieth century has also led to a reappraisal of the standard distribution of physics and chemistry. In J. J. Thomson's work on the electron, as Michael Chayut has emphasized, the chemical considerations were an integral part of the process of discovery.[34] Other historians have similarly reconsidered the role of chemistry in the early history of radioactivity. Because radioactivity became part of nuclear physics in the twentieth century, many historians have assumed that the chemical work of the identification of radioactive elements played a minor role in the advancement of science, that this was only a preliminary phase quickly superseded by the atomic interpretation. Lawrence Badash, in particular, has viewed the experimental research of the Curie laboratory aimed at the isolation and purification of radioactive sources as being well behind Rutherford's physical approach to radioactive phenomena.[35] This standard view is not one of national prejudice; rather, it betrays a spontaneous and implicit hierarchy of physics and chemistry.

The boundaries between physiology and chemistry, for their part, raised passionate debates during the nineteenth century. Nineteenth-century chemists submitted animal compounds to elementary analyses in their laboratories, while the physiologists claimed that they could not understand the processes transpiring in living bodies because they dealt with dead matter.[36] Following Lavoisier's theory of respiration, it was usually admitted among chemists that "an animal," as Dumas put it, "constitutes in effect, a combustion apparatus, from which carbonic acid is continuously disengaged, and in

33. Servos, *Physical Chemistry*; Barkan, *Walther Nernst*; Nye, *From Chemical Philosophy* and *Before Big Science*; Laidler, *World of Physical Chemistry* and "Chemical Kinetics"; Schummer, "Physical Chemistry."

34. Chayut, "Thomson."

35. Badash, *Radioactivity in America*; Malley, From Hyperphosphorescence. For a criticism of the standard view see Boudia, "Marie Curie et son laboratoire."

36. Coley, *From Animal Chemistry* and "Studies in the History of Animal Chemistry"; Holmes, "Elementary Analysis" and *Claude Bernard*; Fruton, *Molecules of Life*.

which, consequently, carbon continously burns."[37] Similary, Liebig assumed that muscular work consumed nitrogenous substances and that the consumption of nitrogen could be measured by the quantity of urea in the urine. The reductive balance-sheet approach of physiological processes was the basis of both Liebig's animal chemistry and his agricultural chemistry. However successful this approach may have been among farmers, it was challenged by physiologists like Claude Bernard.

In addition to the study of these controversies, the emergence of hybrid disciplines such as physiological chemistry and biochemistry has raised an interesting debate over the nature of a scientific discipline. Robert Kohler has argued that physiological chemistry failed to institutionalize as a discipline in the 1850s because university chairs were divided between organic chemistry and physiology; Holmes has challenged this view, arguing that institutional demarcations did not prevent intense activity in this field prior to Felix Hoppe-Seyler's founding in 1877 of the *Zeitschrift für physiologische Chemie.*[38] Underlying this debate is the question of whether a discipline should be best characterized as a cognitive or as an institutional structure, including the establishment of social networks. The emphasis in the current historiography is on the pivotal role of social aspects.[39] Nonetheless, a full understanding of discipline formation will require scholars to address the issue of how a research area emerges and stabilizes as an independent entity.

7. Research Schools

In counterpoint to the intense scholarship of the 1960s and 1970s devoted to the study of scientific revolutions and their heroic figures, the daily practice of chemistry has become an intense if not new historiographical approach. A number of more recent scholars have focused their efforts on the study of research schools, the formation of professional chemists, and industrial practices. In so doing, they have taken up such topics as the historical study of instruments, publication practices (including textbooks), teaching, patenting, laboratory life, and routine analyses by experts.[40] This shift of attention, from the study of heroic figures to the study of everyday, humdrum science and scientists, could have discouraged publishers who know that more profit may be expected from biographies of stellar scientists than from monographs on the blowpipe or a local laboratory. In fact, an impressive number of thematic is-

---

37. Dumas, *Essai de statique chimique;* English trans. from Holmes, *Between Biology and Medicine,* 11.
38. Kohler, *From Medical Chemistry;* Holmes, *Between Biology and Medicine.*
39. Lesch, *Science and Medicine in France;* Bäumer, *Von der physiologischen Chemie.*
40. See, e.g., Hamlin, *Science of Impurity,* and Watson, "Chemist as Expert."

sues of journals and collective volumes have appeared that provide new insights into everyday chemistry.

The emergence of research schools—one of the major new characteristics of nineteenth-century science—is best seen in chemistry, where laboratory classes and work encouraged the formation of research groups around a patron.[41] The notion of research schools was originally used as a helpful analytical unit for illuminating scientific styles. The classic studies here include J. B. Morrell on the comparative strategies of Thomas Thomson's and Liebig's training of chemists, and Joseph Fruton on Emil Fischer in comparison to Franz Hofmeister. More recently, Nye has compared the Ecole Normale Supérieure and the Manchester research groups at the turn of this century.[42] Among other things, the study of research schools can sometimes illuminate how new fields of research became established science or how difficult techniques became routine. As such studies show, there is no natural process of "normalization" in science. For example, Holmes's analysis of Liebig's laboratory at Giessen shows that only a conjunction of various local circumstances allowed a successful alliance of teaching and research, and this for only a few years.[43] Neither in Marburg nor in Leipzig, as Rocke has shown, did Hermann Kolbe manage to convince his advanced students of the validity of his theoretical approach; nor did he ever create a research tradition. In France, by constrast, Charles Adolphe Wurtz formed a cognitive network of atomists (most of them Alsatians) who resisted the dominant antiatomist climate of the French chemical community.[44]

8. The Chemist as a Professional

When and how did "chemist" become a professional occupation? The introduction of chemistry into higher education was the cornerstone of the process of professionalization. In the reform of Prussian university education led by Wilhelm von Humboldt, emphasis was placed on the fundamental sciences and the acquisition of knowledge for its own sake. Science continued to be taught in the philosophy faculty. However, at the urging of Liebig, who defended a nonutilitarian view of science, training in research was included in the university syllabus. At a time when French faculties were giving ex-

41. Geison and Holmes, *Research Schools;* Klosterman, "Research School"; Carneiro, Research School.

42. Morrell, "Chemist Breeders"; Fruton, *Contrasts in Scientific Style;* Nye, "National Styles." Crosland's *Society of Arcueil* (1967) did not describe that society as a research school, a point made only in Hannaway's brilliant review of Crosland's book.

43. Holmes, "Complementarity of Teaching and Research."

44. Rocke, "Group Research in German Chemistry"; Carneiro and Pigeard, "Chimistes alsaciens à Paris."

cathedra lectures on chemistry, Liebig in Giessen was giving practical classes eight hours a day. When Liebig's former students found university and similar appointments, they modeled their instruction in chemistry on his. In time, the German universities, the Royal College of Chemistry in London, and a number of American universities became nurseries of young professional chemists well trained for a variety of careers in pharmacy, chemical industry, or chemical education.[45] Once research had become an accepted part of a professor's obligations, it inevitably became a key requirement for academic appointment and promotion. The tandem of teaching and research thus became the main vehicle of professionalization in chemistry.

However, the question remained as to whether this German-inspired scheme was applicable to other contexts. The institutional background of the modernization of chemistry in imperial Germany, as described by Jeffrey Johnson,[46] for example, was too closely linked to local and national circumstances to be transferrable everywhere. Was there anything like a standard process of professionalization of chemistry typical of our period? The European dimension of this question was recently addressed as part of the European program "Evolution of Chemistry, 1789–1939." Here scholars analyzed, in a series of workshops, how the apprenticeship role in chemistry gave way to that of academic and industrial (chemical engineering) careers as a distinct, self-conscious profession. The resulting collective volume devoted to this issue has fundamentally changed our view of the process of the professionalization of chemistry.[47] For one thing, it has enlarged the field of vision. Whereas such leading countries as Britain, France, Germany, and the United States had been relatively well studied during the 1970s and 1980s,[48] this effort yielded new understanding about the organization of chemistry and the creation of chairs, journals, and learned societies in many other countries and regions (including Catalonia, Piedmont, Poland, Portugal, Russia, and Sweden). Although we still lack a clear view of the process of the institutionalization and professionalization of chemistry on a European-wide scale, local studies of this kind may lead us to question our current view of the development of chemistry, which takes the politically and economically leading

---

45. Gustin, Emergence of the German Chemical Profession; Hufbauer, *Formation of the German Chemical Community*; Roberts, "Establishment of the Royal College"; Lewenstein, "To Improve Our Knowledge."

46. Johnson, *Kaiser's Chemists*.

47. Knight and Kragh, *Making of the Chemist*.

48. Russell, Coley, and Roberts, *Chemistry as a Profession*; Bud, Discipline of Chemistry; Bud and Roberts, *Science versus Practice*; Servos, "Industrial Relations of Science"; Thackray et al., *Chemistry in America*; Hufbauer, *Formation of the German Chemical Community*; R. Turner, "Justus Liebig versus Prussian Chemistry"; Meinel and Scholz, *Die Allianz*; Daumas, *Centenaire de la Société chimique*; Crosland, *In the Shadow of Lavoisier.*

nation-states of the nineteenth century as its model.[49] One possible implication is that the traditional center/periphery distinction may prove inadequate; each country seems to have had its own history and each locality its own dynamics of institutionalization that were dependent on specific political and industrial contexts.

A related issue may be opened to debate during the coming decade. Increasing specialization has long been considered a natural movement, the unavoidable consequence of the advancement of science. Chemistry has provided a paradigmatic example of the whole process of disciplinary specialization. Around the middle of the nineteenth century, national societies were founded in various countries: the Chemical Society of London in 1841, the Société chimique de Paris in 1857, the Deutsche Chemische Gesellschaft in 1867, the Russian Chemical Society in 1868, the American Chemical Society in 1876. These institutions began publishing specialized journals. The chemical community also introduced the international disciplinary conference: the Karlsruhe Congress was the first conference of chemists; organized in 1860, it brought together 140 chemists from thirteen countries. Hand in hand with the disciplinary organization went the development of chemistry courses in universities and technical schools. The categories of "pure" and "applied" chemistry, introduced in 1751 by the Swedish chemist Johan Gottschalk Wallerius, proved crucial, as Meinel has pointed out, for legitimizing the academic status of chemistry, while also enhancing its prestige as a useful science.[50] This distinction successfully served the promotion of chemistry teaching during the nineteenth century in a variety of curricula such as pharmacy, medicine, and agricultural science.

However, it seems that disciplinary specialization was but one possible strategy among several, and one that raised many tensions. Despite the rise of smaller, specialized associations of scientists, institutions open to scientists of all disciplines flourished in the early decades of the century. Chemists appeared on several stages, including that of popularization. The Royal Institution of London, founded in 1799, functioned both as a research center and as an institute for popularizing science, while the Gesellschaft Deutscher Naturforscher und Ärtze, founded in 1822, the British Association for the Advancement of Science, founded in 1831, and the American Association for the Advancement of Science, founded in 1848, all offered annual meetings open to all types of scientists, whether specialist or generalist. The Paris Academy of Sciences, for its part, maintained its multidisciplinary weekly publication throughout the nineteenth century.[51]

49. Rae, "Chemical Organizations," opens the way to further case studies of this kind.
50. Meinel, "Theory or Practice?"
51. Crosland, Science under Control.

Textbooks provided an important vehicle for spreading chemistry in nineteenth-century society. Apart from one isolated study of German textbooks,[52] only several great treatises—notably, those by Lavoisier, Berthollet, and Berzelius—have to date attracted the attention of scholars. A recent workshop (also under the auspices of the European program "Evolution of Chemistry") has begun considering textbooks as valuable sources for the historian of chemistry. As Owen Hannaway has argued, textbooks in chemistry were of special importance in the didactic origins of chemistry in the early modern period.[53] During the nineteenth century, the intensified production of textbooks, stimulated in good measure by the development of school systems as well as by the growth of the publishing business, gradually generated a gulf between creative science and textbook science, a genre defined by its standardization and conservatism:[54] A wide variety of textbooks addressed the needs of medical students, doctors, pharmacists, dyers and farmers, as well as the interests of nonprofessionals. Such textbooks ranged from the extremely popular to the extremely esoteric. Nineteenth-century chemistry was most definitely a fashionable and useful science, an integral part of public culture in industrialized countries. More scholarship is needed, however, to understand more fully the role of chemistry textbooks in the dynamics of the formation of chemistry as a profession. In this regard, historians of chemistry need to ask, among other things, about the extent to which textbooks, with their specific ways of recording results aimed at facilitating understanding and memorization, helped standardize and normalize the language and the practices of chemistry. And they might inquire as to the extent to which writing a textbook stimulated the creativity of the chemist-author. One thinks of Mendeleev's discovery of the periodic law, coming as it did in the course of his effort to organize the chapters of his *Principles of Chemistry*. But are there other such cases? Again, to what extent did textbooks contribute to enrolling recruits in chemistry and shaping the profile of academic and industrial chemists?

9. Industrial Chemistry

From knowledge to production: this traditional sequence long seemed unproblematic in most reference works on chemical technologies.[55] In the

---

52. Haupt, *Deutschsprachige Chemielehrbücher.*

53. Hannaway, *Chemists and the Word.*

54. Bensaude-Vincent, Garcia-Belmar, and Bertomeu-Sanchez, *Émergence d'une science des manuels.*

55. See, e.g., Clow and Clow, *Chemical Revolution;* Musson and Robinson, *Science and Technology;* Haber, *Chemical Industry;* J. Smith, *Origins and Early Development;* and Wengenroth's essay "Science, Technology, and Industry" in the present volume.

1970s, several historians of chemical technology entered the debate on the role of science in the Industrial Revolution initiated by *Technology and Culture* and questioned the tacit connection between the chemical revolution and the contemporary technological advances in the manufacture of soda, dyes, bleaching powder, and gunpowder.[56]

The standard histories of relations between academic and industrial chemistry in nineteenth-century Germany have emphasized the leading role played by academic chemists in the development of dyes and the close relations between universities and industrial firms.[57] The structure of the benzene ring, a fundamental discovery by a German professor, Kekulé, and the basis of the synthesis of the anilin dyes, thus became a symbol of modernity, the turning point from chemical industry to industrial chemistry. Recent studies of dyeing technologies have, however, substantially revised this traditional view. When natural dyes were displaced by synthetic dyestuffs, they were by no means a decaying technology and were already science-based.[58] Much research conducted in order to improve natural dyes had contributed to the development of synthetic dyestuffs.[59] Thus, thanks to a renewed interest in the history of natural products, the stark contrast between empirical and science-based technologies has been qualified.

As far as synthetic dyestuffs are concerned, the imprecise notion of science-based industry needs clarification through more case studies of corporate strategies. As Wolfgang König has pointed out, such a notion may cover many different situations, including the application of fundamental discoveries in technological innovations, the appointment of university graduates to industrial firms, the direct transfer of research results from university to industry, and the research laboratory in an industrial site.[60] Moreover, if we include World War I in our period, a new configuration of relations developed with the mobilization of chemists for chemical warfare. Although collaborations between scientists and the military were rarely operational, they were truly

56. See the essays in *Technology and Culture* 1 (1965) and 17 (1976). See also "Discovery of the Leblanc Process," in which Gillispie deconstructed the rhetoric surrounding the inventor, stressing the variety of circumstances that converged toward the final success of this revolutionary innovation. See also Perrin on Chaptal ("Of Theory Shifts"). More recently, historians of gunpowder production in France have argued that the strategy that enabled Lavoisier to double the production of the Arsenal within a decade rested less on research into the process of nitrification and detonation than on financial measures and, above all, on the development of a pedagogy for training in this technology (Mauskopf, "Gunpowder and the Chemical Revolution" and "Lavoisier and the Improvement of Gunpowder Production"; Bret, "Lavoisier et l'apport de la chimie académique").

57. Haber, *Chemical Industry*.

58. R. Fox and Nieto-Galàn, *Natural Dyestuffs;* Nieto-Galàn, *Colouring Textiles*.

59. See Travis, Hornix, and Bud, "Organic Chemistry and High Technology," and Travis, *Rainbow Makers*.

60. König, "Science-based Industry"; Reinhardt, "Instrument of Corporate Strategy."

efficient in the area of chemistry, particularly in the case of Germany. Chemists were active not only in the explosives and munitions industries, especially with regard to manufacturing poisonous gas, but also in medical applications such as radiology (using Röntgen rays) and in connection with food supply.[61]

In short, the standard view of a mid-nineteenth-century shift from empirical chemical industries to industrial chemistry (i.e., science-based technologies) has predominated because our view of technology has been too often restricted to the paradigmatic examples of artificial soda and synthetic dyestuffs. Once the scope of studies is broadened, many potential patterns of relations between academic and industrial chemistry become visible. The traditional dichotomy between science and technology will become (indeed, has already become) too poor and too rigid to characterize the diverse cultures of chemistry.[62]

10. Spaces of Chemical Knowledge

Despite the deep revisions of standard episodes mentioned above and the new light shed by the growing interest in the everyday practices of chemistry, there remain numerous fields to be explored. Although the laboratory and its activities has been the specific space of chemists from ancient times to the present, study of it has, strangely enough, been much neglected. Over the past decade, historians of physics and of biology have conducted numerous studies of experimental practices, including replications of past experiments.[63] And one can certainly find papers devoted to the instruments of great chemists like Lavoisier or Priestley.[64] In particular, Holmes's recent study of Lavoisier's laboratory notebook for the year 1773 shows the great interpretive potential opened up by minute investigation of experimental practices. The use of instruments in chemistry is the focus of a volume edited by Holmes and Trevor Levere, and a volume edited by Ursula Klein looks at chemists' means of representing invisible objects, emphasizing the crucial role of "paper tools" and the interplay between visual models and experiments in chemistry.[65] Yet few historians have been attracted to the study of

61. Haber, *Poisonous Cloud;* MacLeod, "Chemists Go to War."

62. See, e.g., the case studies collected by the European group devoted to "Strategies of Chemical Industrialization," in Homburg et al., *Chemical Industry in Europe.* For further discussion of this topic, see Wengenroth's essay in this volume.

63. See, e.g., Sibum, "Reworking the Mechanical Value," and Blondel and Dörries, *Restaging Coulomb.*

64. Roberts, "Word and the World"; Golinski, *Science as Public Culture,* 129–51; Bensaude-Vincent, "Between Chemistry and Politics."

65. Holmes, *Lavoisier;* Holmes and Levere, *Instruments and Experimentation;* Klein, *Tools and Modes of Representation.*

the spaces involved in experimental chemistry. Whatever the interest of these novel approaches, the organization of laboratories calls for more direct attention by scholars.

Laboratories have multiple functions, including teaching, research, testing, and control. How and when such diversification occurred is a central yet rarely addressed problem.[66] During the nineteenth century, the term "laboratory" came to refer principally either to a more or less large space for the training of dozens of students or to an academic room (usually much smaller) where one or two scientists conducted experimental investigations.[67] It could furthermore sometimes (if less often) mean the set of portable instruments or scientists working in the field (as with geologists) or in a small room devoted to the testing and control in factories and health services or, later, the big, well-equipped spaces devoted to industrial research. More studies that characterize the experimental practices of chemistry attached to these various spaces are needed. To what extent did industrial pressures and the concern for standardization influence academic laboratories? In a study of balances and spectroscopes, Matthias Dörries has analyzed how the standardization of experimental practices—especially the requirement that chemists in distant locations be able to make comparable and reproducible measurements—necessitated that instruments be defined and analyzed in great detail.[68] Instruments thus became a new area of research and were the collective construction of many scientists. Thanks to the joint efforts of museum curators and scholars, chemical instruments and techniques, already a booming field of inquiry, will likely attract more historians in the near future.[69]

11. The Philosophy of Chemistry

Unlike physics, chemistry during the past decades has not attracted much attention from philosophers, and those few it has attracted are mainly concerned with tracing influences.[70] Yet a number of chemists actively pursue the philosophy of chemistry and have recently founded a journal, *Foundations of Chemistry*.[71] These chemist-philosophers are mainly concerned with the issue of the reduction of chemistry to quantum physics; for them philosophy is

66. See, e.g., Holmes, "Complementarity of Teaching and Research."

67. A recent European Science Foundation workshop provided useful details on these kinds of laboratories; see Meinel, *Research Laboratories.*

68. Dörries, "Balances."

69. See, e.g., Bud and Warner, *Instruments of Science,* and Holmes and Levere, *Instruments and Experimentation.*

70. For instance, Condillac's influence on the chemical nomenclature, on which see Albury, Logic of Condillac, and Beretta, *Enlightenment of Matter.*

71. The journal is edited by Eric Scerri and published by Kluwer Academic.

in part a resource for claiming the autonomy of chemistry. A German group in Marburg, however, has developed a different perspective that it calls "methodical constructivism." Its aim is to reconstruct methodically the languages of chemistry, starting at the level of the vocabulary referring to operations and instruments. In addition, Germany took another initiative in the promotion of the philosophy of chemistry with the establishment of a journal, *Hyle,* whose title (from the Greek for "matter") offers a potentially superb program for philosophical investigations.[72] This boomlet in the philosophy of chemistry should be an invitation to develop interactions with historians of chemistry. For the particular case of the nineteenth century, chemists' epistemic choices as regards atomism and the status of structural formulae are certainly major philosophical issues.[73] When philosophy is brought into historical narratives, it is usually considered as an obstacle. For instance, it is usually assumed that the long resistance of French chemists to atomism was due to the strong influence of Comtean positivism on the French chemical community.[74] The new interest in the philosophy of chemistry may help correct this attitude.

Be that as it may, the entire issue of the influence of philosophy upon chemistry or of the philosophical foundations of chemistry calls for serious rethinking. More efforts are needed to understand historically the nature of the philosophy of chemistry, assuming that such a thing exists. Here, instead of tracing influences, it might be more fruitful to analyze the epistemological attitudes and philosophical choices of chemists as these emerged from their own practices and the controversies in which they were involved. To what extent did, for example, the atomic debates or the triumphant distinction between pure and applied chemistry help shape a positivistic framework for chemistry during the nineteenth century? To turn this question around, to what extent did an ever more dominant positivism contribute to promoting chemistry as a model science in the mid-nineteenth century? Subtle interactions of this kind, when investigated in local and specific contexts, may very well provide fruitful grounds for comparative studies that will allow a deeper understanding of chemistry as an integral part of our culture.

72. *Hyle: An International Journal for the Philosophy of Chemistry,* edited by Joachim Schummer and available on-line: http://www.uni-karlsruhr.de/philosophie/hyle.html. See also Psarros, "Constructive Approach," and Psarros, Ruthenberg, and Schummer, *Philosophie der Chemie.*

73. See, e.g., Kapoor, "Dumas and Organic Classification"; Kim on Dumas's epistemological choices in "Constructing Symbolic Spaces"; and, on Auguste Laurent, Blondel-Megrelis, *Dire les choses.*

74. A few decades ago, however, the philosophical and religious commitments of chemists such as Priestley or Davy were considered the driving forces of their scientific achievements. See, e.g., Farrar, "Nineteenth-Century Speculations"; McEvoy and McGuire, "God and Nature"; McEvoy, "Joseph Priestley"; and Knight, *Transcendental Part of Chemistry.*

12. An Audience for the History of Chemistry

In their ongoing effort to challenge whiggish or presentist accounts of the founding events in chemistry, historians are continuously confronted with the historical narratives produced by professional chemists. Professional historians often alter the traditional images of the founding heroes, whom the chemist-historians seek to commemorate. The more the historians of chemistry contextualize their subjects, the greater may be the danger of isolation from the common views of chemistry's past. Is there a natural hostility or a fatal contradiction between the reconstructions of the past based on archival materials by professional historians and the reconstructions by working chemists?

Whatever the answer to this question may be, chemists and their historians need to bridge the widening gulf between them. This is a legitimate concern since historians without chemists would be deprived of an indispensable source of information, a natural audience for their publications, and financial support. Historians of science need both a close connection with the actual practice of science and the active collaboration and criticism of chemists. Still, the danger of a schizophrenic division of history, as has been repeatedly pointed out,[75] should not encourage unprofessional or second-class publications. On the contrary, it is only through a more professional attitude on the part of historians that the gulf can be bridged.

A number of historians, including Colin Russell, Rachel Laudan, and Alan Rocke, have characterized in fine detail the prejudices that underlie the standard accounts of the evolution of chemistry as written by the nineteenth-century chemist-historians. Similarly, it is not so difficult to identify the nationalistic biases in the French cult of Lavoisier as the founder of chemistry, partly because the subsequent French claim that "chemistry is a French science" sparked a fierce controversy between French and German chemists in the aftermath of the Franco-Prussian War.[76] Two hundred years after the chemical revolution, it is still possible to see that the accounts of this episode are far from neutral. Most accounts were so heavily influenced by the chemistry and the philosophical choices of their writers that they say more about them than they do about the chemical revolution itself.[77]

It is still more difficult to apply this critical approach to our current historiographical practices. Yet it may prove helpful to consider writing the history of chemistry as an activity that needs to be contextualized in exactly the same manner as chemical investigation itself. Historiographical reflection should be an integral part of the agenda of the history of chemistry, even and espe-

75. See Jensen, "History of Chemistry."
76. See Bensaude-Vincent, "Founder Myth."
77. McEvoy, "Positivism, Whiggism"; Jacques, "Lavoisier et ses historiens français."

cially when studies are addressed to the public at large. A promising step in this direction merits notice: In a recent publication, sponsored by the Chemical Heritage Foundation and aimed at instigating a concern for the history of modern chemical science and industry among chemists (industrialist, academic, and schoolteachers alike), a whole section was devoted to reflexive accounts on detailed case studies.[78] There are, for example, many founder myths in the history of chemistry. An intense study of commemorations within the history of chemistry could provide a good starting point for joint efforts between historians and active chemists. In so doing, we might come to understand better why, when, and under what circumstances the history of chemistry has been excluded from teaching, and how it is that most scientists are largely ignorant of past intellectual accomplishments in their fields. By working together to study the daily work and lives of scientists, their institutions, and their philosophical outlooks, chemists and historians stand to gain a more realistic, intimate, and exciting view of the development of modern chemistry.

ACKNOWLEDGMENTS

I am grateful to José Ramon Bertomeu and Ulrike Felt for their bibliographical help. David Cahan did much work on the manuscript and made cogent suggestions for improving it.

78. See Mauskopf, *Chemical Sciences.*

Eight

SCIENCE, TECHNOLOGY, AND INDUSTRY

ULRICH WENGENROTH

❦

THE NINETEENTH CENTURY witnessed the rise of modern industry. From western Europe to Britain to North America agriculture lost its preeminent role in societal reproduction and yielded to industrial manufacturing and technology-intensive services like railroads, steam navigation, and telecommunication. Dramatic changes in the social fabric and the face of landscapes spread from the North Atlantic region throughout the world and bore witness to a fundamental shift in human history. Both the number of people and artifacts grew at an unprecedented pace. This emerging modern world was driven by an unending stream of new products turned out by factories employing radically new technologies, skills, and organization. Technological innovations, being the most tangible results of this new, accelerated mode of reproduction, were soon understood to represent the rationale of nascent industrial society. Never before in history and never within a single lifetime had so much novel material culture been produced. This sudden leap of productive potential puzzled contemporaries and continues to preoccupy historians.

One of the many questions raised by this historical watershed concerns the sources of innovation in nineteenth-century industry. While social and economic historians have concentrated on skills and organization, historians of science and technology have debated the character of novelty in technology. How much did technological innovation owe to recent advances in the sciences? To what extent was nineteenth-century industry science-based? To what extent were developments in science and industrial technology independent of each other? Did science perhaps ultimately benefit more from technology than technology did from science?

The positivistic school, which dominated the history of science in the 1950s and 1960s, thought that industrial technology was applied science and technological innovation not much more than putting the results of scientific research to work. A. Rupert Hall provides a representative summary of this view. In 1962 he declared: "The late eighteenth century was the point in time at which the curve of diminishing returns from pure empiricism dipped to meet the curve of increasing returns from applied science. This point we can fix fairly exactly, and so we may be sure that if science had stopped dead with

[Isaac] Newton, technology would have halted with [John] Rennie, or there-abouts. The great advances of later nineteenth century technology owe everything to post-Newtonian science."[1]

Engineering appeared to be subordinate to enlightened progress in the "hard" sciences, with physics at their center. Recalling this earlier orthodoxy, Ruth Schwartz Cowan drew a parallel to gender relationships in her 1994 pres-idential address to the Society for the History of Technology, "Technology Is to Science as Female Is to Male."[2] The group had been created "from a rib out of the side of the History of Science Society,"[3] at a time when historians of technology had begun to stress the autonomy and originality of technologi-cal knowledge and strategies vis-à-vis both science and economics, the two fields that had claimed to incorporate technology and its history as a subset. The "technology equals applied science" controversy constituted the back-ground against which the newly independent field of history of technology asserted its group identity. If "technology" could be subsumed under "sci-ence," then could history of technology, at least for the nineteenth and twen-tieth centuries, be placed under history of science? In subsequent years, the formal separation of the academic fields of history of science and history of technology has helped to distinguish the differences between science and technology during two centuries of industrialization. In its early years, the so-ciety's journal, *Technology and Culture,* focused more on the characteristics of science, technology, and engineering than on any other single subject.[4] With two special issues in 1961 and 1976,[5] as well as with a great number of scattered articles on science-technology relations over the years, the debate eventually moved from apodictic ontological battles to an elaborate discussion of the specificity of technological knowledge and practice. While the bitterness of secession has never been fully overcome, the sophistication of argument has benefited immensely from this distancing.

However, and regrettably, the distancing has gone to such an extent that by now the two disciplines have nearly lost sight of one another. Courses in the history of technology and the history of science take little notice of what hap-pens on the other side of the newly erected academic fence. And while histo-rians of technology, in their effort to prove the distinctness of technological knowledge and practice, at least continue to discuss science, historians of sci-ence rarely offer more than passing remarks on the field of technology and in-dustry. The peripheral field of the history of scientific instrumentation is

1. Hall, "Changing Technical Act," 511.
2. Cowan, "Technology Is to Science."
3. Staudenmaier, *Technology's Storytellers,* 1.
4. Ibid., chap. 3.
5. Respectively, "Science and Engineering" and Reingold and Molella, "Interaction of Science and Technology."

perhaps the only place where technology is frequently discussed in their publications. In one recently published companion to the history of modern science "technology" and "industry" are not even allotted a chapter and make only an extremely modest appearance in the subject index.[6] Each field remained ignorant of the other during the years of unsettled claims of primacy and priority in the development of the history of modern industrial civilizations.[7] It has taken joint meetings, stimulating input from science studies, and a new generation of historians unscathed by the wars of secession to enter into a new dialogue over the role of science and technology during the "post-Rennie world."

Historians no longer see the historical relationship of science and technology as one of epistemological hierarchy, ennobling one subject over the other. Rather, they see a systemic interrelatedness of the two, which makes it difficult to separate them neatly, if at all. The metaphor of fields interpenetrating each other now seems more appropriate than that of bodies encapsulated in or juxtaposed on one another. It appears as an almost ironic reversal of the earlier debate when some historians of science, inspired by twentieth-century European philosophy, now claim that all science is technology.[8] With the missionary hubris of the Enlightenment ideal approaching dusk and with the reinvention of humanism,[9] the history of science and technology as they existed in practice now attracts more interest than the apologetic story of unfolding truth and progress that dominated much of the earlier literature. Still, this body of literature continues to be an invaluable source for the history of science, technology, and industry in the nineteenth century. Like all sources, it has to be read with its time and ideological background in mind. The same is of course true, and more difficult to recognize, for contemporary historiographical outlooks.

This essay addresses the most important issues of the science-technology-industry triangle during the nineteenth century. Following a general discussion of the "science-technology push," wherein hypotheses such as the "linear model" of "technology equals applied science" are critically treated (section 1), the essay turns to the dispute over the role of science in the Industrial Revolution (section 2). This in turn leads to an assessment of the importance of science to engineering in early-nineteenth-century industry (section 3). While Britain still holds center stage at this point, the emphasis thereafter shifts to the European Continent and America. The issues here include technology transfer and the tools employed by states and industries to catch up

6. Olby et al., *Companion.*
7. Many scholars have made this observation; see, e.g., Laudan, "Natural Alliance."
8. Wise, "Mediations," 253; Shapin and Shaffer, *Leviathan and the Air Pump,* 25.
9. Toulmin, *Cosmopolis.*

with Britain (section 4). The focus is on forms of knowledge and the early development of a school culture in engineering. The institutionalization of technological education as a deliberate effort to promote industrial development emerges as an important clue to understanding the complicated science-technology relationship. It turns out that, well before any measurable impact of the content of science, academic science and industrial technology were heading toward a common language that was instrumental in promoting intensified exchanges between the two fields (section 5). Engineering science, following a lengthy gestation, developed as an autonomous academic subject in the second half of the nineteenth century and played the successful intermediary between the findings of science and industrial application (section 6). A still quite limited number of science-based industries in chemistry, electrical engineering, optics, and mechanical engineering eventually provided the empirical background for evaluating the contribution of science to technology and industry at the end of the nineteenth century (section 7).

1. A Science-technology Push?

Studying the relationship between science, technology, and industry during the nineteenth century can be done from the perspective of any one of these three elements. The outcome of such a study is, however, very likely to reflect this a priori choice. Both the history of science and the history of technology, as they have come down to us, lend themselves to a heroic "push hypothesis," whereby science pushes technology and technology in turn pushes industry toward innovation. This once pervasive "linear model," which continues to be used to influence and legitimize public spending on research and development, has lost much of its persuasiveness over the last decades. If, on the other hand, we conceive of a triangle of science, technology, and industry and do so in terms of systemic interrelatedness, the result is a much more complex and plausible pattern of multidirectional pushes and pulls to and from each of the three elements.

The largest of these elements in terms of numbers of people and amount of material resources involved is certainly industry. In the linear model, industry was at the end of the process, since science needed first to be transformed into technology before it could become applicable to production. This brought the history of industry closer to the history of technology than to the history of science—and much of the history of technology is in fact hard to distinguish from the history of industry or even business history. This state of affairs is also revealed by certain professional preferences. Historians of technology are often to be found at conferences on economic history, business history, and labor history, and their writings often appear in these fields' publications; this is much less the case for historians of science. There exists a continuum between the history of technology and certain subfields of social and economic history that scarcely reaches into the history of science. In view of recent studies of industrial research, where the history of science and the history of industry meet, this restraint now seems outdated. Obviously, academic affiliations die harder than academic orthodoxies.

If the linear model is no longer seen as a fair representation of the science-technology-industry relationship, then examining the demonstrable impact of science and technology becomes the appropriate starting point for a historical investigation. Taking a broad view of nineteenth-century industry, science appears on but a few scattered islands on the seas of industry; nor is modern technology as pervasive as many histories of technology would have us believe. It has long been the common opinion that industry owed little to the content of scientific knowledge during (at least) the first half of the nineteenth century.[10] Looking at technology we do find, to be sure, impressive examples of newly mechanized production sources: textile mills, iron works, machine shops, and the like. At the same time, however, much or rather most of mid-nineteenth-century industry relied, as Raphael Samuel has shown us in a seminal article, on hand labor.[11]

What was technologically new in mid-nineteenth-century industry was not dominant. The organization of labor and space, information about markets, and a new entrepreneurial spirit seem to have been more prominent concerns of most new industrialists. Such issues as factory design and the deployment of labor were regarded as more pressing challenges than the design of machines until well into the twentieth century.[12] This is reflected in two of the most influential texts about early industry, both of which were written by scientists who put the issue of the organization of factory production first. Both the chemist Andrew Ure, in his *Philosophy of Manufactures; or, An Exposition of the Scientific, Moral, and Commercial Economy of the Factory System in Great Britain* (1835), and the mathematician Charles Babbage, in his fundamental work *On the Economy of Machinery and Manufactures* (1835), concentrated on organization much more than on technology, let alone science. These books certainly had more impact on the nineteenth century than Babbage's *Difference Engine* or his *Reflections on the Decline of Science in England* (1830). It should be noticed that the economist's concept of technological progress, as a residual in the production function, includes management, accounting, marketing, human resources, and the like. In the context of a history of science and technology, however, this wider concept seems inappropriate since it blurs the very distinctions that are central to our disciplines.

10. Mason, *History of the Sciences,* 503.
11. Samuel, "Workshop of the World."
12. Biggs, *Rational Factory.*

Economic historians like Robert Fogel, a Nobel Prize winner in economics, have shown that rapid economic growth in the nineteenth century did not
necessarily depend on its most exciting technologies.[13] Fogel demonstrated
how the United States, relying on a somewhat different combination of older
technologies like land transport and river and canal navigation, could have developed equally well without railroads. With a wealth of hydraulic power,
large parts of Switzerland, Italy, and California experienced successful industrialization without the use of coal and steam, the very symbols of nineteenth-
century industry. Paul David, using examples from the late nineteenth century,
has shown us just how contingent irrevocable decisions on technological
development can be. Rather than always promoting the technologically and
economically best possible alternative, market forces might as well lock in
suboptimal technologies and make further development "path dependent."[14]
Irrespective of the inherent qualities of new technologies, complex social
processes decide their relative success. There is no hidden logic in history that
will eventually usher in some optimal state, deriving the full potential from its
stock of scientific and technological knowledge. What John Maynard Keynes
exemplified for employment in the twentieth century was true for technologies and industries in the nineteenth as well: equilibria and optima did not by
themselves converge.

Inevitability, determinism and one-dimensional causality have been the
prime victims of late-twentieth-century social and economic studies. A reductionist view of nineteenth-century industry as the epitomization of accelerated technological progress in the narrower sense of the history of science
and technology is no longer defensible. A whole world has undergone a
transformation, a fundamental historical change. Without science and technology this transformation would undoubtedly have been quite different,
probably unrecognizably different, but the same is true for many other
elements: democracy, nationalism, labor movements, mass culture, service
economies, secularization, religious revival, and so on. Privileging one element over others to explain the dynamics of a systemic context will not help
one to understand either the system or the element in question. The contributions of science and technology to nineteenth-century history are not best
understood by starting a historical investigation with science and technology.
This, however, was legitimately the business of the history of science and the
history of technology. We learn but little about industry in studying just technology and but little about technology in studying just science. This observation alone would make Hall's quotation in the introduction to this essay, and
the school that it stands for, appear untenable. At the same time it helps us un-

13. Fogel, *Railroads and American Economic Growth*.
14. David, "Path Dependence."

derstand why we find so little about technology proper in general studies of nineteenth-century history and so little history of science in histories of nineteenth-century technology. It is not just parsimonial neglect, arrogance, or blindness; rather, it indicates, to a large extent, an appropriately balanced outlook. Science and technology are only two of the many elements of nineteenth-century history; moreover, they hardly existed in any "pure" form.

It is important to realize that industry had many more users than producers of technology. The textile industry, blast-furnace works, gas works, sawmills, breweries, and dairies all used technology to transform raw materials into forms that were not and are not recognized as the products of "technologies." This is even more true in the case of service industries like railroads and shipping companies, which used complex technology to transport things and people. These important industries created not technologies but organizations, ways of doing things rather than ways of producing things. They were buyers of knowledge encapsulated in knowledge-intensive artifacts produced by others. This is the world about which Ure and Babbage wrote their most influential books. Research on making people cooperate or handle machines and materials was more common than research on the machines and materials themselves. The science of industry was a social science with many faces, from Ure through Karl Marx and on to Frederick Winslow Taylor. The "human motor" remained the most important power source of early industry, one which, however, posed social and political challenges. Only slowly and peripherally did an interest in industry's physical dimension grow through the century.[15]

2. Science in the Industrial Revolution

All efforts to understand the changing relationship between science, technology, and industry during the nineteenth century must start with the Industrial Revolution. While there is widespread agreement in the literature today that industry owed very little to science during the turn from the eighteenth to the nineteenth century, this agreement is neither total nor unqualified. In their magisterial study, published in 1969, A. E. Musson and Eric Robinson argued forcefully for an applied-science model for understanding technological innovation in late-eighteenth-century Britain.[16] They documented in great detail the debt of early industrial inventor-engineers to knowledge gleaned from the new science of Robert Boyle and Isaac Newton, making the application of science a significant distinction of inventions created during the Industrial

15. Rabinbach, *Human Motor.*
16. Musson and Robinson, *Science and Technology.*

Revolution and earlier. While they did not argue that science was the single most important contribution to the Industrial Revolution, their account stands out for having linked many inventions to direct scientific input and stressed the existence of a continuum from pure to applied science and eventually to technology (rather than juxtaposing the separate worlds of science and engineering). To Musson and Robinson chlorine bleaching and the Watt steam engine were striking examples of applied science, of innovations that benefited immensely from their authors' embedment in the scientific discourse and the scientific communities of the time.

This position has met strong criticism not only from historians of technology, who would stress the independent nature of technological knowledge, but also from established historians of science who were otherwise favorable to the view that technology increasingly became a form of applied science in the later nineteenth century. A. R. Hall summed up these reservations in a programmatic article, the title of which asked, "What Did the Industrial Revolution in Britain Owe to Science?" Hall argued that inventors during the Industrial Revolution in fact owed very little to contemporary developments in science. The mathematics that they used, often for the first time in engineering, was itself centuries if not millennia old. Moreover, references made to science could be shown to have been of no importance to the actual inventions. On closer scrutiny they appeared to be mere window dressing, intended to attach science's authority to engineering ingenuity. Hall argues that Nicholas Leblanc was guided to a useful reaction not by chemical science but by false analogy, that James Watt's separate condenser did not need a theory of latent heat, that Josiah Wedgwood's use of phlogiston language "was no more than a way of rationalizing what was physically observable." Furthermore, he maintains that there are a number of examples where "scientific clarification postdated the technical improvements which it ought to have preceded." The core of Hall's argument, even if not stated explicitly, is that Musson and Robinson fell victim to a *post hoc ergo propter hoc* fallacy, the original sin of much historical theorizing.[17]

According to Hall, "Men like [John] Smeaton, Wedgwood, Watt, [Thomas] Telford, [Richard] Trevithick, [George] Stephenson, [and] Rennie were above all great technical engineers. Certainly they used more exact and sophisticated experimentation than their predecessors; they also relied far more on the analysis of quantitative data. But these novel characteristics—and their significance was still limited enough—should be regarded as rather incipient modifications of an ancient tradition, partly enforced by the desire to use new materials like cast and wrought iron, than as the effect of a revolution wrought within technology by an infusion of scientific theories and discover-

17. Hall, "What Did the Industrial Revolution in Britain Owe to Science?" (quotes on 141, 145).

ies." He concluded: "The history of the Industrial Revolution in Britain shows amply how ready the technical innovators were to work out new ideas empirically when, as was then often the case, science had little guidance to offer." "Science" here stands for the content of science; as to the methodology of science, the story was quite different. In scientific procedure Hall saw many applications to engineering work. These were "attempts to classify technical processes logically," "the employment of systematic experimentation, usually involving model[s]," and "the treatment of data quantitatively."[18]

Margaret C. Jacob has been the most sophisticated recent defender of the Musson-Robinson linear model.[19] In her much refined argument, which bears witness to the cultural turn history has undergone in the past decades, she again gives Newtonian mechanics and the new chemistry of the eighteenth century a prominent place in the Industrial Revolution. As she states at the very start of her study, she is convinced that "the elements of the natural world encoded in science were not peripheral to industrialization and Western hegemony; rather they were central to it."[20] Moreover: "The late eighteenth-century application of scientific knowledge and experimental forms of inquiry to the making of goods, the moving of heavy objects whether coal or water, and the creation of new power technologies dramatically transformed human productivity in the West."[21] Jacob infers from the strong contemporary interest in science, or rather natural philosophy, that there was a growing "audience for science" and that not only the forms of scientific investigation but also its content became instrumental and indispensable for technological development at the beginning of the nineteenth century. In her view, "the industrial entrepreneur girded with skill in applied science" was already the key figure of early European industrialization.[22]

However, it is hard to find positive evidence for Jacob's assertion. Prominent engineers' proximity to science is demonstrated more by association and mutual respect than by demonstrable input of scientific knowledge. Jacob's recent book largely concentrates on Watt and the steam engine, the centrality of which to economic and industrial development in early nineteenth century England has been previously much debated;[23] it is thus only a narrow segment of the Industrial Revolution that she has drawn back into the applied-science debate. Hall's paradigmatic criticism of Musson and Robinson is still applicable and the issue remains unresolved. While science was indisputably

18. Ibid. (quotes on 148, 151, 146, resp.).
19. Jacob, *Scientific Culture*. This book is largely a continuation of her earlier work; see idem, *Cultural Meaning*.
20. Jacob, *Scientific Culture*, 3.
21. Ibid., 4.
22. Ibid., 177.
23. Tunzelman, *Steam Power* and "Technology in the Early Nineteenth Century."

on the agenda of early nineteenth-century industrial figures and very much influenced the way technological problems were discussed, there is still no conclusive evidence that it had a noteworthy impact on technology beyond the sharing of a common methodology and ideology.

These common characteristics of science and technology have been much less contested, although there has been a hidden science-technology hierarchy in some of the literature. One example is Stephen F. Mason's widely appreciated *History of the Sciences*. Mason maintained: "Whilst the content of scientific knowledge did not have much influence upon the development of industry up to 1850, the method of science did."[24] This would suggest a linear model in the realm of methodology, with science emerging as the source of problem-solving strategies; a scientific approach to technology would thereby have existed long before a measurable scientific input. More recent literature has shied away from this second-order determinism and rather stressed equality and simultaneity in the science-technology relationship. "Twins" has become the common metaphor. To Jacob, science and technology at the time of the Industrial Revolution were "fraternal twins, born into a family particularly eager for profits and improvement: they have different *personae*, different looks, but are still profoundly related."[25] To Edwin Layton, they were, later in the nineteenth century, "mirror-image twins," sharing many of the same values but pursuing different ends and acquiring their own kind of knowledge in the process.[26]

As Ian Inkster has shown, the technology twin was, to be sure, often well educated in the sciences—in Britain more than in any other country prior to 1850[27]—but it was not always those who were best-educated scientifically who stood behind the most striking technological breakthroughs. Textiles, machine tools, and ironmaking developed into the backbone of British industry without noticeable scientific input. Textiles, the leading sector in the early stages of industrialization, thrived on improvements of basic inventions, from John Kay's flying shuttle to Edmund Cartwright's power loom, all of which had been made in the eighteenth century.[28] Even the steam engine is seen by most scholars of the Industrial Revolution as having been a great stimulus for science rather than its offspring.[29] As for chemistry, though it was not the most important industry in the early nineteenth century it does remain the one candidate to come to the rescue of what is left of the linear model. Yet even here opinion is split and in its majority leans toward the view that tinker-

24. Mason, *History of the Sciences*, 503.
25. Jacob. *Scientific Culture*, 9.
26. Layton, "Mirror-Image Twins."
27. Inkster, *Science and Technology in History*, 72.
28. O'Brien, Griffiths, and Hunt, "Technological Change."
29. Inkster, *Science and Technology in History*, 70.

ing was more important as a source of tangible results than any reasoning that could be called scientific. No great advances have been made in the literature on this question since Charles Gillispie's work on the origins of the Leblanc process, wherein he showed that the invention of the process was unaffected by the contemporary chemical revolution of Lavoisier.[30] This does not, however, rule out the utility of wrong assumptions on the path to new technologies. In this respect the twins, nascent science and technology, showed a similar mind-set that was hardly their birthright but often enough in the nineteenth century led to lucky outcomes for both.

3. Engineering in Early Nineteenth-Century Industry

During the nineteenth century engineering became the backbone of industrial development. As Akos Paulinyi has shown, material-forming processes were the common denominator of a revolution in technology that accompanied and pushed the early industrialization of Europe and North America.[31] Textiles and railways, the two leading sectors, relied heavily on inputs from mechanical engineering. Machine tools and heavy machinery for moving objects occupied center stage in the workshop of the world. Together with the rapidly developing skills of machinists, production engineers, and factory managers they were the enablers of an unprecedented advance in artifacts useful in everyday life. The pinnacle of early-nineteenth-century technological knowledge was embedded in the output of machine shops from London through Birmingham to Sheffield.

If early engineers had participated in the rage for Newton's mechanics during the eighteenth century, then there was little of practical value to be found in it. Unlike celestial mechanics, terrestrial machinery was governed by the limitations of friction and other forms of resistance and elasticity as much as by economic considerations, all of which limited a purely mechanical approach to its problems. Tremendous heat losses, reflected in efficiency quotients of less than 0.05, rather than theories of heat were the dominant factor in the design of steam engines. Gathering knowledge of what worked seemed a more attractive and promising strategy than theory building in a mechanical world far too imperfect to benefit from idealizations. Again, Hall insists that Watt's tables of empirical data on steam and his collection of guidelines, approximate figures, and proportions for mechanical problems constitute a more lasting legacy than his acquaintance with Black's notion of latent heat.[32]

30. Gillispie, "Discovery of the Leblanc Process."
31. Paulinyi, "Revolution and Technology."
32. Hall, "What Did the Industrial Revolution in Britain Owe to Science?" 140.

As the Dresden research group directed by Gisela Buchheim and Rolf Sonnemann have stressed in their collective "history of the sciences of technology," the method of ratios, borrowed from architecture, became the foundation of much of nineteenth-century engineering. John Farey made extensive use of this approach in his *Treatise on the Steam Engine* (1827). Ferdinand Redtenbacher and Arthur Morin brought it to the Continent. Academic teachers like Olinthus Gregory of Woolwich, and Thomas Young, professor at the Royal Institution, were instrumental in providing engineers with assistance in solving problems of practical mechanics such as friction, kinematics, force of resistance, and material strength.[33] Most important, according to the Dresden group, were the extensive collections of the values of material substances, specifications on dimensions, formularies, and general technical encyclopedias compiled or edited by experienced engineers like Peter Barlow, John Rennie, William Fairbairn, and—the most theoretically advanced member of this group—Thomas Tredgold.[34] Hans Joachim Braun in his essay on methodological problems of nineteenth-century engineering science has emphasized that, in their effort to accommodate the practical needs of engineers, some authors—Julius Weisbach, for example—reduced the complexity of the mathematical tools to such an extent that their peers accused them of being unscientific.[35] This did not, however, render their books any less useful to the practitioner.

The situation was very different in France, where engineering was taught and researched at a much more abstract level, firmly based on mathematics and theoretical mechanics. This, however, did not help domestic industry. Neither the works of Claude Navier, Gustave Coriolis, Jean Poncelet, and Sadi Carnot nor the *géométrie descriptive* of the school of Gaspard Monge connected to the contemporary problems of engineering in any useful way. In terms of industrial applications, the French scientific effort in engineering during the first half of the nineteenth century fell flat. In his magisterial history of industry in France, Denis Woronoff acknowledges no noteworthy contributions from science until the 1880s. He portrays an almost tragic situation where nothing came of the most heroic efforts to promote industry through science: "All branches of mathematics, physics, and chemistry are being mobilized for the promotion of industry. In France in particular, this idea of applied science, which had been so dear to Georges Cuvier and Jean Antoine Chaptal, this idea of the transfer of fundamental knowledge towards the 'arts,' is dominant. To say to the contrary is to submit to routine. It is to admit, however, that the results are hardly convincing. Not that the bridges are

33. Buchheim and Sonnemann, *Geschichte der Technikwissenschaften*, 182–83.
34. Ibid., 185.
35. Braun, "Methodenprobleme der Ingenieurwissenschaft."

burned or that the inventors do not benefit from a scientific culture. But the era is one of unlikely encounters of theories without prospect and inventions without theory."[36] Notwithstanding brilliant French science, competitive technology and knowledge about it continued to issue from Britain. Going to Britain and learning by doing in British firms proved to be the most successful way to transfer technology to the Continent and catch up with the leader.

4. The Tools for Catching Up

Academic Tools

Notwithstanding the very limited success of scientific education in France, a number of European countries took to teaching science and technology on an academic level with a view to fostering their infant industries. It was an effort to catch up with British industry in the absence of a wealth of factories and workshops, which served as educational institutions for learning by doing. The academic teaching of technology was a second-best solution, with blackboards and laboratories substituting, often poorly, for real factories and machinery. It is no surprise that the educational institutions of the military served as a model. The military was in a similar situation in that it had to educate its junior officers in an art that could not be practiced at will; unlike marauding troops of seventeenth-century mercenaries, politically disciplined eighteenth-century armies had to resort to blackboards and maneuvers. Moreover, the construction of fortifications and the buildup of artillery required specialized education. Such motives lay behind the near simultaneous creation of two Parisian institutions: the École des Ponts et Chaussées (1747), which taught civil engineering, and the École du Genie Militaire (1748).[37]

A second precedent came from the experience of the royal mining schools on the Continent. These schools, among the earliest institutions of technological higher education, had constituted a technological backbone for generating royal income from silver and copper production in early modern times. Two centuries of systematic investigation in mining, metallurgy, and its technology had led to the creation of mining academies in Freiberg (Saxony), Berlin, and Schemnitz (Slovakia) in 1770. These academies, together with the École Polytechnique of the French Revolutionary Army (1794), formed the background and model for the institutions of higher education in Continental Europe during the post-Napoleonic Reform era. Governments throughout Europe thought that the only lasting protection from revolutionary wars, invasions, and occupation in an age of mass mobilization was the erection of a

---

36. Woronoff, *Histoire de l'industrie,* 238.
37. Buchheim and Sonnemann, *Geschichte der Technikwissenschaften,* 68–69.

powerful industry along the lines of the British model. Yet they understood industrialization as a task for the state to organize, promote, and control rather than one to be left to the unhampered initiatives of private enterprise.

Within a few decades a great number of technological schools for higher education were established in central Europe, most of them in the German states, where they began to assume a new form, that of the *Technische Hochschule,* or institute of technology. These institutes continued the tradition of the states' educating, organizing, and controlling their own technological experts. Like the French military schools and the German and Austrian *Bergakademien* (mining academies), the newly created polytechnical schools turned out civil servants. As Peter Lundgreen has stressed, the certificates sought by these schools' students were those that qualified them for state service; with the vast majority heading for positions in the state bureaucracy rather than in industry, there was little need for a separate, academic certificate.[38] Lars Scholl and Cornelis Gispen have found that well into the second half of the nineteenth century autodidacts outnumbered college graduates among engineers in German private industry.[39]

The same is true for France and its model institutions of higher technical education. There, more than anywhere else, engineers in state service were separate from engineers in industry. According to Terry Shinn, the evolution of these distinct occupational categories and the immense gap separating them constitute the principal theme in the development of French engineering between 1750 and 1880.[40] Industry benefited little from the prestigious Parisian *écoles.* Most of the latter's graduates during the nineteenth century served in the artillery, followed by the corps of bridges and roads.[41] Between 1830 and 1880 only 10 percent of the *polytechniciens* eventually ended up in industry. Most of them worked in the mining (27 percent) and chemical (22 percent) industries or with the railway companies (18 percent). While those in mining might actually have been employed as *ingénieurs,* in other industries it is doubtful whether their formal education made them attractive to employers. Jean-Pierre Daviet found that they were employed to organize and rationalize manufacturing rather than to innovate technology. They were, in other words, appreciated as managers rather than as scientists or engineers. The organizational skills acquired in their military careers after graduation made them valuable to industry.[42] This emphasis on factory organization over technology recalls again the themes of Babbage's and Ure's books. As Denis

38. Lundgreen, "Engineering Education," 44.

39. Scholl, *Ingenieure in der Frühindustrialisierung,* 297; Gispen, "Selbstverständnis und Professionalisierung deutscher Ingenieure," 49.

40. Shinn, "From 'Corps' to "Profession,'" 185.

41. Ibid., 186.

42. Daviet, *Un destin international,* 231–32.

Woronoff, the historian of French industry, put it: "their intervention maximized the use of technologies [already] employed."[43]

Technology Transfer

Industry continued to rely on hands-on experience, and early-nineteenth-century government proved instrumental in facilitating it. While academic institutions on the Continent strove for excellence in the sciences and continued to reinforce social stratification, as Fritz Ringer has argued persuasively,[44] royal institutions devoted to promoting trade and industry focused on importing skills and empirical knowledge from Britain and disseminating them among artisans, engineers, and industrialists.

Textile machinery and steam engines ranked first among the tangible products of new technology coveted by Continental governments to promote their infant industries. But the importing, legally or clandestinely, of machine tools had a more lasting effect. Machine tools were of paramount strategic importance for the development of autonomous Continental industries. Machine tools represented the only technology that could be employed to replicate itself and at the same time provide tools for other industries. Machine tools were a sort of *perpetuum mobile* of industrialization. Kristine Bruland, drawing upon her extensive studies of Norwegian industrialization, has well summed up the important differences between the textile industry and mechanical engineering:

> Norwegian cotton and wool entrepreneurs acquired machinery, expertise, information and labor from abroad in "packages" which were put together by British textile engineering firms. The entrepreneurs themselves required commercial and marketing skills: they could, and did, remain relatively lacking in technical expertise. In the engineering industry, by contrast, skill development and competence building were central: this is because engineering is not so concerned with the production of standardized products, but is much more a matter of technical problem solving in which competence is of critical importance.[45]

It was the capacity of a country to build its own machines rather than to mass-produce textiles that was to be decisive in its growth in the nineteenth century. Both British and foreign governments knew this all too well. The tools of and impediments to technology transfer have been an important subject for historians of Continental industrialization.

Akos Paulinyi has summed up the British policy intended to prevent this transfer. While scientific knowledge had always been allowed to circulate

43. Woronoff, *Histoire de l'industrie*, 243.
44. Ringer, *Education and Society in Modern Europe*.
45. Bruland, "Norwegian Mechanical Engineering Industry," 266.

freely, until well into the nineteenth century craftsmen and machinery were understood to embody British industry's prowess. To protect its competitive advantage, Britain banned the export of most production machinery, beginning with knitting frames in 1696 and adding greatly to the catalog between 1782 and 1795. A notable exception to the prohibition was steam engines, which were not thought to be of the same strategic importance. Knowledge embedded in machines was not allowed to be handed over to Britain's competitors until 1842, by which time smuggling had reached such a level as to make the protective efforts of the Board of Trade ridiculous. Skilled workmen, as carriers of tacit knowledge, were not allowed to leave the country until 1824. A more liberal trade policy eventually won wide support: British trade, it was argued, stood to gain rather than lose from prosperous and developing neighbors.[46]

The counter to this thorough protection was state-supported espionage, smuggling, and the illegal recruitment of highly skilled labor. That technology transfer from Britain to the Continent was largely illegal in the early years of industrialization did not prevent its taking place, as numerous authors, including William Henderson on Germany and John Harris on France, have shown.[47] Paulinyi found that after 1815 the illegal export of British machinery became a normal if risky branch of international trade; British machine shops interested in selling as much of their product as possible were more than willing partners. In 1841 British entrepreneurs persuaded a parliamentary commission that the export bans were ineffective. At the same time, they praised the achievements of the institutions established to promote local industry in France, Saxony, Prussia, Switzerland, and Belgium.[48] The Conservatoire des Arts et Métiers in Paris, the Polytechnisches Institut in Vienna, the Gewerbe-Institut in Berlin, and the Polytechnischer Verein in Munich were among the major recipients of these early high-tech imports.

The curricula at these institutes were mainly concerned with replicating "imported" machine tools, as Paulinyi has demonstrated with the example of the Prussian Gewerbe-Institut.[49] This early form of reverse engineering extended knowledge and expertise as much as the stock of precious investment capital. Mechanical engineering, the linchpin of the whole new factory system, was firmly rooted in empirical knowledge. Very much relying on the mind's eye (to borrow Eugene Ferguson's felicitous metaphor), Britain's foremost engineers provided few abstractions of their work in written documents. Continental attempts to replicate this kind of knowledge similarly

46. Paulinyi, "Umwälzung der Technik," 470–73.
47. Henderson, *Britain and Industrial Europe;* Harris, *Industrial Espionage and Technology Transfer.*
48. Paulinyi, "Umwälzung der Technik," 473.
49. Paulinyi, "Technologietransfer," 108.

relied on personal experience. Pietro Redondi has stressed that journals at the time were of great intellectual authority, but, as Woronoff has warned, at least until midcentury they were of limited practical value when it came to putting new technology in action.[50] Reports in the *Verhandlungen des Vereins zur Förderung des Gewerbefleißes in Preußen,* for example, mainly drew attention to innovations that would have to be inspected and re-created based on observation. The same is true of publications from private initiatives like the notable *Bulletin* of the Société d'encouragement pour l'industrie nationale, inspired by Chaptal.

While there existed a number of very instructive manuals for textile manufacture, some of which had even been translated,[51] the technologically more sophisticated branch of mechanical engineering had to await the second half of the nineteenth century before "books" were taken seriously on the shop floor. Earlier literature had argued that the second quarter of the nineteenth century had already seen the "objectification" of engineering knowledge and the "institutionalization" of technology transfer.[52] More recent European research, however, has come to different conclusions. In a seminal article, Paulinyi, who has devoted most of his research to investigating the many forms of eighteenth- and nineteenth-century technology transfer to and through Continental Europe, has concluded "that at least until the mid-1850s personal contact among the carriers of technological knowledge, hands-on experience of new machinery, and availability of 'fine specimens' for reproduction were much more important instruments of technology transfer than printed text."[53] In a similar vein, David Jeremy has observed that, for the United States until 1830, "inanimate sources of technical information were inadequate as a vehicle of technology diffusion."[54] Similarly, Woronoff has written of France: "Between 1780 and 1840 a visit of the factories and mines across the Channel is to the elite of innovative manufacturers what the tour to Rome is to artists, an obligatory passage."[55]

Engineering Schools

In view of the inadequacy of academic institutions for promoting technical skills and immediately useful technological knowledge, several Continental states followed and eventually superseded the British model of establishing engineering schools of an intermediate level. As Charles Day has shown, in

50. Redondi, "Nation et entreprise"; Woronoff. *Histoire de l'industrie,* 241.

51. Paulinyi, "Technologietransfer," 118; Bernoulli, *Betrachtungen;* Ure, *Cotton Manufacture;* Montgomery, *Practical Detail of the Cotton Manufacture.*

52. Redlich, "Frühindustrielle Unternehmer und ihre Probleme," 343.

53. Paulinyi, "Technologietransfer," 128–29.

54. Jeremy, *Transatlantic Industrial Revolution,* 72.

55. Woronoff, *Histoire de l'industrie,* 242.

France the *écoles d'arts et métiers* turned out large numbers of students with good knowledge of elementary mechanics but little of science and mathematics. For most industrial purposes this level of education was quite satisfactory; students actually ended up in factories and workshops, where they soon rose to management positions.[56] An effort to bridge the wide gap between the highly academic École Polytechnique, which had the reputation of offering little knowledge useful for an engineering career in industry, and the *écoles d'arts et métiers* was the creation in 1829 of the École Centrale des Arts et Manufactures, which combined science, mathematics, and mechanics in a curriculum aimed at industrial application.[57] As various case studies have shown, graduates of this new school came closest to the ideal of highly competent innovators in industry.[58]

However, early on the École Centrale showed a drift toward an increasingly academic curriculum, as did many of the German trade schools *(Gewerbeschulen)* or polytechnical schools. According to Wolfgang König, these technical schools had been created in the 1820s and 1830s and were based on the idea of comprehensive technical education *(technische Allgemeinbildung)*, which was seen to serve the needs of industry better than an academic education along the lines of the famous French *écoles*. The German trade schools, like the *écoles d'arts et métiers,* were training institutions for future employees in private industry; civil servants, by contrast, continued to come from the universities or, later, the institutes of technology. In the absence of engineering programs at universities outside Prussia and Hesse-Darmstadt, however, the trade schools in a number of German states also had to meet the need for civil servants.[59]

It is common opinion that outright academicization of technical education had to await the 1870s, when a number of polytechnical schools were renamed as institutes of technology; the latter claimed higher academic status and incorporated a greater share of university-level mathematics and science in their curricula. Until the late 1860s, however, there was little engineering education of an academic character available in the German states. The situation was not so different from, say, northern Italy, where the *scuole tecniche* and the *istituti tecnici* provided a similar level of technical education. As Anna Guagnini has shown, Italy in fact preceded Germany in establishing courses in engineering leading to degrees at university level (at Milan in 1862).[60]

Until the last third of the nineteenth century German industrialization owed little to high-powered scientific or mathematical education, and even

56. Day, "Making of Mechanical Engineers."
57. Weiss, *Making of Technological Man.*
58. Daviet, *Un destin international,* 252–59.
59. König, "Technical Education," 66–67.
60. Guagnini, "Higher Education," 512.

the oft-claimed positive effect of the trade schools on German industrial success has met with some reservations. In a recent synthesis of the German literature, König rather soberly concludes "that the influence of technical education on industrial performance was overestimated by contemporaries and is overestimated by historians today. . . . We are justified only in stating that the German system of technical education was not a constraint on the development of industry."[61] At the same time König stresses the importance of formal technical education for engineers as a means to gaining status; this was the policy of the Association of German Engineers (Verein Deutscher Ingenieure) ever since its creation in 1856. The applicability of Ringer's analysis of the social effects of higher education and motives for pursuing it seems even more sweeping in regard to nineteenth-century technical education than the author himself had anticipated.

In recent literature, technical education in its various forms has been credited with less explanatory power than was earlier attributed to it when it comes to analyzing the different paths of economic growth during the nineteenth century. In his comprehensive overview of the literature, Lundgreen warned: "It is one of the dubious retrospective extrapolations from the present to assume that formal education, such as the academic training of engineers, is somehow necessary, if and when the private economy is about to become industrialized."[62] Lundgreen found that before 1870 a great variety of institutional arrangements for higher technical education existed and that there is no convincing evidence in the literature that any of these forms contributed measurably to any nation's industrial development. Referring to Monte A. Calvert's work on American engineering education,[63] Lundgreen suggests that "we should do well to assume a dualism between an older shop culture and the encroaching school culture that eventually gave way to a universal school culture."[64] But that was not to happen before the end of the nineteenth century.

To sum up, the main conclusions of research into technical education prior to 1870 can best be stated negatively. France did not lack a system of technical schools producing successful technicians and engineers for industry. The German system of technical education cannot be shown to have substantially contributed to the country's rapid industrialization. And with the discussion over French economic performance during the nineteenth century having recently been reopened,[65] it is not at all clear which model was superior and to what extent.

61. König, "Technical Education," 81.
62. Lundgreen, "Engineering Education," 39.
63. Calvert, *Mechanical Engineer in America*.
64. Lundgreen, "Engineering Education," 39.
65. O'Brien and Keyder, *Economic Growth*.

Sobering as it might be, recent revisionism regarding the importance of institutionalized technical education during most of the nineteenth century
presents an unsatisfactory picture. Even if societies in nineteenth-century Europe, and especially Germany, were status-ridden, it seems difficult to believe
that their sustained efforts to establish a system of scientific, mathematical,
and technical education did not have more tangible effects. The situation is
similar to that of a related great topic, the Industrial Revolution. Here again,
many negative conclusions—"nots"—have been reached based on sophisticated testing of longstanding orthodoxies about the importance of foreign
trade, transport innovation, steam power, and investment in factories. Donald McCloskey, surveying twenty years of revisionist research in 1994, concluded that "the task of the next twenty years will be to untie the Nots."[66] His
statement seems equally applicable to the question of science's contributions
to technology and industry during the first seven decades of the nineteenth
century. In the absence of new perspectives on this important relationship,
historians of science and technology might well have to accept the verdict of
ruthlessly quantitative economic history, again well expressed by McCloskey:
"Few parts of the economy used much in the way of applied science in other
than an ornamental fashion until well into the twentieth century. In short,
most of the industrial change was accomplished with no help from academic
science."[67] Joel Mokyr has recently developed a promising new approach using Simon Kuznets's concept of "useful knowledge," positing, rather than the
traditional juxtaposition of scientific and technological knowledge, a kind of
"master catalog" that engineers increasingly could draw on.[68]

5. Creating a Common Language

After many failures simply to translocate British technology, the ability to explain "scientifically" a process became the proof that it had been understood
well enough to justify the expense of introducing it at home. The transition
from "non-verbal technology" to "verbal technology," which according to
Jeremy played such an important role in technology transfer, had only just
begun in the early nineteenth century.[69] But as the language of science gradually displaces drawings (the "alphabet of the engineer," in Marc Isambard
Brunel's phrase),[70] as the dominant tool for analyzing a technological process
or an artifact, the "secret" that had made every investment a risk began to
be revealed. New technology, encapsulated in standardized language and

66. McCloskey, "1780–1860," 253.
67. Ibid., 266.
68. Mokyr, "Knowledge, Technology, and Economic Growth."
69. Jeremy, *Transatlantic Industrial Revolution*, 25.
70. For an analysis of drawings in engineering, see Ferguson, *Engineering and the Mind's Eye*.

stripped of its context, could be transferred in an idealized form. Explanation still failed as often as it succeeded, but knowledge that could not be expressed in terms of science and mathematics seemed no longer legitimate and trustworthy. Hans-Liudger Dienel, in his study of nineteenth-century refrigeration technology, one of the showcases of science-led innovation, insists that one of the foremost "applications" of science in this industry was to give credibility and legitimacy to new technologies.[71]

"Scientific analysis," in this perspective, became the tool of a backward industry in need of government support to catch up quickly with the advanced engineering practice and routine in use elsewhere. It is based both on a formal language, which makes it easier to communicate without personal contact, and on the assumption that all reasoning along scientific lines is valid. Using the restricted code of scientific language very much helped to create a common language among engineers, civil servants, and university experts, who had to cooperate in order to develop an infant industry. Lundgreen has drawn attention to a division of labor between "state engineers . . . and civil engineers along the lines of supervision and execution."[72] The two groups, although educated quite differently, needed a common language, which, given the asymmetry of power, had to be the language of the academically educated supervisors. More research in this direction needs to be done, however, before we can fully understand this important dimension of the science-technology relationship in the nineteenth century.

The revisionist view that scientific and academic technical education at universities, the *Technische Hochschulen,* and the French *écoles* had little material effect on innovativeness in nineteenth-century industry is less puzzling when one considers the necessity of creating a common language among state civil servants, who oversaw large technical projects like railways and mining operations, and practical engineers in industry. As we know from Max Weber, rationalization and bureaucratization in modern societies need objective, transparent, and reproducible procedures. Mikael Hård explicitly put his history of the "scientification of refrigeration and brewing" in a Weberian framework.[73] To Weber, depersonalization and objectification of knowledge are among the great achievements of modern rationality, for they increase the regularity and calculability of action. Alexander Gerschenkron, for his part, has shown that industrialization in Continental Europe, unlike in Britain, relied heavily on modern institutions, from investment banks to nation-states.[74] More recently and more generally, John Staudenmaier, in his analyses of pub-

71. Dienel, "Professoren als Gutachter."
72. Lundgreen, "Engineering Education," 40.
73. Hård, *Machines Are Frozen Spirit.*
74. Gerschenkron, "Economic Backwardness."

lications in *Technology and Culture,* has stressed that "the inventor must communicate the value and the nature of the new design concept to an appropriate audience before his/her idea becomes a real invention. The role of engineering theory in such cases is to provide a language for such communication." And referring to studies by Thomas Hughes, Otto Mayr, Lynwood Bryant, and others, he writes: "These examples suggest that engineering theory is aptly described by the metaphor of language."[75] Hård and Andreas Knie have successfully applied the linguistic concept of competing grammars to a comparative history of German and French diesel engineering in the early twentieth century.[76] The history of nineteenth-century technology would greatly benefit from a similar approach.

Science and mathematics were powerful tools for putting technical projects into writing, incorporating them into the rational-legal system of the modern world and thus making them both intelligible and manageable for political decision makers. In this context, the surprisingly high rate of state recruitment among graduates from academic institutions of technical education would indeed materially contribute to promoting the industrialization of Continental Europe. At the same time, it would reconcile, on the one hand, the strong view of contemporaries and most of the older literature that higher standards of technical education sped the development of industry and, on the other, more recent findings that the students of the new schools did not drive innovation in industry.

In creating a common language as a means of catching up with British industry, state civil servants, scientists, and engineers on the Continent forged an alliance that did not exist to a similar extent in Britain. When industries later appeared that relied more substantially on the academic world, this alliance gave them a head start. Especially Germany, with its twenty-some universities and nine institutes of technology, appeared better prepared for the new scientific age.[77]

6. "Engineering Science" and "Sciences of Technology"

Authors from different backgrounds and with different concerns agree that the "scientification" of engineering and technology did not gather momentum before the mid-nineteenth century. The emergence of a substantial body of engineering theory during the second half of the century, which gave rise to early "science-based industries," has, however, been widely accepted as heralding a new era in the science-technology relationship. The change did

75. Staudenmaier, *Technology's Storytellers,* 110.
76. Hård and Knie, "Grammar of Technology."
77. Lundgreen, "Natur- und Technikwissenschaften."

not, however, simply entail science being adopted by engineering in ever more fields; engineering also developed its own scientific approach.

Klaus Mauersberger, in his work on the formation of technological mechanics, has elaborated on the line between useful knowledge and abstractions detached from shop-floor problems. He has characterized the "theoretic kinematics" of Franz Reuleaux,[78] arguably the most theoretical of nineteenth-century German engineers, as "over-theoretical." In advancing what they saw as a universal language for engineering, Reuleaux and his school failed to understand "that if it was possible to interpret a machine as a kinematic chain, it was still impossible to make out a distinct machine from a pre-given language, which in the end limited his language . . . to kinematic analysis."[79] When Reuleaux's approach ultimately failed, Mauersberger wrote, it was because "engineering practice itself . . . had become a corrective against the onesidedness of a kinematic approach to engineering design."[80] The battle between theorists and practitioners, however, raged on during the last decades of the nineteenth century. It took until the early twentieth century before engineering theory descended "from the lofty heights of theoretical mechanics to the practical demands of engineers."[81] At the same time, Mauersberger acknowledged that, in hindsight, increasing theoretical abstraction in technological mechanics at German universities proved indispensable to establishing the subject as the autonomous discipline of academic engineering, which would bear fruit in the next century.[82]

The Anglo-Saxon world obviously experienced a less agonizing transition from celestial to terrestrial mechanics than did the Germanic. David Channel, in his study of the evolving engineering science of W. J. M. Rankine of Scotland (author of a number of standard textbooks for university-trained engineers during the second half of the nineteenth century), stressed the effort to reconcile theory and practice in "a framework in which scientific laws could be modified so that they could accommodate material bodies."[83] Late-nineteenth-century engineering science as developed by Rankine and others was a distinct body of research and education that could not be reduced to "applied science." As Channel observed, "The concept of efficiency, for example, is in some sense a mathematical measurement of the degree in which ideal theories are modified by actual materials."[84]

78. Reuleaux, *Theoretische Kinematik.*
79. Mauersberger, "Herausbildung," 28.
80. Mauersberger, "Maschinenelemente," 160.
81. Mauersberger, "Herausbildung," 21.
82. Ibid., 29.
83. Channell, "Harmony of Theory and Practice," 52.
84. Ibid.

Building on the work of Calvert,[85] James E. Brittain and Robert C. Mc-Math Jr., in their study on the early development of the Georgia Institute of Technology under Robert H. Thurston, present the advent of school culture and the progressive division of labor in engineering.[86] Thurston was explicit about seeing the technical skills used on the shop floor as an application of the theories mastered by academically trained mechanical engineers. He also strongly advocated the view that the ability to codify knowledge and express technology in mathematical terms dramatically enhanced the precision and productivity of industrial processes. The "Russian" system of education that he had first introduced at the Stevens Institute of Technology, drawing inspiration from an exhibit of the Imperial Technical School of Moscow at the Philadelphia Centennial Exhibition, was to become his model. Showing great sympathy for authoritarian European approaches, Thurston was convinced that only the import of the European system of higher technical education would fully mobilize the intellectual potential of American engineers.

In several articles published during the 1970s, Layton put forward the most influential interpretation of the unfolding of engineering theory in nineteenth-century America.[87] He stressed the epistemological differences between science and engineering theory while acknowledging the similarity in methods and language employed. Science and technology as represented by their respective communities looked to him like mirror-image twins—superficially very much alike yet fundamentally different in their outlook and objectives. In an article programmatically entitled "American Ideologies of Science and Engineering," Layton stated the main differences between the two theoretical approaches: "Engineering theory and experiment came to differ with those of physics because it was concerned with man-made devices rather than directly with nature. Thus, engineering theory often deals with idealizations of machines, beams, heat engines, or similar devices. And the results of engineering science are often statements about such devices rather than statements about nature. . . . By its very nature, therefore, engineering science is less abstracted and idealized; it is much closer to the 'real' world of engineering. Thus, engineering science often differs from basic science in both style and substance."[88]

In the late 1960s, scholars in the Soviet Union, eastern Europe, and East Germany undertook a comprehensive effort to investigate the emergence of engineering theory, the building of school culture in engineering, and the

85. Calvert, *Mechanical Engineer in America.*
86. Brittain and McMath, "Engineers and the New South Creed."
87. Layton, "Mirror-Image Twins," "Technology as Knowledge," "American Ideologies," and "Scientific Technology."
88. Layton, "American Ideologies," 695.

institutionalization of engineering on an academic level.[89] The initiative came from Yuri S. Meleshchenko and his research group at the Leningrad department of the Institute for the History of Science and Technology of the Academy of Science of the USSR. Starting in 1969, they studied the methodological and social problems involved in the history and formation of the "technological sciences." The Russian team first developed a periodization of technological knowledge and science, identifying the mid-eighteenth century through the late nineteenth century as the period of formation of technological science. They saw this as a time in which technology made increasing use of science and mathematics and the technological sciences established themselves as a field encompassing a number of disciplines and separate from science proper. Only in the twentieth century did the "classical" period of technological science begin, with robust and durable forms of interaction between technological science and science proper, which increasingly took the lead in developing new technology. In the late twentieth century, according to this periodization, a unified system of science, technology, and production eventually unfolded, a process characterized as the "scientific-technological revolution."[90]

As presented by the Dresden school, the sciences of technology were to apply the "know-why" approach of science, together with its methodology and (mostly mathematical) language, to the man-made world. The concept of "sciences of technology," then, may help us to better understand the evolving school culture of technology during the nineteenth century and to avoid the inconclusive debate about the line of demarcation between science (of nature) and engineering. The Dresden school acknowledged that an increasing amount of research was being conducted on artifacts and the processing of resources for the production of artifacts and, further, that this research was emancipating itself from the narrow horizons of application. The school has thus focused on the intersection of modernity's theoretical approaches to nature and technology. This acceptance of a true intersection of the sciences of nature and the sciences of technology is different from Layton's mirror-image twin metaphor, which stressed the epistemological differences between science and technology.

While the Russian team did not produce many case studies and mostly concentrated on twentieth-century problems, the historical dimension of its approach was taken up and much refined and elaborated at the Center for the History of the Sciences of Technology at the Dresden University of Technology, created in 1978 by the East German ministry of higher education. Similar smaller-scale efforts were undertaken in Bulgaria, Rumania, and Czechoslo-

---

89. A very useful account in German is Blumtritt, "Genese der Technikwissenschaften."

90. Meleshchenko, "Kharakter i osobennocti nauchno-tekhnicheskoy revolyutsii"; Ivanov, Cheshev, and Volossevich, "Besonderheiten"; and Ivanov and Cheshev, *Entstehung und Entwicklung*.

vakia. In East Germany, the Russian *tekhnichesky nauky* was translated almost literally into *Technikwissenschaften* (sciences of technology), which again was understood to be different from and more precise than the established German notion of *Ingenieurwissenschaft* (engineering science).[91] Unlike the Russians, the Dresden team produced a large number of case studies, mostly dissertations, including many on nineteenth-century subjects. Unfortunately, this wealth of dissertations has remained unpublished, although as of 2000 some results had found their way into the *Dresdner Beiträge zur Geschichte der Technikwissenschaften* (Dresden Contributions to the History of the Sciences of Technology) and, in highly condensed form, into the 1990 volume on the history of *Technikwissenschaft* edited by Buchheim and Sonnemann.[92] The Dresden scholars' work was highly appreciated by their West German peers; the school quickly became the authoritative institution on the history of academic engineering in Germany. In recognition of its academic achievements, the Dresden institute hosted the International Committee for the History of Technology (ICOHTEC) congress in 1986 and presented its approach before the international community of historians of technology.[93]

The sciences of technology, in adopting the cognitive ideal of science (of nature), placed theoretical knowledge above empirically achieved recipes, which nevertheless continued to be a backbone of successful innovation in industry, at least during the nineteenth century. To the Dresden school, the epistemological dividing line thus ran within technology rather than between technology and science. Disciplines rather than fields of engineering emerged as the sciences of technology. Although the Dresden approach is far from complete, it remains the most comprehensive attempt to understand the cognitive dimensions of the changing science-technology-industry interplay during the nineteenth century. It does not yet conclusively answer the question as to how qualitatively, let alone quantitatively, important the evolving sciences of technology were to the industrial prowess of late-nineteenth-century Europe and America. The level of scientific activity in technological fields is coupled rather loosely to industrial success at large. In some industries, however, this coupling appears to have been pertinent to early leadership.

7. Science-Based Industries

Electrical engineering, organic chemistry, optics, and cryogenics—these were fields where scientists played an important role in the development of indus-

---

91. For a programmatic statement by the group leader see Sonnemann, foreword; for a history and assessment of this approach see Hänseroth and Mauersberger, "Dresdner Konzept."

92. Buchheim and Sonnemann, *Technikwissenschaften*. The original dissertations are available in Dresden, Munich, and Berlin.

93. Sonnemann and Krug, *Technology and Technical Sciences.*

trial technology and new products. At the same time these were technologies where British industry proved strikingly less responsive than its rivals in the United States and, especially, Germany. These science-based industries became the showcase of the German innovation system between the end of the nineteenth century and the First World War. Decades of investment in academic institutions for technical education seemed to have come to fruition with a singularly successful cooperation between, on the one side, universities, institutes of technology, and trade schools *(Gewerbeakademien)* and, on the other, industry. Scientists and engineers moved freely between these two realms and built hybrid careers at the interface of science and technology. As Eda Kranakis has shown, this was not only a German phenomenon; it also characterized many other outstanding engineers in the nineteenth century.[94]

The Chemical Industry

The most extensively studied of the late-nineteenth-century science-based industries is organic chemistry, in particular the German dyestuffs industry, which by 1900 was turning out close to 90 percent of world dyestuff production. The reasons for this preeminence range from a favorable patent law to managerial skills in defending an early lead. Prominent in all accounts, however, is the high quality of human capital available to industry. Lutz F. Haber, the historian of the German chemical industry, claims that the "German manufacturers . . . were able to draw on a large reservoir of extremely capable chemists whose enthusiasm for research, often of a painstaking, routine nature, was unmatched in other countries, except Switzerland."[95] Between 1877 and 1892 alone, German universities established seventeen new chemical laboratories; at the same time the number of students in science departments multiplied, giving rise to concerns of an oversupply of scientists.[96] Germany became the center of academic education in chemistry. In contrast, "the native Englishman seeking a chemical education before the rise of the civic universities had to resort to a variety of different ways of scraping an education. One way was to go to Germany. . . . Virtually all the English professors of chemistry of the later part of the century had undergone this experience."[97]

What made German academic chemistry unique in the eyes of many historians was the ease with which professors cooperated with industry and the benefits university science derived from industrial technology. Key figures in this two-way exchange—chemists like Carl Graebe, Carl Liebermann, and

94. Kranakis, "Hybrid Careers."

95. Haber, *Chemical Industry*, 129.

96. Wetzel, *Naturwissenschaften und chemische Industrie*, 129–35.

97. Sanderson, *Universities and British Industry*, 19. Similarly, Beer, *Emergence of the German Dye Industry*, 15.

Adolf Baeyer—had started their careers at the trade schools, which had been explicitly created to foster industry.[98] On the other side, as Anthony Travis has shown, early immunology and chemotherapy benefited greatly from technological developments in industry.[99] The principal agents of this exchange were professors who had research contracts with industry. In a recent book on research in the German dyestuffs industry, Carsten Reinhardt confirms the well-known picture of graduate students and assistants in university laboratories pursuing research consistent with the interests of one of the major chemical companies.[100] Eventually an ideal-typical situation developed wherein the university partner would develop and publish on more scientifically fundamental aspects of joint research programs while the head of the industrial laboratory would concentrate on technologically useful dimensions. Likewise, chemists moving from industry to universities would bring technological projects with them. Conflicts over secrecy in the run-up to patenting were inevitable in this setting and usually followed the preferences of the industrial company concerned. In the end, however, it was a very fertile symbiosis, with science playing a crucial and indispensable part. Since the days of the classic study by John Beer, there has been no controversy over the science-based character of this industry in the literature.[101]

The Electrical Industry

The second important science-based industry was electrical engineering. Coming out of physics laboratories, it had its research and innovation centers in both Germany and the United States. And in Thomas Alva Edison it had the most productive inventor of all times (in terms of the number of patents) and the inventor of "a method of invention," if we accept one of the most frequent characterizations of him. Edison was convinced of the value of scientific input when it came to mechanical and electrical problems.[102] Thomas Hughes, in his comparative study of American inventors, characterizes him as someone who built on and expanded tacit understanding: "In this sense he was an applier of electrical science, but his intimate knowledge of the behavior of electrical devices and his ingenious application of accumulated tacit knowledge distinguished him from most scientists, who tended to be more verbal and theoretical."[103] This did not protect him from wasting much of his accumulated wealth in a futile attempt to separate by magnetic means iron ore at reasonable cost. In this he was similar to another nineteenth-century

98. Borscheid, *Naturwissenschaft, Staat und Industrie,* 127–28.
99. Travis, "Science as Receptor."
100. Reinhardt, *Forschung in der chemischen Industrie,* 332–33.
101. Beer, *Emergence of the German Dye Industry.*
102. Hughes, *American Genesis,* 32–33.
103. Ibid., 68.

genius of tacit knowledge, Henry Bessemer, who grew rich on a technologically brilliant steelmaking process whose chemistry he had not understood, and who lost a fortune on an attempt to apply the technology of his converter plant to stabilizing a seagoing vessel to prevent sea-sickness. Tacit knowledge, which had long worked so well for both men, let them down toward the end of their careers, thus betraying its unpredictability.

Wolfgang König has extensively researched the emergence of electrical engineering as an academic discipline in Germany and its relationship to industry. He confirms the view of much of the earlier literature that mid-nineteenth-century theories of electromagnetic machinery had little impact on or use for practitioners.[104] According to König, developments in the 1860s and 1870s were still governed by trial and error, while the 1880s saw increasing application of empirically won formulae; theoretical penetration of engineering problems, however, had to wait until the 1890s.[105] König's findings corroborate Ronald Kline's earlier case study on the development of the induction motor.[106] If science contributed to solving major engineering problems in the design of electric motors and generators, it did so only at the very end of the nineteenth century. Still, earlier science did provide the language to express and communicate problems and empirically won solutions alike.

As in the discussion of the relation between thermodynamics and the steam engine, König has turned the phrase "science-based industry" on its head, calling electrical engineering before the First World War an "industry-based science."[107] As he has shown, using historical statistics, there was a "shift from theoretically educated professors in the 1880s to professors who possessed practical experience in the 1890s" at the German institutes of technology, which at the time dominated higher education in electrical engineering worldwide.[108] "Whereas in the 1880s the institutes and state bureaucrats saw it as applied science, they later came to see electrical engineering as a practice-oriented science."[109] The more successful electrical industry became in Germany, the less "scientific" was academic education. Robert Fox and Anna Guagnini support this interpretation in their study of industrial research in Europe. They stress the character of the workshop as a laboratory and the noncorrelation of academic achievement and industrial success in electrical engineering.[110]

This situation was very different from the one in the United States as de-

104. Buchheim and Sonnemann, *Technikwissenschaften*.
105. König, *Technikwissenschaften*, 305.
106. Kline, "Science and Engineering Theory."
107. König, *Technikwissenschaften*, 227–96; idem, "Science-based Industry."
108. König, "Science-based Industry," 82.
109. Ibid., 87.
110. R. Fox and Guagnini, "Laboratories, Workshops, and Sites," esp. chap. 4.

scribed by Kline. American technical colleges were expected to "teach basic principles and leave practical training to corporations."[111] Given the contemporary excellence of both national systems in promoting innovations in the electrical industry, there is no reason to believe that either strategy was more successful. By recruiting an increasing number of professors from industry, the German institutes of technology imported practical knowledge in electrical engineering and came to be "more concerned with normal design than with innovation."[112] Only after the turn of the century, with the creation of extensive laboratories along the American model, did they begin to play an important part in applied research. These American industrial laboratories, above all the General Electric Research Laboratory, have been a focal point of the debate on the science-technology-industry triangle before the First World War.[113] The GE laboratory is to the history of science in electrical industry what the Bayer laboratory is to the history of science in chemistry. Both were places where efforts to find a solution to a specific problem might draw on and turn to "basic" research. One might add the optical laboratory of Ernst Abbe at Carl Zeiss in Jena to this category, although its impact on industrial development in general was much more limited.[114]

Mechanical Engineering

Mechanical engineering has certainly not been at the forefront of the science-technology debate in the way that organic chemistry and electrical engineering have. Mechanical engineering was more concerned with standardization, systems of measurement, mass production, single-purpose machines, and the "scientific" exploitation of the human motor and human skills, and mechanical engineers toward the end of the nineteenth century became above all organizational innovators, a phenomenon examined in the books of Alfred Chandler and Philip Scranton on American industry.[115] Successful mechanical industry, big or small, relied almost exclusively on shop culture.

Mechanical engineering worthy of the designation "science-based" was found only on the fringes of the industry. One notable example was late-nineteenth-century refrigeration technology, which was, it seems, single-handedly put on a scientific foundation by Carl Linde. The story of Linde, cryogenics, and the science-technology-industry triangle has been concurrently and independently researched by Hård and Dienel.[116] Linde was close

111. Kline, *Steinmetz*, 167.

112. König, "Science-based Industry," 89.

113. Birr, *Pioneering in Industrial Research;* Wise, *Willis R. Whitney;* Reich, *Making of American Industrial Research;* Carlson, *Innovation as a Social Process.*

114. Cahan, "Zeiss Werke and the Ultramicroscope."

115. Chandler, *Visible Hand;* Scranton, *Endless Novelty.*

116. Hård, *Machines Are Frozen Spirit;* Dienel, *Ingenieure zwischen Hochschule und Industrie.*

to unique among successful inventors in mechanical engineering in that he was both a university professor and an exponent of theoretical engineering, "insisting," in Dienel's words, "on a deductive, scientific penetration of practical technology." Linde was the student of famous school-culture proponents like Franz Reuleaux and Gustav Zeuner, and, as Hård has written, he "first stipulated that the most efficient refrigeration process ought to be identical with the reverse Carnot cycle and then determined what refrigerant it ought to apply. He deduced what an optimal ice-machine should look like and what degree of efficiency would be its possible maximum. The outcome of this exercise was the 'disenchantment' not only of nature, but also of mechanical refrigeration per se." While acknowledging this (self-)description of Linde and his work, Dienel, in contrast to Hård, concludes that "this was a problematic precondition for practical work," and he refutes Linde's assertion that his first experimental prototype of 1873 was already a "complete success." In the end, some of Linde's assumptions turned out to have been wrong, and thermodynamics played no role whatsoever during the agonizing years of engineering design, testing, and development.[117] Nonetheless, Linde was convinced that his extensive experimenting was firmly rooted in theory, and during the famous theory-practice debate in German engineering (in 1895) he was in the theory camp.[118]

Rudolf Diesel, Linde's student and later a refrigeration engineer, took from Linde's lectures on the Carnot cycle the idea of designing an optimal internal-combustion engine. However, according to Diesel's own writings and the most comprehensive book-length study of his invention,[119] Diesel always acknowledged that intuition rather than "science" had been the main source for his invention. Science played an important role in directing his curiosity and testing the consistency of his reasoning, but it did not translate simply into technology. On the other hand, no account of Linde's refrigeration process or the Diesel engine can exclude crucial information from science. Technical thermodynamics as it was introduced into industry in the last decades of the nineteenth century is inconceivable without recourse to school culture. To be sure, it was still not the mainstream of mechanical engineering, but it was important in industrial terms. And as Dienel has shown convincingly, practitioners very soon were ahead of engineering scientists, again restoring the shop-school hierarchy of the steam engine in technical thermodynamics.[120]

117. Dienel, *Ingenieure zwischen Hochschule und Industrie*, 104–5 (quotes), 326; Hård, *Machines Are Frozen Spirit*, 231.

118. König, *Künstler und Strichezieher*, 35–46.

119. Knie, *Diesel, Karriere einer Technik*.

120. Dienel, *Ingenieure zwischen Hochschule und Industrie*, chap. 5.

There are not many cases in engineering where science unequivocally led to technological development. Instead, and more often, we find evidence that science served as window dressing or a means of confirming and expressing in standardized language what had been found out earlier. David Mowery and Nathan Rosenberg, who have surveyed the literature on metallurgy, a central discipline in nineteenth-century technology, have concluded that it "was a sector in which the technologist typically 'got there first,' developing powerful new technologies in advance of systematic guidance by science. The technologist demanded a scientific explanation from the scientist of certain properties or performance characteristics."[121]

In the end it was the scientist's skill at measuring and testing for properties, that is, technical skills developed in the pursuit of scientific experiment, that made him indispensable for industrial enterprise. Materials testing became one of the major fields of employment for scientists, as we learn from the literature on laboratories—initially quite small—in the steel industry.[122] On a larger scale, even governmental research institutions like the Physikalisch-Technische Reichsanstalt, described by David Cahan as "masters of measurement,"[123] turned out to be in good measure service agencies for industry's need to quantify and standardize rather than themselves pushing scientific research ahead. At the same time, however, this development confirms that authority increasingly rested with the institutions of science.

8. Conclusion

The historiography of the relations of nineteenth-century science, technology, and industry has shown that science was but one instrument for innovation, and certainly not the most important one. Craft knowledge, designs for the organizational framework of factories, tinkering, and building on experience (which mushroomed as industrial activity intensified)—all these together continued to have more effect on industrial growth than did science. Notwithstanding Margaret Jacob's insistence, talk of science among early industrialists was not recognized to have been equivalent to making use of its findings. Through most of the nineteenth century, science seems to have been much more a mental than a material foundation of industry. In this respect, it was accepted as one of the most important tools for catching up with British industry. Together with mathematics, science became the universal and legitimate language for conversing about technology, rendering other forms of knowledge "tacit." Debates about appropriate forms of engineering

121. Mowery and Rosenberg, *Technology and the Pursuit of Economic Growth*, 33.
122. Livesay, *Andrew Carnegie*, 114.
123. Cahan, *Institute for an Empire*, chap. 4.

education reflected the issue of theoretical versus practical knowledge, which remained unresolved throughout the century. Even "science-based industries" at the end of the century did not escape critical reassessment of the importance of inputs from science. The linear model is now dead; but it has not yet been successfully replaced by a new orthodoxy.

It would be tempting to run a counterfactual test on the performance of nineteenth-century industry in the absence of science in the way that Robert Fogel did for the American economy without railways. Maybe economic growth would not noticeably have been affected: technology and industry might take a somewhat different path, but not necessarily a slower one. In view of the sobering difficulty of reaching affirmative conclusions concerning the Industrial Revolution, it would not even be a big surprise. The importance of science in the nineteenth century, though, is so firmly rooted in the minds of both the actors and the public that it is hard to draw a fine line between its ideological and its "material" impact. And if science and technology are social activities, then they help constitute the actors' minds. Science was certainly and increasingly the shaper of the minds of engineers, state civil servants overseeing the building of infrastructure, the military, and the public at large when it came to making sense of technological achievements. We do not know, nor can we guess, what society would have looked like had it not begun during the nineteenth century to believe in the power of science to a degree greater—recent revisionist accounts of the history of its contributions to technology and industry would suggest—than was justified. But then, that would not be our history.

ACKNOWLEDGMENT

I thank Joel Mokyr for his helpful comments on an earlier version of this essay.

Nine

THE SOCIAL SCIENCES

THEODORE M. PORTER

❦

1. Ordering the Social

REFLECTIVE ACCOUNTS OF the history or the current state of the social sciences account for much by alleging their immaturity. And it is true that the modern social disciplines are not much older than a century. Their establishment was linked to a gradual redefinition of the methods and intellectual content of social knowledge. But to accept the formation of university specialties as the beginning of true social science, and their limits as its proper boundaries, is to beg a crucial historical question: namely, the relation of social science to a wider terrain of social discourses and practices. Even now, social science is widely dispersed and heterogeneous, with broad, ill-defined margins. The individual academic fields are themselves disunified, and the adjective "social" is incapable of holding together a diverse lot of researchers who work on behavior, cognition, economies, politics, and culture. Practitioners outside the academy are in varying degrees liminal, included within some definitions of social science and excluded by others. National and regional divisions have been powerful historically, and remain so. Until the late nineteenth century, university credentialing was almost unknown in these studies, and the problem of identity correspondingly diffuse.

Indeed, the demarcation of social science scarcely became an issue before about 1890. Modern researchers imposed their conceptions retrospectively on previous centuries, relying on newly formed ideals that defined it more narrowly and in their own preferred image. Purity and objectivity became watchwords of professional social science, and as moral values they helped to shape it, but the social sciences did not, indeed could not, cut their links to politics and administration. Divorced from this more practical domain, the academic social disciplines would wither, and in any case would offer far less to interest the historian. In the nineteenth century, social science was inseparable from policy and reform, their union probably closer than that between science and technology. The success of the university disciplines has supported too narrow an understanding of the history of social knowledge, one that also deflects attention from some of its decisive contemporary features. Outside that ordering are forms of social philosophy, inquiry, and practice with antique roots, which were thoroughly reconfigured and flourished as never before during the nineteenth century.

Thomas Kuhn was impressed in 1958–59 by the discrepancy between the normal shop talk of physicists and the philosophical proclivities of his colleagues at the Stanford Center for Advanced Study in the Behavioral Sciences.[1] He concluded from their arguments about method and foundations that the social disciplines lacked paradigms. It may be doubted that the conversations of economists and psychologists so often turn to philosophy in other venues. At least the disposition to ask deep questions has not often disfigured their historical writings. The prevailing historiography continues to view even the "prehistory" of the social sciences through the lenses of the modern disciplines. It would be pointless to deny that our contemporary fields are continuous in some ways with earlier traditions of thought and inquiry. But these were harder to distinguish from practical activities, including exploration, education, governance, and social reform, than their descendants. Often they were not clearly demarcated from each other, or from natural science, literature, and law (and indeed, these problems of definition have persisted). Although there are many laudable exceptions to the disciplinary whiggishness I describe, the only unifying concept for the history of the social sciences has been the discipline—which is also, ironically, the principal axis of their heterogeneity.[2]

Still, the historiography of social science appears now to be irredeemably disparate. Very few works address systematically the intellectual relations and cultural connections among the social sciences.[3] Fragmentation by discipline is only the most conspicuous of the divides that fracture this field, or rather, this multiplicity. Here I emphasize those arguments and approaches that break down the illusion of disciplinary purity and autonomy. I hold histories of social ideas in high esteem, but I aim to get beyond them to specify the significance of social science for a broader cultural history. I want also to sketch out some ways of thinking about the history of social science that transcend its pervasive divisions and that might provide grounds for a rich and multifaceted historiography rather than independent specialist discourses. I discuss first (section 2) the invention of "social science" in the early nineteenth century, and its bearing on the contemporaneous reshaping of "science" as a privileged cultural category. Section 3 considers how nineteenth-century social science has been refashioned retrospectively into a set of distinct sciences by

1. Kuhn, *Structure of Scientific Revolutions*, x.
2. E.g., Brown and Van Keuren, *Estate of Social Knowledge*.
3. Among the few exceptions: R. Smith, *Fontana History of the Human Sciences*, and T. Porter and Ross, *Cambridge History of Science*, vol. 7, *Modern Social Sciences*.

disciplinary histories. In section 4 I begin a reconsideration of the linkages between social and natural science, focusing first on the greater biological domain (section 5) and then on the statistical (section 6). Next, I turn my attention to its setting within politics (section 7) and as a tool of reform and administration (section 8). Finally, I offer some suggestions on periodizing social science (section 9) and a concluding argument about how to situate it within a more encompassing historicism.

2. Inventing Social Science, Inventing Science

The aspiration to a "science of man," on the model of natural philosophy or natural history, was characteristic of, though not quite invented by, the Enlightenment.[4] Philosophical inquiry into questions of soul, mind, morals, and politics and theoretical or empirical studies of trade, money, the state, and alien peoples go back much further. So too do various practical strategies of defining, investigating, and managing people and their activities, under such names as law, the census, and bookkeeping. By the later eighteenth century, the methods, concepts, and credibility of natural knowledge had been linked to the business of statecraft. In the face of sweeping historical changes in the early nineteenth century, all in the shadow of a vast revolution, the need for a proper science of society was advertised with much urgency. Nevertheless, until almost the end of that century it remained scattered and unsettled, and the gathering of distinct studies like ethnography, political economy, and psychology under the rubric "social science" was rare before 1900. To dismiss the very different assumptions and alliances of the nineteenth century as prehistory is to take as natural our contemporary privileging of disciplined professionalism. Conversely, to recognize a longer history is to grant that social knowledge is, in practice, multifarious, and its current form highly contingent.

If the program for social science that began to take shape in the late eighteenth century was new, it nonetheless had a history. Its advocates did not announce a program of refining bureaucratic and legal practices that had been evolving for millenia. Rather, they situated themselves in a narrative of advancing science. For A. R. J. Turgot, the Marquis de Condorcet, Claude Henri de Saint-Simon, and Auguste Comte, the history of science was precisely the history of progress. The creation of a proper social science would complete this grand project. It was, for Comte, the last and highest phase of the growth of science, the key to human perfectibility and orderly social improvement. The particular filiations implied by these accounts varied. Just prior to the French Revolution, Condorcet identified measurement and mathematics as the distinguishing features of science, and endeavored himself to found a so-

---

4. C. Fox, Porter, and Wokler, *Inventing Human Science.*

cial mathematics.[5] This broad understanding was also advanced by Adolphe Quetelet and by a few of his more scientifically inclined statistical colleagues during the second quarter of the nineteenth century. Comte, by contrast, heaped scorn on the social mathematicians, insisting that each new science required its own distinctive methods. It would be a mistake to pattern sociology after physics or to view it as an extension of its immediate predecessor in the hierarchy of sciences, physiology. Nevertheless, Comte framed his origin story in relation to the natural sciences. Sociology did not grow out of, but negated, the "theological" and "metaphysical" stages of society. It was not a mere improvement of prior social practices but an expression of enduring scientific ideals and new scientific methods in a dark and troubled domain where the torch of knowledge was urgently needed.[6]

The purpose of such arguments, by Comte and others, was to announce a new science. What they achieved was rather different. They did not merely extend science but helped to redefine it, and in a way that was not entirely to their own advantage. After all, their pretensions were not readily accepted. Social science has consistently been regarded as peripheral to the scientific enterprise—not quite literature or the humanities but not a natural science either, and perhaps lacking the objectivity, falsifiability, experimental control, or mathematical rigor that distinguishes "true" science. But to dismiss it so easily is merely facile, and even if in the end we are more impressed by the differences between social and natural science than by the continuities, the experience of the social disciplines illuminates the cultural position of science generally. If it is not science, then it is a doppelgänger, and its own characteristic obsession with objectivity and method may be regarded as a kind of scientific ultramontanism—a more extreme scientism than physics.

Social science, through its failure as much as through its success, helped to create "science" as a privileged category. Observers of recent science, impressed by the proliferation of disciplines, subdisciplines, and interdisciplines, can easily forget that the old regime enjoyed no unity of science. Up-to-date Anglophone historians of early modern science, eschewing anachronism, are left with no encompassing label to designate their object of study but only a multitude of antique names: natural philosophy, natural history, experimental physics, mixed mathematics. These studies were, for some purposes, grouped together under the rubric of science, but the label was unhelpful because it was so broad. After all, politics, theology, and language were also, and perhaps equally, sciences. Champions of social science advanced a definition of science that was conspicuously narrowed and yet intended to include their own studies.

5. Baker, *Condorcet;* Manuel and Manuel, *Utopian Thought.*
6. Pickering, *Comte.*

This campaign was not merely implicit. Much nineteenth-century philosophy of science, including the whole project to clarify "scientific method," was
allied to a campaign to justify social science or to reconstruct it. This includes
the very considerable efforts of Comte, John Stuart Mill, and Karl Pearson.
They laid out idealized standards of scientific procedure, or the scientific attitude, through which they hoped to establish a foundation for social knowledge. Rigorously enforced, any of these versions of scientific method would
have been as destructive of physics and biology as of political economy and
psychology. Only occasionally, however, were they invoked as debating points
within the established natural sciences, to challenge the legitimacy of optics,
morphology, or even meteorology. They were deployed, rather, to police the
boundaries of science, to exclude the more popular, rebellious, and undisciplined claimants to scientific status, as well as artistic or literary efforts like
Friedrich Schelling's *Naturphilosophie* or Johann Wolfgang von Goethe's theory of color. Occasionally, as with biological evolution, method was invoked
to refute theories with disturbing implications for religion or for human dignity. It was, however, social scientists who appealed most often to scientific
method, and social scientists were its ironic victims. Their very marginality
made "science" selective and, in a way, coherent, helping it to attain a cultural
unity even as its disciplinary structure and theoretical content became more
fragmented. To its enemies, science became lifeless and mechanical, or shal-
low and complacent, but each of these epithets named the obverse of a
widely recognized virtue: rigorous, impartial, empirical, solid. Advocates of
social science laid claim to the same admirable characteristics and reinforced
them, whether they gained their prize or not.

3. Histories of Unprofessional Social Science

This ambition shaped the historiography of the social sciences, which has
been largely controlled and often written by practitioners. But who are these
people? The ones who write histories are most often at universities. Few if any
are "social scientists" as such; they are psychologists, economists, sociologists,
and anthropologists. Most disciplinary histories have been written by "basic"
researchers, not by technocrats, and emphasize theoretical traditions. These
writers have rarely been interested in the comparatively unstructured empirical methods of the nineteenth century, and they eschew the union of investigation and reform. Most have been reluctant to see their special topics as part
of broader moral and philosophical discourses, to assess how it matters that
much social science was widely seen in its time as a division of politics, philosophy, or law. They have usually believed in the wider importance of their
subject, even in times long past, but have rarely made this larger story their
focus.

The histories they have written have aimed in many cases to provide their disciplines with a living past. Some of the best were written before the history of science became a specialty, long before there were societies, divisions, or journals devoted to the history of any of the "human sciences." A few classics have outlived their children and grandchildren, and not always or only for bad reasons.

Psychology and Economics

For example, Edwin G. Boring's *History of Experimental Psychology,* first published in 1929 and revised in 1957, is the very model of a disciplinary history. The historical learning of this psychologist was immense, yet his book is no mere compilation but, rather, a treasure chest of opinionated reasonableness. It is not so relentlessly focused on experiments as to leave out the philosophical sources of psychological ideas, but Boring took care not to turn Aristotle, John Locke, and David Hume into Great Psychologists *avant la lettre.* "Psychology" per se was defined by experiment and could be dated back no further than the early nineteenth century. The first true "psychologist," Wilhelm Wundt, arrived still later.

Boring was mostly immune to hero worship, insisting instead on the force of an all-embracing zeitgeist, whose character he consistently failed to specify. His first chapter, on "the rise of modern science," placed psychology in a grand scientific tradition, which he then made concrete by linking psychological experimentation to physiology. He was indulgent with suspect relatives, such as phrenology and mesmerism. "In brief, phrenology was playing its ambiguous role as cause and symptom of the *Zeitgeist,* which was moving mind away from the concept of the unsubstantial Cartesian soul to the concept of the more material neural function. Phrenology was wrong only in detail and in respect of the enthusiams of its supporters."[7] He was skeptical of the radical behaviorism that had become so influential in American psychology in his own day. His impulse to write history came in part from a fear that the behaviorists would lead his field down the primrose path of merely pragmatic applications.[8] His preoccupation with basic science was thus intended not merely or mainly to defend the validity of psychology but as part of a struggle for its soul against John B. Watson.

Joseph Schumpeter, an eminent Viennese economist displaced to Harvard, was a particularly distinguished disciplinary historian. His *History of Economic Analysis* was published posthumously in 1954. He justified the endeavor mainly by way of its benefits to the economist but in a way that encouraged

---

7. Boring, *History of Experimental Psychology,* 58.
8. O'Donnell, "Crisis of Experimentalism"; more generally, Ash, "Self-Presentation of a Discipline."

broadening and self-questioning rather than disciplinary complacency. The "significance and validity" of economic problems and methods, he wrote, can only be grasped in relation to the background that formed them.

> Scientific analysis is not simply a logical consistent process that starts with some primitive notions and then adds to the stock in a straight-line fashion. It is not simply progressive discovery of an objective reality—as is, for example, discovery in the basin of the Congo. Rather it is an incessant struggle with creations of our own and our predecessors' minds, and it 'progresses,' if at all, in a criss-cross fashion. Therefore any treatise that attempts to render 'the present state of science' really renders methods, problems, and results that are historically conditioned and are meaningful only with reference to the historical background from which they spring.[9]

Schumpeter's outlook was interestingly anthropological. Science, he proposed, may be simply defined as "tooled knowledge . . . by the criterion of using special techniques." And indeed, if a "primitive tribe . . . uses techniques that are not generally accessible and are being developed and handed on within a circle of professional magicians" then we have no basis for excluding this magic from science.[10] He argued for a sociology of economics. Yet he concluded his introduction with a move to confine sociology of knowledge and to narrow his subject matter from economics to economic analysis, which eludes Mannheim's sociology because its procedures and methods "are almost as exempt from ideological influence as vision is subject to it." In the execution, Schumpeter punctuated his more technical chapters with "review[s] of the troops," "developments in neighboring fields," and "socio-political backgrounds." Historians of economics, however, have largely endorsed the narrower Schumpeter and treated his book as a reference work on the ideas and methods of the great economists.

Disciplinary history of economics admits a variety of possibilities. The institutionalist Joseph Dorfman, for example, chronicled American economics in five volumes that joined economics to economic history and policy and attended to popular assumptions as well as academic theory. At the other extreme is Mark Blaug's *Economic Theory in Retrospect*. Confronted, he tells us, with the "ultra-Marxist" interpretation of economics as class ideology, he "wondered whether the diametrically opposite thesis—economic theory for economic theory's sake—is not less misleading." "Of course," such an approach "would be limited and inadequate," only a part of the story and perhaps not even the best part, but needful, because it is a part rarely told.[11] While he must have felt himself to be struggling against the current, recon-

9. Schumpeter, *History of Economic Analysis*, 4.
10. Ibid., 7.
11. Dorfman, *Economic Mind*; Blaug, *Economic Theory in Retrospect*, preface to second ed. (1968).

structions of this kind are what economists generally like best, and the book has had several editions. It is consulted and cited far more than other, more historical works like his own on the Ricardians or Terence Hutchison's classic survey of the period from 1870 to 1929.[12]

Of the social disciplines, economics and psychology are the least hospitable to historicism. Paradoxically, this may be why there are specialist societies and journals devoted to history of psychology and "economic thought."[13] In the Anglophone world, at least, history has declined in recent decades in economics and psychology departments at research universities but has taken hold at teaching institutions. A common style of research paper treats the work of a famous author on a problem or topic of current interest. The standard of historical scholarship is generally higher in research monographs than in journal articles, textbooks, and grand, panoramic surveys. These last abound. Many or most might well be characterized, in Robert Young's words, as the work of "part-time historical scholars who mistake secondary sources for primary ones,"[14] and who, one may add, reveal very little in the way of cultural or historical imagination.

The history of economics, especially in the Anglophone literature, is overwhelmingly about Britain, until the United States takes over in the twentieth century. Some, like Schumpeter, consider various discourses, interpreted as protospecialties, such as money, credit, and cycles or equilibrium analysis. Most are preoccupied with theory, often following a canonical succession from Adam Smith, Thomas Robert Malthus, David Ricardo, and John Stuart Mill to William Stanley Jevons, Alfred Marshall, and—at last an author who wrote in French—Léon Walras. The background to economic mathematics follows a different line, with more attention to the French, and is often written to a higher standard.[15] Controversy is not wholly absent from these histories of political economy. The debate between Ricardo and Malthus about gluts and overconsumption, for example, is used to set up an issue that resurfaces more than a century later with John Maynard Keynes. But the diversity of economic theories is more often plotted as historical sequence or as the coexistence of specialties. By contrast, the history of psychology is commonly presented in terms of a competition of rival schools: associationism, mesmerism, phrenology, psychophysics, psychoanalysis, brain physiology, behaviorism, and so on. Beneath it all, psychology is depicted as somehow separating itself from philosophy and acquiring the status of science.

12. Blaug, *Ricardian Economics;* Hutchison, *Review of Economic Doctrines.*

13. E.g., *History of Political Economy* and *Journal of the History of the Behavioral Sciences.*

14. Young, "Scholarship and the History of the Behavioral Sciences." Young argued that the problem of body and mind was the crucial one in early psychology.

15. Ménard, *Formation d'une rationalité économique;* Israel and Ingrao, *Invisible Hand;* Dumez, *Économiste;* Zylberberg, *Économie mathématique.*

Sociology and Political Science

Sociology and political science are almost entirely twentieth-century disciplines, yet each is linked to an antique intellectual tradition. In both, history has become largely alien to the dominant traditions of research, though not to the same extent as in psychology and economics. Political philosophy holds on weakly to its status as one of the basic fields in political science, deserving representation in any balanced department. Social theory is less clearly demarcated in sociology but is readily distinguished from empirical research. In both disciplines, historical writing has been concerned far more with great traditions of thought than with disciplinary structures, research methods, and forms of expertise. The more self-consciously scientific areas of sociology and political science are only rarely the objects of historical investigation.

There have been positivist histories of pre-twentieth-century sociology, beginning with Comte's own, but these have little to offer someone who should resolve in our own day to write a disciplinary history. There was not much before 1890 under the label of sociology that a modern, self-consciously scientific sociologist would care to recognize. The historical self-understanding of sociology, at least as it pertains to all centuries before the twentieth, focuses on a few major thinkers. Raymond Aron's outline of the "stages of sociological thought" comes to the late nineteenth century by way of Montesquieu, Comte, Karl Marx, and Alexis de Tocqueville. Robert Nisbet, drawing inspiration from Arthur Lovejoy, examined how the great thinkers, especially Marx and Tocqueville, framed the "unit ideas" of sociology: community, authority, status, the sacred, and alienation. Sociological historians have not wanted to sever social philosophy from the larger patterns of history but have interpreted their discipline, sometimes quite elegantly, as an effort to come to terms with a new world of industrialized economies and democratic politics. The rise of social theory is identified with modernity, yet often also with a nostalgic conservatism, and is always portrayed as an effort to penetrate its seeming incoherence and identify an intelligible order.[16]

Even for theorists, only the period since 1890 is generally regarded as the living past, which cannot quite be relegated to history at all. Sociology, largely American in its disciplinary origins, looks back to a generation of European founding fathers.[17] Anthony Giddens began his 1971 book on Emile Durkheim, Max Weber, and Marx by announcing that social theory requires radical revision, which "must begin from a reconsideration of the works of those writers who established the principal frames of reference of modern sociology."[18] It was the quasi-historical writings of influential sociologists that first

16. Aron, *Etapes de la pensée sociologique;* Nisbet, *Sociological Tradition.*
17. Wagner, "Science of Society Lost."
18. Giddens, *Capitalism and Modern Social Theory,* vii.

set the great Europeans up in this founding role. Preeminent among them is *The Structure of Social Action* by Talcott Parsons, first published in 1937. Parsons dispatched quickly his dead ancestors. "Who now reads [Herbert] Spencer?" he asked, quoting Crane Brinton. The progress of science had passed Spencer by. Weber and Durkheim, however, had still to be reckoned with, as did Vilfredo Pareto, the third person in the Parsonian trinity. Add to them Marshall, to whom Parsons devoted a full chapter, and the founding generation has representatives from France, Germany, Italy, and Britain. He aimed to discover a unity of conception in the works of this diverse group, one that could provide the foundation for real knowledge.

Parsons was more imperialist than specialist. Economics fell within his ambit, and he was tempted in the second edition to impress Sigmund Freud and Franz Boas as well into the "process of convergent theoretical development" over which he presided.[19] After sorting truth from error in the thinking of his protagonists, he was able to discover surprising harmonies in their conclusions. Weber and Durkheim, for example, started "from opposite poles of thought—Weber from historical idealism, Durkheim from highly self-conscious positivism," yet the "basic conceptual framework" at which each arrived was "almost identical." Weberian charisma mapped nicely onto Durkheim's idea of the sacred, and Weber's "legitimacy" was similar to Durkheim's "moral authority." In the absence of mutual influence, this agreement could only be explained "as a matter of correct interpretation of the same class of facts."[20] Pareto, the neoclassical economist, identified "conceptual elements" that belonged, also, to "the *same* theoretical system." This was, at its core, a "voluntaristic theory of action" curiously similar to Parsons' own and the basis, he thought, for the continued advance of social research.[21]

Sociological theorists like Giddens and Jeffrey Alexander have continued to engage with the classical authors as a means to develop their own social theories. Others have written on the institutionalization of sociology, which, however, was unsystematic and undisciplined until at least 1900.[22] Only a few sociological writers on the history of their discipline have been interested in the empirical side of social research, still fewer in its alliances to administration and reform. Typical is Philip Abrams's book on the early history of British sociology, still the standard treatment of its subject, which he wrote as a study of failure. The British tradition, he argued, was largely empirical, "statistical rather than sociological," and lacked the theoretical sense of "knotty

19. Parsons, *Structure of Social Action*, pp. B–C.
20. Ibid., 669, 670.
21. Ibid., 716, 720. On synthetic narratives in the history of sociology, see Levine, *Visions of the Sociological Tradition*, 35–58.
22. T. Clark, *Prophets and Patrons*.

complexity" that distinguished the genuine sociologies of America and the European continent.[23] Stephen Turner, in a more historicist spirit, has shown that even the big ideas of sociology drew heavily from statistical traditions.[24] The history of empirical social research as such was advocated most effectively by the Viennese emigrant and Columbia University sociologist Paul Lazarsfeld. Anthony Oberschall, his student and research assistant, wrote a model study of empirical sociology in Germany from 1848 to 1914. This was a tradition, he argued, that ultimately failed, but for a time had the active participation and support of some of the founders of German sociology, including Weber and Ferdinand Tönnies.[25] Antoine Savoye has studied the history of empirical sociology in France, and Martin Bulmer has written extensively on social surveys in Britain.[26] Both, like Abrams, have doubts about their protagonists, who were reformers rather than academics and thus lacked the detachment of true social science. The early-twentieth-century *Methodenstreit* at the Verein für Sozialpolitik, where Weber pressed for objectivity against the open political commitments of his senior colleagues, looms large in such accounts.[27] They protest, perhaps too much, and their insistent objectivity provides a clue that the links of empirical research to business and government continue to pose a vexing issue. Still, at least they have portrayed a social science that is more than purely theoretical.

Few political scientists have written on the history of their discipline, and these have usually focused, understandably, on the twentieth century, above all on America. The history of political philosophy, in contrast, embraces a grand tradition beginning with the Greeks and involves a form of writing that has sometimes merged with that of historians. Quentin Skinner, a political philosopher interested mainly in early modern England, has written influentially on the possibilities and advantages of a contextualized intellectual history that reads the great books of political philosophy against lesser contemporary writings.[28] The nineteenth century has not provided so fruitful a stimulus to methodological reflection.

Geography and Anthropology

Geography and anthropology have not buried their own histories to the same extent as other social sciences. American graduate students in anthropology are often required to study the canonical works of their field, or rather of a

23. Abrams, *Origins of British Sociology,* 150, 153.
24. S. Turner, *Search for a Methodology.*
25. Lazarsfeld, "Quantification in Sociology"; Oberschall, *Empirical Social Research in Germany.*
26. Savoye, *Débuts de la sociologie empirique;* Bulmer, Bales, and Sklar, *Social Survey in Historical Perspective;* and, in a more historicist vein, Bannister, *Sociology and Scientism.*
27. For a different view, see Proctor, *Value-Free Science.*
28. Skinner, "Meaning and Understanding."

grand tradition beginning with Aristotle. The historical sensibility in these
fields, however, is often self-critical, or at least critical of the fathers as too
closely tied to merchant voyages and imperialism. *Orientalism,* by the literary
critic Edward Said, helped to define this genre of autocritique. Scholarly studies of the literature of exploration and of ethnographic fieldwork, while
emphasizing that these involved asymmetric relationships of power, have recognized also a genuine spirit of curiosity, sometimes even sympathy. Early
ethnography does not, perhaps, reduce to the tropes of dismissal and roman-
ticization of "the primitive," even if the researcher, with his collections and
notebooks, was inevitably an alien presence.[29] Although blending at their
margins into natural science, these disciplines have if anything been more resistant to scientism than the other social sciences. An early essay by Boas, who
moved from a German background in physics and psychophysics to become
the founding figure in American academic anthropology, is revealing. Citing
Alexander von Humboldt and quoting Goethe, he defended geography as a
descriptive science, one that elevated particulars rather than dissolving them
into general laws.[30]

Although exploration has become an appealing topic throughout the academic humanities, the history of geography as a science has been written almost exclusively by geographers. The field is sufficiently unsettled that
textbook authors are sorely tempted to use history aggressively, pruning and
reinterpreting the geographical tradition to defend certain approaches and
criticize others.[31] Much of the best scholarship has come out of France,
where geography was and remains especially influential. There, according to
Vincent Berdoulay, the field was born of a selective appropriation of German
geography in the wake of the humiliating Franco-Prussian War. It was institutionalized under the Third Republic, whose notably inflexible universities
created chairs in the subject decades before any were devoted to Durkheimian
sociology or (much later still) Walrasian economics. It was supported by a
flourishing collection of French geographical societies, elite organizations
that, in this period, mixed utilitarian and commercial aims with an intense nationalism and colonialism. Geography was neither a research discipline for its
own sake nor a factory for technocrats but a discipline called into being to educate the French nation. Paul Vidal de la Blache, its acknowledged leader, defended it as a synthetic study that extended from physical features of the earth
to anthropology and linked them. Vidalian geography defined a special kind
of nationalism, one that interpreted the genius of the French nation in terms
of regional diversity and of rural traditions deriving from a distinctive rela-

29. Liebersohn, *Aristocratic Encounters;* Stocking, *Ethnographer's Magic;* Kuklick, *Savage Within.*
30. Boas, "Study of Geography."
31. Livingstone, *Geographical Tradition.*

tionship to the soil. Generations of French schoolchildren learned to revere *la France,* "the Hexagon," in these terms.[32]

We might choose Robert Lowie's *History of Ethnological Theory* (1937) as an exemplary disciplinary history of anthropology. As with Parsons's, his history aimed to reshape a discipline, not merely to understand its past. Lowie was a student of Boas, and he identified ethnology with the study of culture. "The modern scientific procedure is to refrain from all subjective pronouncements," he declared. Yet he forthrightly praised some fellow anthropologists and censured others, such as John Lubbock, for their disposition to condemn alien peoples for cruelty or lack of religion. He disapproved also of ethnographers who insisted on offering their aesthetic impressions, thereby confusing their science with literature. Empiricism alone was necessarily inadequate. For example, James G. Frazer's study of totemism, the *Golden Bough,* was a literary rather than an anthropological success, because it failed to follow "the strides of theory." But theory had to prove itself in the details. He was critical of system builders, like Spencer and Lewis Henry Morgan, and of "vigorous intellects," like John McLennan, who let their imaginations run ahead of the facts. Indeed, this history of ethnological theory defended careful study and methodological rigor above all. Boas himself, Lowie argued, stood not for bold speculation but for an independent approach and for originality in framing problems. He also praised Durkheim for a methodology that rivaled the objectivity of the "older sciences." He believed firmly in fieldwork and even suggested that missionaries and traders may often understand a culture better than ethnologists—he was perhaps thinking of Margaret Mead—who have paid only brief visits to their peoples.[33] Lowie celebrated the Boasian sense of fieldwork as the core of ethnographic work and no longer as something gentlemen could harvest from the reports—however disciplined—of travelers.[34] Beyond that, ethnology was for him a matter of attitude. Its development was mainly an achievement of the late eighteenth and nineteenth centuries and, in the early period, above all of Germans. But it was problem enough to study indigenous peoples, and he did not undertake to link the birth of his discipline to a German cultural tradition.[35]

4. Social Science and the Public Culture of Science

Comte had made the founding of social science the culmination of scientific advance, and his admirer George Sarton sometimes wrote in the same spirit on its history. The more rigorous philosophical form of history introduced by

---

32. Berdoulay, *École française;* Lejeune, *Sociétés de géographie en France;* Revel, "Knowledge of the Territory."

33. Clifford, *Predicament of Culture,* 30–31.

34. Stocking, *After Tylor;* Blanckaert, "Histoires du terrain."

35. Lowie, *History of Ethnological Theory.*

Alexandre Koyré after the Second World War made Sarton seem naive and relegated social science to the borderlands of history of science. There it has remained, even as the progress of scientific ideas has given way in the historiography of science to institutions, professions, laboratories, and practices. In this case, however, marginality has not always meant dishonor. Histories of science that attend also to social science have on occasion gained recognition for their distinctively imaginative contextualization. Such ambitions still go somewhat against the grain. Notwithstanding much rhetoric to the contrary, some of the most influential moves in recent historiography of science have tended to maintain or even encourage its traditional internalism, if this word may be stretched to include inward-looking studies of institutions and practices as well as of theories and methods. Attention to social science has been a way to connect science to the larger culture.

To see social science as part of a larger scientific tradition does not exclude, but rather stimulates, other perspectives on the development of social knowledge and practice. The analogy to natural science was not self-evident but a matter of open debate and often of deep personal ambivalence. One plausible and powerful rival was the law—itself divided, in Europe, between strictly codified Roman law and the more "organic" tradition of common law. Donald R. Kelley formulated his wide-ranging study of the legal tradition in terms of a fundamental antagonism between *physis* and *nomos,* nature and law, one that remained at issue in the nineteenth century.[36] Religion was another rival to science as a basis for understanding the human domain. As with the law, some of the most interesting interactions involved not a competition but a synthesis of religious and scientific imperatives.[37] The human sciences, and especially psychology, have often been understood as peeling off from philosophy. Psychiatry and physical anthropology were basically medical fields, and medical language has been pervasive throughout the human sciences. In the century of Honoré de Balzac, Charles Dickens, and Émile Zola, literature too helped give shape to social understandings. Wolf Lepenies has written a powerful and influential interpretation of the history of sociology in France, Britain, and Germany in terms of a set of choices, and much indecision, "between literature and science."[38]

Most studies of the relations between the social and natural sciences have emphasized intellectual exchanges. Historians of science tend now to argue that these interactions were not ruled by a hierarchy of prestige or maturity, but were often and interestingly reciprocal. The shared scientific discourses of nature and society could be rich in ideas and often in technical concepts and methods as well. Some of these episodes have been discussed in terms of in-

36. Kelley, *Human Measure.*
37. E.g., Morgan, "Competing Notions of Competition."
38. Lepenies, *Between Literature and Science.*

fluence, or selective appropriation, but in many cases the mutual shaping goes beyond what a language of influence can reasonably encompass. The real issue may not be the exchange of particular doctrines but a shared situation or overlapping problematic, as in the "common context of biological and social thought" discussed by Young.[39] This was not a mere interaction of specialist disciplines but a shared location in a broad public discourse.

5. Biology and Society

Young coined his phrase to characterize the relation between nineteenth-century social thought, especially Malthusian political economy, and Darwin's theory of evolution. The evidence for an important intellectual filiation between natural selection and population doctrines seems conclusive and includes clear statements by both Darwin and Alfred Russel Wallace. Young aimed to demonstrate a deeper link, one that joined biology and social science to the progress of capitalism and to changes of class relations in nineteenth-century Britain.

Malthus expressed his political aims clearly in successive versions of his *Essay on Population*. The succinct first edition of 1798 was provoked by the utopianism of Condorcet and William Godwin and of the French Revolution. On the basis of modest data and a parable of unchecked reproduction and growing scarcity, Malthus showed how utopia would turn to its opposite in a few generations. Beginning with the much longer and, as it now seems, more sociological "second essay" of 1803, he aimed his sights at the system of English poor laws, whose looseness and generosity seemed to encourage irresponsible breeding. They must surely, he argued, lead to more misery and hunger in the long run than they can relieve in the present. His was a vision that discounted benevolent impulses and seemed to his many critics to depict social life as a Hobbesian struggle. In 1835 the poor laws were in fact made harsher after extensive debates and a famous report that invoked Malthusian arguments. Aristocratic charity seemed to have given way to bourgeois insistence on labor, thrift, and self-reliance, an individualistic world of competition. Marx credited Darwin with having shown, unwittingly, that what British economists lauded as the highest stage of civilization was in fact the normal condition of the animal world. Young, writing in the Marxist tradition, interpreted Darwin's Malthusianism as not the borrowing of a unique insight but, rather, the theoretical appropriation of views that were already very much part of his social world. For this reason, too, he argued that late-century "social Darwinism" cannot be dismissed as a sociological misuse of Darwinian biology. It was rather an ideology shared by biological and social thought.[40]

---

39. Cohen, *Natural Sciences.*
40. Young, *Darwin's Metaphor;* Hamlin, *Public Health.*

The category of social Darwinism continues to be used widely, in reference not only to the most powerful economic and scientific nations but to the rest of the world, and especially to Latin America.[41] There it refers to the activity of governments, like that of Porfirio Diaz in Mexico, that look to an economic and administrative elite to modernize their countries. None of these efforts has involved a biological understanding deriving from Darwin's evolution by natural selection. As Robert Bannister has shown, few intellectuals or writers of any influence looked to survival of the fittest as an important mechanism of social progress.[42] More commonly, as with Spencer, the press of want was taken as necessary to promote moral virtues of industry and self-reliance. The seeming callousness of the present would advance a bright future: these habits would be passed on to offspring, leading in the end to a self-regulating order and the disappearance of the state. Spencer's was a utopian view that Malthus could not have condoned, but there is an unmistakable resemblance to the theodicy with which Malthus concluded his 1798 *Essay*. The pressure of population should not lead to social progress through the death of the poor but provide a stimulus to activity and labor.[43] Malthus and Spencer endorsed self-reliance, not endless struggle or progress through the death of the undeserving.

Spencer's evolutionary biology was not far from the mainstream. An abundant literature now shows that natural selection was rarely identified as the principal mechanism of evolution before the end of the century.[44] Most biologists put more emphasis on Lamarckian use-inheritance or directed variation, often in an explicitly teleological way. Even Darwin relied increasingly on alternative mechanisms of evolutionary change. No more than Malthus can he be regarded as a mouthpiece for a widely shared faith in the natural progress of society through competitive struggle. In the early nineteenth century, these biological and economic doctrines drew from Christianity and not merely from naturalism. Religion was not always linked to charity. Boyd Hilton shows that the comparative harshness of social policy in Britain for about six decades after 1789 owed less to the logic of classical political economy than to the severity of evangelical Protestantism. Or rather, religion gave moral force to economic arguments, construing laissez-faire in terms of submission to divine judgment and limited liability as theologically unsound. The weak British response to the terrible Irish famine of 1846 owed much to evangelicalism. By 1859, the God of the Victorians had become appreciably more benevolent, and a merciless doctrine of human advance through elimination of the poor, or even of "savages," had still less appeal.[45] The wide-

41. For further discussion on social Darwinism, see Robert J. Richards's essay in this volume.
42. Bannister, *Social Darwinism*.
43. James, *Population Malthus*.
44. Bowler, *Non-Darwinian Revolution*.
45. Hilton, *Age of Atonement*.

spread biological racism of the late century was linked to a loose cluster of ideas in which evolutionary doctrines of progress were key. The new racism was as often socialist as liberal and was intellectually more nearly Lamarckian than Darwinian.[46] We should not forget that these biological notions of society remained powerful during the birth of the welfare state.

The interconnection of biological and social thought in the tradition of Malthus and Darwin has often been understood in relation to the Industrial Revolution. Malthus, by this view, articulated the viewpoint of modern capitalism, one that regards the poor instrumentally as "hands" for factory labor and refuses to interfere with the marketplace. Darwin then applied this same remorseless logic to the biological world, thereby reinforcing the social argument for unrelenting competition in an epoch of industrial concentration and robber barons. Interpretations of this kind, morality tales of nefarious ideologies formed of misguided scientific theories, are by now so familiar as to be found in introductory textbooks of several disciplines and even in the popular press.

The idea of reducing the social to the natural, very much at issue in the twentieth century, was uncommon until the latter part of the nineteenth. Anxiety about the imperialism of natural science dates mostly from the period after 1860, when German historical economists began to worry that talk of natural laws of society led to a rigorous determinism, ruling out the possibility of reform.[47] Previously, nature had been widely viewed as something more spontaneous and fluid. Peter Reill shows that historicism arose in the German Enlightenment in alliance with a vitalistic, antireductive natural science, and that the social and natural were very much intertwined in the period from about 1750 to 1820.[48] Alexander von Humboldt's *Kosmos* was continuous with this movement, an effort to unify the physical and the cultural sciences under the rubric of geography. Positivism disrupted these harmonies, in the name not of physical reductionism but of the autonomy of distinct domains of experience. Although Comte held that the sciences developed in succession and that physiology, following physics and chemistry, was a precondition for sociology, he forcefully opposed the unification of the sciences.[49]

The relations between nature and culture in the later nineteenth century were most interestingly at issue in anthropology. George Stocking has given a particularly eloquent account of the relations between biological evolution and the resurgence of speculative histories that ranked contemporary peoples along a single trajectory of human progress. But biological reductionism,

---

46. Gould, *Ontogeny and Phylogeny.*
47. T. Porter, "Lawless Society"; Wise, "How Do Sums Count?"
48. Reill, "Science and the Construction."
49. Heilbron, *Rise of Social Theory.* Hayek, *Counter-Revolution of Science,* criticizes the scientific hubris of Saint-Simonian socialism.

whatever it may have owed to evolutionary theories, derived also from increasingly rigid racial doctrines and from an expanding domain of measurement. Anthropometry was used to support, and sometimes to contest, theories of the separate origins of human races. In the later nineteenth century it was deployed to erect racial hierarchies based especially on the angle of the jaw and the shape and interior volume of the skull. Both in England and in France, a broadly ethnographic approach was challenged in the early 1860s by the formation of societies calling themselves "anthropological," dominated by doctors who were interested in physical characters more than kinship, language, or customs. Although the new anthropology gained for itself a certain rowdy popularity, it did not in the end supplant the ethnographic approach in societies devoted to investigations of this kind. Eventually, "physical anthropology" was domesticated as one of the standard subfields of an anthropological discipline. But for several decades, and especially during the eugenic enthusiasm of the early twentieth century, biology and culture were in competition to account for the customs and achievements of diverse peoples. In America, Boas deployed arguments from physical anthropology against racialist theories and made cultural anthropology a center of resistance to biological explanations of human differences.[50]

6. Statistics in History

The history of statistics has blossomed in recent years as an area of intersection between natural and social science. Calculation has long stood in Western society as a paradigm of rationality and also, since late in the eighteenth century, of unfeeling rigor. Statistics seems to be exemplary in the process of bureaucratization and rationalization described by Weber, a characterization that readily accommodates the linking of natural and social science. Weber was keenly aware that a rationalism of this kind rests on values or habits that cannot themselves be derived rationally. He identified the mentality of calculation particularly with capitalism, whose moral foundation he linked, in the most famous of his studies, to the anxieties of predestinarian Protestantism. By the eighteenth century, he argued, this process of rationalization had broken free of the Calvinist sensibility from which it had sprung and was advancing on its own momentum. Bureaucracy and science, including social science, were its vanguard.

Weber provided for social scientists a way of understanding the modern world, and in particular a way of framing and historicizing their own role in society. Many have celebrated their role as agents of rationality. Others, such as the Frankfurt school critics Max Horkheimer and Theodor Adorno, criti-

50. Stocking, *Victorian Anthropology* and *Race, Culture, and Evolution;* Gould, *Mismeasure of Man.*

cized the quantitative mentality in social science as an ally of capitalism, an agency more of manipulation than of understanding, because it refused to probe beneath the surface of things. Herbert Marcuse, a member of the Frankfurt group, wrote a pioneering historical study of nineteenth-century social science, built around a contrast between the philosophical dialectic of Georg Friedrich von Hegel and the scientific positivism of Comte. He regarded Hegel as unmistakably the more profound thinker, but modern social science had followed the easier path.[51] Like Weber, the Frankfurt critics characterized that path in a very abstract way, one that is still echoed when scholars speak without blushing of "modernity" and "the Enlightenment project."

Rationality and bureaucracy have been central issues in the history of probability and statistics. Lorraine Daston argues that probability was commonly understood in the Enlightenment as the mathematization of reasonableness, or, in Pierre-Simon de Laplace's words, good sense reduced to calculation. By late in the eighteenth century, it was defended as the proper basis for determining the advisability of smallpox inoculation and ordering juries and elections, as well as for calculations about life insurance, annuities, and games of chance.[52] Her study concludes about 1840, after Siméon-Denis Poisson's late attempt to give empirical precision to the combinatorics of juries, when subjective probabilities in general came under attack. By this time, she argues, the ambitions of classical probability ceased to be credible. She links its decline to the end of the Enlightenment, particularly to the French Revolution, which undermined confidence in the ideal of individual rationality on which the probability calculus depended.[53] There may indeed be a sense in which nineteenth-century elites despaired of the general power of reasonableness, now that the unruly masses could disrupt the political order. But they did not lose confidence in their own. Social science in a bourgeois culture could not be a research specialty; the nineteenth century was distinctly less tolerant of recondite political discussion than the eighteenth had been. Now it was part of a public sphere, where politics was taken much too seriously to mix it up with what, to untutored good sense, seemed mathematical fantasies. It is revealing that in the later twentieth century, with the growth of a more detached, academicized social science, most of the topics of Enlightenment probability theory have returned, including models of elections, of juries, and even of general reasonableness, now called judgment under uncertainty.[54] The bourgeois century had more use for empirical statistics than for probabilistic standards of rationality. It used them to try to manage populations in an age of instability.

51. Marcuse, *Reason and Revolution*.
52. Daston, *Classical Probability*.
53. Daston, "Rational Individuals."
54. Gigerenzer et al., *Empire of Chance*, chap. 4.

Nineteenth-century statistics had only a little to do with mathematical probability. It was a social science whose business was facts, which were to be exhaustively collected and honestly presented. In the German and Italian states, statistics became a university discipline, a status that the professors felt obliged to earn with profound writings about its definition, methods, and aims. These were so heterogeneous, and of so little use in the practical business of collecting and presenting numbers, as to create mainly perplexity and consternation. Most "statists" worked in census bureaus or other official statistical offices. In Britain and America, and to a lesser degree elsewhere, there grew up statistical societies of business and professional men who busied themselves investigating the morality, order, and cleanliness of the urban poor. From about 1820 to 1880, statistics was far more representative of social science than sociology, then little more than a word. It was a liberal study, a practical science of administration and reform. Statists of this sort did not claim personal expertise, but allowed the facts to speak for themselves and in this way to reveal objectively the real consequences of government decisions and policies. The increasing power of ideology in a more vigorous political life meant that social science should display its modesty, even though it had become more powerful.[55] In the late eighteenth century, an alliance of mathematicians and high bureaucrats had tried to measure the population of France using samples and calculation. Condorcet looked to such methods as the basis of a new, more rational system of governing.[56] But it took humble nineteenth-century bureaucracies of clerks to establish credible numbers for populations and crime rates.

Statistical science was also more humble in doubting its own capacity to reshape the world. Enlightenment philosophes looked often to kings to put great reforms into effect. The new social science, in contrast, assumed that society had its own dynamic, and that the task of legislators and experts was to anticipate and ease the course of progress. There was a social rationality, collective and historical, which could be uncovered by a science that averted its gaze from the soul of the individual and assumed as its mission to classify and tabulate. The most vocal champion of the new science of mass regularities was Quetelet, the Belgian astronomer turned statistician. His faith in natural order arising out of seeming chaos epitomizes the more contemplative aspect of statistical investigation. It was broadly linked to a liberal policy of patient watchfulness and modest intervention, and it gave reassurance that disorder was superficial and transitory.

Quetelet wanted to be the Newton of a new science of "social physics," and his disposition was enthusiastically mimetic. Yet he created something new: it was precisely this stability of mass phenomena, "statistical laws," that

---

55. T. Porter, *Rise of Statistical Thinking* and *Trust in Numbers*.
56. Brian, *Mesure de l'état*.

inspired physicists, like James Clerk Maxwell, and biologists, like Francis Galton, to comprehend nature as, in some important aspects, analogously statistical. Here, too, in the assemblies of molecules we call gases and the transmission of hereditary traits from parents to offspring, could be found a rational order of the collective, produced spontaneously and yet invisible at the level of individuals. This was not merely a technical point but a matter of genuine philosophical interest. Even theoretical physics could participate in a common context of moral discourse, as was revealed clearly in statistical debates of the 1860s and 1870s about free will, chance, and determinism.[57]

Numbers were an important tool of administration in the nineteenth century, but quantification came also to stand for a moral ideal of precision and honest work in the social and natural sciences.[58] Even in the physical sciences—such as meteorology or geology—numbers did not always lead quickly to mathematical theories, and theories certainly did not depend on a quest for mathematical laws to give them meaning. The only social sciences in which mathematical theory gained general acceptance by the end of the nineteenth century were demography and economics. The former was closely linked to statistical studies of populations. Mathematical economics, in contrast, was almost wholly detached from the empirical, even if Jevons defended it as a natural consequence of the quantitative character of economic phenomena. Jevons was himself an inveterate quantifier, but his economic theory was more nearly linked to a British tradition of mathematical logic. Walras, in Switzerland, and Irving Fisher, in the United States, relied on analogies between economic choices and the mechanical motions of bodies in a force field.[59] In Britain, America, and, most powerfully, Germany, there were historicist revolts against the deductive economics of Ricardo and the classical tradition. Most historical economists opposed laissez-faire, pursuing instead an economics allied to moral commitment and state action. But they believed also in empiricism and, more than that, in quantification as an indispensable tool of economic management. Many came to disdain the abstract mathematical neoclassicism of Jevons, Walras, and Marshall, not least because it made no contact with measurement.[60]

7. Democracy and Exclusivity

Already at the beginning of the nineteenth century, but much more decisively at the end, natural science stood as a model of disciplined knowledge. Comte

---

57. Gillispie, "Intellectual Factors"; T. Porter, *Rise of Statistical Thinking.*
58. Wise, *Values of Precision.*
59. Schabas, *World Ruled by Number;* Mirowski, *More Heat than Light.*
60. T. Porter, "Rigor and Practicality" and *Rise of Statistical Thinking;* Koot, *English Historical Economics.*

and Pearson appealed to this idiom as a critique of democracy, arguing that public opinion deserved no more respect in relation to difficult social questions than to astronomical ones. Yet Comte became intensely critical of what he called scientific narrowness, exalting in its place a broadly educated intellectual elite of men like himself. A generation later, he was known for his religion of humanity as much as for his philosophy of science, and, at least in Britain and France, his doctrine supported the activities of gentlemanly reformers rather than learned specialists.[61]

Before about 1890, the technical expertise of social science was more closely associated with its bureaucratic aspect than with its scientific aspirations. Some of the most forbiddingly academic prose in social science was widely discussed and debated. A particularly rich example is Ricardo's very abstruse *Principles of Political Economy and Taxation* (1817). Ricardian political economy was at the center of a movement of liberal reform during the 1830s and 1840s. The classical theory was also applied to questions of imperial administration and to trade union legislation. Rather more startlingly, there was an influential movement of Ricardian socialism, with active working-class support, drawing freely from arguments that in retrospect appear singularly dry and abstract. This involved, to be sure, an important dimension of critique: namely, that orthodox economics was a mere apology for powerful interests; that it ignored the ethical dimension of economy, reducing workers to mere instruments of production; and that it was so blinded by theory as to ignore the facts of depression and poverty. The Ricardian socialists also argued, however, from a labor theory of value to criticize the exploitation implied by the large shares of production going to land and capital.[62] Marx has been called, with no excess of generosity, a minor Ricardian economist. Although his prose was often morally charged, his critique of classical political economy was directed not against its ethical indifference or its blindness to empirical reality but the insufficiency of its theoretical development. His theory of surplus value in its turn was endlessly debated, often in what seems a highly scholastic way, yet by the 1870s it had become an important part of the ideology of German social democracy. Marxian social science, which eventually was debated at universities, gained its fame and notoriety from its association with labor movements, and maintained an important if unsettled association with praxis during the second half of the nineteenth century.[63]

The sciences of mind, too, flourished in a context of wide public interest and discussion. Efforts to materialize thinking and perception by identifying mind with brain contributed, eventually, to more specialized practices and a

61. Kent, *Brains and Numbers;* Kadish, *Oxford Economists.*

62. Thompson, *People's Science;* Yeo, *Contest for Social Science.*

63. Jorland, *Paradoxes du capital;* Schorske, *German Social Democracy;* Carver, "Marx and Marxism."

narrower discourse. Initially, they created a space for philosophical and even public debate of physiological claims. Phrenology, never a specialist practice, illustrates nicely the resonances of this variety of materialism. It was a popular and subversive science from its beginnings. The idealist views it opposed were linked to the power not of an academic discipline but of churches and the old-regime states with which they were, sometimes uneasily, allied. The phrenological brain was hard to reconcile with an immaterial soul, and this likely was an important aspect of its appeal to middle-class enthusiasts in the early nineteenth century and to workers a bit later. The more gentlemanly men of science who redrew the boundaries of science to exclude it were as conscious of its ideological and religious as of its scientific heterodoxy.

Roger Cooter's study of phrenology in Britain has much in common with Adrian Desmond's later account of radical evolutionary theories[64] and with Noel Thompson's work on popular political economy in Britain. All three authors might be said to derive from a shared historiographical context and to describe a common social and political one. Theirs is a class-based, broadly Marxian account of Britain in its first phase of industrialization, and of the part science played in this dynamic. Cooter urges the most far-reaching social implications, bestowing on phrenology something of the antirevolutionary influence that Elie Halévy assigned to Methodism. After about 1825, phrenology attracted increasing interest from the working classes, who also found advantage in its radical tendencies. Its greatest popular success, however, came after 1835, when George Combe's *Constitution of Man* (1828) became a bestseller. Combe linked phrenology to self-improvement and the division of labor, reversing its prior association with radical materialism.[65] Its effectiveness in subduing working-class radicalism remains to be shown, but it was unmistakably part of popular discussion, in no way the preserve of an academic elite. Indeed, it was, intellectually as well as socially, antielitist. The phrenological brain, as Steven Shapin has noted, bore its decisive marks on the surface, and the science was sold as a practical tool that anyone could use to assess the character of friends and strangers.[66]

Phrenology gained support for a variety of reasons. Even by comparison with the moral and political sciences, its standing was tenuous among the elite of science at universities and academies in the capitals of Europe. Its unregulated, democratic character goes some way toward explaining their disapproval: its methods of investigation were loose and unstructured, and its evidence largely anecdotal. Intellectual standards, however, are in this case hard to distinguish from religious and political views. Rival philosophies of

---

64. Desmond, *Politics of Evolution*.
65. Cooter, *Cultural Meaning of Popular Science*.
66. Shapin, "Phrenological Knowledge."

the mind, especially idealist ones, affirmed the integrity and responsibility of the self. Comte, who was intensely concerned about the standards of knowledge, disliked the individualism of traditional moral science, and he was outspoken in his preference for phrenology over introspective psychologies.[67] Despite his support, and despite the intensification of research in cerebral localization,[68] phrenology had become *merely* popular, a pseudoscience, by midcentury. Mesmerism, by origin closer to physics than psychology, was transformed eventually into a mental science, but never an orthodox one, and became even more suspect after it acquired links to spiritualism. Still, it could not be quelled, and it is part of the tradition that gave rise to psychologies of the unconscious and the irrational.[69] Freud, who drew heavily from unorthodox biological theories, had no connection to the introspective psychology of the universities. He inspired a movement like those that supported mesmerism, and his success owed little in the first instance to acceptance by scientists. Rather, psychoanalysis succeeded by attracting patients and through successful popularization in books and articles, not least by Freud himself. Although always controversial, it eventually gained academic standing through its success with a highbrow public.[70]

It would be wrong to infer from popular interest in doctrines like socialist economics and phrenology that social science meant a democracy of learning. But it was not the preserve of experts and professionals, either. During much of nineteenth century, biologists and even physicists had to struggle to establish their authority over the domains they claimed. Elite practitioners of social science were still less able to limit their field to a community of the learned. And even they did not generally think of social science as something technical and remote from ordinary experience.

Of the fields of knowledge discussed in this essay, political economy came closest to being a specialized discourse at the beginning of our period. By 1776, the year of Smith's *Wealth of Nations,* it stood for a refined body of learning. But who was an economist? In Germany, economic doctrines were taught to future bureaucrats at universities, and economic expertise was offered to princes. In France, political economy gradually penetrated schools of law and engineering. In Britain, it was increasingly taught at colleges and universities, beginning in 1805 with Malthus's appointment at the East India College in Haileybury, yet still was closely linked to legislation and reform, "that noble science of politics," and was as much moral as philosophical.[71]

67. Goldstein, "Advent of Psychological Modernism."
68. Young, *Mind, Brain, and Adaptation.*
69. Goldstein, "Psychology"; Ellenberger, *Discovery of the Unconscious;* Winter, *Mesmerized.*
70. Sulloway, *Freud;* Shakow and Rapoport, *Influence of Freud.*
71. Collini, Winch, and Burrow, *Noble Science of Politics;* Winch, *Riches and Poverty.* See also Elie Halévy's classic *Growth of Philosophic Radicalism.*

The triumph of a mathematical neoclassicism, which was largely inaccessible to amateurs, has been linked to the professionalization of economics.[72] But this proceeded very slowly. For another half century after 1870, most British students of economics were pursuing commercial studies.[73] The United States, as usual, introduced economics as an academic profession, in roughly the period 1880 to 1920. There too it was initially a policy science.[74] In Britain until the 1940s, it remained less an academic specialty than a tool of governance, the shared wisdom of responsible, educated men.

And also, to a degree, of women. There were, in fact, many women in nineteenth-century social science, and not always in subordinate roles. Their prominence furnishes some indication of its standing as an improving discourse rather than a technical science. At the British Association for the Advancement of Science, women were allowed only in section F, statistics. The gentlemen of science were ambivalent about this section. From the moment of its creation in 1833, they worried that it might sponsor unseemly political wrangling. The statists, defensively, adopted the most extreme doctrine of empiricism, declaring that theirs was a science of facts and that opinions had no place in their deliberations. Section F, nevertheless, became a very popular one, attracting the participation of local citizens wherever the association met, and sponsoring some notable debates. Members of the natural-science sections came to regard it as a concession to the public. Political arguments at scientific meetings, to be sure, were not exactly open opportunities for the participation of women, even if women were in the audience. Neither, it seems, was the most theoretical of the social sciences, political economy, which involved women such as Harriet Martineau and Jane Marcet as popularizers. Much social science, however, was linked to philanthropy and charity. It aimed to provide for orphans, widows, and the sick, for the inhabitants of prisons, asylums, and poor houses. It aimed to bring the light of education, as well as windows, to the miserable basements that housed the impoverished, to teach them the virtues of thrift, order, and sanitation. This form of social science, as Eileen Yeo points out, was reasonably compatible with accepted roles of upper-class women. It was charity according to the precepts of enlightened maternalism, combining benevolence with stern reason, hence promoting a middle-class ordering of the working-class household.[75] It was, of course, also practiced by men, often men of wealth and standing.[76]

Some women in social science gained considerable fame in the late nine-

72. Coats, *Sociology and Professionalization of Economics,* esp. papers 5 and 7, originally published in 1967 and 1972; Stigler, "Marginal Utility Theory"; Maloney, *Marshall.*

73. Tribe, "Political Economy to Economics."

74. Church, "Economists as Experts."

75. Yeo, *Contest for Social Science.*

76. Chambelland, *Musée social.*

teenth century: Beatrice (Potter) Webb in Britain; Jeanne Weill ("Dick May") in France; Jane Addams, among others, in the United States. Women had a prominent role in the American Social Science Association, and more generally in U.S. social research and social reform in the late nineteenth century. Their activities have, of late, attracted scholarly interest. The broad trajectory of women's roles in American social science reflects familiar patterns. For about three decades after the Civil War, they found new careers and activities open to them in social research and social reform. Some of the scholarship seems to imply that ameliorative social science was somehow natural to women and that the push for objectivity led to their exclusion. But we should not fail to note that for men, too, social science in this period was understood as a charitable activity, tied to morality, or as the indispensable instrument of reform and the promotion of public welfare.[77] Academic professionalization toward the end of the century involved new recruitment patterns as well as a revision of research values. It tended to consign women to special fields like social work, home economics, and child psychology.[78] There were, however, some very prominent women in early twentieth-century anthropology.

8. Powers and Practices

Michel Foucault did not invent the analysis of social knowledge as power, but this interpretation has come to be closely identified with his influence. Yet there is no Foucaultian school of historians of social science. It is perhaps a more impressive tribute to his contemporary reputation among scholars that his writings have inspired historical work on the social sciences by a wide range of scholars from many disciplines. Others have addressed such topics without relying consciously on Foucault. These works form, collectively, a rich and fascinating literature, one that engages more important issues and provides a more faithful picture of the role of social knowledge in the nineteenth century than can studies of specialized ideas and methods.

Foucault's *Discipline and Punish* identifies the prison as a model for the exercise of power in the last two centuries. In place of a regime based on public spectacles of physical suffering, the prison meant confinement under the watchful gaze of the authorities. In the Panopticon of Jeremy Bentham, which Foucault made famous, the prisoners are all exposed to surveillance from a single, central point, but cannot know when they are being observed. The French title, *Surveiller et punir,* emphasizes this visibility and eternal watchfulness, the heart of modern power. It is paired with discipline, with strict routines of performance, regulated, most often, by the clock. "At the

77. Rodgers, *Atlantic Crossings.*
78. Silverberg, *Gender and American Social Science;* Rossiter, *Women Scientists in America.*

heart of the procedures of discipline" is the examination, which "manifests the subjection of those who are perceived as objects." Through it, all are classified, normal and abnormal, then transferred to the cells of a table for a more economical overview—one that, as Foucault stressed, could be quantified. And these are the principles not only of prisons, but also of schools, barracks, hospitals, and factories. "The carceral network constituted one of the armatures of this power-knowledge that has made the human sciences historically possible."[79]

Foucault has never won so devoted a following in history or the social disciplines as in English and art history, but he has inspired, and also validated, a scrutiny of the social sciences that is not limited to theories and ideas. The applied social sciences became more interesting because nefarious; collections of data achieved new dignity as representations; statistical and financial measures were now loci of power; and official or scientific categories imposed themselves on the individuals they were set up to describe. A look beyond straightforward histories of the social disciplines to the wider, more amorphous field of historical writing about social knowledge reveals that numbers and calculations have become the liveliest topic of current research. This work makes social science central to some key developments of modern history.

François Ewald, Foucault's assistant for many years at the Collège de France, has called for moving beyond the history of ideas to "a history of practices and of the types of rationality in which they are reflected." His book on insurance and the welfare state (L'état providence), links the history of statistical reason to an important fin-de-siècle shift in France in conceptions of risk and responsibility. Until almost 1900, the social ideal was upheld mainly by conservatives like Frédéric Le Play. They wanted to leave the health and security of workers to their patrons—who would act not bureaucratically, according to rules, but personally, in response to individual need. Liberals, in opposition, maintained the sanctity of personal responsibility. Between socialization and individuality no compromise was possible, Ewald argues, but the new form of reasoning developed by statisticians like Quetelet bypassed this debate. It redefined risk in collective terms. Whatever behaviors might have contributed to this or that injury to a worker in a mine or factory, a certain rate of accidents follows naturally from its material and social organization. Based on this understanding, the French parliament at last approved, in 1898, accident insurance for workers that did not require a showing of fault. The presumption that employers should pay implied also, somewhat paradoxically, that they were no longer accountable for individual mishaps but only for the accident rate. At the same time, this altered radically the implica-

79. Foucault, *Discipline and Punish,* quotes on 184–185, 305.

tions of taking risks, since the financial costs of misfortune were to be borne by others. The consequences of the new technologies of risk were far-reaching. Insurance soon became a social good, defining in crucial respects the relation of the state to its citizens. Foucault had declared the importance of a new regime of norms, defined largely in terms of external standards, by which people were assessed and judged themselves, in the world of modern prisons and schools. In the case of social insurance, the norms emerged from within, from the statistics themselves.[80]

Ian Hacking has written about statistics in relation to philosophical questions of chance and determinism, but through research on public health officers and census bureaus. Bureaucratic activities of this kind, he argues, brought about the taming of chance.[81] As with Ewald, such investigations interest Hacking because they made social life appear more predictable, hence manageable, and because statistics became a fount of social norms. He also emphasizes the ways that statistics created new objects, such as crime or mortality rates, and contributed to the definition of new kinds of people, defined by occupation, ethnicity, behaviors, and medical conditions. Alain Desrosières, also writing about bureaucratic statistics, has shown how quite different conceptions of, for example, the professions, have become accepted in public discourse and inscribed in government regulatory structures in different countries. Even if they set out merely to describe social reality, these statistical investigations tended almost inevitably to reshape it.[82]

The power of social statistics is not only a matter of administrative control but also one of hearts and minds. Benedict Anderson's *Imagined Communities* includes, in its second edition, a chapter on censuses in Southeast Asia and how they helped to form a sense of nation in Indonesia, a heterogeneous state of many islands. Censuses could also, however, undermine claims for unified nationhood if, as in Italy for example, they persistently revealed sharp differences between north and south. Shaded maps impressed millions of eyes with the disparities of enlightenment between the north and south of France, whose boundary was marked by a line from Saint-Malo to Geneva.[83] Or, as in India, statistics could enshrine one category of differences and undermine another. As Bernard Cohn explains, the first "modern censuses" there involved comparisons of caste levels across regions, with the intention of finding equivalents and establishing an "all-India system of classification of castes."

80. Ewald, *Etat providence,* quote on 25.

81. Hacking, *Taming of Chance;* T. Porter, *Rise of Statistical Thinking.*

82. Hacking, *Taming of Chance* and "Making up People"; Desrosières, "How to Make Things" and *Politics of Large Numbers;* T. Porter, *Trust in Numbers;* Anderson, *American Census;* Mespoulet, *Statistique et révolution;* Curtis, *Politics of Population;* Favero, *Misure del regno.*

83. Anderson, *Imagined Communities;* Patriarca, *Numbers and Nationhood;* Palsky, *Des chiffres et des cartes.*

British elites often admired the differentiation of status implied by this institution. Others disapproved of its rigidity. Whatever the attitude of the colonizers, their investigations helped to standardize caste in India and to objectify it for the Indians themselves as a fundamental aspect of their culture and society.[84]

Most scholarship on the uses of political economy in the nineteenth century is concerned with ideologies and with the formation of economic policy, especially in Britain. Some notable topics include the parliamentary overhaul of the poor laws and of factory regulation in the 1830s, the adoption of free trade in 1844, writings on trade unions after about 1860, discussion of the colonies, especially India, and continued debates about trade toward the end of the nineteenth century.[85] The quantitative side of economics was more closely associated with commerce and banking but often also with natural science. Those, like Jevons, who believed financial panics to be cyclical, often sought explanations in the periodic phenomena of nature.[86] The relation of energy to quantitative economics also went beyond analogy. Norton Wise shows how "work" was simultaneously economic and physical in mid-nineteenth-century Britain. Some political economists hoped that energy measures could give quantitative specificity to a labor theory of value. They also determined the energetic output in relation to the cost of various foods that might be given the poor, to minimize expenses, and they compared humans to machines in terms of output of work. They calculated whether more work would be derived if waste matter were used as fertilizer to grow crops for the consumption of laborers than if it were burned to power engines.[87] Nineteenth-century railway engineers in France developed a quantitative economics, a systematic accounting that joined engineering calculations to financial and even utilitarian ones, as a guide to the planning of public works.[88] Russian economists before and after 1917 gathered statistics and calculated in an effort to understand and reform peasant agriculture.[89] Historians of economics have given short shrift to these forms of political economy, preferring to emphasize a more abstract and theoretical tradition, but this is unjustified even as whig history. Statistics and calculation have never ceased to be central to the work of economists, especially to their activities in business and government. Calculations of energy, of money, and of "utility" have figured not

84. Cohn, "Census, Social Structure, and Objectification"; Dirks, *Castes of Mind;* Cannadine, *Ornamentalism,* chap. 9.

85. E.g., Winch, *Classical Political Economy* and *Economics and Policy;* Barber, *British Economic Thought and India;* Berg, *Machinery Question;* Cowherd, *Political Economists.*

86. Morgan, *History of Econometric Ideas;* Klein, *Statistical Visions in Time.*

87. Wise, "Work and Waste."

88. T. Porter, *Trust in Numbers;* Ekelund and Hébert, *Secret Origins of Modern Microeconomics.*

89. Stanziani, *Economie en revolution.*

just in utopian fantasies, but of concrete strategies for measuring and managing work.[90]

Historians of psychology have been more alert than economists to the interaction of management and measurement in their discipline. This begins, perhaps, with the development of the "personal equation," in which astronomers developed tools, which in retrospect are called psychological, for managing and coordinating a labor force of observers and (human) computers.[91] The history of mental testing has been a lively field for some three decades, involving historians and psychologists in a shared discussion. Here, psychology involves issues of clear importance for a larger history of schooling and the recruitment and formation of elites.[92] Mental testing also helped to shape the psychological discipline, especially in America. Kurt Danziger's exemplary book, *Constructing the Subject,* shows that the growth of the science, including its standards of knowledge, cannot be separated from psychological practices in the public domain. His story, though mostly a North American one, begins in Germany, where so many Americans received their advanced degrees in the late nineteenth century. Psychologists were drawn particularly to the laboratory of Wilhelm Wundt. There they encountered a distinctive set of social relations. Crucially, his experimental subjects were no less expert than the experimenter, for Wundt saw introspection as requiring skill and training. His version of the laboratory, however, proved in the end less influential than Francis Galton's biometric form of inquiry. This latter, as Danziger shows, was nicely suited to the situation of psychology in the American educational system, where age-graded students came to form the raw material for experiments, and where psychological tests were extensively used by administrators to examine and to sort them. After the early twentieth century, the psychological subject was normally a statistical population.[93] Even psychophysics came to be linked to education, as we learn from a remarkable paper by Trudy Dehue. The field that Fechner invented to demonstrate the unity of mind and matter, and that Wundt developed as the heart of an experimental, introspective psychology, was deployed in the early twentieth century to assess learning. Blind testing and randomization, the crucial features of R. A. Fisher's immensely influential scheme of experimental design, were introduced and first used systematically in educational psychophysics.[94]

90. Rabinbach, *Human Motor.*

91. Schaffer, "Astronomers Mark Time."

92. Kevles, "Testing the Army's Intelligence"; Samelson, "Putting Psychology on the Map"; Sokal, *Psychological Testing;* Carson, "Army Alpha."

93. Danziger, *Constructing the Subject;* Coon, "Standardizing the Subject"; Morawski, *Rise of Experimentation.*

94. Dehue, "Deception, Efficiency, and Random Groups." On mathematical statistics in nineteenth-century social science, see Stigler, *History of Statistics* and *Statistics on the Table.*

As late as 1900, the orientation of social science was predominantly a practical one. The familiar modern disciplines had only recently entered the burgeoning American system of universities. Everywhere, specialization was almost unknown, and the aspiration to scholarly detachment was uncommon. Even in Germany, where the sciences of state and, later, society had been taught regularly at universities since the eighteenth century, the orientation was synthetic. An Enlightenment-era education in the cameral sciences might join political economy to mining, agriculture, or ceramics, depending on the local economy.[95] David Lindenfeld shows in his excellent book on the sciences of state that the Humboldtian research ideal arrived very late to these studies. They were, in the words of a contemporary, *Brotstudium* for administrators. In the 1860s, they began to be split between a social science—drawing from economics, history, and statistics—and a "legal positivism" that had less to do with science. Even so, the historical economist Gustav Schmoller continued for decades afterward to teach economics in connection with demography, ethnography, and technology. Commentators as diverse as Schumpeter and Fritz Ringer have demeaned the "state sciences" as mixed up or backward-looking. Lindenfeld defends them for their thorough engagement with practical issues such as land reform, guild regulation, and, later, Bismarck's social insurance reforms.[96]

In Germany, the professoriate had great prestige on account of its high position in the civil service as well as the respect that attached to learning. Its capacity to assume a role in matters of administration was not merely an effect of academic expertise. At the other extreme, in Britain, where social investigation took place mainly in voluntary institutions like the statistical societies and the National Association for the Promotion of Social Science, specialized expertise had still less to do with the standing of social science. The latter body, active between 1857 and 1886, attracted members of the very highest social standing and devoted itself to serious political issues. Its scientific claims were based less on a rhetoric of social laws than on the methodical study of empirical facts. Its members did not disdain theoretical writing, but they also controlled the sort of knowledge that comes from an active involvement in the affairs of government. It may be usefully contrasted with an imitator, the American Social Science Association, which never attracted so exalted a membership and which began to expire not long after its founding in 1865. But to criticize these bodies for their failure to pursue the ideals of scientific detachment that began, around 1900, to be enunciated more forcefully in the academy is to apply an anachronistic standard. The shift to objectivity was inspired in part by the unwillingness of universities to defend faculty who be-

95. Tribe, *Governing Economy*.
96. Lindenfeld, *Practical Imagination;* Ringer, *Decline of the German Mandarins*.

came known for controversial political positions. It indicated weakness as well as strength.[97]

9. The Problem of Periodization

Social science was so heterogeneous a thing in the nineteenth century that one could scarcely expect any consensus on periodization. Indeed there is not. There is general agreement on only one thing: that there occurred a landmark shift in its character between about 1870 and 1890. Historians interested in sociology or in France tend to assign great significance to the French Revolution in 1789. Foucault identified a great rupture around 1800, a shift from a classical, typologizing episteme to a denser and more historical one, though, strangely, the treatment of power in *Discipline and Punish* is hard to distinguish from the classifying mentality of the old regime in his *Order of Things*. A recent collection urges the importance of continuities reaching from the high Enlightenment to the mid-nineteenth century.[98] It is tempting, especially for central Europe, to define 1848 as an important transition.

Any viable periodization of social science must reflect its social history. This means, in the first instance, relating it to the economic and political transformations of modern Europe. Without denying the existence of intellectual traditions that trace back to antiquity, it makes little sense to talk of social science before the Enlightenment. We can connect its origins with the growth of a public sphere, though not necessarily a "bourgeois" one,[99] and hence with the upsurge in political activity in much of Europe and North America in the late eighteenth century. The French Revolution, rather more than the American, made the project of social understanding appear more urgent, the processes of social change more dangerous. Social science continued, perhaps even more than before, to involve a faith in progress, but now mixed often with a nostalgic dimension. Still, the social science of the period up to 1830 was predominantly liberal and largely theoretical or journalistic in character.

The year 1830 marks the beginning of a transition to a new period, first in Britain, France, and the Low Countries. The United States, where Jacksonian democracy inspired Tocqueville's ambivalent reflections, represents a variant. In the German lands, a date in the 1840s makes more sense. It was then that a compelling consciousness of social changes driven by the growth of industry, and soon by railroads and steamships, added to postrevolutionary

97. Goldman, "Peculiarity of the English?" and "Experts, Investigators, and the State"; Szreter, *Fertility, Class, and Gender.*

98. Heilbron, Magnusson, and Wittrock, *Rise of the Social Sciences.*

99. Habermas, *Structural Transformation.*

political nervousness. Progress became complicated because the loyalty of working people to middle-class political ideals appeared increasingly doubtful. Men and women of substance began speaking in many languages of "the social question." The formation of statistical societies in England marked the real beginning of social science as an institutionalized genre. The institutions were necessary because social science was now more interventionist, more administrative. The urban poor and factory workers were the leading objects of its attention. It was obsessed with dirt, darkness, and disease, with ignorance and superstition, crime and revolution. It was no longer mainly a site of ideological opposition to status-based regimes. Social science interpenetrated with the rapidly expanding bureaucracies whose business it was to try to regulate industry, trade, agriculture, and labor, or simply to keep records and, often, to publish them.

Most of the social disciplines look back to founding figures around 1890. In economics, the origins of marginalist neoclassicism are dated to about 1870, and Marshall's treatise of 1890 marks its completion. The founding of Wundt's laboratory gives psychology also a somewhat earlier date, though again there is a slow process of consolidation. Freud, always viewing himself as a world-historical figure, attached the date 1900 to his *Interpretation of Dreams* (1899). Sociology has Durkheim and Weber—canonized by Parsons in the 1930s—and a cohort of Americans at Columbia and then Chicago. Anthropology looks to Boas in America and to the British expedition to the Torrey Straits in 1898. Political science, which in its academic form is rather distinctively American, has a founding generation in the same period, which urged the need for a strong American state.[100] Geography developed gradually under the French Third Republic after 1870 and was then exported to America.

How should these dates be interpreted? It is tempting to deflate them. The social sciences began to assume their modern institutional form in the American research university at just this time;[101] perhaps the Americans invented a discontinuity, and a generation of founding fathers, as a mythical marker of their new status. But if the institutions themselves are to be given credit for this new form of intellectual production, why is the mythologized generation mostly European rather than American? And how does the establishment of social science relate to other historical transitions in a period widely regarded as epochal? After 1900, there was a new physics of the quantum and relativity, a new biology of Mendelian genetics and statistics. In the arts, the 1890s are associated with the birth of modernism. In government, this is the period of so-

100. Seidelman, *Disenchanted Realists;* Farr, "Estate of Political Knowledge."
101. Geiger, *To Advance Knowledge;* Oleson and Voss, *Organization of Knowledge.*

cial democracy and the "new liberalism."[102] We find here the origins of the welfare state (and with it of such bureaucratic interventions as eugenics) and also of an irrational mass politics of nationalism and anti-Semitism. What is often called the second industrial revolution, associated with scientific technologies like organic chemicals and networks of electric power, took off between 1870 and 1890. The period before World War I, despite a modest retreat from free trade, was a time of unprecedented economic interdependence. Most social science in this period involved the effort to come to terms with the economic changes, class tensions, and political instabilities of the fin de siècle. The American professionalizing project, too, has to be understood partly in these terms, as is apparent for fields such as sociology and economics. Even ethnology, though formally concerned with remote and "backward" societies, involved an implicit program for European reform, as Henrika Kuklick shows. Investigation of the "primitive" meant a search for origins, perhaps for an enduring core of religious values but also for elements of savagery that, it seemed, had not yet been expunged from Western societies.[103]

The transformation of social science around 1890 was intellectual as well as institutional, European as well as American. And it was unmistakably part of a larger history, whose economic, social, political, and cultural dimensions were interwoven with the scientific. A rich body of scholarship addresses the rise of modernism. H. Stuart Hughes has characterized it in terms of intellectual discontinuity: a generational revolt against positivism, a discovery of the unconscious and the irrational. Carl Schorske provides an especially rich and convincing account in his study of Vienna, which he calls a hothouse of modernism. Liberalism, allied to a faltering industrialization, came late to Austria, and began almost immediately to break down as mass politics turned vicious. Artistic modernism expressed despair at the prospects of liberalism and disenchantment with politics. Yet Freud, for example, saw the irrational in psychology as in some way mirroring the threatening mass movements of his time. His therapies aimed at self-understanding and were in some ways a displacement of politics, which remained liberal if not optimistic. The aspiration of European social thought to science meant penetrating to the causes beneath an unattractive surface, and gaining some control of them.[104]

There was some shift in European social science toward a posture of objectivity in the years around 1900, in part because politics seemed often so un-

102. On social democracy, the new liberalism, and social thought, see Kloppenberg, *Uncertain Victory*, and Collini, *Liberalism and Sociology*.

103. Kuklick, *Savage Within*; Kuper, *Invention of Primitive Society*.

104. Hughes, *Consciousness and Society*; Schorske, *Fin-de-Siècle Vienna*. More generally, Ross, *Modernist Impulses*, esp. introduction.

inviting. This was simultaneously an affirmation of higher values, of truth, and an expression of despair. The easy victory of academic social science in the United States over a voluntaristic, mugwump variety reflects the comparative weakness of American intellectual elites. Dorothy Ross's authoritative account of the social sciences in America identifies a cultural anxiety that mirrors, in a way, the European experience, as the rise of social discontent in the 1890s threatened the traditional exceptionalism of American political thought. There is, she implies, an element of escapism in the abandonment of history and the move to scientism in this period. Mary Furner identifies a more specific set of conflicts that exposed the weakness of American social science and inspired a retreat to objectivity. Thomas Haskell is probably correct to insist on the harmony between specialist, university-based social science and modernity, but this was a distinctively American version of the modern, and one that remained inchoate even in the early twentieth century.[105]

University-based disciplines of social science at the end of the nineteenth century were not, however, uniquely American. Mitchell Ash gives a fascinating analysis of German psychology, growing up uneasily within philosophy faculties and moving toward the holist antireductionism of the gestalt, which was in some ways antimodern but not straightforwardly antiliberal. German anthropology in the same period was associated with the new empire abroad and with a naturalizing project—a new antihumanism—at home. In place of the self-cultivation of European historicism it offered the naturalistic study of alien peoples, taken to be unprogressive and thus outside of history.[106] In England, the general acceptance by university faculty of Marshall's neoclassical economics involved a curiously unstable mix of detached, technical science and continued aspirations to serve the public good.[107]

In the United States, too, one should not exaggerate the extent or the speed with which specialized university disciplines detached themselves from administration and social reform in the name of science. Early in the century, a vigorous social-science utopianism, aspiring to design a brave new world, flourished in the settlement houses, Taylor societies, and foundations, drawing from religious traditions and from utopian dreams that were powerful even in the academy.[108] The antagonism between objectivity and commitment was sharpest in relation to leftist politics, and even the advocacy of socialist measures became more acceptable during the Depression. Still, the

---

105. Ross, *American Social Science;* Furner, *Advocacy and Objectivity;* Haskell, *Professional Social Science.*

106. Ash, *Gestalt Psychology;* Zimmerman, *Anthropology and Antihumanism.*

107. Maloney, *Marshall.*

108. Alchon, "Self-Applauding Sincerity."

model of social science that organizes it into disciplines and locates it above all in universities was established for the first time in the United States near the end of the nineteenth century. This is the heart of the transition around 1890 from a social science of administration and reform to a set of specialist disciplines located in universities. But it was neither an autonomous set of institutional developments nor a passive response to society's needs. The new social science was an intellectual program to comprehend and manage the sharpened social tensions that engulfed Europe as well as America, and much of the rest of the world too.

10. Beyond the Histories of Academic Specialists

If it is anachronistic and unduly narrow to divide the intellectual terrain of the social into timeless specialties, this does not mean that it is illegitimate to identify traditions of social thought and action. Many of the very best historians of social knowledge, such as Stocking and Danziger, have written as historians of anthropology or psychology. They have, however, recovered a past that is not merely a prologue to the present and have sought out rich new ways to contextualize their subject matter. They have also treated their subjects as actors, not only thinkers, and refused to draw a line between knowledge and intervention.

My criticism here has been directed primarily at the tendency of disciplinary historians, and sometimes of intellectual historians and historians of science, to take current academic boundaries as natural and to ignore the broader circumstances of social science. But it is equally unsatisfactory to treat social science as little more than a reinscription in scientific language of prevailing values. Too many social and cultural historians have reduced it to a mouthpiece of the dominant classes, adding its authority to ideologies of racism, individualistic struggle, or masculine domination. If social science is not closed off from the world, neither is it merely passive. Its discourses have some autonomy, some customs and rules of their own, and its practitioners do not merely articulate ideas but modulate or redirect their speech in a competition for prestige and resources. This perspective has most often been attached to Pierre Bourdieu's metaphor of the intellectual field, which however does little to structure an understanding of this kind. In his own work, Bourdieu tends to speak reductively of the "legitimation of domination," which needs to be "unveiled" by a strangely detached sociology that is his own.[109] Still, historians of social science need to recognize, with him, that speaking and writing are a form of practice, pursued by human actors who have their own purposes and ambitions, and that science has its "habitus," which re-

109. Bourdieu, *Noblesse d'état*, 7–15.

sponds without capitulating to pressures from the political and social order. Its relation to the world is not simply one of harmony or subordination, nor of identification with a particular faction, but often of tension and anxiety.

Still, the greatest need of the history of social science at present is not to recognize its autonomy. It is to find new and richer ways to understand its interactions with the societies in which it works. To write histories as if social science were or should have been practiced in isolated disciplines, committed to the disinterested pursuit of knowledge, having put moral commitments aside, is not merely complacent, not merely an effort to impose present values on the past. It is to defend a vision that remains deeply problematical and one-sided. Claims for objectivity, ironically, are also claims for a greater role in democratic political life. The social sciences depend on this role as detached and independent experts, and cherish the myths that support it. Social scientists are now found in many agencies of government, in all areas of business life, in foundations, charities, schools, militaries, and prisons. University researchers work as consultants or, at the very least, train students who will do so. These forms of involvement, and not only specialist publications by university faculty, define the place of social science in the world, not merely as a by-product but as the intrinsic outcome of two centuries of history. We need a more subtle historical appreciation of this complex, multifaceted tradition that gradually ordered the social during the nineteenth century and, in so doing, became social science.

Ten

INSTITUTIONS AND COMMUNITIES

DAVID CAHAN

❦

1. A Historiographical Paradox, Characterization, and Scope

THE HISTORY OF scientific institutions and communities is marked by a historiographical paradox: A large number of scholars—indeed most—routinely express their interest in the subject, yet relatively few pursue it in a systematic and concerted manner. A comparison of the *Isis Guide to the History of Science* (1992), with its very long list of scholars declaring their interest in the study of institutions and communities, and the annual *Isis Current Bibliography* issued by the History of Science Society, with its very short list of publications under the section "Communities and Institutions," points to this contradiction.

So does the relative (and nearly absolute) lack of systematic and concerted attention to institutions and communities within general reference works in the history of science. For example, René Taton's edited volume *Science in the Nineteenth Century,* a tome of more than six hundred pages that appeared first in French in 1961 and then in English in 1965, provides nothing on communities and only a rather general and brief discussion of institutions.[1] This shortcoming cannot simply be ascribed to the French or to the relative scholarly immaturity of historians of science circa 1960. A generation later, a British editorial team produced the *Companion to the History of Modern Science,* an impressive volume of nearly eleven hundred pages containing sixty-seven individual, often highly sophisticated, articles devoted to a wide array of topics concerning the history of modern science. Yet the *Companion* contains not a single historical article devoted to nineteenth-century institutions or communities; indeed, the topic of "communities" is not even listed in the volume's index, while "institutions" gets exactly one reference.[2] Given this neglect in

1. The closest this volume comes to a discussion of communities and institutions is Taton, "Causes of Scientific Progress in Western Europe," and Cohen, "Science in the United States." Cf., too, Taton, "Emergence and Development."

2. Olby et al., *Companion*. To be sure, the volume does contain an unhistorical article by Pinch, "Sociology of the Scientific Community," which is essentially a plea for the sociology of scientific knowledge. In the same volume see also W. H. Brock's instructive historical survey, "Science Edu-

general reference works, it should come as little surprise to learn that there is no reference work for scientific institutions and communities equivalent to, say, the *Dictionary of Scientific Biography* or the *Dictionary of the History of Science*.[3] Some sort of *Companion to the History of Scientific Institutions and Communities* would be a most useful and welcome volume.

This neglect of nineteenth-century institutions and communities within general reference works is all the more disconcerting in that there do exist several surveys that form good starting points for more advanced investigations. As long ago as 1904, John Theodore Merz, one of the founding figures of the history of science, devoted three long and penetrating chapters in the first volume of his magisterial four-volume study, *A History of European Thought in the Nineteenth Century,* to what he called the "scientific spirit" in France, Germany, and England. Merz's text and, above all, footnotes still provide newcomers to the field with a learned introduction to nineteenth-century scientific institutions and communities.[4] So too do such far more recent and general studies as Donald Cardwell's *Organisation of Science in England,* Colin Russell's *Science and Social Change, 1700–1900,* Robert Fox and George Weisz's edited volume *The Organization of Science and Technology in France, 1808–1914* (in particular, their fine introductory essay, "The Institutional Basis of French Science in the Nineteenth Century"), Harry Paul's *From Knowledge to Power: The Rise of the Science Empire in France, 1860–1939,* Joseph Ben-David's chapter on Germany in his *Scientist's Role in Society,* and Robert V. Bruce's *Launching of Modern American Science, 1846–1876.* These and other books, along with Merz's *History,* provide entrée to the systematic and coherent study of nineteenth-century scientific institutions and communities for their respective countries.[5] Nonetheless, we sorely need a comprehensive, one-volume work that surveys the origins, development, and growth of nineteenth-century institutions and communities, roughly in the way that James McClellan has done for eighteenth-century scientific societies.[6]

The reason for the paradox, then, lies neither in a lack of sincerity on the

---

cation"; Emerson, "Organisation of Science," which offers (quite limited) discussion of scientific institutions for the early modern period; and several other related instructive historical articles: Schroeder-Gudehus, "Nationalism and Internationalism"; Morrell, "Professionalisation"; Shapin, "Science and the Public"; and Outram, "Science and Political Ideology."

3. Gillispie, *Dictionary of Scientific Biography;* Bynum, Browne, and Porter, *Dictionary of the History of Science.*

4. Merz, *History of European Thought,* 1:89–301.

5. Although the present essay is limited to discussions of developments in America, Britain, France, and Germany, mention might nonetheless be made of James R. Bartholomew's fundamental study *The Formation of Science in Japan,* which concentrates on the formation of the Japanese scientific community and institutions in the late nineteenth and early twentieth centuries.

6. McClellan, *Science Reorganized.*

part of scholars nor in a lack of introductory, panoramic accounts of nineteenth-century scientific institutions and communities. Moreover, studies in the social history of nineteenth-century science often provide much (if not always systematic) information about institutions and communities, biographical studies sometimes do so as well, and even conceptually oriented studies occasionally provide useful, if often dispersed, information about the institutions and communities in which their subjects worked and lived or their concepts developed. Rather, the reason seems to lie in the difficulty of characterizing institutions and communities in a way that provides the historian with a useful means for pursuing systematic analysis. Placing their study within some sort of conceptual framework is no easy matter.

For present purposes, scientific institutions and communities may be characterized as consisting of populations of individuals who share similar cognitive interests and values that serve to provide them with a collective social identity and to advance individual scientific careers and group needs. Such populations are naturally composed of individual scientists and their variegated associates, yet they only become institutions and communities when those individuals—perhaps only few in number—act in concert over an extended period of time and perceive themselves as bound together in some particular professional manner. When such individuals hold more or less similar cognitive interests and values and act more or less together as a group, we often speak of a "community," though scholars sometimes use the related (and usually undefined) notions of "discipline" or "school." Like individuals, institutions and communities usually have multivalent and changing identities and functions, and therein lies one of the greatest difficulties in studying them. Moreover, institutions and communities are not limited to formal, named social organizations or physical structures: societies, associations, institutes, laboratories, departments, journals, and the like. They include also the informal groupings of scientists ("networks," "invisible colleges," and, again, "schools") that, for instance, facilitate the transfer of scientific knowledge, skills, and values from individual to individual and generation to generation. Hence, in principle a scientific institution or community may be an academic or nonacademic unit within a state or private structure; a physical building housing laboratories, classrooms, and the like; or, more abstractly, a set of social and professional relations among individuals. However diverse such institutions and communities may be, to one degree or another they all share a collective identity.

This essay seeks to portray something of the state of study of nineteenth-century scientific institutions and communities and to draw attention to neglected or insufficiently examined areas of study. It does so by highlighting certain studies that accord with (if sometimes only implicitly) the above-mentioned characterization. It makes no pretense of covering the entire field—a goal that neither its limited length nor the author's competence allows. It offers apologies *en avant* for topics and studies to which it does not refer—the development of botanical gardens, zoos, museums, and astronomical observatories, scientific expeditions, the popularization of science, "research schools," Humboldtian science, and journals, to name a few—or of which it was unaware.

The essay's scope is further limited by two other major conditions: First, it restricts its attention to empirical studies of the major scientific institutions and communities of nineteenth-century Britain, France, Germany, and the United States. The theoretical analysis of scientific institutions and communities—work done largely by sociologists and, to a lesser extent, philosophers of science—is not the subject of this essay. Nor is the increasingly active scholarly area of science in the colonial world: here a separate essay would be warranted. These four countries constituted the heartland of nineteenth-century science, even as, toward the end of the century, Canada, other European nations, and countries in South America and Asia (above all, Japan) showed a marked increase in the number, size, and complexity of their scientific institutions and communities. Scholarly work to date has concentrated principally on the Big Four, predominantly on British and, to a lesser but rapidly increasing extent, American scientific institutions and communities. This pronounced emphasis—or possibly overemphasis—on Anglo-American science has done an injustice to the dynamic developments of nineteenth-century science in France and Germany, not to mention other developing areas within and beyond Europe.

Second, the essay concentrates principally on institutions and communities devoted to science, as opposed to those devoted to medicine and technology. To be sure, the latter were often enough admixed or associated with more strictly scientific institutions and communities; the hybrid nature of numerous institutions and communities is undeniable.[7] Here again, limitations of space and competency require drawing limits to this essay's scope.

Within these limits, then, the essay discusses a series of issues and topics that have comprised the heart of research on nineteenth-century scientific institutions and communities. It first addresses the efforts and largely unmet need to measure the demography of nineteenth-century scientific institutions and communities (section 2). It then turns to studies of the enlarged role of universities as a setting for the scientific enterprise (section 3) and of the importance of new types of nonacademic institutions (section 4). Section 5 takes up analyses of national academies of science and scientific societies. Finally,

---

7. On the complexities of understanding the relations during the nineteenth century between science and medicine, and science and technology, see the essays by Michael Hagner and Ulrich Wengenroth in this volume.

Section 6 argues for what, adapting a well-known notion of Benedict Anderson's, I shall call the imagined community of science. The latter's contrast with the concrete institutions and communities noted in sections 2–5 forms one of this essay's major claims: namely, that what are usually seen as discrete, individualized institutions and communities came during the nineteenth century to constitute an imagined yet real scientific community.

2. The Emergence of Scientific Communities and Prosopography

Whatever characterization one may give or meanings one may attach to the term "scientific community," it is well to remember that it refers to an evolving historical phenomenon: The scientific community of, say, seventeenth-century England or France or Europe as a whole was, at best, markedly different in terms of size, composition, structure, and even function from that of the nineteenth century or of today. Historians of early modern science are well aware of this distinction. For example, although John Christie, in his analysis of the origins and development of the Scottish scientific community in the seventeenth and eighteenth centuries, speaks of a "scientific community," he notes that the term "is divested of many of its modern connotations." Nonetheless, he argues that by 1760 this community had reached "maturity" in the senses of self-regeneration and "communal structure."[8] Similarly, Roger Hahn points out that "the very phrase 'scientific community' is an anachronism in the Old Regime because there was no category corresponding to the term." Nonetheless, Hahn too uses it throughout his study of the Paris Academy of Sciences under the ancien régime, thereby implicitly indicating its (partial) social reality and the difficulty of finding a better term for characterizing the collective social dimension of the scientific enterprise in early modern France.[9]

From the nineteenth-century point of view, it seems not unfair to say that these and other scholars of the social history of early modern science are largely referring to the formation, by the late Enlightenment, of some sort of protocommunity (or better, protocommunities) of natural philosophers and

8. Christie, "Origins and Development" (first quote on 122–23, second on 134). Cf. "Eighteenth-Century Scientific Community," in which Gascoigne does not hesitate to speak of a "scientific community."

9. Hahn, *Anatomy of a Scientific Institution,* 85 (quote), 86, 107–8, 264, 271, 274, 285, 287–88, 301, 304, and 307. McClellan also uses the term "community" freely throughout his study, yet he argues, contra Hahn, that eighteenth-century scientific societies provided career possibilities to at least the better scientists. (McClellan, *Science Reorganized,* 233–51). See also Hahn, "Scientific Careers" and "Scientific Research as an Occupation," in which Hahn argues against science as a socioprofessional class, in either eighteenth- or twentieth-century terms, and rejects the existence of a scientific community in eighteenth-century France in terms of a socially cohesive class.

natural historians rather than to a community of scientists in the, say, post-1850 sense. Scholars broadly agree that by the late Enlightenment the traditional categories of natural philosophy and natural history had begun to dissolve and slowly reconstitute themselves into individual scientific disciplines, which in turn began to feel the need for and gradually became the crucial elements that constituted "the scientific community" or "scientific communities."[10] It is worth further recalling the semantic point that there were, literally speaking, no "scientists" in the eighteenth century—the French normally spoke of an *érudit* or a *savant*. And though it is true that William Whewell coined the (English) word "scientist" in 1834,[11] the term only began to find general usage during the second half, if not the last third, of the century. Hence it may be legitimately asked if, even in principle, there could have been a community of scientists before, say, 1850. To be sure, there were individuals known as "astronomers," "chemists," or "mathematicians" in the early modern world, yet most specialist scientific roles—one thinks above all of "biologist" and "physicist"—were post-1800 neologisms and new social-historical categories that reflected the rise of specialized and professionalized scientific activity.[12] Widespread usage of these and other neologisms and the flowering of specialist communities did not occur until around the mid-nineteenth century. Moreover, the eighteenth century was not so much one of national or large-scale communities as of local clubs, groups, and similar small-scale societies or corporations. As members of these diverse local social entities, eighteenth-century natural philosophers and natural historians simply did not belong to the same kinds of social or sociological categories as nineteenth-century disciplinary scientists, who were members of specialized, often national or even international scientific organizations. Understood in these terms, there were no scientific communities before the mid-nineteenth century.[13] It was only then, as scientific education and certification began to be more broadly instituted and standardized, as disciplines and specialist institutions began to take shape on a sufficiently wide scale, and as improved or new means of communication and transportation were devel-

10. Schaffer, "Scientific Discoveries"; Gascoigne, "Eighteenth-Century Scientific Community." For the German version of this "great transition," see R. Turner, "Great Transition." Cunningham and Williams, "De-centering the 'Big Picture,'" argues for the origins of science between 1760 and 1848.

11. Hahn, "Scientific Careers" and "Scientific Research as an Occupation"; Ross, "Scientist."

12. For further discussion of this point see the essays in this volume by Robert J. Richards, on biology, and by Jed Buchwald and Sungook Hong, on physics.

13. Lepenies, *Between Literature and Science,* 2; Daston, "Objectivity and the Escape from Perspective," 608. The importance of clubs as fundamental social units of science continued into the nineteenth century, as Gay and Gay, "Brothers in Science," illustrates.

oped, that modern scientific communities could begin to take meaningful shape and function on a national and international scale.[14]

How, then, can we ascertain more precisely the emergence and growth of scientific communities in the nineteenth century? One potential way is to undertake prosopographical studies of them or at least of their parts. Such collective biographical studies of scientific communities (or of individual institutions, societies, and the like) would, among other things, establish their size and growth rates (i.e., their changing demography); develop appropriate categories of social structure and analysis, above all as concerns scientific education and career development (e.g., issues about "amateur" versus professional, mobility, and reward systems); help pinpoint and isolate scientific "elites" from scientific "masses" and, in so doing, help differentiate knowledge producers, users, disseminators, and general supporters; and, finally, correlate scientific with nonscientific concerns (e.g., religion). The general aim of such studies should be to develop as precisely as possible notions and analysis of, and data concerning, the emergence and changing social structures of nineteenth-century scientific communities and their associated institutions.[15]

Conducting prosopographical studies should help us sharpen our rather vague notion of the "professionalization" of science: While most scholars are agreed that there was an increasing and widespread professionalization of science in the course of the nineteenth century, there is nonetheless much uncertainty as to what precisely that means. Full-time devotion to and pay for scientific work; advanced, well-defined educational credentials; and research and publishing standards certainly lay at the heart of professionalization. But even these elements had different meanings in different countries and at different points in time. Moreover, while science became a profession in the course of the century, it was not one in quite the same way as the classical professions of theology, law, and medicine. Finally, there is an inherent teleological danger in invoking the term "professionalization." In short, the explanatory power of the notion of professionalization seems limited; at the very least, it must be used with great care.[16]

14. Merz, *History of European Thought,* 1:19–20, 303–5; Gillispie, "Science and Technology." Martin Rudwick has argued that "talk of a geological 'community' " in the 1830s and 1840s is "misleading, not least because it suggests anachronistically a strong-boundaried professional group marked by standardized training and certification, with only the uninitiated lay public outside." (Rudwick, *Great Devonian Controversy,* 418.)

15. Stone, "Prosopography," provides an excellent general introduction to and analysis of the potential strengths, limitations, and dangers of prosopography. Gascoigne, "Eighteenth-Century Scientific Community," is a "research note" providing a prosopographical analysis on the basis of scientists listed in the *Dictionary of Scientific Biography.*

16. I thank Theodore Porter for his remarks to me on this issue.

This is certainly not the first call to undertake prosopographical research in the history of science. During the mid-1970s, Steven Shapin and Arnold Thackray carefully laid out some of the potential intellectual rewards and methodological strategies of prosopographical research as it applies to the history of science, and they discussed some of the tools that scholars might use to illuminate the social evolution of the scientific enterprise in Britain between 1700 and 1900. One of the potential advantages of prosopographical analysis, they rightly noted, is that it can help historians avoid anachronistic categories and teleological thinking. Another is that it can be used to correlate socioeconomic status and intellectual performance and to test related causal hypotheses. Shapin and Thackray themselves offered a preliminary, three-level prosopographical analysis of the British scientific community: they divided it up among those "who *published* a scientific paper, book or pamphlet"; those who did not publish but who were formally affiliated with some scientific society or institution; and those "who patronized, applied or disseminated scientific knowledge or principles, but who themselves neither published science, taught science, nor actively associated themselves with scientifically-oriented institutions."[17] Nathan Reingold proposed a similar tack: he suggested that the scientific community—he was referring specifically to that of nineteenth-century America—consisted of three groups: researchers, practitioners ("individuals wholly or largely employed in scientific or science-related occupations" but who published little or nothing), and cultivators, i.e., those who possessed and applied scientific knowledge or otherwise supported researchers and practitioners, but who rarely published, since for them science was a pleasure or hobby rather than a profession.[18] During the course of the century, the cultivators, Reingold claimed, gradually became extinct, while the number of practitioners expanded significantly and the researchers led the way in forming an American scientific community.[19] Both Shapin and Thackray's and Reingold's approaches offer potentially useful analytical categories for future prosopographical studies.

Reingold's suggestive conceptual-analytical essay offered no statistical

17. Shapin and Thackray, "Prosopography as a Research Tool," quotes on 12–13. Thackray, "Natural Knowledge," exemplifies its use in understanding the origins and development of the Manchester Literary and Philosophical Society. For a critique of Thackray's account see Glen's letter to the editor of the *American Historical Review,* followed by Thackray's rejoinder. Shapin, "Pottery Philosophical Society" and "Property, Patronage, and the Politics of Science," demonstrate prosopography's use. Pyenson, "'Who the Guys Were,'" provides another call for the use and a broader survey of prosopography in the history of science, as well as a critique of Shapin and Thackray's analysis, while Cantor criticizes Shapin in Cantor, "Critique of Shapin's Interpretation." Inkster, "Cultural Enterprise," effectively extends Thackray's study to a nineteenth-century provinical town.

18. Reingold, "Definitions and Speculations."

19. Ibid., 31–34 (quote on 31), 40.

documentation. However, the prosopography of the emergent American scientific community has received a strong start from scholars such as Donald deB. Beaver, Clark Elliott, Robert V. Bruce, and, to a lesser extent, George Daniels, each of whom has provided some statistical data.[20] These and other scholars have relied on publications, society memberships, and the like to establish their databases and have sought to characterize the community as a whole. Daniel Goldstein, by contrast, has recently used the Smithsonian Institution's lists of correspondents to argue that the American natural history community was numerically larger, geographically broader, and socioeconomically more diverse than scholars had previously imagined.[21] Others, including Sally Gregory Kohlstedt in her studies on amateur scientists and Ronald and Janet Numbers in theirs on science in the Old South, have made similar points about the size and diversity of the nineteenth-century American scientific community.[22] Such work indicates that prosopographical analysis must include a broad set of categories and resources, not merely publication figures or listings from a modern reference work like the *Dictionary of Scientific Biography*.

Nearly all prosopographical studies to date have concerned the American and British communities during the eighteenth and nineteenth centuries.[23] This emphasis probably reflects not only the cultural backgrounds and interests of Anglo-American historians of science but also the relative numbers of scientists active in Britain and America. A French observer in the 1830s, for example, reluctantly acknowledged that the Geological Society of London had more members than all the comparable groups on the Continent combined, which has led Martin Rudwick to conclude that "the world of geology was widely perceived as being dominated by the English."[24] While there remains much to do in the way of prosopographical research on the American and British communities, there is even more to do for their counterparts in France, Germany, and elsewhere.

Prosopographial study of national scientific communities over an extended period of time is a daunting task. John Lankford's highly quantitative study of the origin, growth, and development of the American astronomical

20. Beaver, "American Scientific Community"; Elliott, "American Scientist, 1800–1863," "American Scientist in Antebellum Society," and *Biographical Dictionary of American Science*, appendixes, 209–347; Bruce, "Statistical Profile of American Scientists"; Daniels, *American Science in the Age of Jackson*, chap. 1 and app. 1.

21. Goldstein, "'Yours for Science.'" For an extensive list of similar, critical studies, see ibid., n. 7, which refers to work by a number of scholars on nineteenth-century American women scientists and regionalism.

22. Kohlstedt, "Nineteenth-Century Amateur Tradition"; Numbers and Numbers, "Science in the Old South."

23. One noteworthy exception is Gascoigne, "Historical Demography of the Scientific Community."

24. Rudwick, *Great Devonian Controversy*, 28.

community between 1859 and 1940 shows, however, what can be done.[25] One of his book's great strengths is its precision of social description. In avoiding the impressionistic characterizations that one gets in all too much social history of science, it is a model of its kind.

Where Lankford provides his readers with a sense of the *longue durée,* Paul Forman, John Heilbron, and Spencer Weart have given a synchronic view of the physics discipline circa 1900 in terms of its demographic and material components. They provide highly useful data, including growth rates, on the number of academic physicists in Europe and the United States, and on their personal income, laboratory expenditures, new plant, and productivity.[26] Such information provides both important measures for assessing the dimensions of national and international physics communities at the turn of the century and potential benchmarks for comparing the physics community with those in other contemporary disciplines. Similarly, Lewis Pyenson and Douglas Skopp have done a demographical analysis of the social and educational backgrounds of physicists in Germany at the turn of the century.[27] Della Dumbaugh Fenster and Karen Hunger Parshall, for their part, have quantitatively profiled the American mathematical community circa 1900. They provide telling measures of its membership and the latter's participation, geographical distribution, education, and research interests, as well as the role of foreign participants, which, taken together, lead to a precise account of American mathematicians' emergence as a research community. Moreover, like Goldstein they show that the American scientific community was geographically broader than has often been appreciated.[28] Finally, Thackray, Jeffrey Sturchio, Thomas Carroll, and Robert Bud have offered copious data—"historical indicators," as they call them—concerning the diachronic development of the American chemical community between 1876 and 1976.[29]

Improved understanding of the formation of national scientific communities in the nineteenth century above all requires improved understanding of

25. Lankford, *American Astronomy.* Other, comparable quantitative studies include Kevles, "Physics, Mathematics, and Chemistry Communities"; Rossiter, *Women Scientists in America;* and Cahan, "Institutional Revolution."

26. Forman, Heilbron, and Weart, "Physics *circa* 1900." Forman et al. also provide some data for Japan. Though reference to studies on the formation of the Japanese scientific community is beyond this essay's scope, attention should nonetheless be drawn to the important work, for physics, of Koizumi, "Emergence of Japan's First Physicists," and, more broadly, to Bartholomew, *Formation of Science in Japan;* both studies provide important data and analyses.

27. Pyenson and Skopp, "Educating Physicists."

28. Fenster and Parshall, "Profile of the American Mathematical Research Community."

29. Thackray et al., *Chemistry in America.* For an exemplary study of the formation of a national disciplinary community, see Hufbauer, *Formation of the German Chemical Community.* See also Golinski, *Science as Public Culture,* esp. 145–52, 236–87, and Giuliani, *Il Nuovo Cimento,* which provides a wealth of data on the Italian physics community and its formation.

the demographics of the formation of disciplinary entities. The sociology of science has for some time now been focused on—and, in the view of some, been obsessed with—the sociology of scientific knowledge. A renewed interest in the historical sociology of science is needed. With the widespread availability and low cost of powerful computers and database systems, historians and sociologists of science now have at their disposal the instruments needed to undertake serious research into the demography and prosopography of nineteenth-century scientific institutions and communities. Such studies promise not only quantitative precision but also a better qualitative sense of the vitality of scientific institutions and communities.

3. The University as the Enlarged Setting of Science

While scholars may debate the role and importance of universities in fostering science during the early modern period, there is widespread agreement that during the nineteenth century they became the principal institutional setting for science. To be sure, the existence and expansion of the universities was not coextensive with that of the natural sciences: This model of the university as the dominant institution of science and the heart of its communities is largely due to the expansion of university-based science in Germany after 1810 and in a number of other countries (above all, the United States) during the last third of the century. Britain and, especially, France followed rather different institutional patterns.

The ideological hallmark of German universities after 1810 was, as R. Steven Turner has convincingly argued, the demand for autonomy—"*Einsamkeit und Freiheit,*" in the language of Wilhelm von Humboldt and his followers—via the gradual elimination of any religious, social, or economic orientation. Universities were (in principle at least) to be places where science was pursued for its own sake, independent of any nonscientific pressures. On this interpretation (which, Turner notes, has dominated subsequent historiography), science and its institutions grew in the German universities on the basis of institutional autonomy, and not (for the most part) as a result of large-scale social forces, the influence of special interests with technological or other (especially ideological) nonscientific goals, or government policy.[30]

Ben-David and Turner are the two major modern theoretical interpreters of this development.[31] Ben-David argued that the decentralization of the German university system led to competition within it. Young scientists and

30. R. Turner, "German Science." For general background on the German universities see McClelland, *State, Society, and University,* and R. Turner, "Universitäten," as well as the literature cited in the latter.

31. Stichweh, *Zur Entstehung des modernen Systems,* makes a modern, systems-theoretical argument for the autonomy of German science.

medical researchers exploited this system by creating and concentrating on specialized disciplines that did not involve utilitarian matters; the result (until the 1850s) was institutional innovation, new scientific roles, disciplinary development, and new facilities—at least for those who were most successful. After 1850, Ben-David maintained, the German universities became increasingly inflexible.[32]

Turner complemented (and implicitly criticized) Ben-David's interpretation by stressing the role of what he called the "research imperative" as the driving force behind university change in Prussia (and by implication elsewhere in Germany). He argued that whereas eighteenth-century German university professors were expected merely to transmit the known by lecturing and writing general textbooks, their nineteenth-century successors were increasingly oriented toward publishing specialized research articles and books that provided novel results. Between 1815 and 1848 a transformation occurred within the Prussian professoriate that led to specialized research publications becoming the basis for university appointments and promotions. Moreover, Turner argued, in contrast to Ben-David, that the state played a critical role: It helped create a research ideology, contributing to a transformation of the professoriate's values from those appropriate to local universities to those befitting national, disciplinary-oriented scientific communities. To Ben-David's analysis of the importance of competition, Turner added the importance of ideology as manifested in the state's appointment and promotion decisions.[33]

Turner argued that the "research imperative" first saw the light of day in the philological and historical seminars and that after 1830 it spread to the natural sciences, where it helped shape those emerging disciplines.[34] In imitation of the humanistic seminars, several German universities established new, homologous seminars for mathematics and the natural sciences. Yet these seminars were also, as Kathryn M. Olesko's study of Franz Neumann's physics seminar at Königsberg (1834–76) argues, devoted to science education. During its first two decades, Neumann certainly used his seminar as a platform for conducting pathbreaking research and for training outstanding students, several of whom became important researchers in the next generation of German physicists. Nonetheless, Olesko clearly shows that its principal purpose and function was to provide science education through discipline and prac-

---

32. Ben-David, "Scientific Productivity," "German Scientific Hegemony and the Emergence of Organized Science" (in his *Scientist's Role in Society*), and "Organization, Social Control, and Cognitive Change"; Ben-David and Zloczower, "Universities and Academic Systems."

33. R. Turner, "Growth of Professorial Research," "*Bildungsbürgertum* and the Learned Professions," "Prussian Universities," and "Prussian Professoriate."

34. W. Clark, "Dialectical Origins of the Research Seminar," traces the philological seminar's origins between 1738 and 1838.

tice: Neumann drilled his students in the protocols of the instrumentational and quantitative techniques used in experimental physics. His seminar was not principally concerned with doing research, as previous historical accounts and hagiography had proclaimed; rather, it sought to equip future secondary-school science and mathematics teachers with techniques and values that they would in turn impart to their pupils.[35]

We will need many more such in-depth studies of other seminars in other sciences and at other universities—such as Gert Schubring's for the Bonn natural sciences seminar[36]—if we are to understand their role in leading to the rise of science institutes. Moreover, as F. L. Holmes has shown in the case of Justus von Liebig's laboratory at Giessen, teaching and research could complement one another,[37] a point that Fox and Anna Guagnini have recently stressed in their synthesizing study of the rise of applied research in nineteenth-century European academic and industrial life.[38]

After midcentury, the natural-science seminars were gradually replaced by institutes. At first, these were institutes in name only or, in the case of physics, nothing more than the old cabinets that housed equipment and apparatus and that, in some cases, belonged to the seminar itself. But starting in the 1860s, German universities gradually acquired independent buildings that housed teaching and research laboratories for physics, professors' offices, storage rooms, and so on. I have argued that this dramatic new development between 1865 and 1914 constituted an "institutional revolution" in German physics.[39] These independent institutes became the envy of foreign visitors, some of whom sought to imitate the Germans, and became a key sign of the professional formation and independence of the physics discipline.

New buildings were likewise gradually acquired for German scientific institutes in numerous other disciplines. A number of scholars have begun chipping away at the issue of disciplinary developments. Jeffrey Johnson has portrayed the successes and failures of chemistry institutes at German universities through three successive generations, from circa 1850 to 1914, pointing especially to the roles of size, entrepreneurial leadership, hierarchy, and, in general, institutional inertia and flexibility.[40] Robert Kohler has outlined the disciplinary development of physiological chemistry in Germany be-

35. Olesko, *Physics as a Calling*. See also Olesko, "Commentary."

36. Schubring, "Rise and Decline." See also idem, "Entwicklung des Mathematischen Seminars" and "Mathematische Seminar der Universität Münster."

37. Holmes, "Complementarity of Teaching and Research."

38. R. Fox and Guagnini, "Laboratories, Workshops, and Sites," esp. 99–139.

39. Cahan, "Institutional Revolution." See also Jungnickel and McCormmach, *Intellectual Mastery of Nature*, on German physics institutes.

40. Johnson, "Academic Chemistry in Imperial Germany." See also idem, "Hierarchy and Creativity in Chemistry"; R. Turner, "Justus Liebig versus Prussian Chemistry"; Meinel, *Chemie an der Universität Marburg*; Gustin, "Emergence of the German Chemical Profession."

tween 1840 and 1900.[41] Richard Kremer has studied three early physiology institutes in Prussia (Bonn, Berlin, and Breslau), looking in particular at the contexts, interests, and rhetoric that called for their greater support.[42] Arleen Tuchman has detailed the institutional development of experimental physiology at Heidelberg, showing physiology's importance to medicine there and, hence, its socioeconomic importance to the state of Baden.[43] Similarly, Timothy Lenoir has outlined the origins of Carl Ludwig's new institute for physiology in Leipzig, also arguing that the state's science policy (in particular, Saxony's drive to advance clinical medicine and industrialization) was responsible for the new institute.[44] Lynn Nyhart has sketched the development of zoology institutes in the 1880s and 1890s, arguing that "the new buildings represent nicely the intellectual transformation from natural history to scientific zoology."[45] She has also systematically tracked the institutional development of morphology (above all, its professoriate) within the German university system, in a way that is comparable to Christa Jungnickel and Russell McCormmach's extensive treatment of the institutional development of theoretical physics in Germany.[46] And Kurt-R. Biermann has traced the institutional development of mathematics at the University of Berlin.[47] These and other studies have begun to portray systematically the growth of university-based scientific institutes in Germany.[48] Still, our knowledge remains fragmentary and all too often is concerned with origins rather than long-term growth and development. We need to follow institutes over time and to integrate these studies of local institutes by using interpretative issues like competition, the "research imperative," science education, modernization, and the like.

A half century after the German universities began establishing their seminars and institutes, along with a research ethos that was gradually diffused throughout the German higher-education system, American colleges and universities started to lay the foundation for graduate-level instruction and research in the sciences. To be sure, in the years between the revolution and the Civil War scientific institutions had been established and an American scientific community had begun to take shape, as John Greene has shown in his survey of the Jeffersonian period and George Daniels has shown, to a lesser

41. Kohler, *From Medical Chemistry*, 9–39.

42. Kremer, "Building Institutes."

43. Tuchman, *Science, Medicine, and the State* and "From the Lecture to the Laboratory."

44. Lenoir, "Science for the Clinic." Alan Rocke has argued against Lenoir's claim that Saxony was motivated by socioeconomic considerations and has shown that the state, instead, simply sought to achieve "academic excellence." See Rocke, *Quiet Revolution*, 265–86 (quote 439, n. 41).

45. Nyhart, "Natural History and the 'New' Biology," 435.

46. Nyhart, *Biology Takes Form;* Jungnickel and McCormmach, *Intellectual Mastery of Nature.*

47. Biermann, *Mathematik und ihre Dozenten.*

48. In general, see the articles concerning nineteenth-century science institutes (esp. those by Gregory, Schubring, Holmes, Cahan, and Johnson) in Olesko, *Science in Germany.*

extent, for the Jacksonian.[49] Nonetheless, as Bruce and Kohlstedt have both forcefully argued, it was largely between 1846 and 1876 that American science was "launched" and the American scientific community first established.[50]

During and after the Civil War, a series of new state universities were established under the Morrill Land-Grant College Act (1862), a number of new private higher educational institutions were founded (among them Boston Tech, later the Massachusetts Institute of Technology, in 1861, Cornell in 1868, and the Johns Hopkins University in 1876), and a number of older colleges (above all, Harvard, through its Lawrence Scientific School, and Yale, through its Sheffield Scientific School) began devoting more assets to the natural sciences and stressing electives. The Reconstruction period in American history meant educational as well as social, political, and economic reconstruction: the United States now began building its modern higher-educational and scientific research system.

Laurence Veysey's classic study *The Emergence of the American University* (1965) brilliantly portrayed the general transformation between 1865 and 1910 of a key set of older American colleges as well as new state and private institutions into modern universities. Among other things, he analyzed "rival conceptions of American higher learning," showing how piety gave way to utility and how research values and practices became institutionalized at a number of key universities.[51] He (and others) also showed that a number of leading American scholars and scientists who attended or visited German universities between the 1860s and the 1890s adopted or adapted some of the values, practices, and institutional structures that they thought they had seen or experienced at German universities: a stress on "pure science," freedom to teach and learn as one thought best, and empirical research; the construction of laboratories and institutes; and so on.[52] Nowhere was this more true than at Hopkins, which from its founding until the mid-1890s set the pace and tone for research in American universities.[53] As Robert Kargon, Owen Hannaway, Karen Hunger Parshall, and Jane Maienschein have respectively shown, such key scientific figures as Henry Rowland in physics, Ira Remsen in chemistry,

---

49. Greene, *American Science in the Age of Jefferson;* Daniels, "Process of Professionalization" and *American Science in the Age of Jackson.*

50. Bruce, *Launching;* Kohlstedt, *Formation of the American Scientific Community.* More recently, Kohlstedt has questioned this periodization; see Kohlstedt, "Parlors, Primers, and Public School."

51. Veysey, *Emergence of the American University,* quote on 18. See also Ben-David, *Scientists's Role in Society,* 139–68, on American universities; Shils, "Order of Learning"; the essays in Oleson and Voss, *Organization of Knowledge;* Guralnick, "American Scientist in Higher Education"; Geiger, *To Advance Knowledge,* 1–57; and Reingold, "Graduate School and Doctoral Degree."

52. Veysey's older account of the German influence on American institutions should be tempered by the more recent and nuanced essays in Geitz, Heideking, and Herbst, *German Influences.* See also Frank, "American Physiologists in German Laboratories."

53. On Hopkins, see Veysey, *Emergence of the American University;* and Bruce, *Launching,* 335–38. The older, standard histories are French, *History of the University,* and Hawkins, *Pioneer.*

James Joseph Sylvester in mathematics, and Henry Newell Martin and William Keith Brooks in biology helped set the research visions and practices for Hopkins during its first forty years and, through the work of their leading students, for their respective disciplines.[54] Still, as Hannaway and Larry Owens have argued for graduate education at Hopkins, and as Stanley Guralnick has stressed for the antebellum college system and Kohler for the Ph.D.-granting institutions from 1850 to 1900, graduate research institutions were in good measure built on the values and structures of their collegiate bases.[55] This general point—the importance of understanding the development of the American educational system for understanding the development of the American research system—is one that resonates with the analyses of Olesko, Holmes, and others concerning Germany.[56]

If Hopkins led in institutionalizing new (and deepening older) ideals and practices of university science, it was not, of course, alone. Yet generally speaking, we sorely lack the sorts of modern, critical studies dedicated to other institutions and their individual scientific leaders that we have for Hopkins or dedicated to a single discipline at a number of universities. Philip Pauly has provided an example of the latter in his study of the rise of academic biology at Harvard, Hopkins, Chicago, Columbia, and Pennsylvania in the late nineteenth century, and Kohler has written on building the discipline of biochemistry in America.[57] But we need more detailed studies like, for example, Mary P. Winsor's of Harvard's Museum of Comparative Zoology, Jane Maienschein's on Charles Otis Whitman and the establishment of biology at Chicago, and Donald Osterbrock's on the Yerkes Observatory.[58] We also need more studies of the emergence and growth of individual departments, such as Lawrence Aronovitch has provided for physics at Harvard and

54. Kargon, "Henry Rowland"; Hannaway, "German Model of Chemical Education"; Parshall, "America's First School"; Maienschein, Transforming Traditions, chaps. 1, 2. On Martin and Hopkins, see also Owens, "Pure and Sound Government."

55. Hannaway, "German Model of Chemical Education"; Owens, "Pure and Sound Government"; Guralnick, Science and the Ante-Bellum American College; Kohler, "Ph.D. Machine."

56. Cf., too, Lankford, American Astronomy, 383–88. Lankford argues that, in terms of numbers of observatories and personnel, American astronomy of the late nineteenth and early twentieth centuries surpassed its British, French, and German counterparts because it was tied to higher education and not, like them, to national observatories.

57. Pauly, "Appearance of Academic Biology"; Kohler, From Medical Chemistry. Nineteenth-century American physiology has recently been well served through essays by Frank, "American Physiologists in German Laboratories"; Fye, "Growth of American Physiology"; and, on institutional developments at Columbia and Harvard, Laszlo, "Physiology of the Future." Similarly, for the institutionalization of American astronomy, see Lankford, American Astronomy.

58. Winsor, Reading the Shape of Nature; Maienschein, "Whitman at Chicago" and Transforming Traditions, passim; Osterbrock, Yerkes Observatory. See also Benson, "From Museum Research to Laboratory Research," on the rise of biological laboratory research at a series of American institutions in the late nineteenth and early twentieth centuries.

Gary Cochell for mathematics at Cornell.[59] Such studies should not, of course, be limited to the institutional development of the sciences at private universities: state universities naturally deserve attention, too, especially such major schools as California, Illinois, Michigan, Minnesota, and Wisconsin. M. Eugene Rudd and I have sought to provide just such a study of the institutionalization of science at a state university: we show how the physicist DeWitt Bristol Brace, who had studied with Rowland in Baltimore and Hermann von Helmholtz in Berlin, established the Department of Physics at the University of Nebraska in 1888, how during the next seventeen years he transferred the culture, values, and institutional structures and practices of science that he had absorbed at Hopkins and Berlin to the Great Plains, and how by 1900 he had made Nebraska into one of America's leading centers of physics.[60] Similar studies of other figures and institutions can and should be undertaken.

If the universities came to hold center stage in the German scientific community after 1810 and in the American community late in the century, the institutional development of science in Britain was far less closely tied to that of the universities. During the first half, if not two-thirds, of the nineteenth century scientific research was at best only weakly present in British universities.[61] The strength of British scientific research originated in the ambitions and entrepreneurial activities of individual scientists, not through institutionalized support. Though Cambridge and Oxford could boast of a number of talented mathematicians and theoretically oriented scientists, the ancient universities provided very limited research facilities for experimental work. Not until after the reforms of midcentury, or even the 1870s, did university-based science come to Britain in full force.[62] There were of course some exceptions: Harvey Becher, for example, has shown that scientific activities and reforming efforts at Cambridge prior to 1850 were more extensive than many then and since have supposed, and Nicolaas Rupke has argued much the same for William Buckland's activities in pre-1850 Oxford.[63] Moreover, Susan Faye Cannon has argued for the presence and importance of a network of early Victorian scientists at Cambridge.[64] Yet until 1850, British science, as Jack Morrell argued in a fundamental article, found its leitmotif in individualism and was characterized by an unsystematic, disorderly nature.[65] Research, espe-

59. Aronovitch, "Spirit of Investigation"; Cochell, "Early History."

60. Cahan and Rudd, *Science at the American Frontier.*

61. Merz, *History of European Thought,* chap. 3, argues for the weakness of university-based British science.

62. Cardwell, *Organisation of Science,* 36–69, 92–98, 136–41.

63. Becher, "Voluntary Science"; Rupke, "Oxford's Scientific Awakening." For a comparative study, see D. Wilson, "Educational Matrix."

64. Cannon, "Cambridge Network."

65. Morrell, "Individualism and the Structure of British Science." See also Merz, *History of European Thought,* 1:251, 271–72; Morrell, "Patronage of Mid-Victorian Science"; and Becher, "Voluntary Science."

cially of the experimental sort, done in the ancient universities prior to 1870 must be attributed far more to the efforts of individual researchers than to the priorities of the institutions. Indeed, it was not until after the rise of the new London-based universities and their Scottish counterparts, and the findings of a series of commissions of inquiry, that Oxford and Cambridge fully re-formed themselves.[66]

But especially after 1870, both Cambridge and Oxford significantly strength-ened and expanded their science facilities and faculties. The age of the indi-vidualist in British science was gradually replaced by that of university-based science, with the state, as Roy MacLeod has argued in a series of essays, in-creasingly the source of financial support.[67] Historians of science have been rather good at documenting and analyzing the improvements in the institu-tional state of science at Oxbridge. Gerald Geison has shown how Michael Foster built up physiology at Cambridge, and has described the broader rise of laboratory research in biology there.[68] Gerrylynn Roberts has treated the introduction of chemistry into the Cambridge natural science tripos.[69] Ro-mualdas Sviedrys has studied the rise of physics laboratories in Britain, in-cluding Cambridge.[70] Cambridge's Cavendish Laboratory, devoted to physics and financed with new money not subject to the colleges' control, has been treated by a number of scholars.[71] Oxford, for its part, has recently received careful attention from Fox on the university museum, Rupke on geology in the first half of the century, and Roger Hutchins on John Phillips and astron-omy, while physics has received attention from Graeme Gooday, though the moribund nature of the Clarendon Laboratory under Robert B. Clifton has otherwise failed to elicit a full-scale historical treatment.[72]

With the exceptions of University College and King's College London (which in 1837 became the first constituent members of the new University of London) and Owens College (which became Victoria University in 1880 and

66. Cardwell, *Organisation of Science*, 86–98. For an exhaustive treatment of nineteenth-century Oxford see the essays in Brock and Curthoys, *History of the University of Oxford*, vol. 6, *Nineteenth-Century Oxford*.

67. See, e.g., Cardwell, *Organisation of Science*, 84–98, and the essays in MacLeod, *Public Science and Public Policy*.

68. Geison, *Michael Foster*.

69. Roberts, "Liberally-Educated Chemist." For a social analysis of the natural science tripos examinees at Cambridge, see MacLeod and Moseley, "'Naturals' and Victorian Cambridge."

70. Sviedrys, "Rise of Physical Science" and "Rise of Physics Laboratories."

71. In addition to the items cited in n. 70, see Cavendish Laboratory, *History of the Cavendish Laboratory*; Crowther, *Cavendish Laboratory*; Falconer, "Thomson and 'Cavendish Physics'"; and Kim, Emergence of the Cavendish School and "J. J. Thomson." For the Cavendish as a metrological institution, see Schaffer, "Late Victorian Metrology" and "Physics Laboratories."

72. R. Fox, "University Museum"; Rupke, "Oxford's Scientific Awakening"; Hutchins, "John Phillips"; Gooday, "Precision Measurement"; Morrell, "Research in Physics," 263–66.

Manchester University in 1904), new colleges did not emerge in Britain until the 1870s and 1880s and did not become universities until at least 1900.[73] Edinburgh, Britain's third university, sought, as Morrell has shown, to remake its reputation after 1860 as a leading science and medical center, a reputation that it had held in the late eighteenth century but had lost after about 1820. However, a lack of sufficient government resources prevented it, at least during the mid-Victorian era, from regaining its former eminence or establishing the level of research concentration found at German and, increasingly, American universities.[74] Glasgow after 1870, however, thanks in part to the vigorous leadership of William Thomson and his desire to commercialize scientific knowledge, reformed and promoted itself into a national university with strong scientific and engineering components.[75]

Nineteenth-century French science and its relationship to France's university system took a rather different course from those followed in Germany, the United States, and Britain. The fundamental difference that makes comparing France with the other great powers of science problematic is, on the one hand, the role of an alternative set of institutions—the *grandes écoles*—in shaping the scientific, technological, and state bureaucratic elite and, on the other, the limited functions of a marginalized university system.

From the mid-eighteenth century onward, and above all following the Revolution, France had two systems of higher education in the sciences: that of the universities and that of the *grandes écoles*. Although an individual instructor might well teach in both systems—one thinks of the *cumul*—there were key differences between the two: The universities, refounded by Napoleon in 1808, were open to everyone who had a baccalaureate degree; following two years of general work, students there concentrated on a single science. The *grandes écoles*, by contrast, only took students who were one to three years beyond the baccalaureate and had passed a rigorous competition; it prepared them more broadly than the universities did for professional life—that is, they were not as specialized. Hence, the *grandes écoles* were far more selective in admission, more prestigious and affluent, and had a different notion of scientific culture than the universities. While the *écoles* had (and have) a strong elitist and hierarchical nature (both in terms of student recruitment and with re-

73. Roderick and Stephens, "Scientific Studies and Scientific Manpower," provides an overview of its subject. On the architecture of science at British universities, see Forgan, "Architecture of Science." One might also include here the establishment of the Royal College of Chemistry in 1845, on which see Roberts, "Establishment of the Royal College," and Bud and Roberts, *Science versus Practice*, 16, 48, 51–3, 63, 66, 69, 71–6, 79, 98–100, 117, 137–8.

74. Morrell, "University of Edinburgh," "Science and Scottish University Reform," and "Patronage of Mid-Victorian Science."

75. Sviedrys, "Rise of Physics Laboratories," 409–15; C. Smith, "'Nowhere but in a Great Town.'" See also C. Smith and Wise, *Energy and Empire*.

spect to one another and the universities), they have also been marked by a reluctance to change. Along with the École Normale Supérieure, which only began training university professors after 1880, the key *écoles* for science and technology in the nineteenth century included the École des Ponts et Chaussées, the École de Génie Militaire de Mézierès, the École des Mines, the École Polytechnique, and the École Centrale des Arts et Manufactures. These and similar institutions, above all the Polytechnique, stressed mathematics and largely neglected the experimental sciences. They became the training grounds for the elite of the technocratic bureaucracies.[76] The Polytechnique, the most important of these institutions for the sciences proper, has been studied as a whole and from a social historical viewpoint by Terry Shinn and, more recently, by scholars in two edited volumes containing studies of some key figures, subjects, and themes.[77] Craig Zwerling has studied the emergence at midcentury of the École Normale Supérieure, as well as the social origins and career patterns of its graduates. He shows that the increased need for a well-trained governmental bureaucracy and the growing industrialization of France led to the enhanced support of science there, for the school trained the lycée teachers who would be responsible for increased science instruction at the secondary level; at the same time, under Louis Pasteur's leadership, research was given greater emphasis.[78] (To be sure, Antoni Malet has qualified Zwerling's account of Pasteur's role and compared the École Normale Supérieure with the École Polytechnique, showing their changing roles during the course of the century.)[79]

The fundamental background study of the nineteenth-century French university system is by George Weisz, who grounds his study of the emergence of modern universities in French political history and offers an extended social analysis of the system but also has much to say about the universities' specific involvement with the sciences and research.[80] Similarly, Victor Karady has studied the marketplace for scientists in nineteenth-century France, in regards both to the universities and to the *grandes écoles*. He

---

76. For a particularly succinct summary of the differences of these two systems, see Dhombres, "Tradition française"; for the *grandes écoles*, see Shinn, "Specialized Institutions" and "Science, Tocqueville, and the State." Grattan-Guinness, "*Grandes Ecoles, Petite Université*," discusses the institutional distinction insofar as it bore on mathematics from the late eighteenth to the mid-nineteenth centuries. See also R. Fox and Weisz, "Institutional Basis of French Science."

77. Shinn, *Savoir scientifique*; Belhoste, Dahan Dalmedico, and Picon, *Formation polytechnicienne*; Belhoste et al., *France des X*. See also Bradley, "Scientific Education versus Military Training," "Facilities for Practical Instruction," "Scientific Education for a New Society," and "Financial Basis of French Scientific Education."

78. Zwerling, *Emergence of the Ecole Normale Supérieure* and "Emergence of the Ecole Normale Supérieure."

79. Malet, "Ecole Normale."

80. Weisz, *Emergence of Modern Universities*.

shows that there was a distinct improvement in career expectations due to a lagging number of candidates relative to the increasing number of posts available.[81] More to the point, Fox and Shinn have each provided fundamental studies of the French university system's developments in the natural sciences. Fox has shown and analyzed the transformation of French scientists from a small group of *notables* dependent on individual patronage, in the period 1800–1830, into, by the close of the century, a much larger, less personal, less elitist, and far more professional and discipline-oriented group. Fox also showed how French scientists' political involvement, especially in the 1850s and again in the late 1870s, helped transform the higher education system. They emphasized their specialized skills, created increased facilities for science (laboratories, journals, societies, and the like), and found new, nonstate patrons (industrialists) and audiences, for all of which they paid an intellectual and cultural price. In the course of the century, research became an increasingly important function of French university science, and faculty members were transformed from independent intellectual *savants*, who devoted their time to giving highly oratorical lectures and state examinations to students and well-polished lectures before cultivated dilettantes, into professors pursuing research-oriented careers in specialized disciplines, often with applied-science orientations. Starting in the late 1870s, and especially after 1885, the science faculties grew and substituted dependency on local, industrial interests for dependency on the state.[82]

Shinn examined the expansion of science, and especially research, at four of the ten French science faculties (Paris, Besançon, Lyon, and Montpellier) within the university system, and in so doing studied, as did Fox, the affect of the rise of applied science after 1870 on "pure" science at the faculties. Shinn found that between 1808 and 1845 research in mathematics, physics, and chemistry grew, but that from 1846 to 1875 it stagnated and compared unfavorably with that at the *grandes écoles* and the Collège de France. During the latter period, the quality of facilities and morale declined and scientific growth in France was stunted. Like Fox, Shinn pointed to a variety of factors to explain these shortcomings: a weak emphasis on research, greater emphasis on teaching at a low intellectual level, the burden of giving state examinations, strict governmental surveillance, and too much time devoted to *haute vulgarisation*. Then during the last quarter of the century, as the political elites of the Third Republic came increasingly to believe in the social and material progress to be derived from science, they sought allies among scientists, leading to an unprecedented expansion of the faculties. To no small extent, the French reformers modeled their system of science faculties on that of Ger-

81. Karady, "Educational Qualifications."
82. R. Fox, "Science, the University, and the State."

many: the emphasis was on high-quality research, not popularization before refined audiences. Yet because of the heavy-handed influence of local economic interests, between 1900 and 1914 research in pure science, Shinn found, declined sharply and that in applied science stagnated.[83]

In addition to studying the strong role of the *grandes écoles* and the weak role of the university science faculties in higher education and research, scholars of nineteenth-century French science have been preoccupied with a number of other issues, two of which deserve at least brief mention: those of Paris versus the provinces and, relatedly, of centralization versus decentralization. Ever since the nineteenth century, all scholars of nineteenth-century French science have recognized the dominant role of Paris within the French system. Most recently, Fox and Mary Jo Nye have shown that, at least during the second half of the century, as local economic and cultural interests in science increased, there emerged a remarkable vitality to scientific enterprises in such places as Mulhouse, Nancy, Grenoble, Lyon, Toulouse, and Bordeaux.[84] Trying to understand local efforts, be they in the provinces or in Paris, seems to be a more promising approach to understanding "French science," which was never as monolithic as some older scholarly studies implied, than broadly trying to assess the role of centralization in French scientific life. Although Ben-David's provocative claim that Parisian centralization led to a decline in French science has hovered over the analyses of all French scientific institutions of the nineteenth century, there is nothing inherently positive or negative about centralization in regards to its effects upon scientific growth.[85] The issue of Paris versus the provinces, or of centralization versus decentralization, points, moreover, to the fact that historiographical studies of nineteenth-century French science are cast in a distinctly more political and social, and a less strictly institutionalist, setting than studies of their American, German, and British counterparts.[86]

83. Shinn, "French Science Faculty System." Similarly, Lundgreen, "Organization of Science and Technology," cogently argues that after 1860 the French and German institutions of science and technology became multifunctional and hence similar to one another.

84. R. Fox, "Presidential Address"; Nye, *Science in the Provinces.*

85. See, e.g., Ben-David, *Scientist's Role in Society,* chap. 6; Gilpin, *France in the Age of the Scientific State,* 77–123; R. Fox, "Scientific Enterprise"; and Outram, "Politics and Vocation." Critical treatments of the issues of centralization and decline in nineteenth-century French science (with citations of the relevant literature) may be followed in Crawford, "Competition and Centralisation"; Paul, "Issue of Decline" and "Science française"; and Nye, "Scientific Decline."

86. This point is made by R. Turner, "German Science," 29–30, citing Nye, "Recent Sources and Problems." Outram, "Politics and Vocation," esp. 29, explicitly says that understanding the role of French institutions requires situating them within their broad political and intellectual contexts.

4. Nonacademic Institutions as the Setting of Scientific Life

Alongside the universities (and, in France, the *grandes écoles*), a heterogeneous set of nonacademic institutions emerged as settings of nineteenth-century scientific life. These institutions complemented their academic brethren and merit attention here.

The nonuniversity institutional scene was simplest in Germany, for there the universities overwhelmingly dominated scientific life. (The institutes of technology [*technische Hochschulen*], devoted principally to educating engineers, do not come within the purview of this essay.) Science in Germany was, moreover, a state affair, and its legal foundation and financial support derived from the individual *Länder* (state provinces). After the founding of the Reich in 1871, however, Berlin officials managed to circumvent this constitutional nicety as occasion necessitated, gradually building up a select set of national scientific and technological research and testing institutes. In particular, in 1887 the Physikalisch-Technische Reichsanstalt was founded in Berlin. It was a new sort of national scientific institution, an imperial one devoted to pure science as well as technological research and testing. As I have argued in *An Institute for an Empire*, the Reichsanstalt's hybrid constitution led to its concentration on physical measurement (especially metrological work) and testing. In this way, it satisfied the needs of diverse groups: the physics community, industrial technology, the state, and society at large. It became one of Germany's largest, best-manned, and best-equipped scientific institutions, arguably the crown jewel among its pre–World War I scientific institutions.[87]

The Reichsanstalt's success in physics and physics-related technology soon inspired a number of German chemists to try to establish a Chemisch-Technische Reichsanstalt. By 1911, these efforts had been transformed and led to the founding of the Kaiser-Wilhelm-Gesellschaft, which comprised a series of individual scientific institutes; these institutes ultimately became Germany's most prestigious.[88] A number of scholars have offered book-length studies of the Gesellschaft's origins: Günter Wendel has given a heavy-handed Marxist approach; Lothar Burchardt has focused on science policy and, especially, finance; and Jeffrey Johnson has provided a broader, and more satisfying, account by considering the Gesellschaft in terms of a modernizing Germany.[89]

87. Cahan, *Institute for an Empire.*

88. For good overviews of and much detail on the early Gesellschaft and its associated historiography, see vom Brocke, "Kaiser-Wilhelm Gesellschaft" and "Kaiser-Wilhelm- / Max-Planck-Gesellschaft."

89. Wendel, *Kaiser-Wilhelm-Gesellschaft;* Burchardt, *Wissenschaftspolitik;* Johnson, *Kaiser's Chemists.* See also vom Brocke's work, cited in the previous note, and several of the articles concerning the pre-1918 history of the Gesellschaft in Vierhaus and vom Brocke, *Forschung im Spannungsfeld von Politik und Gesellschaft.*

The founding of the Reichsanstalt and the Gesellschaft reflect the increasing inability of the German university system in the late nineteenth and early twentieth centuries to meet the needs perceived by leaders of the German scientific community and the German governments for advanced scientific research.

While the Gesellschaft had no direct counterpart in America, Britain, or France, the Reichsanstalt did serve in part as a model for the British National Physical Laboratory, founded in 1900, and the American National Bureau of Standards, founded in 1901. Unlike the Reichsanstalt, though, neither the British nor the American laboratories had "pure" scientific aims. Edward Pyatt has written a detailed, official, and uncritical history of the British institution, while Russell Moseley has provided a more critical account of its history prior to 1914, placing special emphasis on the role of the British government's science policy.[90] Rexmond Cochrane has written an official history, comparable to Pyatt's, covering the first half century of the American institution.[91] As for France, it never managed to establish a national metrological laboratory.[92]

French scientists (or the state) did, however, establish two particularly creative, nonuniversity institutions. The story of the first of these, the Society of Arcueil, has been told by Maurice Crosland, who used it as a means to provide "a view of French science at the time of Napoleon I," to quote his subtitle. Crosland provided a wealth of detail about the institutional and community scenes of French science, above all physics and chemistry, and argued that the society, although it lasted formally only from 1807 to 1813, is a key to understanding French science during the first half of the nineteenth century. In particular, the fact that the society had to resort to the private means of Claude-Louis Berthollet and Pierre-Simon de Laplace reflected the difficulties and limitations of gaining state support for research: the state bureaucracy and the hierarchy of the scientific establishment, along with the state's financial weakness, prevented or impeded such support.[93] Hannaway has provided an excellent critique of Crosland's study, pointing out that the society's leaders were themselves the central figures in the French scientific establishment and arguing that the society's identity, the source of its coherence, and its influence arose not from being a society or group but rather from being a private and embryonic "research school."[94]

---

90. Pyatt, *National Physical Laboratory;* Moseley, Science, Government, and Industrial Research and "Origins and Early Years."

91. Cochrane, *Measures for Progress.*

92. To be sure, the Laboratoire Central d'Electricité in Paris did address testing and metrological needs in regards to electricity.

93. Crosland, *Society of Arcueil.* Crosland also argues that the society represented a Newtonian style of science, an aspect of his study that goes beyond the present essay.

94. Hannaway, review of Crosland, *Society of Arcueil.* The entire, thorny topic of "research

The second notable nonuniversity institution is the Muséum d'Histoire Naturelle, France's principal institution in the nineteenth century for the study of the biological sciences, above all natural history. Camille Limoges's account remains the basic, and certainly the best, introductory study of its development from its re-creation in 1793 (out of the old Jardin du Roi) until 1914. He traces the museum's fluctuating reputation; provides statistics concerning its professorial staff, financing, and productivity; analyzes its successive periods of strength (1800–40), decline (1840–70), and crisis (1870–1914); and argues that it was gradually marginalized as an institution, within both the French and the international scientific communities.[95] Several other scholars have also analyzed the museum's history, though mostly with other purposes in mind: Toby Appel has discussed it within the context of her excellent analysis of the debate over animal structure between Georges Cuvier and Etienne Geoffroy Saint-Hilaire;[96] Dorinda Outram has considered it in terms of the multiple meanings of "space";[97] and Claude Schnitter has studied Edmond Frémy's attempt after 1850 to strengthen the museum as a science-teaching center, an attempt abandoned after 1890 by his successors.[98] More recently, an edited volume has appeared containing no fewer than thirty-two articles on various aspects of the museum during the century following its re-creation. These articles treat a variety of figures and issues, above all questions concerning experimentation, exhibitions, and the collection, observation, and classification of plants and animals. The museum's narrower institutional history, however, in particular its birth, development, and relations with other institutions and the world at large, receives lesser attention in this long and handsome volume.[99]

Apart from the National Physical Laboratory, the major nonacademic British institutions that have attracted scholars' attentions are the Royal Institution and the mechanics' institutes. Both have evoked radically different interpretations by their historians. Morris Berman has provocatively argued that the Royal Institution, which was Britain's major laboratory during the first half of the nineteenth century, reflected the changing nature of science, which had previously been largely an activity of the leisure classes. According to Berman, the Royal Institution was founded and exploited by the more industrious members of the aristocracy as a tool to advance their agricultural

schools" is both related to and transcends the present essay. For a good overview and analysis of this topic see Geison and Holmes, *Research Schools,* a special issue of *Osiris,* esp. Servos, "Research Schools and Their Histories."

95. Limoges, "Development of the Muséum d'Histoire Naturelle."
96. Appel, *Cuvier-Geoffroy Debate.*
97. Outram, "New Spaces in Natural History." See also idem, *Cuvier.*
98. Schnitter, "Développement du Muséum national d'histoire naturelle."
99. Blanckaert et al., *Muséum au premier siècle.*

interests and to further exploitation of their estates. By the 1820s, however, the institution's leaders, including Michael Faraday, began using it, in Berman's view, to help construct an orderly society and to turn sociopolitical problems into technical ones. Berman's aim was not so much to write a history of the institution as to raise questions about the changing nature and function of science as society became increasingly industrialized in the late eighteenth and early nineteenth centuries. He thus sought to understand the changing ideology of science as it sought to accommodate itself to an industrializing, class-based, and professionalizing society, i.e., to a new set of social interests and values.[100]

In contrast to Berman the political and intellectual radical, Gwendy Caroe has offered a straightforward, old-fashioned narrative that is as sympathetic and uncritical of the institution as Berman's is hostile and hypercritical. Caroe gladly concedes that the institution was initially a "tool of capitalism" but characterizes it as "a willing servant rather than a tool during that decade, for after all it was the Proprietors' Institution; they owned it, they payed the piper and called the tune." Unlike Berman, she sees nothing sinister in the institution's emergence and development; for her, its cooperation with the capitalist system was perfectly natural.[101] Notwithstanding these two very different interpretations, we still lack a solid, critical study of the Royal Institution during the nineteenth century.

The mechanics' institutes, created to instruct certain members of the British working classes, have, by contrast, received attention from an array of scholars, including historians of education and technology. Cardwell has provided an introductory treatment of them as part of his survey of the organization of nineteenth-century science in England, and Ian Inkster has usefully reviewed the literature on the institutes (through 1975) and stressed the importance of studying them in their particular social and cultural contexts.[102] Like Berman in his treatment of the Royal Institution, Shapin and Barry Barnes have sought to reinterpret the mechanics' institutes in terms of the social control they effected: they claim that the institutes' founders aimed to provide "scientific education for certain members of the working class" in order to "render them, and their class as a whole, more docile, less troublesome, and more accepting of the emerging structure of industrial society."[103] Inkster, by contrast, has argued for a broader interpretation, noting that their

100. Berman, *Social Change and Scientific Organization;* see also idem, "Early Years of the Royal Institution."

101. Caroe and Caroe, *Royal Institution,* 49.

102. Cardwell, *Organisation of Science,* 39–44, 71–5, 82–5; Inkster, "Social Context of an Educational Movement."

103. Shapin and Barnes, "Science, Nature, and Control," 32. Ibid., n. 1, provides references to studies of mechanics' institutes by historians of education.

founders' purposes were sometimes radical rather than oriented toward control of the workers.[104] John Laurent, also in contrast to Shapin and Barnes, has argued that the institutes in northern England later in the century provided scientific education that "was used by working-class people for their own purposes—namely, the development of an alternative social and economic philosophy." Where Shapin and Barnes interpret the institutes as providing control of the working classes, Laurent, like Inkster, sees them as being a resource for workers' self-control and enlightenment (i.e., socialist thought).[105] Finally, Gordon Roderick and Michael Stephens stress how inadequate the institutes were in terms of contributing to the scientific training of workers for industrial purposes.[106]

5. From Academies to Specialized Societies

From the 1660s to the 1790s the Royal Society of London and the Paris Academy of Sciences were the dominant institutions supporting scientific research in their respective nations and served as the models of institutionalized science in other states. The various revolutions and restorations from the 1790s to 1830 led to the closing or reorganization of numerous academies and to a general sense of crisis concerning them; the fate of the Paris Academy during these years is merely the best-known instance. While national and regional academies continued to be valued for the prestige, honors, and rewards that they bestowed upon individuals and upon their nation or region as a whole, and while they continued to function to a degree as intellectual, social, and political centers for the gathering of scientific elites and the publication of scientific journals and monographs, after 1800 they gradually lost much of their influence over the direction of scientific research. In part, this was due to the increased importance of universities in conducting such research. Indeed, academies in places like Berlin, Göttingen, Vienna, Bologna, Turin, and Edinburgh owed a good deal of their well-being to the professors at local universities who published in their respective academies' journals.[107] While nineteenth-century academies were not moribund in terms of research, they principally served to communicate results and distribute awards, honors, and prestige for work done elsewhere. Their importance for scientific research was thus primarily indirect.

104. Inkster, "Social Context of an Educational Movement." Inkster, "Science and the Mechanics' Institutes," argues that scientific culture, as much as the educational movement, was responsible for the institutes.

105. Laurent, "Science, Society and Politics," 586.

106. Stephens and Roderick, "Science, the Working Classes, and Mechanics' Institutes"; see also Roderick and Stephens, Scientific and Technical Education.

107. Mazzolini, "Nationale Wissenschaftsakademien," 247–50.

This decline in terms of direct research support is reflected in the amount of historiographical attention accorded nineteenth-century national academies. While there have been any number of studies of individual academies of science and of academies as a whole during the eighteenth century, studies of nineteenth-century national academies have been quite limited in number. (There has, however, been increasing interest in writing the history of regional and municipal academies, neither of which I discuss here.) The only general survey of national academies of science—more precisely, of the characteristics of European national academies—for the nineteenth century is Renato G. Mazzolini's useful account.[108] Moreover, with one or two exceptions we lack critical, book-length studies of individual nineteenth-century academies; most studies tend to be of the sympathetic, not to say hagiographic, variety and to concentrate on the achievements of a select set of a given academy's individual members.

The major exception in this regard is Crosland's *Science under Control: The French Academy of Sciences, 1795–1914,* which provides a detailed and comprehensive study of its subject and serves, in effect if not intent, as the successor study to Hahn's. Crosland's theme is twofold: the French government's control of the academy and the latter's control of science within France. In the course of pursuing his theme, Crosland seeks to demystify the academy's operations, to contextualize the actions of scientists (both the elite and the lesser ranks) within French society, and to present the public image the academy projected of itself and of science within France. There is virtually no aspect of the academy—from its membership lists and its buildings to its governmental and international relations—that Crosland's extensive study does not thoroughly discuss.[109]

The Royal Society was treated in Henry Lyons's old administrative and financial history[110] but has found its principal historian for the nineteenth century in Marie Boas Hall. The main title of Hall's book, *All Scientists Now,* suggests its theme: the gradual transformation of the society after 1847 into a meritocratic, professional organization. By 1870, Hall shows, the society had become a more or less completely professional group of scientists. In her view, its successful reform led to increased prestige and influence. In contrast to Crosland, Hall largely eschews administrative details (especially after 1850) and, unfortunately, any sort of statistical or sociological analysis; she also has relatively little to say concerning the larger British scientific and historical

108. Ibid. See also Grau, *Berühmte Wissenschaftsakademien,* which is strong in bibliographical references.

109. Crosland, *Science under Control.* See also idem, "French Academy of Sciences." On the Academy's continuation of its prerevolutionary tradition of involvement with technological innovation, see Davis, "Artisans and Savants."

110. Lyons, *Royal Society.* See, however, Miller, "Between Hostile Camps."

contexts. Instead, she concentrates on internal politics and personalities. Her study resembles an official, authorized history without literally being one.[111] MacLeod, on the other hand, has critically treated the society's history by using a political model, rather than one focused on professionalization, to interpret its changes between 1830 and 1848 in terms of a broad British context. Unlike Hall, MacLeod argues that, even when reformed the society remained dominated by an elite, albeit a different, more performance-oriented one, which had little interest in rank-and-file scientists and did not encourage specific fields of science.[112] As for the post-1848 society, Ruth Barton has provided one of the few critical studies.[113]

Though all of the major German states had academies of science, most attention has focused on the Berlin (Prussian) academy, since it was the oldest and most prominent and is often enough regarded as the German counterpart to the Royal Society and Paris Academy. The classic work here, Adolf Harnack's turn-of-the-century three-volume hagiography, scarcely treats the post-1870 period.[114] More recently, works by Werner Hartkopf and Gert Wangermann have provided much useful information on the academy, and Conrad Grau has provided an overview.[115] In addition, Jürgen Kocka has recently edited a volume on the academy during the imperial period, which includes interpretative essays concerning structural changes to the institution as a whole; its setting within and relation to its larger social, institutional, German-national, and international contexts; and leading personalities, disciplines, and scientific topics within the academy.[116]

The American National Academy of Sciences, as an honorary and advisory body, was of little scientific importance during the nineteenth century; there is, nonetheless, all too little in the way of historical analysis. The starting points are Cochrane's narrative (and uncritical) book-length account, A. Hunter Dupree's more analytical article, and Dupree's discussion in his study of science in the federal government.[117]

Studies of patronage and the prestige that accrued to the academies as a re-

111. Hall, *All Scientists Now*.

112. MacLeod, "Whigs and Savants."

113. Barton, "'Influential Set of Chaps.'"

114. Harnack, *Geschichte der Königlich Preußischen Akademie der Wissenschaften zu Berlin*.

115. Hartkopf and Wangermann, *Dokumente;* Hartkopf, *Berliner Akademie der Wissenschaften;* Grau, *Preussische Akademie der Wissenschaften*. See also the first volume (of three) of Stern et al., *Berliner Akademie der Wissenschaften,* a study that is Marxist-Leninist but informative.

116. Kocka, Hohfeld, and Walther, *Königlich Preußische Akademie der Wissenschaften*.

117. Cochrane, *National Academy of Sciences;* Dupree, "National Academy of Sciences" and *Science in the Federal Government*. See also National Academy of Sciences, *A History*. Oleson and Brown's edited collection on American scientific and learned societies in the pre–Civil War period, *Pursuit of Knowledge in the Early American Republic,* contains several essays concerning America's state and municipal academies of science.

sult of their awarding prizes and grants and enhancing status and titles have
largely concerned Britain and France, especially their respective national
academies. MacLeod has provided a fine series of informative and analytical
studies on patronage and the government's increasing involvement in Victorian science.[118] Gerard Turner has edited an important collection of essays on
the patronage of science in the nineteenth century, which, in addition to
MacLeod's study of science and the treasury, includes instructive studies by
Morrell on Edinburgh, Cardwell on Manchester, and William Brock on state
patronage.[119] For France, Fox has written an excellent survey of patronage
from 1800 to 1870, arguing that it was "individual initiative, rather than government sponsorship, which brought French science its successes early in the
century, and it was the lack of such initiative which brought French science to
its unhappy state during the Second Empire."[120] On a more general level,
Paul's survey of "the rise of the science empire in France 1860–1939" and
Nye's of five provincial French scientific communities together nicely complement Fox's work for the period after 1860.[121] As for the Paris Academy of
Sciences, Elisabeth Crawford has systematically studied its prize system in the
second half of the century, while Crosland and Antonio Gálvez have argued
that it transformed at least part of its prize system into a grant system.[122] The
reward system of nineteenth-century science culminated, in a sense, in the introduction of the Nobel Prizes in 1901, concerning which Crawford has written the fundamental study.[123]

As already noted, from the late Enlightenment onward the academies were
gradually marginalized as new specialized societies and their associated disciplines assumed ever larger shares of roles occupied by the academies and individual patrons during the early modern period: as supporters of research, as
meeting places for scientists, and, to a lesser extent, as the means of communicating results. Eric Hobsbawm's old but still superb analysis of the late eighteenth and early nineteenth centuries as an age of dual revolutions—the
industrial in Britain and the political in France—deserves extension to the
institutional realm of science. There, too, the situation was revolutionary.

118. See, esp, MacLeod, "Science and the Civil List," "Royal Society and the Government
Grant," "Of Medals and Men," "Support of Victorian Science," "Resources of Science," and "Sci-
ence and the Treasury." These and related pieces are collected in MacLeod, *Public Science and Public
Policy.*

119. Morrell, "Patronage of Mid-Victorian Science"; Cardwell, "Patronage of Science"; Brock,
"Spectrum of Science Patronage."

120. R. Fox, "Scientific Enterprise," 10.

121. Paul, *From Knowledge to Power;* Nye, *Science in the Provinces.*

122. Crawford, "Prize System of the Academy of Sciences"; Crosland, "From Prizes to
Grants"; Crosland and Gálvez, "Emergence of Research Grants"; Crosland, *Science under Control*
(which thoroughly treats the academy's patronage).

123. Crawford, *Beginnings of the Nobel Institution.*

The foundings of the Lunar Society of Birmingham circa 1768, the Manchester Literary and Philosophical Society in 1781, and the Linnean Society of London in 1788 were harbingers of the emergence of provincial and specialized, discipline-oriented scientific societies, which would flourish in the nineteenth century. Once the hiatus caused by the wars and political upheavals of 1789 to 1815 was overcome, scientific societies experienced unprecedented growth and diversity.[124] By 1850, numerous new, specialized, and often provincial, societies had been established; after 1850, many transformed themselves into national, disciplinary societies. This development has presented scholars, as Robert Schofield indicated more than a generation ago, with new opportunities—and a compelling need—to write monographic histories of scientific societies, not least for the nineteenth century.[125]

In 1760, Britain and Ireland together had only a dozen formal scientific organizations; by 1870, there were 125, including 59 specialist societies.[126] Apart from sheer growth in numbers, the history of British scientific societies shows at least four general characteristics. First, as numerous scholars have noted, many were provincial in origin and location.[127] Second, the British scientific community at the same time produced many societies that were or became national in scope—e.g., the Linnean Society, the Royal Astronomical Society, and the Chemical Society of London.[128] Third, British science was also club science. Several scholars have pointed to the importance of the X-Club, for example, as a fraternity or network of Victorian scientists, while Hannah and John W. Gay have argued for the importance of fraternal culture in the world of science clubs (the Universal Brotherhood of the Friends of Truth, the Red Lion Club, the B-Club, and the Chemical Club of the late nineteenth century).[129] Fourth, as studies by Rudwick and by Adrian Desmond effectively remind us, much of British scientific life in the nineteenth century was located

124. McClellan, *Science Reorganized*, 253–54. McClellan notes that specialized societies were not unknown to the eighteenth century and argues that to a certain extent the academies themselves were specialized societies (ibid., 256–57). Yet their number, sizes, and importance in the post-1815 period dwarfed anything known to the eighteenth century. On the political context of French science from 1793 to 1830, and the lack of dramatic improvements in institutions during this period, cf. Outram, "Politics and Vocation."

125. Schofield, "Histories of Scientific Societies."

126. Morrell, "Patronage of Mid-Victorian Science," 354, citing Thackray, "Reflections on the Decline," 29.

127. E.g., Thackray, "Natural Knowledge"; Schofield, *Lunar Society of Birmingham;* Orange, "Rational Dissent and Provincial Science"; Morrell, "Early Yorkshire Geological and Polytechnic Society" and "Bradford Science."

128. On the Chemical Society, see Bud and Roberts, *Science versus Practice,* 48–51, 65–69, 100–8.

129. Jensen, "X-Club"; MacLeod, "X-Club"; Jensen, "Interrelationships within the Victorian 'X Club'"; Barton, "'Influential Set of Chaps'" and "'Huxley, Lubbock, and Half a Dozen Others'"; various essays in F. Turner, *Contesting Cultural Authority;* Gay and Gay, "Brothers in Science."

in, if not dominated by, London, which served as the natural center of the British scientific community, as the inevitable destination point of all foreign scientists who visited Britain, and as Britain's imperial scientific hub.[130]

To give some particular examples: Several historians of geology have studied the origins and early years of the Geological Society of London (founded 1807).[131] Shapin has examined the Pottery Philosophical Society as a means of analyzing the cultural uses of provincial science.[132] Both J. Bastin and Lynn Barber have studied the Zoological Society of London (founded 1826); more recently, Desmond has extensively reexamined its origins and early years, stressing its ideological and class functions.[133] Adrian Rice, Robin Wilson, and J. Helen Gardner have studied the emergence, structure, membership, and activities of the London Mathematical Society from its founding in 1865 until 1900, and in so doing have helped us to understand the British mathematical community as a whole and its place within the international mathematical community.[134] Sophie Forgan, to cite a final British example, has shown the potential for interpreting the context, image, and function of scientific societies by looking at their buildings' architecture.[135] Although these examples suggest that much has been done since Schofield called for such studies, there remains an enormous unmet need, not least because most studies concentrate on societal origins rather than subsequent growth and development.

In France, to quote Fox, "most societies were, at best, marginal to the national system of research and education." Fox has nicely portrayed the emergence of French disciplinary scientific societies out of the larger set of *sociétés savantes,* which were largely provincial and local and which encompassed the entire spectrum of knowledge—scientific, humanistic, and applied. Most members of these societies were provincial amateurs and autodidacts; they had little to do with research. During the 1850s and 1860s,

130. For good, short overviews for the first four decades of the century see Morus, Schaffer, and Secord, "Scientific London," and Hays, "London Lecturing Empire." See also Rudwick, *Great Devonian Controversy* and "Charles Darwin in London"; Desmond, *Politics of Evolution* and "Making of Institutional Zoology"; Morrell, "London Institutions and Lyell's Career"; Hays, "Science in the City" and "London Institution"; Kurzer, "Chemistry and Chemists"; and Forgan, "'But Indifferently Lodged.'"

131. Rudwick, "Foundation of the Geological Society"; Laudan, "Ideas and Organizations"; R. Porter, *Making of Geology,* 145–50; Weindling, "Geological Controversy"; Secord, *Controversy in Victorian Geology,* chap. 1.

132. Shapin, "Pottery Philosophical Society."

133. Bastin, "First Prospectus of the Zoological Society" and "Further Note on the Origins"; Barber, *Heyday of Natural History;* Desmond, "Making of Institutional Zoology."

134. Rice, Wilson, and Gardner, "From Student Club to National Society"; Rice and Wilson, "From National to International Society."

135. Forgan, "Context, Image, and Function."

the slow professionalization of French scientists and their increased desire to do research manifested themselves in part through the rise of new, national, disciplinary-oriented societies, which by about 1870 had overtaken the *sociétés savantes*. Still, even as late as 1900 most of the national disciplinary societies had only limited involvement with research and publication. And most, Fox tellingly notes, were established long after their British counterparts.[136]

The first French disciplinary societies included the Société Géologique (1830), which was closely modeled on its British predecessor;[137] the Société Entomologique (1832); the Société Météorologique (1852); the Société Botanique (1854); the Société Chimique de Paris / France (1857), of which Ulrike Fell has given a nice account that in part illustrates the process by which a disciplinary society developed;[138] the Société Mathématique de France (1872), of which Hélène Gispert has given a particularly good account that also does much to illustrate the state of the French mathematical community;[139] the Société Française de Physique (1873), concerning which we have no study; and the Société Zoologique de France (1876), whose early history Fox has treated.[140] Notwithstanding their shortcomings, these societies helped establish French disciplinary scientific communities and a national French scientific community.

Our knowledge of German societies—both provincial and national—is particularly weak. Walter Ruske's old and uncritical account of the important Deutsche Chemische Gesellschaft needs a successor.[141] On the more positive side, Jungnickel and McCormmach's study of theoretical physics in nineteenth-century Germany examines in some detail how physical societies there served both a local and a disciplinary function.[142] Lynn Nyhart has recently studied the case of the German zoologist Karl Möbius in Hamburg, thereby showing how the study of the civic realm and its local scientific societies and needs can offer an important route to understanding how science functions within culture.[143] Similarly, Olesko has treated Helmholtz within the civic culture of Königsberg and its societies.[144] Finally, on Berlin as a cen-

136. R. Fox, "*Savant* Confronts His Peers." See also Paul, *From Knowledge to Power,* 267–85.

137. Rudwick, *Great Devonian Controversy,* 27–30.

138. Fell, "Profession of Chemistry in France."

139. Gispert, "France mathématique." See also Gispert and Tobies, "Comparative Study of the French and German Mathematical Societies," which compares the Société Mathématique de France and the Deutsche Mathematiker-Vereinigung.

140. R. Fox, "Early History of the Société Zoologique" and "Société Zoologique de France."

141. Ruske, *100 Jahre Deutsche Chemische Gesellschaft.*

142. Jungnickel and McCormmach, *Intellectual Mastery of Nature.* See also R. Turner, "Growth of Professorial Research." On the Berlin Physical Society, see Schreier, Franke, and Fiedler, "Geschichte der Physikalischen Gesellschaft."

143. Nyhart, "Civic and Economic Zoology."

144. Olesko, "Civic Culture."

terpoint of German science, a volume by Hubert Laitko et al., *Wissenschaft in Berlin,* is particularly informative.[145] That said, we nonetheless lack scholarly studies of many important German scientific societies as well as an overview of all such societies.

Though no longer the most current work on the subject, Ralph Bates's pioneering survey of scientific societies in the United States still provides useful entrée. Bates documented the fact that between 1790 and 1865 numerous state and municipal academies of science were founded in the northern and eastern states, along with a number of specialized societies, and that the south lagged in institutional development.[146] More recently, studies by Joseph Ewan and Lester Stephens have helped refine societal developments for the pre–Civil War south.[147] Between 1866 and 1918, the institutional growth documented by Bates spread throughout the country, and American scientists experienced "the triumph of the ideal of specialization."[148]

More specifically, many of the major American national disciplinary societies were formed between the 1880s and circa 1910. Although chemists began organizing in New York in 1876, the American Chemical Society only became national in the 1890s.[149] Toby Appel provides a fine treatment of the founding (1883) and first four decades of the American Society of Naturalists, which, though not itself a disciplinary society, spawned the Naturalists, the American Physiological Society (1887), the American Association of Anatomists (1887–88), the American Morphological Society (1890; subsequently the American Society of Zoologists), the Botanical Society of America (1894), and the Society for Plant Morphology and Physiology (1897).[150] Appel has also given a thorough treatment of the founding of the American Physiological Society (1887), showing that it was shaped by the two worlds—biology and medicine—to which it was attached and that it was one of the first American national disciplinary societies.[151] At the same time the Geological Society of America (1888), the American Mathematical Society (1888), the American Psychological Association (1892), the American Physical Society (1899), and

145. See Laitko et al., *Wissenschaft in Berlin,* 96–303.

146. Bates, *Scientific Societies in the United States,* 28–136. See also Dupree, "National Pattern of American Learned Societies."

147. Ewan, "Growth of Learned and Scientific Societies"; Stephens, "Scientific Societies in the Old South."

148. Bates, *Scientific Societies in the United States,* 28–136 (quote at 136).

149. Browne, "Half-Century of Chemistry in America"; Skolnick and Reese, *Century of Chemistry;* Browne and Weeks, *History of the American Chemical Society.*

150. Appel, "Organizing Biology." On the Naturalists, see also Conklin, "Fifty Years of the American Society of Naturalists."

151. Appel, "Biological and Medical Societies." See also Brobeck, Reynolds, and Appel, *History of the American Physiological Society.*

the American Astronomical Society all emerged.[152] In brief, during the past two decades scholars have started to recount and analyze the history of nineteenth-century American scientific societies.

6. Conclusion: Imagined Scientific Communities

This essay has focused on historical studies devoted to the demography of nineteenth-century science, the enlarged role of universities as the setting for the scientific enterprise, the importance of new types of nonacademic institutions, and the declining importance of national academies of science relative to that of specialized societies. Seen in isolation, these studies give at best a discontinuous image of a scientific community made up of individual institutions, some local, some national, and some disciplinary in orientation. Often enough, these institutions and their individual members had little or nothing to do with one another. Nonetheless, this image is, I believe, ultimately misleading.

In his well-known and seminal study, *Imagined Communities,* Benedict Anderson sought to explain the origins and spread of nationalism. He argued that the notion of nationhood was (and is) essentially an imagined one in that the vast majority of a nation's members never meet or even know one another, yet each imagines him / herself as being in more or less close political communion with the other members. For Anderson, the nation as an "imagined political community" is characterized by its limited, sovereign, and, notwithstanding a measure of inequality and exploitation, communal features. Language, printing, consciousness, patriotism, racism, the use and abuse of history, and so on, he argued, help establish and maintain a sense of nationhood among otherwise independent individuals or ethnic groups.[153]

Anderson's conception of an imagined community may be useful for conceiving "scientific nations," communities that, like many of those analyzed by Anderson, arose starting in the late eighteenth century and that, in many cases, achieved noticeable maturity during the nineteenth. As in the populations Anderson described, most members (scientists) would not have known most other members personally, yet they imagined themselves as belonging to one and the same community. By adopting, over the course of the nine-

---

152. On geology, see the undocumented, narrative account of Eckel, *Geological Society of America.* On mathematics, see Archibald, *Semicentennial History of the American Mathematical Society,* and Parshall and Rowe, *Emergence of the American Mathematical Research Community.* On physics, see Phillips, "American Physical Society." On astronomy, see Osterbrock, "AAS Meetings before There Was an AAS"; DeVorkin, "Pickering Years"; and Rothenberg and Williams, "Amateurs and the Society."

153. Anderson, *Imagined Communities,* quote on 6.

teenth century, the term "scientist" (or *"scientifique,"* *"Naturwissenschaftler,"* *"scienziato,"* or the like), individuals effectively declared their membership in a larger group and gave themselves a professional and social identity on both the general communal and the disciplinary planes. Anderson's conception, by allowing us to move from the particular individual or local community to a general notion of community, that is, to recognize simultaneously the existence of numerous, more or less independent, local scientific communities within the larger, posited community, thus allows us to talk about *the* scientific community. For the idea of a scientific community is, *pace* some philosophers and even some sociologists of science, more than that of individuals bound together to create new knowledge through cognitive and material exchange; it is also the social and cultural sense of belonging to a certain social group, that is, of having a certain social identity.

This notion of an imagined scientific community may pertain to three levels of development during the course of the nineteenth century: the rise of disciplinary, of national, and of international scientific communities. For each, in its own way, helped bind specialists and socially isolated local scientists to one another. By way of closing, I shall touch briefly on each of these.

A number of scholars, above all McCormmach and Charles Rosenberg, have provided thoughtful discussions concerning the general nature, rise, and importance of studying disciplines.[154] Rudolf Stichweh, for his part, has given an extremely abstract, sociological analysis (in terms of systems theory) but one that gains some embodiment through his account of the rise of the physics discipline in Germany between 1740 and 1890.[155] On a more empirical level, Jungnickel and McCormmach's study of the rise of theoretical physics in the German-speaking world, though not explicitly discipline-oriented, in effect portrays the discipline's rise there.[156] Similarly, Robert Silliman has provided an account of the emergence of the physics discipline in France in the early nineteenth century, and Nye has studied the disciplinary development of chemistry there and elsewhere and has also written an introductory synthetic account of the disciplinary development of chemistry and physics between 1800 and 1940.[157] Jan Golinski has argued convincingly for the emergence of

154. See, esp., Rosenberg, "On the Study of American Biology and Medicine" and "Toward an Ecology of Knowledge"; and McCormmach, editor's forward. See also Reingold, "Definitions and Speculations"; Lemaine et al., *Perspectives on the Emergence of Scientific Disciplines;* and Stichweh, "Sociology of Scientific Disciplines."

155. Stichweh, *Zur Entstehung des modernen Systems.*

156. Jungnickel and McCormmach, *Intellectual Mastery of Nature.*

157. Silliman, "Fresnel and the Emergence of Physics"; Nye, *From Chemical Philosophy,* esp. 13–31, and *Before Big Science. Pace* Kohler and Nye, Matthias Dörries has argued that while the disciplinary boundaries of midcentury French chemistry and physics are perhaps rationally distinguishable, in practice they were not. He thus questions, but does not fully discard, the usefulness of the metaphor of disciplines. (See Dörries, "Easy Transit.")

the discipline of chemistry in Britain between 1760 and 1820, stressing the importance of the new discipline as the crystallization point for the formation of the British chemical community.[158] Paul Farber has shown the emergence of ornithology as a discipline between 1760 and 1850.[159] Daniel Kevles' well-known and unmatched study of the formation of the American physics community, while largely devoted to the twentieth century, has much to offer on how the discipline of physics initially took shape in late nineteenth-century America.[160] Much to the point, John Servos's excellent book on the development of physical chemistry in America is explicitly devoted to the discipline's formation there.[161] Likewise, Kohler has explicitly sought to understand the formation of biochemistry as a discipline.[162] Pauly, as already noted, has traced the development of academic biology as a discipline in late-nineteenth- and early-twentieth-century America.[163] Similarly, Fenster, Parshall, and David Rowe have shown the emergence of the American mathematical discipline during this same period.[164] And, to cite one last example, Nyhart has argued that, in Germany, zoology as a discipline developed only gradually during the course of the nineteenth century but that from the late 1860s onward university zoology students began pursuing careers as zoologists.[165] These and other empirical accounts of discipline formation during the nineteenth century see or illustrate the emergence of a scientific community as a natural by-product.

The notion of a scientific community becomes more abstract, more imagined, on the national level, where it has been embodied in general scientific societies. The first of these, the Gesellschaft Deutscher Naturforscher und Ärzte (Society of German Natural Scientists and Medical Doctors), founded in 1822, still lacks a worthy scholarly account of its history during the nineteenth century.[166] The Gesellschaft not only helped acquaint German scientists with one another and organize the widely dispersed body of German scientists, many of whom envied the Royal Society of London and the Paris Academy of Sciences as organizing centers; it also, ironically, inspired imita-

158. Golinski, *Science as Public Culture*, esp. 284–87.

159. Farber, *Emergence of Ornithology.*

160. Kevles, *Physicists.*

161. Servos, *Physical Chemistry.*

162. Kohler, *From Medical Chemistry.*

163. Pauly, "Appearance of Academic Biology" and "General Physiology and the Discipline of Physiology." See also his more recent *Biologists and the Promise of American Life.*

164. Fenster and Parshall, "Profile of the American Mathematical Research Community"; Parshall and Rowe, *Emergence of the American Mathematical Research Community;* Parshall, "Eliakim Hastings Moore."

165. Nyhart, "Disciplinary Breakdown" and *Biology Takes Form.*

166. Schipperges, *Versammlung Deutscher Naturforscher und Ärzte.* Daum, *Wissenschaftspopularisierung,* 119–37, provides a useful overview and cites the relevant secondary literature.

tors in Britain, France, and elsewhere. One of these, the British Association
for the Advancement of Science, founded in 1831, has been the subject of numerous studies, most extensively by Morrell and Thackray, who have provided a detailed, book-length account of its origins and early years.[167] The
German and British institutions in turn helped inspire the American Association for the Advance of Science, founded in 1848, which Kohlstedt has studied
as part of her work on the formation of the American scientific community at
midcentury.[168] As for France, we have no study of the Association Française
pour l'Avancement des Sciences, founded in 1872.[169] Each of these general scientific societies, like their counterparts elsewhere, represented a particular
national scientific community.

The concept of a scientific community becomes still more abstract when
applied on the international plane, yet such a community had also begun to
manifest itself by midcentury.[170] To be sure, perhaps the earliest organized international congress had occurred in the late 1790s, when French leadership
sought to establish an international metric standard, while the first sustained
international cooperation occurred during the 1830s and 1840s, when German
scientists led an effort to obtain worldwide geomagnetic data.[171] But after
midcentury, international congresses and cooperative efforts, which also included establishing literature review journals and large-scale bibliographies,
started to become more routine throughout the various disciplines of science
and technology. In a sense, this trend culminated in the publication of the
Royal Society's *Catalogue of Scientific Papers* (1800–1900), the formation of the
International Association of Academies (1899), and the Great Exhibition of
1900 in Paris, where some fifteen international congresses were held. These
meetings, bibliographies, and associations, together with increased travel and
correspondence and the establishment of the Nobel Prizes, were proof to scientists that they belonged to an international social unit. Together with their
disciplinary and national counterparts, they helped constitute that imagined
and relatively new phenomenon, the scientific community.

167. Morrell and Thackray, *Gentlemen of Science*. This work provides a useful list of other secondary sources. See also MacLeod and Collins, *Parliament of Science*. Burchfield, "British Association and Its Historians," is an appreciative essay review of these and related works.

168. Kohlstedt, *Formation of the American Scientific Community*. See also Bruce, *Launching*, 251–
68.

169. See, however, R. Fox, "*Savant* Confronts His Peers," 272–76.

170. In general, see Gregory, *International Congresses and Conferences;* Schroeder-Gudehus,
"Caractéristiques des relations scientifiques internationales" and "Tendances de centralisation";
and Crosland, "History of Science" and "Aspects of International Scientific Collaboration."

171. Crosland, "Congress on Definitive Metric Standards"; Cawood, "Magnetic Crusade"; Dörries, "Standardisation de la balance de torsion." See also Kanz, *Nationalismus und internationale
Zusammenarbeit*.

Eleven

SCIENCE AND RELIGION

FREDERICK GREGORY

❧

A HISTORIAN WHO attempts to write about science and religion in the nineteenth century faces formidable challenges. The common reference to conflict between science and religion in this era, for example, frequently carries
with it assumptions, not the least of which can involve a hypostatization of
"science" and "religion." Some have resisted this temptation by countering
that "science" and "religion" did not conflict with each other in this period;
people did. According to this approach, both words refer to changing activities that comprise whatever their practitioners at a given historical moment
say they do; hence, historians should not fix religion and science in their
minds as static sets of principles that stand in opposition to each other.[1]

One cannot, however, expect that readers of a historical account, or indeed
historians themselves, will always refrain from appealing to what they regard
as ordinary meanings of the terms science and religion. In spite of an appreciation of the challenge described above, scholars still often refer to the interaction of science and religion on the presumption that readers will understand
what they mean by these terms.

There is another, related complication: suspending one's own convictions
about the nature of science or about religious belief is often easier said than
done. Such convictions, of course, threaten to pull one's analysis out of its historical context and into the present. Indeed, most of those interested in the
subject of science and religion, now as in past generations, have not been historians, nor have they exhibited much concern with history. They have been
concerned instead with what influence, if any, each domain exerts on the
other. When a historical issue like Darwinian natural selection *has* arisen, the
question most frequently pursued has been whether it could be made compatible with some current understanding of theism, not how it affected, say,
Victorian religion.

This essay focuses on the way in which the history of the relations of sci-

---

1. Cunningham, "Getting the Game Right," 370, 381–82. A book by Brooke and Cantor devoted
to the idea that "science" and "religion" are changing activities that cannot be fixed over time is discussed in section 5 below.

ence and religion during the nineteenth century has been represented in historical writing. It does not, for example, discuss how writers in the nineteenth century treated science and religion in general; rather, it examines how authors have portrayed issues that arose then. How have historians and others portrayed the many topics that individuals in the nineteenth century claimed cut across both religious and scientific beliefs? Which episodes have scholars chosen to write about and which have received less or inadequate attention? What questions have historians brought to their work? What generalizations, if any, have persisted? Are any agendas clearly identifiable in the treatments given to the subject? These and related queries form the basis for this exploration.

While the analysis presented here is intended to provide specific responses to the questions just raised, there is as well an overall claim being made. Just as science and religion do not possess eternal meanings that hover above history, neither is the *history* of the relations between them fixed or frozen. As this essay will demonstrate, different historians writing at different times and places have written very different histories of the relations between science and religion in the nineteenth century.

The treatment below is itself a history of historical writing about science and religion in the nineteenth century. As such, it mainly concerns the histories that have appeared since 1970, but it begins (section 1) with the works of historians who wrote in the nineteenth century itself. From our vantage point the motivations for these histories—philosophical, secular, and religious— are clearly visible and tell us much about the age in which they were written. The new century entertained challenges that soon led to crisis, and with it came a different attitude about the historical interaction of science and religion. Section 2 considers the humbler historical treatments of science and religion from the years bracketed by the two world wars, in particular, the resurrection of teleology into science. The period after World War II, dealt with in section 3, produced histories with new questions that began in earnest the process of expanding the historiographical horizon by searching for new grounds of orientation. By the 1970s and 1980s, discussed in section 4, historians had problematized the meanings of "science" and "religion," and their histories reflected the change: the new fashion became that of the social history of science and religion. The final section deals with works written as the century comes to a close, which reflect new agendas attuned to the cultural values of the latest fin de siècle.

1. From Comte to Merz: In the Service of Larger Agendas

The earliest historians of science were themselves too immersed in the events of the period to have included much about the history of the relations of nineteenth-century science and religion. Auguste Comte, whom George Sarton regarded as one of the first to appreciate the value of the history of science, identified grandiose historical patterns that had, in his philosophical scheme, culminated in a new positivistic religion in his day. But among his heroes from the nineteenth century there were no religious thinkers whatsoever, and none of the scientists listed was associated publicly with religious issues.[2]

Nor do we find many nineteenth-century developments in science and religion in William Whewell's *History of the Inductive Sciences* (1837). When religious considerations did impinge on what was for him a current scientific issue, as they appeared to in his coverage of the question of the transmutation of species, Whewell's approach was governed by his appreciation of Kantian philosophy. Like Kant, he took pains to clarify what belonged to religion, what to science. Faced with groups of species "which have, in the course of earth's history, succeeded each other at vast intervals of time," Whewell concluded that either older species had transmuted into newer ones or that "we must believe in many successive acts of creation and extinction of species, . . . acts which, therefore, we may properly call miraculous." While he rejected transmutation because he regarded variation as confined within definite limits, he conceded that one could not account for the origin of species within either physiological or geological science. Whewell observed that Charles Lyell may have described successive creations of species as a part of the economy of nature, but Lyell had nowhere indicated in what department of science the hypothesis belonged. Whewell himself noted that geology and theology "may conspire, not by having any part in common, but because, though widely diverse in their lines, both point to a mysterious and invisible origin of the world," but his view on where the origination of species belonged was unambiguous: "The bare conviction that a creation of species has taken place, whether once or many times, is a tenet of Natural Theology rather than of Physical Philosophy."[3]

Whewell in fact took his place as a primary actor in another debate that had been underway since just before the beginning of the century. Thomas Paine's attack on Christianity in 1796 had produced a number of responses in which natural theologians defended their commonly accepted belief in extraterrestrial life. The best known of these responses was the *Discourses on the Christian Revelation* of 1817 by the Scottish cleric Thomas Chalmers. In 1853 Whewell's anonymously published book, *Of the Plurality of Worlds*, compli-

2. Sarton, "Auguste Comte," 328, 355–57, 339.

3. Whewell, *History of the Inductive Sciences*, 2:564, 570, 574. Quotations are taken from the third edition (1897), in which additions to the earlier editions are clearly marked. Allusions here and below refer therefore to the first edition (1837).

cated the situation by disagreeing with Chalmers. As an Anglican cleric Whewell was motivated by his need to preserve Christianity, as a philosopher by his need for consistency. This episode, as we shall see in section 4, attracted the attention of historians in the 1980s.

Comte and Whewell were not the only ones to utilize the history of science and religion in support of larger intellectual agendas. Other nineteenth-century figures also wrote large-scale histories of science and religion to illustrate new insights that they wished to communicate to their age. Most focused on events that had taken place prior to 1800; some included contemporary material, which, placed alongside older historical episodes, served only to illustrate further the writers' initial theses. The best-known historical writing about science and religion from the latter half of the century came from writers who were either promoting or criticizing the interpretation of history as a battlefield on which a war between science and religion had been waged.

John William Draper's *History of the Conflict between Religion and Science* (1874) followed his *History of the Intellectual Development of Europe* (1862) and the three-volume *History of the American Civil War* (1867–70). The son of an itinerant Methodist preacher in England, Draper had come to America in 1832 at the age of twenty-one. He held a certificate in chemistry from University College in London, a new model university set up by and for dissenters. In 1836 he completed a medical degree at the University of Pennsylvania, enabling him to land a teaching position in chemistry and natural philosophy at Hampton-Sydney College in Virginia. Three years later he took what would become a lifelong position as professor of chemistry and botany on the medical faculty of New York University.[4]

Convinced that the task of sound scientific method was, as his biographer Donald Fleming puts it, "to banish mystery from the world, to drench the world in intellectual sunshine without shadow,"[5] Draper chose the Roman Catholic Church as his enemy. He saw the pope and his church to be the greatest stumbling block to the eventual universal reign of the scientific spirit. While most of the *History* focused on episodes prior to the nineteenth century, Draper's book also portrayed clashes between science and religion from his own era. One such had to do with differences between the biblical and scientific accounts of the age of the earth. Draper conceded that the scientific view of a vast geological history he portrayed had been arrived at "in the present century," noting that geology had not had to endure the vindictive opposition that astronomy had encountered.[6]

Most significant, however, were the issues having to do with the origins

---

4. For details on Draper's early life see Fleming, *John William Draper,* 7–19.
5. Ibid., 46.
6. Draper, *History of the Conflict,* 200.

of the solar system and of species. Draper reviewed the conclusions of nineteenth-century astronomers bearing on the origin of solar systems from an original nebulous condition in order to avoid invoking the direct action of God when explaining the contours of the universe. He then proceeded to "the great theory of Evolution," which, he argued, forced upon us "the conclusion that the organic progress of the world has been guided by the operation of immutable law—not determined by discontinuous, disconnected, arbitrary interventions of God." This was the same tack he had taken some fourteen years earlier as the principal speaker in the famous Oxford session (1860) of the British Association for the Advancement of Science, at which Thomas Henry Huxley and Bishop Samuel Wilberforce had crossed swords. Draper's goal in reviewing clashes between science and religion in the nineteenth century was to persuade the reader of what he regarded as the lesson of the history of science:

> The history of science is not a mere record of isolated discoveries; it is a narrative of the conflict of two contending powers, the expansive force of the human intellect on one side, and the compression arising from traditionary Faith and human interests on the other.[7]

Two years after Draper's work appeared, Andrew Dickson White, president of Cornell University, published a book entitled *The Warfare of Science* (1876). The volume grew out of a lecture he had given in defense of his controversial proposal that the new land-grant university not affiliate with any religious group. Twenty years later White's volume had itself grown into the full-scale historical treatise for which he is best known, the two-volume *History of the Warfare of Science with Theology in Christendom* (1896). White undertook the work with an agenda different from Draper's.[8] Draper praised science but set it in opposition to religion, not theology. He did not find it necessary to include any defense of Christianity. White chose to distinguish dogmatic theology from religion, suggesting that any conflict be drawn between science and dogma, not science and religion. Reconciliation between science and religion *was* possible for White; his goal was "to affirm a rational, nonmythical religion and at the same time to preserve those religious truths which he regarded as absolutes."[9] Unfortunately for him, the reception of his history was no less controversial than his nonsectarian vision of education

7. Ibid., vi. See also 238–43, 246. For an account of Draper's participation in the British Association meeting see Fleming, *John William Draper,* chap. 7.

8. White, *History of the Warfare,* 1: ix, gives the impression that Draper's *History of the Conflict* came out after White's *Warfare of Science* (1876). Since White spent most of 1877 and 1878 traveling in Europe he may not have read Draper's book until after his own was completed. Cf. Altschuler, *White,* 116.

9. Altschuler, *White,* 204.

had been. His division of historical figures into those who had been in the right versus those in the wrong was widely rejected by the religiously orthodox, especially since the limited coverage of nineteenth-century battles invariably depicted the most recent scientific treatments as heroically correct.[10]

Not all histories of science and religion written in the nineteenth century shared the aim of discrediting religion or even of forcing it to reform itself in the light of scientific advances. Otto Zöckler's impressive two-volume *Geschichte der Beziehungen zwischen Theologie und Naturwissenschaft* ("History of the Relations between Theology and Natural Science") appeared in 1877–79, coincident with White's appointment as American ambassador to Germany by President Rutherford B. Hayes. Zöckler, a conservative Lutheran theologian, culminated his historical study with a major section dealing with the nineteenth century, as White's later work would do. But Zöckler's purpose was the opposite of White's: his primary concern was to oppose Darwinism. In the process he made it clear that he rejected the approach taken by Draper and White's first volume. He wrote:

> It would perhaps be more timely and our work would raise more interest in the eyes of many had we wished to write a history of just the *conflict* between theology and natural science. ... A conflict-history *[Conflicts-Geschichte]* would be popular, but it would not be true.[11]

Repeating the position he had first expressed in an 1861 review of the species question, Zöckler depicted Darwin's work as a major contribution that commanded respect. But his historical sketch of its background concluded with the confession that there could be no reconciliation between natural selection and Christianity. Zöckler concluded that treating human behavior in terms of animal behavior in general, as more and more Darwinists were wont to do, relativized human ethics, excused sin, ennobled war, undermined the legal system, and dismissed religiosity as an artificial confusion of the human spirit.[12] His motive for writing the history of science and religion was to expose the obvious error of such conclusions.

Perhaps because of their specific intentions, none of the authors mentioned so far attempted to provide an overview of the nineteenth century itself. Such an attempt came, however, in the work of John Theodore Merz, an electrical engineer with a more than avid taste for history. Unlike all scholars

---

10. This was particularly evident in his depiction of proper methodology. See, for example, his approval of the scientific methodology of the Higher Criticism of biblical texts, which flourished in the nineteenth century. (White, *History of the Warfare*, 2:327–92; also Altschuler, *White*, 208f.)

11. Zöckler, *Geschichte der Beziehungen*, 1:1–2. For more on Zöckler see "Otto Zöckler, the Orthodox School, and the Problem of Creation," chap. 4 in Gregory, *Nature Lost*. I hereby correct an error made there on p. 149: Zöckler's reference to White is to the 1876 *Warfare of Science*, not the later *History of the Warfare*. Zöckler's reference to Draper is found in 1:12, n. 1.

12. Zöckler, *Geschichte der Beziehungen*, 2:794–95.

before him, Merz attempted to summarize the entire history of nineteenth-century scientific and philosophical thought. His four-volume analysis of the century's scientific and philosophical accomplishments was published between 1904 and 1912 and is still read with great profit.[13] From these volumes, and from his *Religion and Science: A Philosophical Essay* (1915), there also emerged an interpretation of the relationship between science and religion in the nineteenth century. Merz's approach is intriguing not because he depicts specific episodes in the interaction between science and religion but because he provides a sweeping and general analysis of the subject. Already in *A History of European Thought in the Nineteenth Century* one detects the theme that will later figure prominently in his essay on religion and science: Merz's need to locate the philosophical and the religious on a higher plane than the purely scientific. The important scientific work of the century, he thought, lay in the realm of practical application; indeed, scientists had shown reluctance to deal with fundamental questions extending beyond the purview of scientific thought. There had been no one like Newton or Leibniz who regarded science and religion as equally important and whose attention had been equally occupied by scientific and religious matters.[14]

If for Merz science had become largely a means to practical ends, it was philosophical speculation that carried some thinkers beyond science's limits and drove them to inquire systematically into the deepest foundations and ultimate ends of the world.[15] Merz extended this depiction of science and philosophy to the arena of our deepest interest, religion. Here one encountered, "in the quiet and serious moments of life," individual concerns and personal convictions that were, after all analysis was said and done, simply embraced.[16] Merz's understanding of religion was not at all confined to ecclesiastical doctrines but broadly focused on ultimate questions of meaning. He believed that this religious interest had given rise to the philosophical tradition that shaped the nineteenth century: the German idealistic thought that arose at the end of the eighteenth century and flourished at the beginning of the nineteenth. Introducing themes that would be developed in volumes 3 and 4, which specifically treated philosophical (including religious) thought, he wrote at the end of his second volume:

> The features peculiar to that period [of German idealism] are still strongly marked on the philosophical countenance of the age: neither the lights nor the shadows thrown by the great luminaries which appeared on the philosophical horizon a century ago have as yet died away.[17]

13. Merz, *History of European Thought.*
14. Ibid., 2:742.
15. Ibid., 1:64. Cf. 1:68.
16. Ibid., 1:69, 71.
17. Ibid., 2:751–52.

Assessing the impact of German thought since Kant on the relations between science and religion, Merz concluded that where a century ago religious interest had expressed itself in concern with the problem of faith and knowledge, after a century of practical scientific achievement the problem had become one of belief and unbelief.[18] As over the century scientists had become more concerned with investigating the purely mechanical order of things, speculation had been increasingly discarded. The effect was to introduce the new problems of pessimism, agnosticism, and what he called indifferentism. At the end of the nineteenth century the question of the independent reality of the moral and spiritual factors of life was, in Merz's view, undecided:

> Knowledge or no knowledge? Certainty or no certainty? Something that is intrinsically good and valuable or a semblance and passing illusion? Philosophically expressed, it is the problem of the Absolute. As to the existence of this the beginning of the century harbored no doubt, the end of the century, the present age, has no certainty.[19]

For Merz himself the existence of an ultimate reality that could be experienced was not in doubt. In a tone reminiscent of the attempt by *Naturphilosophen* of the early nineteenth century to reconfigure cognition aesthetically, he introduced in his last volume the notion that he would rely on in his essay on religion and science: the firmament of the soul. With this notion Merz hoped to transcend what he regarded as the partial field of consciousness that comprised mere sensation and to point to a larger construction that gives the emotional and volitional background of the finite world a superior impress of reality. Here was a greater reality, in which the definite reality of external things and persons had its setting. It could be experienced (though not thought) by artistic and religious minds.[20]

2. Beyond "Warfare": Resurrecting Teleology in Science, 1919–1939

Merz had completed his history before the Great War. Its tone, as we have seen, was nevertheless of a distinctly different pitch from the aggressive and confident preoccupation with the warfare of science and religion that was indicative of the end of the century. Merz looked respectfully to history as the source of challenges not yet met, not as a time of error and prejudice thankfully past. As such, Merz's work stands as a transition to the postwar perspective that marked especially those Europeans who had wandered into hell and stumbled out with a sense of humility and anxiety.

It is a great irony that the so-called warfare historiography that had domi-

---

18. Ibid., 4:609.
19. Ibid., 4:614. Cf. also 4:610, 613.
20. Ibid., 4:788–89. Cf. Merz, *Religion and Science*, 21ff.

nated the history of science and religion in the latter years of the nineteenth century began to unravel thanks to the West's experience of war itself. Historiography from 1919 to 1939 reveals an awareness that the old boldness had been premature. Many historians, their lives directly affected by the disruption of war, sought urgently to reassure themselves that what Western civilization was enduring was not what it seemed: the meaningless grinding on of a world order that was oblivious to human purpose. Such reassurance was not easy to come by. A common theme of postwar reflections on science was a rejection of Newtonian determinism in favor of what Lorraine Daston has called "history of science in an elegiac mode," a pining for a resurrection of teleology in science as a means to help recover a world lost to modernity.[21]

James Simpson, for example, consciously evoked memories of the Great War when he observed, in his *Landmarks in the Struggle between Science and Religion* (1925), that Draper's and White's strictures against representatives of Christianity had softened in the intervening years. The nineteenth century had served as the point of attack for the "warfare" historians of science and religion since it was here that positivism, the historiographical embodiment of determinism, was born. Simpson's postwar analysis of natural law made it clear that determinism was insufficient.[22] His coverage of science and religion in the nineteenth century naturally centered on evolution. It was a spotty review from Jean-Baptiste Lamarck to William Bateson that concluded by pointing out the need to add neo-Lamarckian factors to natural selection. What resulted was a view of evolution that preserved, elegiacally, a teleology of the whole.[23]

The same tone characterized a volume edited by Joseph Needham entitled *Science, Religion, and Reality* (1925). Charles Singer portrayed the nineteenth century cursorily at the end of his essay, "The Historical Relations of Religion and Science," without an apology for the reign of law that science had confirmed in the preceding era. He did, however, claim that scientists had always sought to escape the tyranny of determinism, hinting that for him the escape could be made by merging determinism with Spinozist pantheism, construing the human soul as part of a far greater world-soul.[24]

Needham's volume also contained a contribution by the Italian philoso-

---

21. Daston, "History of Science," 522–31. Although Daston is concerned here with E. A. Burtt's *Metaphysical Foundations of Modern Science* (1924), she includes under the rubric of an elegiac mode "some of the most enduring works on the history of modern science," mentioning specifically those of Alexandre Koyré and Alfred North Whitehead (522). Peter Bowler discusses the revival of beliefs in the purposefulness of the universe and in the progress of civilization that characterized many of the discussions of science and religion in early-twentieth-century Britain in Bowler, *Reconciling Science and Religion*.

22. Simpson, *Landmarks in the Struggle*, vii–viii, 144–63, esp. 160.

23. Ibid., 183.

24. Singer, "Historical Relations," 146ff.

pher Antonio Aliotta that focused on the nineteenth century. Typical of the writers of the day, Aliotta viewed the history of science and religion as a strictly intellectual subject. His survey of developments in science and religion over the century began where Merz's had—with Kant's limiting of reason to the phenomenal realm and his confinement of religion within the bounds of the moral. After reviewing the attempts of the romantics to move beyond Kant by reconfiguring the meaning of both religion and reason, Aliotta rehearsed the developments in science and the study of religion that had produced a flowering of positivism until around 1870.[25] After that time, Aliotta explained, other developments, the most important of which came from thermodynamics, were regarded as having undermined the mechanical view. He never mentioned Max Planck, Albert Einstein, or other subsequent developers of physical theory. It was the irreversibility involved in energy conversions, a result he attributed to the work of Sadi Carnot and Rudolf Clausius, that remained unexplained by mechanical theory. Irreversibility, Aliotta argued, must resist a classical explanation since mechanics, as ordinarily understood, was the study of reversible phenomena. This undermining of mechanics resulted in a philosophical crisis, marked variously by the appearance of Herbert Spencer's agnosticism, Albrecht Ritschl's and Wilhelm Herrmann's radical dualism, and William James's pragmatism.[26]

Aliotta undertook his historical summary of science and religion in the nineteenth century for prescriptive, not descriptive, ends. He regarded attempts like that of James to overcome subjectivism at the turn of the century as inherently irrational and he wished to resist them. All approaches to religion that relied ultimately on the exercise of the will denied what, to Aliotta, was an indisputable fact: that the intellectual was an integral part of the religious. To understand the relationship between science and religion that the nineteenth century had produced, Aliotta first declared that the old correspondence theory of truth had to be abandoned.[27] Yet he could offer only a vague replacement: a progressive realization of superior harmonies, a process that was immune to the disintegration of intellectual content due to scientific advance. Assuming the elegiac mode indicative of his time, he claimed that such a process guaranteed ever new means to capture the harmony that had always been the goal of older dogmas and rituals.[28]

The best-known historical survey of the sciences to appear in this era, George Sarton's multivolume *Introduction to the History of Science* (1927), in fact never reached the nineteenth century. But another survey, of particular relevance here, did. Sir William Dampier began his *History of Science and Its Relations with Philosophy and Religion* (1929) with the observation that scientists used to assume naively that they were dealing with ultimate reality. Now they were coming to understand, he thought, that their fundamental concepts gave only "pictures drawn on simplified lines, but not reality itself."[29] Dampier's treatment of nineteenth-century science and religion, which focused on evolution, reflected this more cautious attitude. While conceding Darwin's achievement, Dampier cautioned against "this overestimation of the possibilities of mechanism," adding that the acceptance of evolution had "produced the illusion that an insight into the method by which the result had been obtained had given a complete solution to the problem [of explaining all activities of organisms]."[30]

Four years earlier Dampier's personal friend Alfred North Whitehead had similarly urged caution. In his chapter "Religion and Science," in *Science and the Modern World* (1925), Whitehead opposed regarding the subject of science and religion as one inevitably involving conflict. He conceded that conflict was "what naturally occurs to our minds when we think of the subject," but he observed that the relationship between the two was far more complex. For one thing, he noted, "both religion and science have always been in a state of continual development."[31] It was an insight, as we shall see, that would not bear fruit in the historiography of science and religion for another half century.

Implications of this new humility about science and its history, based especially on the observation that it was continually in a state of flux, emerged more clearly with the publication of Arthur Lovejoy's *Great Chain of Being* (1936). Lovejoy's study of what he called the "unit idea" of a great chain of being culminated in the early nineteenth century, where, he suggested, a profound transformation had occurred. The German romantics, he argued, had changed the fundamental assumption upon which scientific explanation of the world had rested since Plato; namely, that the match between human reason and the natural world involved static reason and a static natural world. Friedrich Schelling, in particular, had come to believe that the world was a dynamic, changing entity and that human reason must become dynamic in order to understand it. Schelling's contribution was to have introduced "a radical evolutionism into metaphysics and theology" in the attempt "to revise even the principles of logic to make them harmonize with an evolutional conception of reality."[32]

Lovejoy argued that his study of the history of ideas had confirmed "the

29. From the preface to the first edition of Dampier, *History of Science*, vii.
30. Ibid., 316.
31. Whitehead, *Science and the Modern World*, 162. Whitehead noted that of the two, "science is even more changeable than theology." (163)
32. Lovejoy, *Great Chain of Being*, 325.

primacy of the non-rational."[33] By "non-rational" he meant something very specific, different from what later postmodern historians of science and religion would mean when they appealed to the category. In Lovejoy's view the history of science in general, and the history of science and religion in particular, were no longer as restricted as they had been in the nineteenth century in their understandings of what "science" and "religion" entailed. After Lovejoy, historians did not hesitate to reinvestigate even classical episodes from the history of nineteenth-century science and religion in order to uncover new understandings of their complex relationship.

3. The Postwar Years: A Search for New Ground

Two events marked a renewal of interest in history of science during the years immediately following World War II: the founding of the British Society for the History of Science in 1947 and the appearance of Herbert Butterfield's classic work *The Origins of Modern Science* (1949). Although Butterfield barely made it into the nineteenth century in his book, others soon turned their attention to episodes from science and religion in the era.

No work is better known from these years than Charles Gillispie's *Genesis and Geology* (1951). Gillispie opposed the notion that one could describe the relationship between science and religion or between science and theology in Britain in the decades before Darwin's *Origin of Species* as one of warfare. What conflict he could see he attributed to a quasi-theological frame of mind *within* science. His work took a different tack from his predecessors', since his primary concern was not to find out where historical naturalists had made valid or invalid claims. He did not, as he put it, "fret about epistemology."[34] Instead, he sought to understand why his subjects thought as they did. His answer was to show that scientific beliefs were conditioned by nonscientific social and religious opinion. Never had there been, wrote a reviewer in the *Quarterly Review of Biology,* "a more convincing demonstration of the conditioning of scientific beliefs by prevailing social and religious opinion; and, that being so, never a more convincing demonstration of the profound importance to scientists of any time or climate of opinion of a study of the history of science that reveals to what extent every scientist is a creature of his social milieu."[35] By the 1970s, two decades after Gillispie's book first appeared, historians of the relations of science and religion would virtually all agree that a responsible account of the subject had to include a treatment of the social context.

In addition to its role as a harbinger of the social history of the relations of science and religion, Gillispie's study—aided, no doubt, by William Irvine's

---

33. Ibid., 333.
34. Gillispie, *Genesis and Geology,* ix–x.
35. Glass, review of Gillispie, *Genesis and Geology,* 150.

Apes, Angels, and Victorians and John Greene's *Death of Adam* later in the decade—helped launch the modern "Darwin industry." Most of Greene's study dealt with the pre–nineteenth-century background to evolution, but the relationship between science and religion, approached through the history of ideas, defined the overall theme of the work. Not surprisingly, as we shall see in section 4, a discussion of the relations between science and religion has never been far away in the voluminous literature on the history of evolution that has appeared in the last half century.

The reissue in 1955 of A. D. White's classic volumes on the warfare of science with theology was not indicative of the trend in the historiography of science and religion. Instead, the field took its direction from Edward A. White's *Science and Religion in American Thought*, which appeared a year after Gillispie's book. White devoted three chapters to nineteenth-century interpretations of the relationship between science and religion, including those of Draper, A. D. White, John Fiske, and James. White laid out his own assumptions for the reader, although he promised to keep them out of the discussion itself. He approached his topic, he avowed, with a belief in transcendent meaning.[36] In an attempt to be as fair as possible to the historical figures he covered, he tried to portray what had motivated them on their terms. His effort to understand historical figures became most evident in his coverage of post–nineteenth-century events; specifically, his explanation of what was at stake for fundamentalists in the controversies of the 1920s revealed that he indeed knew how they had felt.[37]

By 1960 John Dillenberger could criticize the frequent reissuing of A. D. White's *Warfare*. It was no longer an acceptable scholarly book, wrote Dillenberger in his *Protestant Thought and Natural Science*, because its author was so suspicious of theology. A proper motivation for the historian was to assume "no single interpretation," and especially to reject interpretations that had either defied or capitulated to science.[38] Dillenberger aimed not to give answers but simply to find the genuine issues behind controversies like that over evolution in the nineteenth century. His reading of that episode uncovered, in fact, a concern of enormous magnitude. "The Darwinian impact," he wrote, "was the final threat to all vertical and depth dimension within man and the cosmos. It marked the culmination of a period in which no adequate symbols were left for expressing and thinking about the classical Christian heritage."[39] True to his promise, Dillenberger offered no answer to the problem.

Walter F. Cannon's studies in the early 1960s of science and religion in Britain continued the attempt to understand in their own terms the chal-

---

36. White, *Science and Religion*, vii.
37. Ibid., 117.
38. Dillenberger, *Protestant Thought*, 14–15.
39. Ibid., 253.

lenges the Victorians had faced. Beyond his careful exposition of efforts to incorporate miracles into a worldview increasingly affected by the claims of science,[40] Cannon added a sociological concern to the historiography of nineteenth-century science and religion: how did the associations that individuals formed as scientists and as laity compare?

Among the first to broaden the vision of the relations between physical science and religion in the nineteenth century was Erwin Hiebert. Perhaps because he broke new ground for historians of the so-called hard sciences, Hiebert felt it necessary to justify his 1966 venture into what he regarded as new historiographical territory: "If history is a record of man's ideas and behavior in place and time, then it may be as legitimate to study the interaction between thermodynamic thought and religion as it is to study the internal history of thermodynamics *per se*."[41] Like Gillispie before him, Hiebert was interested in how religious beliefs had affected the evaluation of physical science and how, specifically, the first and second laws of thermodynamics had been used to answer questions of a fundamentally religious cast.

It would not be long before historians of science and religion no longer felt it necessary to justify the historical investigation of the impact each domain had exerted on the other. If Gillispie's focus on the social context of the relationship between Genesis and geology had signaled the beginning of the social history of the relations of science and religion, it was Thomas Kuhn's *Structure of Scientific Revolutions* (1962) that would provide the catalyst for the discipline of the history of science as a whole to shift its attention toward social history. Kuhn's book strongly resonated with the growing sense that science was fundamentally susceptible to cultural influence; indeed, the enormous success that Kuhn's book enjoyed meant that its message transcended the confines of the history of science. Since religion was a principal force in shaping the paradigms of a given time and place, the task of the historian of science and religion took on new significance. By the start of the 1970s "internalist" historians of science found themselves on the defensive. The history of science and religion was increasingly regarded as of primary importance in new attempts to understand life and thought in the nineteenth century.

4. The Rise of the Social History of Science and Religion: The 1970s and 1980s

It has been argued that the rapid expansion of the history of science as a field in the 1970s was largely due to the growth of social history. If this is true, it should come as no surprise that historical studies in the history of the rela-

40. Cannon, "Problem of Miracles."
41. Hiebert, "Uses and Abuses," 1049.

tions of science and religion also proliferated during this decade. The field became much more open than ever before to investigations of all sorts, especially those that uncovered social attitudes about science that reflected widely shared sentiments.

William Brock, for example, investigated the values exhibited by the scientific community in Britain in the 1860s in its response to a declaration signed by some scientists that science could not contradict Scripture. "To what extent," Brock asked, "was 'freedom of expression' in matters of science seen as a 'test case' for freedom of political and religious expression generally?"[42] Leaning on then recent studies of Victorian social scientific networks, Brock concluded that the great majority of members of the scientific community refused to institute a "club law" to intimidate those who thought differently.[43] In a similar vein Stephen Brush analyzed the public debate of the 1870s surrounding the "prayer test," in which John Tyndall had challenged Britain to focus prayer on the sick for several years to see if it could produce any measurable effect. Brush concluded that religious and scientific beliefs of the time shared many features. Religious people, like scientists, had changed their beliefs because of empirical observation, and scientists, like theologians, had succumbed to dogmatism on occasion.[44]

The primary focus of attention for historians of the 1970s studying the nineteenth-century interactions of physical science and religion was Lord Kelvin. Two shorter studies dealing with the theological context of Kelvin's science were accompanied by Joe D. Burchfield's *Lord Kelvin and the Age of the Earth* (1975). A multifaceted study, Burchfield's book demonstrated how the reckoning of geological time, a subject that was intrinsically related to a religious understanding of humankind's past, involved a diversity of issues with implications for the social perception of science: disciplinary hierarchy within the sciences, professionalization of science, the role of scientific exploration, the growing confidence in quantification, and the eroding social authority of the theologian relative to that of the scientist on questions of cosmogony.[45]

This interest in the balance of authority between representatives of the religious community and other intellectuals in nineteenth-century society was very evident in historiography during the 1970s. In America there had been great confidence at the beginning of the nineteenth century that science was among the theologian's most useful allies. Daniel LeMahieu showed in 1976

42. Brock, "Scientists' Declaration," 40.

43. Ibid., 60. The studies referred to included MacLeod, "Victorian Scientific Network," and Jensen, "X-Club."

44. Brush, "Prayer Test," 563.

45. Burchfield, *Lord Kelvin and the Age of the Earth*. The shorter studies were Wilson, "Kelvin's Scientific Realism," and C. Smith, "Natural Philosophy and Thermodynamics."

just how powerful that conviction was in his study of the mind of William Paley, the greatest English-language exponent of the argument from design.[46] Theologians in America shared Paley's view at the beginning of the century, argued Herbert Hovencamp in *Science and Religion in America: 1800–1860* (1978), but by the time of the Civil War confidence in it had collapsed.[47] The British church historian Owen Chadwick, however, cautioned, not always dispassionately, that it was a mistake to attribute the secularization that occurred in the nineteenth century to science; specifically, he argued that it was not Darwin so much as it was Marx and the Higher Criticism of the Bible that had produced a secularized European mind.[48] My own book on German scientific materialism analyzed the popularization of science as it affected religious questions. Materialism had been historically associated with atheism. When German scientific materialists invoked the authority of recent science in their denunciation of all immaterial entities, including the human soul, they proved to be enormously influential in German society.[49] New attention to France and to Catholic thinkers came with Harry Paul's *Edge of Contingency* (1979), which became the standard work on, to use its subtitle, French Catholic reaction to scientific change from Darwin to Duhem.[50]

A considerable amount of the work on the relations of science and religion in the 1970s was, however, due to the burgeoning Darwin industry. Evolution was not completely absent from any of the works from the 1970s already mentioned; it was, however, the focus of many other studies that attempted, ironically, to downplay in one way or another the role of Darwin and Darwinism where religion was concerned. At the beginning of the decade Robert Young attempted to correct the impression that the *Origin of Species* had come into the theological world "like a plough into an ant-hill."[51] Not only was Darwin and Darwinism but one phase in a wider movement that had been underway for some time, but also "the idea of opposing theology could not have been farther from the minds of the main evolutionists." The debate over evolution produced, according to Young, "an adjustment within a basically theistic view of nature rather than a rejection of theism"; furthermore, evolution supplied an alternative belief in progress as utopian and optimistic as belief in the Christian afterlife was for believers.[52] Five years after Young's study Michael

---

46. LeMahieu, *Mind of William Paley.*

47. Hovencamp, *Science and Religion in America.*

48. Chadwick's *The Secularization of the European Mind* was given as the Gifford Lectures of 1973–74 and possessed an apologetic tone.

49. Gregory, *Scientific Materialism.*

50. Paul, *Edge of Contingency.* Paul had earlier examined the response to Darwin among Catholics in his "Religion and Darwinism."

51. The phrase was taken from Henkin, *Darwinism in the English Novel.* See Young, "Impact of Darwin," 17.

52. Young, "Impact of Darwin," 21. Cf. also 27, 31.

Ruse documented the variety of ways of relating science and religion that had long been evident in British society, concluding that "in many respects the various attitudes taken towards the science-religion relationship were the same before and after the *Origin*."[53] Historians should not, Ruse implied, regard Darwin's *Origin* as the watershed event in that relationship that it had frequently been touted to be.

Other historians reinforced the claim that English-speaking society, at least, was quite flexible in its thinking about how evolution and religion were to be understood. Preparing the way for assimilating evolution in America was, according to Ronald Numbers, Laplace's nebular hypothesis.[54] In Laplace's speculation the element of design could be transferred to Newtonian law functioning as a creative agent, and so the idea of creation by natural law could become a part of natural theology. Thus, the notion of the physical evolution of the cosmos proved widely acceptable. Frank Turner's *Between Science and Religion* revealed six English intellectuals who responded in their own creative ways to the issues that flourished in Darwin's wake. None of the six fit into the two camps historians commonly depicted as waging a conflict between scientific naturalism and Christianity; rather, each opted to define an intermediate stance that preserved the ideals and values they were determined to uphold.[55]

In *The Post-Darwinian Controversies* (1979) James Moore consolidated the growing discontent with the lingering notion of science and religion in conflict with a frontal attack on the whole notion of a warfare between science and religion in the nineteenth century. After reviewing the militant historiography and rehearsing the historiographical toll it had taken, Moore presented his own reading of the controversies.[56] The image of warring camps did not accurately depict the participants in the controversies Moore described; it did not, for example, account for the religious believers who objected to Darwinism for scientific reasons, nor did it represent the theological liberals who embraced a neo-Lamarckian evolution so universal in scope that it was worthy of an omnipotent creator.[57] Many found Moore's additional claim—that the high view of God's sovereignty embedded in some distinctly orthodox theological traditions permitted members to accept Darwin's natural selection by diluting the cognitive dissonance between it and teleology—to be theoretically intriguing but historically inaccurate.[58] In spite of such disagreements,

---

53. Ruse, "Relationship between Science and Religion," 522.

54. Numbers, *Creation by Natural Law*.

55. F. Turner, *Between Science and Religion*.

56. Moore, *Post-Darwinian Controversies*, 19–100.

57. The first group Moore identified as Christian anti-Darwinists, the second, resulting from the blending of Darwinism with romanticism, as Christian "Darwinisticists."

58. These were labeled by Moore as Christian Darwinists. Moore, *Post-Darwinian Controversies*, 15, 336.

however, Moore's impressive tome soon became known as the standard refutation of "warfare" historiography.

The 1980s brought even more studies dealing with nineteenth-century science and religion than had the previous decade. In 1985 Ronald Numbers published a useful summary of the literature on the warfare thesis, focusing on historiography among American historians.[59] While most of the works he cited dealt with the era prior to the nineteenth century, notable exceptions included works by Numbers himself on Ellen G. White *(Prophetess of Health)*, Bert Lowenberg on evolution in New England ("The Controversy over Evolution in New England"), Lester Stephens on evolution in the South *(Joseph LeConte)*, and, on nineteenth-century religious reactions to Darwinism, Windsor Hall Roberts *(The Reaction of American Protest Churches to the Darwinian Philosophy, 1860–1900)* and Stow Persons ("Evolution and Theology in America").

With the age-old modality of conflict now considered suspect, historians found their attention drawn "beyond war and peace," to quote the title of an essay in *Church History* by Numbers and David Lindberg (1986). The authors opposed traditional categories as "misleading, even pernicious,"[60] and took much the same tone in their introduction to *God and Nature,* an edited volume of the same year. While Lindberg and Numbers conceded that it may still be necessary for historians to acknowledge the presence of conflict in certain contexts, especially in the nineteenth century, they were impressed with what they identified as "a growing inclination among historians to take science down from its traditional pedestal and treat it as mere ideological property, intrinsically no different from any other kind of knowledge."[61] When this was done, as it had been in recent work by Turner and Martin Rudwick,[62] the element of conflict in the late-nineteenth-century debates could be seen to result as much from a shift in social authority and the social uses of science as from anything else.[63] Over the course of the century the scientist replaced the clergyman as the acknowledged spokesperson wherever nature's truth, the proper method for finding it, or its role in education were concerned.

Darwin's own religious position continued to fascinate historians in the 1980s. Dov Ospovat put to rest the assumption that Darwin's creative development was essentially over by the time of the second sketch of his theory in 1844 by chronicling the stages in his intellectual journey away from teleol-

---

59. Numbers, "Science and Religion," 59–80.

60. Lindberg and Numbers, "Beyond War and Peace," 353.

61. Lindberg and Numbers, *God and Nature,* 9.

62. F. Turner, "Victorian Conflict"; Rudwick, "Senses of the Natural World."

63. For example, Nicolaas Rupke's study of William Buckland, published in 1983, noted the strategic advantages of Buckland's Mosaic geology for promoting geology at Oxford. Rupke, *Great Chain of History.*

ogy.[64] Ospovat showed that the journey took place over two decades after Darwin's voyage around the world, a finding of obvious import in assessing the role of his religion in his science. Neal Gillespie had treated the nature of Darwin's theism in 1979, and Frank Burch Brown did so again in 1986.[65] At the end of the decade James Moore explained how the death of Darwin's daughter Annie could be seen as the focal point of the explanation for Darwin's giving up Christianity.[66] Moore followed that essay not only with a prizewinning biography of Darwin, written with Adrian Desmond, but also with a little book investigating in definitive fashion the legend of Darwin's deathbed conversion.[67]

In commemoration of the centenary of Darwin's death the British Society for the History of Science sponsored a conference in 1982 at the Linnean Society to examine issues in the relationship between science and religion. In the volume that resulted John Brooke addressed a theme that had preoccupied him and numerous others for some time and one on which Brooke became the acknowledged expert. Brooke was impressed with the enormous flexibility and resilience of natural theology, including that of the nineteenth century.[68] Its general agenda—to reinforce a concept of providence—was so powerful that it not only survived Darwin's achievement, but continued, sometimes in secular guises, to function as an effective means of articulating how nature fit into an understanding of human meaning. Brooke concluded from his exhaustive examination of the historical forms and functions of natural theology that "science itself has probably placed fewer constraints on how the natural world is to be interpreted than we are often tempted to think."[69] Among other works on natural theology two from 1988 deserve special mention: Pietro Corsi's study of ecclesiastical politics in Baden Powell's journey from theological conservative to ultraliberal (Science and Religion: Baden Powell and the Anglican Debate) and Jon Roberts's detailed mining of the responses to Darwin among American Protestant thinkers (Darwinism and the Divine in America).

One specific debate within natural theology, that over the plurality of worlds, was the subject of a magisterial study by Michael Crowe. Although

64. Ospovat, Development of Darwin's Theory.

65. Gillespie, Charles Darwin and the Problem of Creation; Brown, "Darwin's Theism."

66. Moore, "Of Love and Death."

67. Desmond and Moore, Darwin; Moore, Darwin Legend.

68. The commemorative volume, Darwinism and Divinity, was edited by John Durant. Brooke's essay was entitled "The Relations between Darwin's Science and His Religion." Brooke's prize-winning book Science and Religion includes a bibliographic essay; the portion pertaining to chapter 6, "The Fortunes and Functions of Natural Theology," lists six other articles on natural theology by Brooke (not including that referred to in n. 69), along with a host of other studies.

69. Brooke, "Science and the Fortunes of Natural Theology," 19–20.

the book's title set the termini of his study as 1750 and 1900, *The Extraterrestrial Life Debate* explored its subject beyond these confines. One episode from the 1800s centered on the question of human uniqueness in the universe and its implications for God's redemptive plan. Crowe described how Thomas Paine's attack on the "conceit" of Christians who believed that Christ's death applied to extraterrestrial as well as terrestrial life precipitated a crisis at the beginning of the century among natural theologians, who scrambled to denounce Paine and his position. The Reverend William Whewell only made things worse at midcentury when he essentially agreed with Paine that one could not hold to the uniqueness of Christ without also holding to the uniqueness of life on earth. So complicated were the implications of commonly held scientific and religious assumptions that the debates resembled, in Crowe's words, "a night fight in which the participants could not distinguish friend from foe until close combat commenced."[70] Crowe's study reinforced the message promulgated by Brooke and others that natural theology was indeed a rich historical lode well worth mining by the student of the relations of science and religion.[71]

Natural theology was not the only subject mined with profit in the 1980s. The upsurge of historical research dealing with the science of the Romantic period also held significance for the history of religion and science. Gone was the older judgment that dismissed Schelling's nature philosophy as an interlude in the progressive development of experimental science. In a spate of new works concerned with "Romantic" science, historians uncovered a host of ways in which the generation following the French Revolution had left lasting marks on the scientific community. Where science and religion were concerned, there could be no thought of appealing to conflict, much less to warfare, to depict the Romantic attitude; hence the general erosion of the categories of conflict and warfare in the historiography of science and religion may well have contributed to an escalation of interest in the science of the Romantic era.

Trevor Levere's work on Coleridge was among the first in the decade to draw out the specifically religious dimension of romantic science. Coleridge, Levere pointed out, progressed from the pantheism he had learned from Schelling and other German *Naturphilosophen* to a transcendental vision of science realized in a specifically Trinitarian Christianity.[72] Ironically, the embeddedness of religious interest in the conception of nature of the German and English Romantics meant that virtually all the new studies of the period

---

70. Crowe, *Extraterrestrial Life Debate*, 558.

71. Brooke himself had written on Whewell and the debate over pluralism in 1977. See Brooke, "Natural Theology and the Plurality of Worlds."

72. Levere, *Poetry Realized in Nature*.

became relevant for the historian interested in the relations between science and religion. My own work was among the few treatments that explicitly singled out the relationship between science and religion in this literature.[73] The specific challenge for German theologians in this period was whether or not Kant's radical separation of religion from natural science could be overcome. Most so-called Romantic scientists insisted that their conception of nature informed their understanding of religion. Even Jakob Fries, a Kantian philosopher of science and religion, altered his mentor's system in such a way that nature again became relevant to religion. Fries's central work, *Wissen, Glaube und Ahndung* (1805), was translated and published in English for the first time in 1989.[74]

In the 1980s scholars continued to assess the larger issue of religious change in the nineteenth century, including the role played by science. Ueli Hasler examined, in Marxist terms that were falling rapidly out of fashion, how German theology had accommodated itself over the course of the century to what he called the "bourgeois conception of nature."[75] In the middle of the decade, James Turner eloquently declared that the blame for the growth of unbelief in America lay squarely with the theologians who had too quickly deferred to scientists' claim that the universe was tailored to our measurements. They were the ones, he wrote, "who had most loudly insisted that knowledge of God's existence and benevolence could be pinned down as securely as the structure of the frog's anatomy—and by roughly the same method."[76] Bernard Lightman led the way in investigating British unbelief in his individual volumes dealing with agnosticism and the Victorian crisis of faith.[77]

Treatment of nineteenth-century science and religion during the previous two decades had also occurred in projects designed to make materials from the history of science more accessible to students and the public. In Britain, the Open University Press published in 1981 a series of seven substantive booklets for a third-level course entitled "Science and Belief: From Darwin to Einstein."[78] Although some of the material was clearly drawn from the twentieth century, much of it summarized key episodes in the relations of science and religion during the nineteenth century, and all of it was authored by respected historians of science.

73. Gregory, "Theology and the Sciences," 69–81. Part 1 of my *Nature Lost?* has considerable material dealing with the first half of the nineteenth century.

74. Fries, *Knowledge, Belief, and Aesthetic Sense.*

75. Hasler, *Beherrschte Natur.*

76. J. Turner, *Without God, Without Creed,* 193.

77. Lightman, *Origins of Agnosticism;* Lightman and Helmstadter, *Victorian Faith in Crisis.*

78. Booklet Series, "Science and Belief."

In America, Jacob Bronowski's television series and book, both titled *The Ascent of Man*, became notable for its treatment of science and religion in its coverage of the seventeenth century. The opposite, however, was true for the nineteenth century. Even the treatment of evolution and religion in the century was understated. For 1973, Bronowski's conclusion that "the theory of evolution is no longer a battleground," while revealing about his approach to covering the topic in the nineteenth century, evinced no awareness of how deep and long-lasting the issues evoked by evolution were. A decade later Carl Sagan's television series and book, *Cosmos,* showed no particular awareness that some historians of relations between science and religion were taking scientific knowledge down from its pedestal. Sagan presented in somewhat condescending fashion the nineteenth-century inference, from evidence of design in the natural world, of the existence of a designer; this he regarded as a pardonable conclusion but clearly unnecessary in light of what Darwin had discovered. In the 1980s James Burke produced a television series and book, entitled *The Day the Universe Changed,* for British and later for American audiences. While Burke geared his approach totally to the vogue of social history, he emphasized the ideological dimension of Darwinian evolution over its significance as an issue in science and religion.

If Bronowski and Sagan regarded religious issues associated with nineteenth-century science with limited patience, so did other natural scientists writing in the 1980s. Ernst Mayr acknowledged openly in his *Growth of Biological Thought* how religion had historically conditioned the way in which the cosmos and the living world were perceived. His study, which explicitly stressed the development of modern biology, was intentionally whiggish. As a consequence, past religious explanations of natural phenomena fell, not surprisingly, by the wayside as the narrative proceeded. This was also true, but to a lesser extent, in the popular historical work of Stephen Jay Gould. Gould appeared to delight in forcing his readers to question standard portraits of the religious enemies of progress in science. His essays, hardly confined to the nineteenth century, nevertheless included a great deal from that era. Gould defended catastrophists, for example, as "the hard-nosed empiricists of their day, not the blinded theological apologists."[79] He made the general point that historical treatments can take "the most notorious of textbook baddies and try to display their theory as both reasonable in its time and enlightening in our own."[80] However, as sympathetic as he was to the motivations of nineteenth-century natural theologians, Gould's history was ultimately meant to teach a lesson about science. When it came to matters like the bla-

---

79. Gould, *Ever Since Darwin,* 150.
80. Ibid., 202.

tantly amoral arrangements nature could make between parasite and host, Gould was clear: moral answers "do not, and cannot, arise from the data of science."[81]

Philosophers also occasionally argued about the relations of science and religion in the nineteenth century. As was the case among scientists, the philosopher's motivation in turning to history was not necessarily that of the historian; historical treatments of science and religion could be expected to be guided by a philosophical interest. Edward Manier sought in 1978 to get at Darwin's methodology and the meaning of his theory. He did so by examining the philosophical, religious, political, and ethical works read and analyzed by the young Darwin and watching for their influence in his later use of language and his establishing standards for explanation in his scientific work.[82] A year after Manier's book came another from philosopher/historian Michael Ruse.[83] In *The Darwinian Revolution* Ruse blended social history with philosophy to produce a synthetic treatment of beliefs—religious, scientific, and philosophical—as they changed over the century. Ruse emphasized that the Darwinian revolution was not a single thing. Where religion was concerned, he pointed to the numerous ways it had helped the evolutionary cause, ending his book less certain, in fact, of why it had failed to suppress evolution.

Social histories of the relations of science and religion from the 1970s and 1980s by and large retained the balance implied by Kuhn's thought between the context in which an episode was embedded and the content of the episode itself. In the larger discipline of history of science, however, some scholars came to emphasize the context of a discovery almost to the exclusion of the discovery itself, as if the social context were all that mattered. In spite of this difference, those historians of the relations of science and religion from these two decades who were willing to identify themselves with the social history of science agreed that treatment of the social context was possible and important. The assumptions on which this agreement was based would be challenged in the 1990s.

5. At Century's End: The Turn to Cultural and Biographical Studies

In the final decade of the century, indeed, of the millennium, historical scholarship on the relations of science and religion in the nineteenth century has continued to thrive and to respond to changing historiographical currents. As the pendulum of interest swung in the 1990s to cultural history, with its suspi-

---

81. Gould, *Hen's Teeth and Horses Toes*, 42–3.
82. Manier, *Young Darwin*.
83. Ruse, *Darwinian Revolution*.

cions of appeals to structures purported to inhere in the social context, historians of science and religion have taken account of and in some cases reflected the trend. To be sure, there is no unanimity that science past and present must always be seen as an expression of culture and never as something *beyond* culture. Still, there has been less debate among historians of science and religion on this issue than in the parent discipline of history of science.

An unmistakable inclination in this decade has been toward the use of biography to study nineteenth-century science and religion. Geoffrey Cantor's study of Michael Faraday stands, in Frank Turner's view, as "the best book-length study of the relationship between religious thought and scientific activity in the life of a Victorian scientist."[84] Yet not all reviewers were as happy with Cantor's attempts to establish links between Faraday's strict biblical religion and his ideas and behavior as a scientist and as a member of British society, especially since, due to the paucity of explicit connections, Cantor had to rely on "resonances" between these spheres. Cantor's exhaustively detailed study of the Sandemanian sect to which Faraday belonged is enormously suggestive in its identification of metascientific principles that, coming from Faraday's private religion, may have "mediated" his public science. The manner in which Faraday embraced his religion and science has long seemed cognitively dissonant to many historians. Cantor's book helps us understand why such dissonance has been the historian's problem, not Faraday's.

Two other biographical studies, while not devoted specifically to the theme of science and religion in the life of the principal figure, do include it. Kenneth Caneva's explicitly contextual approach to the life and work of J. Robert Mayer includes an important observation about the influence of religion in Mayer's science. Consonant with his desire to oppose the materialism of his day, Mayer wished to dissociate force completely from matter. By seeing force as pure cause Mayer could conclude that force was transformable but not destructible, for causes disappear as their effects emerge. Mayer's conclusions about conservation, then, were facilitated in part by his religious opposition to materialism.[85] Likewise Gerald Geison, in his study of the private science of Louis Pasteur, suggests that Pasteur's opposition to materialism was a central motivation in his public debates over spontaneous generation.[86]

Materialism has also captured the attention of other historians, not just those dealing with the German and French debates. In a recent study by Stephen Kim, Tyndall's embrace of scientific naturalism is seen as an expression of his religious faith. While the subtitle of *John Tyndall's Transcendental Materialism* refers to "the conflict between science and religion," Kim's work

84. F. Turner, review of Cantor, *Michael Faraday.*
85. Caneva, *Robert Mayer,* 43–45.
86. Geison, *Private Science of Louis Pasteur,* 121–25.

does not utilize the older warfare historiography. Rather, the reference is to Tyndall's "conflicted" position as a religious materialist.[87] Yet another relevant biographical study is Robert Gustafson's work on James Woodrow, who as a businessman, professor, academic administrator, scientist, and editor devoted much attention to formal consideration of the relationship between science and religion.[88]

Two works in the 1990s have viewed the interaction of religion and science in the nineteenth century as a process involving loss. In my work on natural science and the German theological traditions I examined how neo-Kantian theologians of Albrecht Ritschl's school lost the capacity to incorporate nature into the theological enterprise because they had willingly surrendered nature to the natural scientist. Not all German theologians, of course, abandoned nature; but those who had compensated by redefining theology's task in moral terms (as Kant had done), specifically forbidding it to deal with nature. I sought to show how three other theological traditions—an exclusive conservatism, an inclusive liberalism, and a radical pantheistic naturalism—strove to retain nature, each in a different way. The loss of nature, I suggested, was but one expression of a more general change being experienced throughout the culture and in science itself; specifically, it was part of the loss of confidence that had reigned in the nineteenth century.[89]

Paul Croce's study of science and religion in the era of William James chronicles much the same loss in America. Exhibiting once again the recent penchant for biography, Croce focuses on the cultural circle of the young James. (He is preparing a second volume that will deal with the mature James.) Croce depicts the loss of certainty that James experienced as he absorbed the meaning of Darwin's *Origin* as interpreted by C. S. Peirce. The conclusion of the young James was not unlike that of the Kantians: he retained a craving for sure foundations of experience, but, ironically, his study of science brought him epistemological uncertainty. The combination produced a remarkably open mind that would radically reinterpret the meaning of religious experience.[90]

While, as noted above, most work from the 1990s exhibited a healthy respect for the assumptions of cultural history, some historians of science and religion continued to view the nineteenth century as but one stage in the journey toward a *proper* understanding of how the two domains should be related.

---

87. Kim, *Tyndall's Transcendental Materialism.*

88. Gustafson, *Woodrow.*

89. Gregory, *Nature Lost?* Walter Conser has examined, unsuccessfully in the view of one reviewer, how science and religion were dealt with by American theologians who followed in Schleiermacher's train, the liberal German tradition in my study. See Conser, *God and the Natural World,* and the review by Stiling.

90. Croce, *Science and Religion.*

For example, some still see the century through the lens of conflict. Peter Addinall's interesting summary of the parallel reactions to Hume's devastating critiques in the "conservative" natural theological tradition of Paley and the "liberal" tradition of Kant assumes that science and religion are permanent entities that must inevitably conflict and that Kant's position was essentially the correct one. Addinall assumes, therefore, that his task is to show why Kant's outlook can serve as a profitable foundation for religion.[91]

The work of Christopher Kaiser is couched within a larger agenda of what he calls creational theology, by which he means "a historical worldview and moral stance based on the biblical belief that nature and humanity are created by a wise, powerful God who intends good for them."[92] Rejecting creationist science, which presumes an incompatibility between scientific work and a creational belief, he examines challenges to and expressions of the so-called Newtonian mechanical world picture in the nineteenth century. In his historical analysis Kaiser's work walks the line between expostulating the claim that the creational tradition has historically influenced physics (including, incidentally, twentieth-century physics) and a recommendation of the fundamental insight of creational theology for a balanced worldview in the present.

David Livingstone, though sympathetic to a religious worldview, offers more dispassionate treatment of the century's science and religion. Livingstone is well aware of the pitfalls of conflict historiography. His religious sensitivity becomes evident from the subject matter he chooses to investigate, not from the science he wishes to judge or the theology he wants to recommend. He has reexamined the response to Darwin in the English-speaking evangelical tradition to find out if modern devotees of "creation science" have been correct in identifying themselves as its offspring. He found that, with regard to the leading evangelical theologians concerned with science prior to World War I, they have not.[93] Livingstone's many other recent works on science and religion have either touched or focused on the nineteenth century with impressive methodological astuteness. He has not always persuaded—his notion of the "marriage" of science and religion at the end of the nineteenth century has, for example, been criticized—but his work has been invariably engaging.[94] Livingstone and Mark Noll have made available the writings of the Princeton theologian Charles Hodge on science and reli-

91. Addinall, *Philosophy and Biblical Interpretation.*

92. Kaiser, *Creational Theology,* 8.

93. Livingstone, *Darwin's Forgotten Defenders.*

94. Livingstone, *Preadamite Theory;* Russell, review of Livingstone, *Preadamite Theory,* 555. See also Livingstone, Hart, and Noll, *Evangelicals and Science.* Other recent histories written with a specific religious tradition in view include Jaki, *Scientist and Catholic,* and, for the nineteenth century, Paul, *Science, Religion, and Mormon Cosmology,* chaps. 4–6.

gion, an effort to some extent balanced by the reissue by Alan Barr of several works by Huxley, including several essays and a few selected letters bearing directly on Huxley's views on science and religion.[95]

The Dutch series known as Brill's Studies in Intellectual History has recently produced a volume that reassesses the impact of religion on geology in the years before Darwin's *Origin*. Appreciative of critiques from the 1970s of the focus among historians on catastrophism versus uniformitarianism, J. M. I. Klaver specifically looks at the interaction of religious and scientific ideas in Lyell's thought, comparing Lyell's religious position with that of other contemporary geologists. His approach, which consists heavily of an *explication de texte,* concludes that Lyell's writing "was not anti-religious and his religious sentiments were not essentially different from those of his contemporary colleagues."[96] Klaver is particularly concerned to correct the impression that Lyell's *Principles of Geology* should be regarded as a forerunner to the *Origin of Species* and that Lyell stood as Darwin's champion against the more orthodox reactions of geologists like Adam Sedgwick and Whewell.

Several recent works treat science and religion in the nineteenth century as a means of understanding culture. In two cases there is an open embrace of an agenda; that is, the authors explore history in order to understand and to judge today's culture. Both histories deal with the role of women in past and present science and see the relationship between religion and Western science as the pivotal point on which their agenda is balanced. David Noble and Margaret Wertheim each argue that modern science owes more than a little of its goals, its institutional structure, and its social attitudes to religion.[97] Neither draws heavily on the nineteenth century in establishing their claims that the religious nature of the scientific enterprise helps explain the historical and current underrepresentation of women in science. Wertheim, however, argues that it was in the nineteenth century that science usurped from religion the role of savior. Its earlier usurpation of nature had not been seen by most as essentially threatening to Christianity. The threat became severe, however, when the promise of salvation came to be more effectively associated with science than it was with the church.[98] Not surprisingly, both books have evoked considerable criticism, not only from those displeased with their downplaying of scientific achievement for itself, but also from historians unhappy with the open embrace of a political master narrative.

In his recent treatment of the Bridgewater Treatises, Jon Topham seeks to

95. Hodge, *What Is Darwinism;* Huxley, *Major Prose.*
96. Klaver, *Geology and Religious Sentiment,* xiv.
97. Noble, *World Without Women;* Wertheim, *Pythagoras' Trousers.*
98. Wertheim, *Pythagoras' Trousers,* chap. 7.

change how we regard these works on natural theology.[99] In reexamining how and for whom this series of treatises was produced, who actually read them, and where they were read, Topham finds it necessary to reject any demarcation between "high brow" and "low brow" opinion, even to abandon the categorization of the works as "popular science" since that characterization suggests a "top-down" notion. Topham's approach helps us assess more accurately the relative cultural authority of both science and religion in the nineteenth century.

Three recent works reinforce the interest of contemporary historians in the cultural history of the relations of science and religion. First, in his study of the origins of nineteenth-century historical consciousness, Thomas Howard portrays theology as an expression of culture.[100] In particular, he explores how Jacob Burckhardt's desire to move beyond modernity was conditioned by his theological study and how Burckhardt yet remained ambiguous about providence. As it was for James and for the neo-Kantian theologians at the end of the century, Burckhardt's challenge to respect both science and religion was real and remains recognizable in our own day.

Second, a volume edited by Ronald Numbers and John Stenhouse contains studies by eleven scholars who propose to investigate how differences in location, race, gender, and religion conditioned the way Darwin's message was received.[101] Of the essays that deal with religion (most do), Livingstone's comparison of Darwin's reception in Princeton, Belfast, and Edinburgh reveals that even apparently culturally homogenous locations possessed subtle differences of great significance, while a regional study of Darwinism in the American South by Numbers and Lester Stephens challenges other oversimplifications that have been made. Jon Roberts shows that one cannot lump American denominations together if one seeks to understand the motivations of Protestant leaders, eschewing both social and psychological factors in favor of a reconsideration of the theological issues themselves. Other studies address other religious traditions and places as remote from Darwin's England as Australia and New Zealand. The volume as a whole aims to expand the questions historians ask, not least by forcing us to adjust the categories used in dealing with the reception of Darwin.

Third and finally, the 1995–96 Gifford Lectures, delivered at Glasgow University by Brooke and Cantor, register an eloquent protest against the ubiquitous use of master narratives by various sorts of writers to get what they want out of history.[102] Such uses and abuses of history are extremely common, but

---

99. Topham, "Beyond the 'Common Context,'" 233–62.
100. Howard, *Religion and the Rise of Historicism*.
101. Numbers and Stenhouse, *Disseminating Darwinism*.
102. Brooke and Cantor, *Reconstructing Nature*.

the historiography of the relations of science and religion is a particularly apt locus to make their point since convictions about the relationship between these fields are usually strongly held. Brooke and Cantor show what results when scientists or their detractors who have not done their historical home-work nonetheless create histories that appear to confirm their positions. Their sword is double-edged: modern critics and defenders of science alike feel the sting of the authors' careful historical scholarship. The fact is that the history of human attitudes about science and religion defies easy summation; history is far too messy to confirm any mind-set, however modern. If we wish to be historically accurate we must concede that the very meanings of "science" and "religion" constantly undergo change. However much we in the present may think they transcend time, an empirical investigation of history, like that presented by Brooke and Cantor, shows otherwise. The authors themselves choose to speak of the "engagement" of science and religion in order, in their words, "to capture the many different ways in which the relationship between the two has been presented."[103] A healthy portion of the book deals with material from the nineteenth century that illustrates this diversity.

As is evident from the rich and changing historiography discussed above, historical studies of the relations between science and religion have over the years produced some of the most innovative and intriguing work in the field of history of science as a whole. Of course there remain questions to be answered and new relations between science and religion to be explored. To date, for example, there has been very little written about the history of non-Western religion and science. Since the relationship between non-Western religion and science in general has received little scholarly attention, it is no surprise that its history has been of no interest, especially as that history pertains to the nineteenth century. Even in treatments of the history of the relations between Western religion and science there has been relatively little comparative work done. What differences exist, for example, between Protestantism, Catholicism, and Judaism in their respective receptions of and responses to new scientific ideas and institutions, especially as these affect social conventions and beliefs?

As mentioned above, gender as a dimension of the relations between science and religion has been pursued. Nevertheless, the work that has been done has been carried out within the context of a clear master narrative that has openly guided the historical inquiry. New ways in which gender might inform historians of the relations between science and religion will no doubt emerge as issues of gender continue to inform, as they are now doing, the separate studies of science and religion.

103. Ibid., 7.

The relationship between science and religion has always attracted historians. There can be little doubt that its exploration will continue and that questions that no one has yet even formulated will dictate fresh investigations into its history. By simply writing new histories future historians will confirm the theme of this essay: that the history of the relations between science and religion is, to paraphrase Bronowski, "too complex, too delicate, too rich, and too profound to catch in the butterfly net of our interpretations."[104]

104. Paraphrased from Bronowski's characterization of our ability to capture nature with our senses. See Bronowski, *Ascent of Man,* 394.

BIBLIOGRAPHY

Abbreviations

AHES   *Archive for History of Exact Sciences*
AS   *Annals of Science*
BHM   *Bulletin of the History of Medicine* (title varies slightly)
BJHS   *The British Journal for the History of Science*
ESH   *Earth Sciences History*
HS   *History of Science*
HSPS   *Historical Studies in the Physical and Biological Sciences* (title varies slightly)
RHS   *Revue d'Histoire des Sciences* (title varies slightly)
SHPS   *Studies in History and Philosophy of Science*
SSS   *Social Studies of Science* (title varies slightly)
TC   *Technology and Culture*

Abbri, Ferdinando. 1991. "Tradizioni chimiche e meccanismi di defesa: G. A. Scopoli e la 'Chimie nouvelle.'" *Archivo di Storia della Cultura* 4:75–92.

Abrams, Philip. 1968. *The Origins of British Sociology, 1834–1918.* Chicago: University of Chicago Press.

Ackerknecht, Erwin. 1953. *Rudolf Virchow: Doctor, Statesman, Anthropologist.* Madison: University of Wisconsin Press.

———. 1967. *Medicine at the Paris Hospital, 1794–1848.* Baltimore: Johns Hopkins University Press.

Addinall, Peter. 1991. *Philosophy and Biblical Interpretation: A Study in Nineteenth-Century Conflict.* Cambridge: Cambridge University Press.

Aftalion, Fred. 1987. *Histoire de la chimie.* Paris: Masson. English trans.: *A History of the International Chemical Industry* (Chemical Heritage Foundation, University of Pennsylvania Press, 1991).

Ager, Derek V. [1973] 1981. *The Nature of the Stratigraphical Record.* London: Macmillan Press.

Alborn, Timothy L. 1988. "The 'End of Natural Philosophy' Revisited: Varieties of Scientific Discovery." *Nuncius* 3:227–50.

Albury, William Randall. 1972. The Logic of Condillac and the Structure of French Chemical and Biological Theory (1780–1800). Ph.D. diss., Johns Hopkins University.

Albury, William Randall, and David R. Oldroyd. 1977. "From Renaissance Mineral Studies to Historical Geology in the Light of Michel Foucault's *The Order of Things*." *BJHS* 10:187–215.

Alchon, Guy. 1998. "The 'Self-Applauding Sincerity' of Overreaching Theory: Biography as Ethical Practice, and the Case of Mary van Kleeck," in Silverberg, *Gender and American Social Science,* 293–325.

Aliotta, Antonio. 1925. "Science and Religion in the Nineteenth Century," in Needham, *Science, Religion and Reality,* 149–86.

Allchin, D. 1992. "Phlogiston after Oxygen." *Ambix* 39:110–16.

Altschuler, Glenn C. 1979. *Andrew D. White: Educator, Historian, Diplomat.* Ithaca, N.Y.: Cornell University Press.

Anderson, Benedict. [1983] 1991. *Imagined Communities: Reflections on the Origin and Spread of Nationalism.* London: Verso.

Anderson, Margo J. 1988. *The American Census: A Social History.* New Haven: Yale University Press.

Anglin, W. S. 1994. *Mathematics: A Concise History and Philosophy.* New York: Springer-Verlag.

Anon. 1901. "The Passing of the Century." *Chicago Tribune,* 1 January, p. 18.

Appel, Toby A. 1987a. "Biological and Medical Societies and the Founding of the American Physiological Society," in Geison, *Physiology in the American Context,* 155–76.

———. 1987b. *The Cuvier-Geoffroy Debate: French Biology in the Decades before Darwin.* New York: Oxford University Press.

———. 1988. "Organizing Biology: The American Society of Naturalists and Its 'Affiliated Societies,' 1883–1923," in Rainger, Benson, and Maienschein, *American Development of Biology,* 87–120.

Archibald, Raymond Clare, ed. 1938. *A Semicentennial History of the American Mathematical Society: 1888–1938.* 2 vols. New York: American Mathematical Society.

Arnold, D. H. 1983–84. "The Mécanique Physique of Siméon-Denis Poisson: The Evolution and Isolation in France of His Approach to Physical Theory." *AHES* 28:255–367; 29:37–94, 287–307.

Aron, Raymond. 1967. *Les étapes de la pensée sociologique.* Paris: Gallimard.

Aronovitch, Lawrence. 1989. "The Spirit of Investigation: Physics at Harvard University, 1870–1910," in James, *Development of the Laboratory,* 83–103.

Ash, Mitchell G. 1983. "The Self-Presentation of a Discipline: History of Psychology in the United States between Pedagogy and Scholarship," in Graham, Lepenies, and Weingart, *Functions and Uses,* 143–89.

———. 1995. *Gestalt Psychology in German Culture, 1890–1967: Holism and the Quest for Objectivity.* Cambridge: Cambridge University Press.

Aspray, William, and Philip Kitcher. 1987. "An Opinionated Introduction." In *New Perspectives on the History and Philosophy of Mathematics,* edited by William Aspray and Philip Kitcher, 3–57. Minneapolis: University of Minnesota Press.

Ausejo, Elena, and Marinao Hormigón, eds. 1993. *Messengers of Mathematics: European Mathematical Journals (1800–1946).* Madrid: Siglo XXI de España Editores.

Bachelard, Gaston. 1934. *Le nouvel esprit scientifique.* Paris: Vrin.

———. 1938. *La formation de l'esprit scientifique.* Paris: Vrin.

———. 1949. *Le rationalisme appliqué.* Paris: Presses Universitaires de France.

Badash, Lawrence. 1972. "The Completeness of Nineteenth-Century Science." *Isis* 63: 48–58.

———. 1979. *Radioactivity in America: Growth and Decay of a Science*. Baltimore: Johns Hopkins University Press.

Bakel, A. H. A. C. van. 1994. "Über die Dauer einfacher psychischer Vorgänge: Emil Kraepelins Versuch einer Anwendung der Psychophysik im Bereich der Psychiatrie." In *Objekte, Differenzen und Konjunkturen: Experimentalsysteme im historischen Kontext*, edited by Michael Hagner, Hans-Jörg Rheinberger, and Bettina Wahrig-Schmidt, 83–105. Berlin: Akademie.

Baker, Keith M. 1975. *Condorcet: From Natural Philosophy to Social Mathematics*. Chicago: University of Chicago Press.

Ball, W. W. R. [1888] 1901. *A Short Account of the History of Mathematics*. 4th ed. London: Macmillan.

Bannister, Robert 1979. *Social Darwinism: Science and Myth in Anglo-American Social Thought*. Philadelphia: Temple University Press.

———. 1987. *Sociology and Scientism: The American Quest for Objectivity, 1880–1940*. Chapel Hill: University of North Carolina Press.

Barber, Lynn. 1980. *The Heyday of Natural History*. London: Jonathan Cape.

Barber, William. 1975. *British Economic Thought and India, 1600–1858*. Oxford: Clarendon.

Barkan, Diana. 1999. *Walther Nernst and the Transition to Modern Physical Science*. Cambridge: Cambridge University Press.

Barnes, Barry, and David Bloor. 1982. "Relativism, Rationalism, and the Sociology of Knowledge." In *Rationality and Relativism*, edited by Martin Hollis and Steven Lukes, 21–47. Oxford: Blackwell.

Bartholomew, James R. 1989. *The Formation of Science in Japan: Building a Research Tradition*. New Haven: Yale University Press.

Bartholomew, Michael. 1976. "The Non-Progress of Non-Progression: Two Responses to Lyell's Doctrine." *BJHS* 9:166–74.

Barton, Ruth. 1990. " 'An Influential Set of Chaps': The X-Club and Royal Society Politics, 1864–85." *BJHS* 23:53–81.

———. 1998. " 'Huxley, Lubbock, and Half a Dozen Others': Professionals and Gentlemen in the Formation of the X Club, 1851–1864." *Isis* 89:410–44.

Bassett, Michael E. 1979. "100 Years of Ordovician Geology." *Episodes*, no. 2:18–21.

Bastin, J. 1970. "The First Prospectus of the Zoological Society of London: New Light on the Society's Origins." *Journal of the Society for the Bibliography of Natural History* 5:369–88.

———. 1973. "A Further Note on the Origins of the Zoological Society of London." *Journal of the Society for the Bibliography of Natural History* 6:236–41.

Bates, Ralph S. 1965. *Scientific Societies in the United States*. 3d ed. Cambridge, Mass.: MIT Press.

Bäumer, Beatrix. 1996. *Von der physiologischen Chemie zur frühen biochemischen Arzneimittelforschung*. Stuttgart: Deutscher Apotheker Verlag.

Beaulieu, Liliane. 1993. "A Parisian Café and Ten Proto-Bourbaki Meetings (1934–1935)." *Mathematical Intelligencer* 15:27–35.

Beaver, Donald deB. 1966. The American Scientific Community, 1800–1860: A Statistical-Historical Study. Ph.D. diss., Yale University.

Becher, Harvey. 1986. "Voluntary Science in Nineteenth-Century Cambridge University to the 1850s." *BJHS* 19:57–87.

Beer, Gillian. 1983. *Darwin's Plots: Evolutionary Narrative in Darwin, George Eliot and Nineteenth-Century Fiction*. London: Routledge & Kegan Paul.

Beer, J. J. 1959. *The Emergence of the German Dye Industry*. Urbana: University of Illinois Press.

Belhoste, Bruno. 1991. *Cauchy: 1789–1857*. Trans. Frank Ragland. New York: Springer-Verlag. Originally published as *Cauchy, 1789–1857: Un mathématicien légitimiste au XIXe siècle* (Paris: Belin, 1984).

Belhoste, Bruno, Amy Dahan Dalmedico, Dominique Pestre, and Antoine Picon, eds. 1995. *La France des X: Deux siècles d'histoire*. Paris: Economica.

Belhoste, Bruno, Amy Dahan Dalmedico, and Antoine Picon, eds. 1994. *La formation polytechnicienne 1794–1994*. Paris: Dunod.

Bell, Eric Temple. 1931. *The Queen of the Sciences*. Baltimore: Williams & Wilkins.

———. 1937. *The Handmaiden of the Sciences*. Baltimore: Williams & Wilkins.

———. [1937] 1965. *Men of Mathematics*. Reprint, New York: Simon and Schuster.

———. [1940] 1945. *The Development of Mathematics*. 2d ed. New York: McGraw-Hill.

———. 1951. *Mathematics, Queen and Servant of Science*. New York: McGraw-Hill.

Ben-David, Joseph. 1960. "Scientific Productivity and Academic Organization in Nineteenth-Century Medicine." *American Sociological Review* 25:828–43.

———. 1977. "Organization, Social Control, and Cognitive Change in Science." In *Culture and Its Creators: Essays in Honor of Edward Shils*, edited by Joseph Ben-David and Terry Nicholas Clark, 244–65. Chicago: University of Chicago Press.

———. 1984. *The Scientist's Role in Society: A Comparative Study*. Englewood Cliffs, N.J.: Prentice-Hall, 1971. Rev. ed., with a new introduction, Chicago: University of Chicago Press.

Ben-David, Joseph, and Avraham Zloczower. 1962. "Universities and Academic Systems in Modern Societies." *European Journal of Sociology* 3:45–84.

Benfey, Theodor O. 1992. *From Vital Force to Structural Formulas*. Reprint, Philadelphia: Beckman Center for the History of Chemistry.

———, ed. 1981. *Classics in the Theory of Chemical Combination*. 1963. Reprint, Malabar, Fla.: Robert E. Krieger.

Bensaude-Vincent, Bernadette. 1982. "L'éther, élément chimique: Un essai malheureux de Mendeleev en 1904." *BJHS* 15:183–87.

———. 1983. "A Founder Myth in the History of the Sciences? The Lavoisier Case," in Graham, Lepenies, and Weingart, *Functions and Uses*, 53–78.

———. 1986. "Mendeleev's Periodic System of Chemical Elements." *BJHS* 19:3–17.

———. 1992. "Between Chemistry and Politics: Lavoisier and the Balance." *Eighteenth-Century: Essays and Interpretation* 33:217–37.

———. 1993. *Lavoisier: Mémoires d'une révolution*. Paris: Flammarion.

Bensaude-Vincent, Bernadette, and Ferdinando Abbri, eds. 1995. *Lavoisier in European Context: Negotiating a New Language for Chemistry*. Canton, Mass.: Science History.

Bensaude-Vincent, Bernadette, Antonio Garcia-Belmar, and José Ramon Bertomeu-Sanchez. 2002. *L'émergence d'une science des manuels: Les livres de chimie en France (1789–1852)*. Paris: Éditions archives contemporaines.

Bensaude-Vincent, Bernadette, and Anne Rasmussen, eds. 1997. *La science populaire dans la presse et l'édition, XIXe et XXe siècles*. Paris: CNRS Editions.

Bensaude-Vincent, Bernadette, and I. Stengers. 1993. *Histoire de la chimie*. Paris: Editions la découverte. English trans.: *A History of Chemistry* (Cambridge, Mass.: Harvard University Press, 1997).

Benson, Keith R. 1988. "From Museum Research to Laboratory Research: The Transformation of Natural History into Academic Biology," in Rainger, Benson, and Maienschein, *American Development of Biology*, 49–83.

Berdoulay, Vincent. 1981. *La formation de l'école française de géographie*. Paris: Bibliothèque Nationale.

Beretta, Marco. 1993. *The Enlightenment of Matter: The Definition of Chemistry from Agricola to Lavoisier*. Canton, Mass.: Science History.

Berg, Maxine. 1980. *The Machinery Question and the Making of Political Economy, 1815–1848*. Cambridge: Cambridge University Press.

Berman, Morris. 1972. "The Early Years of the Royal Institution 1799–1810: A Reevaluation." *SSS* 2:205–40.

———. 1978. *Social Change and Scientific Organization: The Royal Institution, 1799–1844*. Ithaca, N.Y.: Cornell University Press.

Bernal, J. D. [1953] 1970. *Science and Industry in the Nineteenth Century*. Bloomington: Indiana University Press.

———. [1954] 1971. *Science in History*. 3d ed. 4 vols.: vol. 1, *The Emergence of Science*; vol. 2, *The Scientific and Industrial Revolutions*; vol. 3, *The Natural Sciences in Our Time*; vol. 4, *The Social Sciences: Conclusion*. Cambridge, Mass.: MIT Press.

Bernard, Claude. [1865] 1949. *An Introduction to the Study of Experimental Medicine*. Trans. Henry Copley Greene. New York: Schuman.

Bernoulli, Christoph. 1825. *Betrachtungen über den wunderbaren Aufschwung der gesamten Baumwoll-Fabrikation nebst Beschreibung einiger der neuesten englischen Maschinen*. Basel: Neukirch.

Berry, William B. 1968. *Growth of a Prehistoric Time Scale Based on Organic Evolution*. San Francisco: W. H. Freeman.

Biermann, Kurt-R. 1959. *Johann Peter Gustav Lejeune Dirichlet: Dokumente für sein Leben und Wirken*. Berlin: Akademie-Verlag.

———. [1973] 1988. *Die Mathematik und ihre Dozenten an der Berliner Universität, 1810–1933: Stationen auf dem Wege eines mathematischen Zentrums von Weltgeltung*. Reprint, Berlin: Akademie-Verlag.

———. 1977. *Briefwechsel zwischen Alexander von Humboldt und Carl Friedrich Gauß*. Berlin: Akademie-Verlag.

———. 1979. *Briefwechsel zwischen Alexander von Humboldt und Heinrich Christian Schumacher*. Berlin: Akademie-Verlag.

———. 1990. *Carl Friedrich Gauss: Der "Fürst der Mathematiker" in Briefen und Gesprächen*. Leipzig: Urania Verlag.

Biggs, Lindy. 1996. *The Rational Factory: Architecture, Technology, and Work in America's Age of Mass Production*. Baltimore: Johns Hopkins University Press.

Biot, J. B. 1816. *Traité de physique expérimentale et mathématique*. 4 vols. Paris. Deterville.

Birkhoff, Garrett, and Saunders MacLane. 1997. *A Survey of Modern Algebra*. 5th ed. Wellesley, Mass.: A. K. Peters.

Birr, Kendall. 1957. *Pioneering in Industrial Research: The Story of the General Electric Research Laboratory*. Washington, D.C.: Public Affairs Press.

Bischof, Gustav. 1854–59. *Elements of Chemical and Physical Geology.* Trans. Benjamin H. Paul and J. Drummond. London: Cavendish Society.

Blanckaert, Claude. 1996. "Histoires du terrain entre savoir et savoir-faire." In *Le terrain des sciences humaines (XVIIIe–XXe siècle),* edited by Claude Blanckaert, 9–55. Paris: Harmattan.

Blanckaert, Claude, Claudine Cohen, Pietro Corsi, and Jean-Louis Fischer, coordinators. 1997. *Le muséum au premier siècle de son histoire.* Paris: Editions du Muséum National d'Histoire Naturelle.

Blaug, Mark. 1958. *Ricardian Economics: A Historical Study.* New Haven: Yale University Press.

———. 1978. *Economic Theory in Retrospect.* 3d ed. Cambridge: Cambridge University Press.

Blondel, Christine, and Matthias Dörries, eds. 1994. *Restaging Coulomb: Usages, controverses et réplications autour de la balance de torsion.* Florence: Leo S. Olschki.

Blondel-Megrelis, Marika. 1996. *Dire les choses: Auguste Laurent et la méthode chimique.* Paris: Vrin.

Bloor, David. 1981. "Hamilton and Peacock on the Essence of Algebra," in Bos, Mehrtens, and Schneider, *Social History,* 202–32.

Blum, Ann S. 1987. " 'A Better Style of Art': The Illustrations of the *Paleontology of New York* [by James Hall]." *ESH* 6:72–85.

Blumtritt, Oskar. 1988. "Genese der Technikwissenschaften: Ein Resümee methodologischer Konzepte." *Technikgeschichte* 55:75–86.

Boas, Franz. [1887] 1996. "The Study of Geography." In *Volksgeist as Method and Ethic: Essays on Boasian Ethnography and the German Anthropological Tradition,* edited by George Stocking, 9–16. Madison: University of Wisconsin Press.

Bolzano, Bernhard. 1905. "Rein analytischer Beweis des Lehrsatzes, dass zwischen je zwey Werthen, die ein entgegengesetztes Resultat gewähren, wenigstens eine reelle Wurzel der Gleichung liege." In *Ostwald's Klassiker der exakten Wissenschaften,* no. 153:3–36. Ed. P. E. B. Jourdan. Leipzig: W. Engelmann.

Bonah, Christian. 1995. *Les sciences physiologiques en Europe: Analyses comparées du XIXe siècle.* Paris: Vrin.

Bonney, Thomas G. 1895. *Charles Lyell and Modern Geology.* London: Cassell.

Bonola, Roberto. 1955. *Non-Euclidean Geometry: A Critical and Historical Study of Its Development.* Trans. H. S. Carslaw. New York: Dover.

Booklet Series. 1981. "Science and Belief: From Darwin to Einstein." 7 booklets. Milton Keynes: Open University Press.

Booth, Christopher. 1993. "Clinical Research," in Bynum and Porter, *Companion Encyclopedia,* 1:205–29.

Borck, Cornelius. 1997. "Herzstrom: Zur Dechiffrierung der elektrischen Sprache des menschlichen Herzens und ihrer Übersetzung in klinische Praxis," in Hess, *Normierung der Gesundheit,* 65–86.

Boring, Edwin G. [1929] 1957. *A History of Experimental Psychology.* New York: Appleton-Century-Crofts.

Borscheid, Peter. 1976. *Naturwissenschaft, Staat und Industrie in Baden, 1848–1914.* Stuttgart: Klett.

Bos, Henk, and Herbert Mehrtens. 1977. "The Interaction of Mathematics and Society in History: Some Exploratory Remarks." *Historia Mathematica* 4:7–30.

Bos, Henk, Herbert Mehrtens, and Ivo Schneider, eds. 1981. *Social History of Nineteenth-Century Mathematics*. Boston: Birkhäuser.

Bottazzini, Umberto. 1986. *The Higher Calculus: A History of Real and Complex Analysis from Euler to Weierstraß*. New York: Springer.

Boudia, Soraya. 1997. Marie Curie et son laboratoire: La radioactivité en France, 1896–1914. Thesis, Université Paris VII.

Bourbaki, Nicolas. 1950. "The Architecture of Mathematics," *American Mathematical Monthly* 57:221–32.

———. 1994. *Elements of the History of Mathematics*. Trans. John Meldrum. Berlin: Springer-Verlag. Originally published as *Eléments de l'histoire des mathématiques* (Paris: Masson, 1984).

Bourdieu, Pierre. 1989. *La noblesse d'état: Grandes écoles et esprit de corps*. Paris: Editions de Minuit.

Bowler, Peter J. 1976. *Fossils and Progress: Paleontology and the Idea of Progressive Evolution in the Nineteenth Century*. New York: Science History.

———. 1983. *The Eclipse of Darwinism: Anti-Darwinian Evolutionary Theories in the Decades around 1900*. Baltimore: Johns Hopkins University Press.

———. 1984. *Evolution: The History of an Idea*. Berkeley: University of California Press.

———. 1988. *The Non-Darwinian Revolution: Reinterpreting a Historical Myth*. Baltimore: Johns Hopkins University Press.

———. 1998. *Life's Splendid Drama: Evolutionary Biology and the Reconstruction of Life's Ancestry, 1860–1940*. Chicago: University of Chicago Press.

———. 2001. *Reconciling Science and Religion: The Debate in Early-Twentieth-Century Britain*. Chicago: University of Chicago Press.

Boyer, Carl B. 1939. *The Concepts of the Calculus: A Critical and Historical Discussion of the Derivative and the Integral*. New York: Columbia University Press; revised as *The History of the Calculus and Its Conceptual Development*. New York: Dover, 1949.

———. 1956. *History of Analytic Geometry*. New York: Scripta Mathematica.

———. 1968. *A History of Mathematics*. New York: Wiley.

Boyle, R. W. 1993. "Geochemistry in the Geological Survey of Canada, 1842–1952." *ESH* 12:129–41.

Bradley, Margaret. 1975. "Scientific Education versus Military Training: The Influence of Napoleon Bonaparte on the École Polytechnique." *AS* 32:415–49.

———. 1976a. "The Facilities for Practical Instruction in Science during the Early Years of the École Polytechnique." *AS* 33:425–46.

———. 1976b. "Scientific Education for a New Society: The École Polytechnique, 1795–1830." *History of Education* 5:11–24.

———. 1979. "The Financial Basis of French Scientific Education and Scientific Institutions in Paris, 1790–1815." *AS* 36:451–91.

Brain, Robert M. 1996. The Graphic Method: Inscription, Visualization, and Measurement in Nineteenth-Century Science and Culture. Ph.D. diss., University of California, Los Angeles.

Braun, Hans-Joachim. 1977. "Methodenprobleme der Ingenieurwissenschaft, 1850 bis 1900." *Technikgeschichte* 44:1–18.

Braun, Marta. 1992. *Picturing Time: The Work of Etienne-Jules Marey (1830–1904)*. Chicago: University of Chicago Press.

Brent, Joseph. 1993. *Charles Sanders Peirce: A Life*. Bloomington: Indiana University Press.

Bret, Patrice. 1996. "Lavoisier et l'apport de la chimie académique à l'industrie des poudres et salpêtres." *Archives internationales d'Histoire des sciences* 46:57–74.

Brian, Eric. 1994. *Le mesure de l'état: Administrateurs et géometres au XVIIIe siècle*. Paris: Albin Michel.

Brianta, Donna. 2000. "Education and Training in the Mining Industry, 1750–1860: European Models and the Italian Case." *AS* 57:3:267–300.

Brigaglia, Aldo, Ciro Ciliberto, and Edoardo Sernesi. 1994. *Algebra e geometria, 1860–1940: Il contributo italiano*. Palermo: Circolo Matematico di Palermo.

Brittain, James E., and Robert C. McMath. 1977. "Engineers and the New South Creed: The Formation and Early Development of Georgia Tech." *TC* 18:175–201.

Brobeck, John R., Orr E. Reynolds, and Toby A. Appel, eds. 1987. *History of the American Physiological Society: The First Century, 1887–1987*. Bethesda, Md.: American Physiological Society.

Brock, M. G., and M. C. Curthoys, eds. 1997. *The History of the University of Oxford*. Vol. 6, *Nineteenth-Century Oxford, Part I*. Oxford: Clarendon Press.

Brock, William H. 1976a. "The Scientists' Declaration: Reflexions on Science and Belief in the Wake of *Essays and Reviews*, 1864–1865." *BJHS* 9:39–66.

———. 1976b. "The Spectrum of Science Patronage," in G. Turner, *Patronage of Science*, 173–206.

———. 1979. "Chemical Geology or Geological Chemistry?" in Jordanova and Porter, *Images of the Earth*, 147–78.

———. 1983. "History of Chemistry." In Corsi and Weindling, *Information Sources in the History of Science and Medicine*, edited by Pietro Corsi and Paul Weindling, 317–46. London: Butterworth.

———. 1990. "Science Education," in Olby et al., *Companion*, 946–59.

———. 1992. *The Fontana History of Chemistry*. London: Fontana. Also published as *The Norton History of Chemistry* (New York: Norton, 1994).

———. 1997. *Justus von Liebig: The Chemical Gatekeeper*. Cambridge: Cambridge University Press.

Broman, Thomas H. 1991. "J. C. Reil and the 'Journalization' of Physiology." In *The Literary Structure of Scientific Argument: Historical Studies*, edited by Peter Dear, 13–42. Philadelphia: University of Pennsylvania Press.

———. 1996. *The Transformation of German Academic Medicine, 1750–1820*. Cambridge: Cambridge University Press.

Bronowski, Jacob. 1973. *The Ascent of Man*. Boston: Little Brown.

Brooke, John Hedley. 1968. "Wöhler's Urea and Its Vital Force: A Verdict from the Chemists." *Ambix* 15:84–114.

———. 1977. "Natural Theology and the Plurality of Worlds: Observations on the Brewster-Whewell Debate." *AS* 34:221–86.

———. 1981. "Avogadro's Hypothesis and Its Fate: A Case Study in the Failure of Case-Studies." *HS* 19:235–73.

———. 1985. "The Relations between Darwin's Science and His Religion," in Durant, *Darwinism and Divinity*, 40–75.

———. 1989. "Science and the Fortunes of Natural Theology: Some Historical Perspectives." *Zygon* 24:3–22.

———. 1991. *Science and Religion: Some Historical Reflections.* Cambridge: Cambridge University Press.

———. 1995. "Chemists in Their Contexts: Some Recent Trends in Historiography." In *Thinking about Matter: Studies in the History of Chemical Philosophy,* 9–27. Aldershot: Variorum.

Brooke, John Hedley, and Geoffrey Cantor. 1998. *Reconstructing Nature: The Engagement of Science and Religion.* New York: T. & T. Clark.

Brouzeng, Paul. 1987. *Duhem, 1861–1916: Science et Providence.* Paris: Belin.

Brown, Frank Burch. 1986. "Darwin's Theism." *Journal of the History of Biology* 19:1–45.

Brown, JoAnne, and David van Keuren, eds. 1988. *The Estate of Social Knowledge.* Baltimore: Johns Hopkins University Press.

Browne, Charles A., ed. 1926. "A Half-Century of Chemistry in America, 1876–1926: A Historical Review Commemorating the Fiftieth Anniversary of the American Chemical Society." *Journal of the American Chemical Society* 48, pt. 2, no. 8A.

Browne, Charles A., and Mary Elvira Weeks. 1952. *A History of the American Chemical Society: Seventy-Five Eventful Years.* Washington, D.C.: American Chemical Society.

Browne, Janet. 1995. *Charles Darwin Voyaging.* Princeton, N.J.: Princeton University Press.

Bruce, Robert V. 1972. "A Statistical Profile of American Scientists, 1846–1876." In George Daniels, ed. *Nineteenth-Century American Science; A Reappraisal,* 63–94. Evanston, Ill.: Northwestern University Press.

———. 1988. *The Launching of Modern American Science, 1846–1876.* Ithaca, N.Y.: Cornell University Press.

Bruland, Kristine. 1999. "The Norwegian Mechanical Engineering Industry and the Transfer of Technology, 1800–1900." In *Technology Transfer and Scandinavian Industrialisation,* edited by Kristine Bruland, 229–93. New York: Berg.

Brush, Stephen G. 1970. "The Wave Theory of Heat." *BJHS* 5:145–67.

———. 1974. "The Prayer Test." *American Scientist* 62:561–63.

———. 1976. *The Kind of Motion We Call Heat: A History of the Kinetic Theory of Gases in the Nineteenth Century.* 2 vols. Amsterdam: North-Holland Pub. Co.

———. 1979. "Nineteenth Century Debates about the Inside of the Earth: Solid, Liquid, or Gas?" *AS* 36:224–54.

———. 1988. *The History of Modern Science: A Guide to the Second Scientific Revolution, 1800–1950.* Ames: Iowa State University Press.

———. 1996a. *Nebulous Earth: The Origin of the Solar System and the Core of the Earth from Laplace to Jeffreys.* Cambridge: Cambridge University Press.

———. 1996b. "The Reception of Mendeleev's Periodic Law in America and Britain." *Isis* 87:595–628.

———. 1996c. *Transmuted Past: The Age of the Earth and the Evolution of the Elements from Lyell to Patterson.* Cambridge: Cambridge University Press.

Bryan, Bettina. 1996. Wilhelm Erb's Electrotherapeutics and Scientific Medicine in Nineteenth-Century Germany. Ph.D. diss., University College, University of London.

Bucciarelli, Louis L., and Nancy Dworsky. 1980. *Sophie Germain: An Essay in the History of the Theory of Elasticity.* Dordrecht: Reidel.

Buchheim, Gisela, and Rolf Sonnemann, eds. 1990. *Geschichte der Technikwissenschaften.* Leipzig: Edition Leipzig.

Buchwald, Jed Z. 1977. "William Thomson and the Mathematization of Faraday's Electrostatics." *HSPS* 8:101–36.

———. 1980. "Optics and the Theory of the Punctiform Ether." *AHES* 21:245–78.

———. 1985. *From Maxwell to Microphysics: Aspects of Electromagnetic Theory in the Last Quarter of the Nineteenth Century.* Chicago: University of Chicago Press.

———. 1989. *The Rise of the Wave Theory of Light: Optical Theory and Experiment in the Early Nineteenth Century.* Chicago: University of Chicago Press.

———. 1993. "Design for Experimenting," in Horwich, *World Changes,* 169–206.

———. 1994. *The Creation of Scientific Effects: Heinrich Hertz and Electric Waves.* Chicago: University of Chicago Press.

———. 1999. "How the Ether Spawned the Micro-world," in Daston, *Biographies of Scientific Objects,* 203–25.

———, ed. 1996. *Scientific Credibility and Technical Standards in Nineteenth and Early Twentieth Century Germany and Britain.* Dordrecht: Kluwer Academic.

Bud, Robert. 1980. *The Discipline of Chemistry: The Origin and Early Years of the Chemical Society of London.* Ph D. diss., University of Pennsylvania.

Bud, Robert, and Gerrylynn K. Roberts. 1984. *Science versus Practice: Chemistry in Victorian Britain.* Manchester: Manchester University Press.

Bud, Robert, and Deborah Warner, eds. 1998. *Instruments of Science: An Historical Encyclopedia.* London: Garland.

Buffetaut, Eric. 1987. *A Short History of Vertebrate Palaeontology.* London: Croom Helm.

Bulmer, Martin, Kevin Bales, and Kathryn Kish Sklar, eds. 1991. *The Social Survey in Historical Perspective, 1880–1940.* Cambridge: Cambridge University Press.

Burchardt, Lothar. 1975. *Wissenschaftspolitik im Wilhelminischen Deutschland: Vorgeschichte, Gründung und Aufbau der Kaiser-Wilhelm-Gesellschaft zur Förderung der Wissenschaften.* Göttingen: Vandenhoeck & Ruprecht.

Burchfield, Joe D. 1975. *Lord Kelvin and the Age of the Earth.* New York: Science History.

———. 1982. "The British Association and Its Historians." *HSPS* 13:1:165–74.

Burdach, Karl Friedrich. 1800. *Propädeutik zum Studium der gesammten Heilkunst.* Leipzig: Dyt'schen Buchhandlung.

———. 1810. *Die Physiologie.* Leipzig: Weidmann.

Burkhardt, Richard. 1977. *The Spirit of System: Lamarck and Evolutionary Biology.* Cambridge, Mass.: Harvard University Press.

Butcher, Norman E. 1983. "The Advent of Colour-Printed Geological Maps in Britain." *Proceedings of the Royal Institution of Great Britain* 55:149–61.

Büttner, M., and E. Kohler, eds. 1991. *Geosciences / Geowissenschaften: Proceedings of the Symposium of the XVIIIth International Congress of History of Science at Hamburg-Munich, 1.–9. August 1989.* Vol. 5, pt. 3. Bochum: Universitätsverlag Dr. N. Brockmeyer.

Bynum, William F. 1980. "Health, Disease and Medical Care," in Rousseau and Porter, *Ferment of Knowledge,* 211–53.

———. 1994. *Science and the Practice of Medicine in the Nineteenth Century.* Cambridge: Cambridge University Press.

Bynum, William F., E. J. Browne, and Roy Porter, eds. 1981. *Dictionary of the History of Science.* Princeton, N.J.: Princeton University Press.

Bynum, William F., S. Lock, and Roy Porter, eds. 1992. *Medical Journals and Medical Knowledge: Historical Essays.* London: Routledge.

Bynum, William F., and Roy Porter, eds. 1988. *Medical Fringe and Medical Orthodoxy, 1750–1850*. London: Croom Helm.

———. 1993. *Companion Encyclopedia of the History of Medicine*. 2 vols. London: Routledge.

Bynum, William F., Roy Porter, and Michael Shepherd, eds. 1985–88. *The Anatomy of Madness*. 3 vols. London: Tavistock.

Cadbury, Deborah. 2000. *The Dinosaur Hunters: A Story of Scientific Rivalry and the Discovery of the Prehistoric World*. London: Fourth Estate.

Cahan, David. 1985. "The Institutional Revolution in German Physics, 1865–1914." *HSPS* 15:2:1–65.

———. 1989. *An Institute for an Empire: The Physikalisch-Technische Reichsanstalt, 1871–1918*. Cambridge: Cambridge University Press.

———, ed. 1993. *Hermann von Helmholtz and the Foundations of Nineteenth-Century Science*. Berkeley: University of California Press.

———. 1996. "The Zeiss Werke and the Ultramicroscope: The Creation of a Scientific Instrument in Context," in Buchwald, *Scientific Credibility*, 67–115.

Cahan, David, and M. Eugene Rudd. 2000. *Science at the American Frontier: A Biography of DeWitt Bristol Brace*. Lincoln: University of Nebraska Press.

Cajori, Florian. 1890. *The Teaching and History of Mathematics in the United States*. Washington, D.C.: U.S. Government Printing Office.

———. [1893] 1985. *A History of Mathematics*. New York: Macmillan. Reprint, New York: Chelsea.

Calvert, Monte A. 1967. *The Mechanical Engineer in America, 1830–1910: Professional Cultures in Conflict*. Baltimore: Johns Hopkins University Press.

Caneva, Kenneth L. 1978. "From Galvanism to Electrodynamics: The Transformation of German Physics and Its Social Context." *HSPS* 9:63–159.

———. 1993. *Robert Mayer and the Conservation of Energy*. Princeton, N.J.: Princeton University Press.

Canguilhem, Georges. [1968] 1994. *Etudes d'histoire et de philosophie des sciences concernant les vivants et la vie*. 7th ed. Paris: Vrin.

———. [1972] 1989. *The Normal and the Pathological*. Trans. Carolyn R. Fawcett. New York: Zone Books.

Cannadine, David. 2001. *Ornamentalism: How the British Saw Their Empire*. Oxford: Oxford University Press.

Cannon, Susan Faye. 1978a. "The Cambridge Network," in Cannon, *Science in Culture*, 29–71.

———. 1978b. *Science in Culture: The Early Victorian Period*. New York: Science History.

Cannon, Walter F. 1960–61. "The Problem of Miracles in the 1830s." *Victorian Studies* 4:6–32.

Cantor, Geoffrey N. 1975. "A Critique of Shapin's Interpretation of the Edinburgh Debate." *AS* 32:245–56.

———. 1991. *Michael Faraday: Sandemanian and Scientist. A Study of Science and Religion in the Nineteenth-Century*. New York: St. Martin's.

Cardwell, Donald S. L. [1957] 1972. *The Organisation of Science in England*. Rev. ed. London: Heinemann.

———. 1971. *From Watt to Clausius: The Rise of Thermodynamics in the Early Industrial Age*. Ithaca, N.Y.: Cornell University Press.

————. 1976. "The Patronage of Science in Nineteenth-Century Manchester," in G. Turner, *Patronage of Science*, 95–113.

Carlson, W. Bernhard. 1991. *Innovation as a Social Process: Elihu Thomson and the Rise of General Electric, 1870–1900*. Cambridge: Cambridge University Press.

Carneiro, Ana. 1992. The Research School of Chemistry of Adolphe Wurtz, Paris, 1853–1884. Ph.D. diss., University of Kent at Canterbury.

Carneiro, Ana, and Natalie Pigeard. 1997. "Chimistes alsaciens à Paris au 19e siècle: Un réseau, une école?" *AS* 54:533–46.

Caroe, Gwendy, with Alban Caroe. 1985. *The Royal Institution: An Informal History*. London: John Murray.

Carson, John. 1993. "Army Alpha, Army Brass, and the Search for Army Intelligence." *Isis* 84:278–309.

Carver, Terrell. 2001. "Marx and Marxism," in T. Porter and Ross, *Cambridge History of Science*.

Castelnuovo, Guido. [1938] 1962. *Le origini del calcolo infinitesimale nell'era moderna*. Bologna: Nicola Zanichelli. Reprint, Milan: Feltrinelli.

Cavendish Laboratory. 1910. *History of the Cavendish Laboratory 1871–1910*. London: Longmans.

Cawood, John. 1979. "The Magnetic Crusade: Science and Politics in Early Victorian Britain." *Isis* 70:493–518.

Cerbai, Ilaria, and Claudia Principe. 1996. *Bibliography of Historic Activity on Italian Volcanoes*. Pisa: Istituto di Geochronologia e Geochimica Isotopica. Report no. 6/96.

Certeau, Michel de. [1980] 1984. *The Practice of Everyday Life*. Trans. Steven Rendall. Berkeley: University of California Press.

Chadarevian, Soraya de. 1993. "Graphical Method and Discipline: Self-Recording Instruments in Nineteenth-Century Physiology." *SHPS* 24:267–91.

Chadwick, Owen. 1975. *The Secularization of the European Mind*. Cambridge: Cambridge University Press.

Chaigneau, Marcel. 1984. *Jean-Baptiste Dumas, chimiste et homme politique*. Paris: Guy le Prat.

Chambelland, Collette, ed. 1998. *Le musée social en son temps*. Paris: Presses de l'Ecole Normale Supérieure.

Chambers, Robert. 1994. *Vestiges of the Natural History of Creation and Other Evolutionary Writings*. Ed. James A. Secord. Chicago: University of Chicago Press.

Chandler, Alfred D., Jr. 1977. *The Visible Hand: The Managerial Revolution in American Business*. Cambridge, Mass.: Belknap.

Chandler, Bruce, and Wilhelm Magnus. 1982. *The History of Combinatorial Group Theory: A Case Study in the History of Ideas*. New York: Springer-Verlag.

Channell, David. 1982. "The Harmony of Theory and Practice: The Engineering Science of W. J. M. Rankine." *TC* 23:39–52.

Charraud, Nathalie. 1994. *Infini et inconscient: Essai sur Georg Cantor*. Paris: Anthropos, Diffusion Economica.

Chartier, Roger. 1982. "Intellectual History or Sociocultural History? The French Trajectories." In *Modern European Intellectual History: Reappraisals and New Perspectives*, edited by Dominick LaCapra and Steven L. Kaplan, 13–46. Ithaca, N.Y.: Cornell University Press.

————. 1987. *The Cultural Uses of Print in Early Modern France*. Trans. L. G. Cochrane. Princeton, N.J.: Princeton University Press.

————. 1994. *The Order of Books: Readers, Authors, and Librarians in Europe between the Fourteenth and Eighteenth Centuries*. Trans. L. G. Cochrane. Cambridge: Polity Press.

Chayut, Michael. 1991. "J. J. Thomson: The Discovery of the Electron and the Chemists." *AS* 48:527–44.

Chester, D. K., A. M. Duncan, J. E. Guest, and C. R. J. Kilburn. 1985. *Mount Etna: The Anatomy of a Volcano*. Stanford, Calif.: Stanford University Press.

Chorley, Richard J., Antony J. Dunn, and Robert P. Beckinsale. 1964. *The History of the Study of Landforms or the Development of Geomorphology*. Vol. 1, *Geomorphology before Davis*. London: Methuen, John Wiley & Sons.

Christie, John R. R. 1974. "The Origins and Development of the Scottish Scientific Community, 1680–1760." *HS* 12:122–41.

Church, Robert L. 1974. "Economists as Experts: The Rise of an Academic Profession in the United States, 1870–1920." In *The University in Society*, 2 vols., edited by Lawrence Stone, 2:571–609. Princeton, N.J.: Princeton University Press.

Ciardi, Marco. 1995. *L'atomo fantasma: Genesi storica dell'ipotesi di Avogadro*. Florence: Leo S. Olschki.

Cimino, Guido, and François Duchesneau, eds. 1997. *Vitalisms from Haller to the Cell Theory*. Florence: Leo S. Olschki.

Clark, Linda. 1984. *Social Darwinism in France*. Tuscaloosa: University of Alabama Press.

Clark, Terry Nichols. 1973. *Prophets and Patrons: The French University and the Emergence of the Social Sciences*. Cambridge, Mass.: Harvard University Press.

Clark, William. 1989. "On the Dialectical Origins of the Research Seminar." *HS* 27:111–54.

————. 1995. "Narratology and the History of Science." *SHPS* 26:1–71.

Clarke, Edwin, and L. S. Jacyna. 1987. *Nineteenth-Century Origins of Neuroscientific Concepts*. Berkeley: University of California Press.

Clifford, James. 1988. *The Predicament of Culture: Twentieth-Century Ethnography, Literature, and Art*. Cambridge, Mass.: Harvard University Press.

Clow, A., and N. Clow. 1952. *The Chemical Revolution: A Contribution to Social Technology*. London: Butchwork Press. Reprint, Philadelphia, 1992.

Coats, A. W. 1993. *The Sociology and Professionalization of Economics: British and American Economic Essays*. Vol. 2. London: Routledge.

Cochell, Gary G. 1998. "The Early History of the Cornell Mathematics Department: A Case Study in the Emergence of the American Mathematical Research Community." *Historia Mathematica* 25:133–53.

Cochrane, Rexmond C. 1966. *Measures for Progress: A History of the National Bureau of Standards*. Washington, D.C.: U.S. Government Printing Office.

————. 1978. *The National Academy of Sciences: The First Hundred Years, 1863–1963*. Washington, D.C.: Academy.

Cohen, H. Floris. 1994. *The Scientific Revolution: A Historiographical Inquiry*. Chicago: University of Chicago Press.

Cohen, I. B. 1965. "Science in the United States," in Taton, *Science in the Nineteenth Century*, 563–70.

————, ed. 1994. *The Natural Sciences and the Social Sciences*. Dordrecht: Kluwer.

Cohn, Bernard S. 1987. "The Census, Social Structure, and Objectification in South

Asia." In *An Anthropologist among the Historians, and Other Essays,* 224–54. Delhi: Oxford University Press.

Coleman, William. 1964. *Georges Cuvier, Zoologist: A Study in the History of Evolution Theory.* Cambridge, Mass.: Harvard University Press.

Coleman, William, and Frederic Holmes, eds. 1988. *The Investigative Enterprise: Experimental Physiology in Nineteenth-Century Medicine.* Berkeley: University of California Press.

Coley, N. G. 1973. *From Animal Chemistry to Biochemistry.* Amsterdam: Bucks, Hulton Educational Publications.

———. 1996. "Studies in the History of Animal Chemistry and Its Relation to Physiology." *Ambix* 43:164–88.

Collie, Michael. 1991. *Huxley at Work, with the Scientific Correspondence of T. H. Huxley and the Rev. Dr. George Gordon of Birnie, Near Elgin.* Basingstoke: Macmillan Professional and Academic.

———. 1995. "George Gordon (1801–1893), Man of Science." *Archives of Natural History* 22:29–49.

Collie, Michael, and Susan Bennett. 1996. *George Gordon: An Annotated Catalogue of His Scientific Correspondence.* Aldershot: Scolar Press.

Collie, Michael, and John Diemer. 1995. *Murchison in Moray: A Geologist on Home Ground, with the Correspondence of Roderick Impey Murchison and the Rev. Dr. George Gordon of Birnie.* Philadelphia: American Philosophical Society.

Collingwood, Robin G. 1939. *An Autobiography.* London: Oxford University Press.

———. 1946. *The Idea of History.* Oxford: Clarendon Press.

Collini, Stefan. 1979. *Liberalism and Sociology: L. T. Hobhouse and Political Argument in England.* Cambridge: Cambridge University Press.

Collini, Stefan, Donald Winch, and J. W. Burrow. 1983. *That Noble Science of Politics.* Cambridge: Cambridge University Press.

Collins, Harry M. 1981. "The Place of the 'Core-Set' in Modern Science: Social Contingency with Methodological Propriety in Science." *HS* 19:6–19.

———. 1992. *Changing Order: Replication and Induction in Scientific Practice.* London: Sage, 1985. 2d ed., Chicago: Chicago University Press.

Conklin, Edwin G. 1934. "Fifty Years of the American Society of Naturalists." *American Naturalist* 68:385–401.

Conser, Walter. 1993. *God and the Natural World: Religion and Science in Antebellum America.* Columbia: University of South Carolina Press.

Cook, Karen S. 1995. "From False Starts to Firm Beginnings: Early Colour Printing of Geological Maps." *Imago Mundi* 47:155–72.

Cooke, Roger. 1984. *The Mathematics of Sonya Kovalevskaya.* New York: Springer-Verlag.

Coolidge, Julian Lowell. 1963. *A History of Geometrical Methods.* New York: Dover.

Coon, Deborah. 1993. "Standardizing the Subject: Experimental Psychologists, Introspection, and the Quest for a Technoscientific Ideal." *TC* 34:757–83.

Cooter, Roger. 1984. *The Cultural Meaning of Popular Science: Phrenology and the Organisation of Consent in Nineteenth-Century Britain.* Cambridge: Cambridge University Press.

———, ed. 1988. *Studies in the History of Alternative Medicine.* Basingstoke: Macmillan.

Corbin, Alain. [1991] 1995. "A History and Anthropology of the Senses." In *Time, Desire and Horror: Towards a History of the Senses,* 181–95. Trans. Jean Birrell. Cambridge: Polity Press.

Corsi, Pietro. 1988. *Science and Religion: Baden Powell and the Anglican Debate, 1800–1860.* Cambridge: Cambridge University Press.

———. 1989. *The Age of Lamarck: Evolutionary Theories in France, 1790-1830.* Berkeley: University of California Press.

———. 2001. *Lamarck: Genèse et enjeux du transformisme, 1770–1830.* Paris: CNRS.

Cowan, Ruth Schwartz. 1996. "Technology Is to Science as Female Is to Male: Musings on the History and Character of Our Discipline." *TC* 37:572–82.

Cowherd, Raymond. 1977. *Political Economists and the English Poor Laws.* Athens: Ohio University Press.

Cox, R. C., ed. 1982. *Robert Mallet, 1810–1881.* Dublin: Institution of Engineers of Ireland.

Cranefield, Paul F. 1957. "The Organic Physics of 1847 and the Biophysics of Today." *Journal of the History of Medicine* 12:407–23.

Crary, Jonathan. 1990. *Techniques of the Observer: On Vision and Modernity in the Nineteenth Century.* Cambridge, Mass.: MIT Press.

Crawford, Elisabeth. 1980. "The Prize System of the Academy of Sciences, 1850–1914," in R. Fox and Weisz, *Organization of Science and Technology,* 283–307.

———. 1984. *The Beginnings of the Nobel Institution: The Science Prizes, 1901–1915.* Cambridge: Cambridge University Press; Paris: Editions de la Maison des Sciences de l'Homme.

———. 1988. "Competition and Centralisation in German and French Science in the Nineteenth and Early Twentieth Centuries: The Theses of Joseph Ben-David." *Minerva* 26:618–26.

———. 1996. *Arrhenius: From Ionic Theory to the Greenhouse Effect.* Canton, Mass.: Science History.

Croce, Paul Jerome. 1995. *Science and Religion in the Era of William James: Eclipse of Certainty, 1820–1880.* Chapel Hill: University of North Carolina Press.

Crosland, Maurice. 1967. *The Society of Arcueil: A View of French Science at the Time of Napoleon I.* Cambridge, Mass.: Harvard University Press.

———. 1969. "The Congress on Definitive Metric Standards, 1798–1799: The First International Scientific Conference?" *Isis* 60:226–31.

———. 1977. "History of Science in a National Context." *BJHS* 10:95–113.

———. 1978a. "Aspects of International Scientific Collaboration and Organisation before 1900," in Forbes, *Human Implications,* 1–15.

———. 1978b. "The French Academy of Sciences in the Nineteenth Century." *Minerva* 16:73–102.

———. 1979–80. "From Prizes to Grants in the Support of Scientific Research in France in the Nineteenth Century: The Montyon Legacy." *Minerva* 17:355–80.

———. 1992. *Science under Control: The French Academy of Science, 1795–1914.* Cambridge: Cambridge University Press.

———. 1994. *In the Shadow of Lavoisier: The Annales de chimie and the Establishment of a New Science.* Oxford: Alden Press.

Crosland, Maurice, and Antonio Gálvaz. 1989. "The Emergence of Research Grants within the Prize System of the French Academy of Sciences, 1795–1914." *SSS* 19:71–100.

Crosland, Maurice, and Crosbie Smith. 1978. "The Transmission of Physics from France to Britain, 1800–1840." *HSPS* 9:1–61.

Crowe, Michael J. 1967. *A History of Vector Analysis: The Evolution of the Idea of a Vectorial System.* Notre Dame, Ind.: University of Notre Dame Press.

————. 1975. "Ten 'Laws' Concerning Patterns of Change in the History of Mathematics." *Historia Mathematica* 2:161–66.

————. 1986. *The Extraterrestrial Life Debate, 1750–1900: The Idea of a Plurality of Worlds from Kant to Lowell.* Cambridge: Cambridge University Press.

Crowther, James Gerald. 1974. *The Cavendish Laboratory 1874–1974.* London: Macmillan.

Cunningham, Andrew. 1988. "Getting the Game Right: Some Plain Words on the Identity and Invention of Science." *SHPS* 19:365–89.

Cunningham, Andrew, and Nicholas Jardine, eds. 1990. *Romanticism and the Sciences.* Cambridge: Cambridge University Press.

Cunningham, Andrew, and Perry Williams. 1993. "De-centering the 'Big Picture': The *Origins of Modern Science* and the Modern Origins of Science." *BJHS* 26:407–32.

————, eds. 1992. *The Laboratory Revolution in Medicine.* Cambridge: Cambridge University Press.

Curtis, Bruce, 2001. *The Politics of Population: State Formation, Statistics, and the Census of Canada, 1840–1875.* Toronto: University of Toronto Press.

Cuvier, Georges. [1810] 1970. *Rapport historique sur les progrès des sciences naturelles depuis 1789.* Amsterdam: Israel.

Cuvier, Georges, and Alexandre Brongniart. 1808. "Essai sur la géographie minéralogique des environs de Paris." *Annales du Muséum d'Histoire Naturelle* 11:293–326.

Dampier, William C. [1929] 1971. *History of Science and Its Relations with Philosophy and Religion.* 4th ed. Reprint, Cambridge: Cambridge University Press.

Daniels, George H. 1967. "The Process of Professionalization in American Science: The Emergent Period, 1820–1860." *Isis* 58:151–66.

————. 1994. *American Science in the Age of Jackson.* New York: Columbia University Press, 1968. Reprint, Tuscaloosa: University of Alabama Press.

Danziger, Kurt. 1990. *Constructing the Subject: The Historical Origins of Psychological Research.* Cambridge: Cambridge University Press.

Darnton, Robert. 1979. *The Business of Enlightenment: A Publishing History of the Encyclopédie, 1775–1800.* Cambridge, Mass.: Harvard University Press.

Darrigol, Olivier. 1993. "The Electromagnetic Revolution in Germany as Documented by Early German Expositions of 'Maxwell's Theory.'" *AHES* 45:189–280.

————. 2002a. "Between Hydrodynamics and Elasticity Theory: The First Five Births of the Navier-Stokes Equation." *AHES* 56: 95–150.

————. 2002b. "Turbulence in 19th-Century Hydrodynamics." *HSPS* 32:207–62.

Darwin, Charles R. 1839. "Observations on the Parallel Roads of Glen Roy, and of Other Parts of Lochaber, with an Attempt to Prove That They Are of Marine Origin." *Philosophical Transactions of the Royal Society of London*, pt. 1, 39–81.

————. [1839] 1952. *Journal of Researches into the Geology and Natural History of the Various Countries Visited by H.M.S. Beagle.* New York: Hafner.

————. 1859. *On the Origin of Species.* London: Murray.

————. 1871. *The Descent of Man and Sex in Relation to Selection.* 2 vols. London: Murray.

————. 1958. *The Autobiography of Charles Darwin, 1809–1882.* New York: Harcourt, Brace and World.

Daston, Lorraine. 1987. "Rational Individuals vs. Laws of Society: From Probability to Statistics," in Krüger, Daston, and Heidelberger, *Probabilistic Revolution*, 295–304.

————. 1988. *Classical Probability in the Enlightenment.* Princeton, N.J.: Princeton University Press.

————. 1991. "History of Science in an Elegiac Mode." *Isis* 82:522–31.

————. 1992. "Objectivity and the Escape from Perspective." *SSS* 22:597–618.

————. 1999. "Objectivity versus Truth." In *Wissenschaft als kulturelle Praxis, 1750–1900,* edited by Hans Erich Bödeker, Peter H. Reill, and Jürgen Schlumbohm, 17–32. Göttingen: Vandenhoeck & Ruprecht.

————, ed. 1999. *Biographies of Scientific Objects.* Chicago: University of Chicago Press.

Daston, Lorraine, and Peter Galison. 1992. "The Image of Objectivity." *Representations* 40:81–128.

Daston, Lorraine, and H. Otto Sibum, eds. 2002. "Scientific Personae." *Science in Context* 15.

Dauben, Joseph W. [1979] 1990. *Georg Cantor: His Mathematics and Philosophy of the Infinite.* Cambridge, Mass.: Harvard University Press. Reprint, Princeton, N.J.: Princeton University Press.

————. 1984. "Conceptual Revolutions and the History of Mathematics: Two Studies in the Growth of Knowledge." In *Transformation and Tradition in the Sciences: Essays in Honor of I. Bernard Cohen,* 81–103. Cambridge: Cambridge University Press. Reprinted in Gillies, *Revolutions in Mathematics,* 49–71.

————. 1985. Introduction, in Dauben, *History of Mathematics,* xvii–xxxv.

————. 1988. Review of Purkert and Ilgauds, *Georg Cantor. Isis* 79:700–702.

————. 1992. "Revolutions Revisited," in Gillies, *Revolutions in Mathematics,* 72–82.

————. 1998a. "*Historia Mathematicae:* Journals of the History of Mathematics." In *Journals and History of Science,* edited by Marco Beretta, Claudio Pogliano, and Pietro Redondi, 1–30. Florence: Leo S. Olschki.

————. 1998b. "Marx, Mao, and Mathematics." *Proceedings of the International Congress of Mathematicians, Berlin.* Special issue of *Documenta Mathematica* and *Jahresbericht der Deutschen Mathematiker-Vereinigung* 3:390–98.

————. 1999. "*Historia Mathematica:* Twenty-Five Years / Context and Content." *Historia Mathematica* 26:1–23.

————. In press. "Mathematics and Ideology: The Politics of Infinitesimals / Marx, Mao and Mathematics: Non-standard Analysis and the Cultural Revolution." *Proceedings of the III Simposio Internacional Galdeano.* Zaragoza: University of Zaragoza.

————, ed. 1985. *History of Mathematics from Antiquity to the Present.* New York: Garland; rev. CD-ROM edited by Albert C. Lewis, Providence, R.I.: American Mathematical Society, 2000.

Dauben, Joseph W., and Christoph J. Scriba, eds. 2002. *Writing the History of Mathematics: Its Historical Development.* Basel: Birkhäuser.

Daum, Andreas. 1998. *Wissenschaftspopularisierung im 19. Jahrhundert: Bürgerliche Kultur, naturwissenschaftliche Bildung und die deutsche Öffentlichkeit, 1848–1914.* Munich: R. Oldenbourg.

Daumas, Maurice. 1957. *Centenaire de la Société chimique de France (1857–1957).* Paris: Masson.

David, Paul A. 1993. "Path Dependence in Dynamic Systems with Local Network Externalities: A Paradigm for Historical Economics." In *Technology and the Wealth of Nations,* edited by D. Foray and C. Freeman, 208–31. London: Pinter.

Davies, Gordon L. 1969. *The Earth in Decay: A History of British Geomorphology 1578–1878.* London: Macdonald Technical and Scientific.

Daviet, Jean-Pierre. 1988. *Un destin international: La Compagnie de Saint-Gobain, de 1830 à 1939.* Montreux-Paris: Éditions des archives contemporaines.

Davis, John L. 1998. "Artisans and Savants: The Role of the Academy of Sciences in the Process of Electrical Innovation in France, 1850–1880." *AS* 55:291–314.

Davison, Charles. 1927. *The Founders of Seismology.* Cambridge: Cambridge University Press.

Day, Charles R. 1978. "The Making of Mechanical Engineers in France: The Écoles d'Arts et Métiers, 1803–1914." *French Historical Studies* 10:439–60.

Dean, Dennis R. 1989. "James Hutton's Place in the History of Geomorphology," in Tinkler, *History of Geomorphology,* 73–84.

———. 1991. "Robert Mallett and the Founding of Seismology." *AS* 48:39–67.

———. 1992. *James Hutton and the History of Geology.* Ithaca, N.Y.: Cornell University Press.

———. 1999. *Gideon Mantell and the Discovery of Dinosaurs.* Cambridge: Cambridge University Press.

Debru, Claude, ed. 1995. *Essays in the History of the Physiological Sciences.* Clio Medica, no. 33. Dordrecht: Kluwer.

Degler, Carl. 1991. *In Search of Human Nature: The Decline and Revival of Darwinism in American Social Thought.* Oxford: Oxford University Press.

Dehue, Trudy. 1997. "Deception, Efficiency, and Random Groups: Psychology and the Gradual Origination of the Random Group Design." *Isis* 88:653–73.

Demeulenaere-Douyère, Christiane, ed. 1995. *Il y a deux cents Lavoisier: Actes du colloque organisé à l'occasion du bicentenaire de la mort d'Antoine Laurent Lavoisier, le 8 mai 1794.* Paris: Académie des Sciences.

Demidov, Sergei S., Menso Folkerts, David E. Rowe, and Christoph J. Scriba, eds. 1992. *Amphora: Festschrift für Hans Wussing zu seinem 65. Geburtstag.* Basel: Birkhäuser.

Den Tex, Emile. 1989. "Helicoidal and Punctuated Cyclicity in Terrestrial Processes and in Geological Thinking." In *Pathways in Geology: Essays in Honour of Edwin Sherbon Hills,* edited by R. W. Le Maître, 408–17. Melbourne: Blackwell Scientific.

———. 1990. "Punctuated Equilibria between Rival Concepts of Granite Genesis in the Late Eighteenth, Nineteenth, and Early Twentieth Centuries." *Geological Journal* 25:215–19.

———. 1996. "Clinchers of the Basalt Controversy: Empirical and Experimental Evidence." *ESH* 15:37–48.

Desmond, Adrian. 1985a. *Archetypes and Ancestors: Palaeontology in Victorian London.* London: Blond and Briggs, 1982. Reprint, Chicago: University of Chicago Press.

———. 1985b. "The Making of Institutional Zoology in London 1822–1836." *HS* 23:153–85, 223–50.

———. 1989. *The Politics of Evolution: Morphology, Medicine, and Reform in Radical London.* Chicago: University of Chicago Press.

Desmond, Adrian, and James R. Moore. 1991. *Darwin: The Life of a Tormented Evolutionist.* New York: Norton.

Desrosières, Alain. 1991. "How to Make Things Which Hold Together: Social Science, Statistics, and the State," in Wagner, Wittrock, and Whitley, *Discourses on Society,* 195–218.

———. 1998. *The Politics of Large Numbers: A History of Statistical Reasoning.* Cambridge, Mass.: Harvard University Press.

DeVorkin, David H. 1999. "The Pickering Years," in DeVorkin, *American Astronomical Society's First Century,* 20–36.

————, ed. 1999. *The American Astronomical Society's First Century.* Washington, D.C.: American Astronomical Society.

Dewey, James, and Perry Byerly. 1969. "The Early History of Seismometry (to 1900)." *Bulletin of the Seismological Society of America* 59:183–227.

Dhombres, Jean. 1993. "Une tradition française dans l'enseignement supérieur scientifique?" In *Le Università e le Scienze: Prospettive Storiche e Attuali. Relazioni presentate al convegno internazionale Bologna, 18 settembre 1991,* edited by Giuliano Pancaldi, 49–57. Università degli Studi, Bologna: Aldo Martello.

Dhombres, Jean, and Jean-Bernard Robert. 1998. *Joseph Fourier, 1768–1830: Créateur de la physique mathématique.* Paris: Belin.

Dienel, Hans-Liudger. 1993. "Professoren als Gutachter für die Kälteindustrie, 1870–1930: Ein Beitrag zum Verhältnis von Hochschule und Industrie in Deutschland." *Berichte zur Wissenschaftsgeschichte* 16:165–82.

————. 1995. *Ingenieure zwischen Hochschule und Industrie: Kältetechnik in Deutschland und Amerika, 1870–1930.* Göttingen: Vandenhoeck & Ruprecht.

Dierig, Sven. Forthcoming. *Experimentierplatz in der Moderne: Physiologie in Berlin, 1840–1900.* Göttingen: Wallstein.

Dieudonné, Jean, ed. 1978. *Abrégé d'histoire des mathémqtiques: 1700–1900.* Paris: Hermann.

Digby, Anne. 1994. *Making a Medical Living: Doctors and Patients in the English Market for Medicine, 1720–1911.* Cambridge: Cambridge University Press.

Dillenberger, John. 1960. *Protestant Thought and Natural Science.* Nashville, Tenn.: Abingdon.

Dinges, Martin, ed. 1996. *Medizinkritische Bewegungen im Deutschen Reich (ca. 1870–ca. 1933).* Stuttgart: Steiner.

Dirks, Nicholas B. 2001. *Castes of Mind: Colonialism and the Making of Modern India.* Princeton: Princeton University Press.

Donovan, Arthur. 1975. *Philosophical Chemistry in the Scottish Enlightenment.* Edinburgh: Edinburgh University Press.

————. 1990. "Lavoisier as Chemist and Experimental Physicist: A Reply to Perrin." *Isis* 81:270–72.

————. 1993. *Antoine Lavoisier: Science, Administration, and Revolution.* Oxford: Blackwell.

————, ed. 1988. *The Chemical Revolution: Essays in Reinterpretation. Osiris,* 2d ser., vol. 4.

Dorfman, Joseph. 1946–59. *The Economic Mind in American Civilization.* 5 vols. New York: Viking Press.

Dörries, Matthias. 1994a. "Balances, Spectroscopes, and the Reflexive Nature of Experiment." *SHPS* 25:1–36.

————. 1994b. "La standardisation de la balance de torsion dans les projets européens sur le magnétisme terrestre," in Blondel and Dörries, *Restaging Coulomb,* 121–49.

————. 1998. "Easy Transit: Crossing Boundaries between Physics and Chemistry in Mid-Nineteenth Century France," in C. Smith and Agar, *Making Space for Science,* 246–62.

Dott, Robert H., Jr. 1974. "The Geosynclinal Concept." In *Modern and Ancient Geosynclinal Sedimentation,* edited by Robert H. Dott Jr. and R. H. Shaver, 1–13. Special Publication no. 19, in honor of Marshall Kay. Tulsa, Okla.: Society of Economic Paleontologists and Mineralogists.

————. 1979. "The Geosyncline—First Major Geological Concept 'Made in America.'"

In *Two Hundred Years of Geology in America: Proceedings of the New Hampshire Bicenten-nial Conference on the History of Geology,* edited by Cecil J. Schneer, 239–74. Hanover, N.H.: University Press of New England.

———. 1985. "James Hall's Discovery of the Craton." In *Geologists and Ideas: A History of North American Geology,* edited by Ellen T. Drake and William M. Jordan, 157–67. Geological Society of America Centennial Special Volume 1. Boulder, Colo.: Geological Society of America.

———. 1997. "James Dwight Dana's Old Tectonics: Global Contraction under Divine Direction." *American Journal of Science* 297:283–311.

———. 1998. "Recognition of the Tectonic Significance of Volcanism in Ancient Orogenic Belts," in Morello, *Volcanoes in History,* 123–31.

Draper, John William. 1874. *History of the Conflict between Religion and Science.* New York: D. Appleton.

Du Bois-Reymond, Emil. 1848. *Untersuchungen über die thierische Elektricität.* Vol. 1. Berlin: Reimer.

———. [1877a] 1912. "Kulturgeschichte und Naturwissenschaft," in Du Bois-Reymond, *Reden,* 1:567–629.

———. [1877b] 1912. "Der physiologische Unterricht sonst und jetzt," in Du Bois-Reymond, *Reden,* 1:630–53.

———. 1912. *Reden.* 2 vols. 2d ed. Leipzig: Veit.

Duchesneau, François. 1997. "Vitalism and Anti-Vitalism in Schwann's Program for the Cell Theory," in Cimino and Duchesneau, *Vitalisms,* 225–51.

Duden, Barbara. [1987] 1991. *The Woman beneath the Skin: A Doctor's Patients in Eighteenth-Century Germany.* Trans. Thomas Dunlap. Cambridge, Mass.: Harvard University Press.

Dudich, Endre, ed. 1984. *Contributions to the History of Geological Mapping: Proceedings of the Tenth INHIGEO [International Commission on the History of Geological Sciences] Symposium, 16–22 August 1982.* Budapest: Akademiai Kiado.

Dumez, Hervé. 1985. *L'économiste, la science, et le pouvoir: Le cas Walras.* Paris: Presses Universitaires de France.

Dupree, A. Hunter. 1957. *Science in the Federal Government: A History of Policies and Activities to 1940.* Cambridge, Mass.: Harvard University Press.

———. 1976a. "The National Academy of Sciences and American Definitions of Science," in Oleson and Brown, *Pursuit of Knowledge,* 33–69.

———. 1976b. "The National Pattern of American Learned Societies, 1769–1863," in Oleson and Voss, *Organization of Knowledge,* 342–63.

Durand-Delga, Michel. 1996. "Jules Marcou, précurseur français de la géologie nord-américaine." *Science, culture, société* 13:59–83.

Durand-Delga, Michel, and Richard Moreau. 1994. "Un savant dérangeant: Jules Marcou (1824–1898), géologue français d'Amérique." *Travaux du Comité Français d'Histoire de la Géologie (COFRHIGÉO),* 3d ser., 8:55–82.

Durant, John, ed. 1985. *Darwinism and Divinity: Essays on Religious Belief.* Oxford: Blackwell.

Duren, Peter, Harold M. Edwards, and Uta C. Merzbach, eds. 1988. *A Century of Mathematics in America.* Vol. 1. Providence, R.I.: American Mathematical Society.

———, eds. 1989. *A Century of Mathematics in America,* Vol. 2. Providence, R.I.: American Mathematical Society.

Duren, Peter, Richard A. Askey, Harold M. Edwards, and Uta C. Merzbach, eds. 1989. *A Century of Mathematics in America,* Vol. 3. Providence, R.I.: American Mathematical Society.

Eagan, William E. 1986. "The Multiple Glaciation Debate: The Canadian Perspective, 1880–1900." *ESH* 5:144–51.

Eble, Burkard. 1836. *Versuch einer pragmatischen Geschichte der Anatomie und Physiologie vom Jahre 1800–1825.* Vienna: Gerold.

Eccarius, Wolfgang. 1976. "August Leopold Crelle als Herausgeber wissenschaftlicher Fachzeitschriften." *AS* 33:229–61.

———. 1977. "August Leopold Crelle als Förderer bedeutender Mathematiker." *Jahresbericht der Deutschen Mathematiker-Vereinigung* 79:137–74.

———. 1980. "Der Gegensatz zwischen Julius Plücker und Jakob Steiner im Lichte ihrer Beziehungen zu August Leopold Crelle." *AS* 37:189–213.

Eckel, Edwin B. 1982. *The Geological Society of America: Life History of a Learned Society.* Geological Society of America, Memoir 155. Boulder, Colo.: Geological Society of America.

Edwards, Charles Henry. 1979. *The Historical Development of the Calculus.* New York: Springer.

Edwards, Harold. 1974. *Riemann's Zeta Function.* New York: Academic Press.

———. 1977. *Fermat's Last Theorem: A Genetic Introduction to Number Theory.* New York: Springer-Verlag.

———. 1984. *Galois Theory.* New York: Springer-Verlag.

Eisele, Carolyn. 1979. *Studies in the Scientific and Mathematical Philosophy of Charles S. Peirce: Essays by Carolyn Eisele.* Ed. R. M. Martin. The Hague: Mouton.

———, ed. 1976. *The New Elements of Mathematics,* by Charles S. Peirce. 4 vols. in 5 bks. The Hague: Mouton.

Eiseley, Loren. [1958] 1961. *Darwin's Century: Evolution and the Men Who Discovered It.* New York: Doubleday & Co.

———. 1979. *Darwin and the Mysterious Mr. X.* New York: Dutton.

Ekelund, Robert, and Robert Hebert. 1999. *Secret Origins of Modern Microeconomics: Dupuit and the Engineers.* Chicago: University of Chicago Press.

Elena, Alberto. 1988. "The Imaginary Lyellian Revolution." *ESH* 7:126–33.

Elkana, Yehuda. 1974. *The Discovery of the Conservation of Energy.* Cambridge, Mass.: Harvard University Press.

Ellenberger, François. 1982. "Marcel Bertrand et 'l'orogenèse programmé.'" *Geologische Rundschau* 71:463–74.

Ellenberger, Henri. 1970. *The Discovery of the Unconscious: The History and Evolution of Dynamic Psychiatry.* New York: Basic Books.

Elles, Gertrude L., and Ethel M. R. Wood. 1901–18. *A Monograph of British Graptolites: Historical Introduction.* Ed. Charles Lapworth. London: Palæontographical Society.

Elliott, Clark A. 1970. The American Scientist, 1800–1863: His Origins, Career, and Interests. Ph.D. diss., Case Western Reserve University.

———. 1975. "The American Scientist in Antebellum Society: A Quantitative View." *SSS* 5:93–108.

———, ed. 1979. *Biographical Dictionary of American Science: The Seventeenth through the Nineteenth Century.* Westport, Conn.: Greenwood.

Emerson, Roger L. 1990. "The Organisation of Science and Its Pursuit in Early Modern Europe," in Olby et al., *Companion*, 960–79.

Engelhardt, Dietrich von. 1979. *Historisches Bewußtsein in der Naturwissenschaft von der Aufklärung bis zum Positivismus*. Freiburg: Alber.

———, ed. 1997. *Forschung und Fortschritt: Festschrift zum 175jährigen Jubiläum der Gesellschaft Deutscher Naturforscher und Ärzte*. Stuttgart: Wissenschaftliche Verlagsgesellschaft.

Engelhardt, Wolf von, and Jörg Zimmerman. 1988. *Theory of Earth Science*. Trans. Lenore Fischer. Cambridge: Cambridge University Press.

Ewald, François. 1986. *L'état providence*. Paris: Grasset.

Ewan, Joseph. 1976. "The Growth of Learned and Scientific Societies in the Southeastern United States to 1860," in Oleson and Brown, *Pursuit of Knowledge*, 208–18.

Eyles, Victor A. 1953. "Scientific Thought of the Early Nineteenth Century." *Nature* 171:714.

Eyles, Victor A., and Joan M. Eyles. 1938. "On the Different Issues of the First Geological Map of England and Wales." *AS* 3:190–212 and plates.

Falconer, Isolbel. 1989. "J. J. Thomson and 'Cavendish Physics,' " in James, *Development of the Laboratory*, 104–17. —

———. 2000. "Corpuscles to Electrons." In *Histories of the Electron: The Birth of Microphysics*, edited by Jed Z. Buchwald and Andrew Warwick. 77–100. Cambridge, Mass.: MIT Press.

Farber, Paul Lawrence. 1982. *The Emergence of Ornithology as a Scientific Discipline, 1760–1850*. Dordrecht: Reidel.

———. 1994. *The Temptations of Evolutionary Ethics*. Berkeley: University of California Press.

Farr, James. 1991. "The Estate of Political Knowledge: Political Science and the State," in Brown and van Keuren, *Estate of Social Knowledge*, 1–21.

Farrar, W. V. 1965. "Nineteenth-Century Speculations on the Complexity of Chemical Elements." *BJHS* 2:297–323.

Fasbender, Philine. 2001. *Physiologie der Funktionen: Untersuchungen zu Johannes Müllers Vorlesung "Physiologie des Menschen" (1826/27). Mit einer Edition des Vorlesungstextes*. Medical diss., University of Göttingen.

Favero, Giovanni. 2001. *Le Misure del Regno: Direzione di statistica e municipi nell'Italia liberale*. Padua: Il Poligrafo.

Feffer, Stuart M. 1996. "Ernst Abbe, Carl Zeiss, and the Transformation of Microscopical Optics," in Buchwald, *Scientific Credibility*, 23–66.

Fell, Ulrike. 1998. "The Profession of Chemistry in France: The Société Chimique de Paris de France, 1870–1914," in Knight and Kragh, *Making of the Chemist*, 15–38.

Fennema, Elizabeth, and Gilah C. Leder. 1990. *Mathematics and Gender*. New York: Teachers College Press.

Fenster, Della Dumbaugh, and Karen Hunger Parshall. 1989a. "A Profile of the American Mathematical Research Community, 1891–1906," in Knobloch and Rowe, *History of Modern Mathematics*, 179–227.

———. 1989b. "Women in the American Mathematical Research Community, 1891–1906," in Knobloch and Rowe, *History of Modern Mathematics*, 229–61.

Ferguson, Eugene. 1992. *Engineering and the Mind's Eye*. Cambridge, Mass.: MIT Press.

Ferrari, Graziano, ed. 1992. *Two Hundred Years of Seismic Instruments in Italy, 1731–1940*. Bologna: Storia-Geofisica-Ambiente.

Fichte, Johann Gottlieb. 1794. *Einige Vorlesungen über die Bestimmung des Gelehrten*. Jena: Gabler.

Figlio, Karl. 1976. "The Metaphor of Organisation: An Historical Perspective on the Biomedical Sciences of the Early Nineteenth Century." *HS* 14:17–53.

———. 1977. "The Historiography of Scientific Medicine: An Invitation to the Human Sciences." *Comparative Studies in Society and History* 19:262–86.

Fink, Karl. 1900. *A Brief History of Mathematics*. Trans. Wooster W. Beman and David E. Smith. Chicago: Open Court. Originally published as *Kurzer Abriss einer Geschichte der Elementar Mathematik* (Tübingen: H. Laupp, 1890).

Fischer, Nicholas. 1982. "Avogadro, the Chemists, and Historians of Chemistry." *HS* 20:77–102, 212–31.

Fischer-Homberger, Esther. 1979. *Krankheit Frau und andere Arbeiten zur Medizingeschichte der Frau*. Bern: Huber.

Fiske, Thomas Scott. 1988. "The Beginnings of the American Mathematical Society," in Duren, Askey, and Merzbach, *Century of Mathematics*, 1:13–17.

Fissell, Mary. 1991. "The Disappearance of the Patient's Narrative and the Invention of Hospital Medicine." In *British Medicine in an Age of Reform*, edited by Roger French and Andrew Wear, 92–109. London: Routledge.

Fleming, Donald. 1950. *John William Draper and the Religion of Science*. Philadelphia: University of Pennsylvania Press.

Floud, Roderick, and Donald McCloskey, eds. 1994. *The Economic History of Britain since 1700*. 2d ed. Vol. 1, *1700–1860*. Cambridge: Cambridge University Press.

Fogel, Robert W. 1964. *Railroads and American Economic Growth: Essays in Economic History*. Baltimore: Johns Hopkins University Press.

Forbes, E. G., ed. 1978. *Human Implications of Scientific Advance: Proceedings of the Fifteenth International Congress of the History of Science, Edinburgh, 10–15 August 1977*. Edinburgh: Edinburgh University Press.

Forgan, Sophie. 1986. "Context, Image, and Function: A Preliminary Enquiry into the Architecture of Scientific Societies." *BJHS* 19:89–113.

———. 1989. "The Architecture of Science and the Idea of a University." *SHPS* 20: 405–34.

———. 1998. " 'But Indifferently Lodged . . .': Perception and Place in Building for Science in Victorian London," in C. Smith and Agar, *Making Space for Science*, 195–215.

Forman, Paul, John L. Heilbron, and Spencer Weart. 1975. "Physics circa 1900: Personnel, Funding, and Productivity of the Academic Establishments." *HSPS* 5:1–185.

Foster, Mike. 1994. *Strange Genius: The Life of Ferdinand Vandeveer Hayden*. Schull, Ireland: Roberts Rinehart.

Foucault, Michel. 1970. "La situation de Cuvier dans l'histoire de la biologie." *RHS* 23: 62–69.

———. 1973. *The Order of Things: An Archaeology of the Human Sciences*. New York: Vintage.

———. 1979. *Discipline and Punish: The Birth of the Prison*. Trans. Alan Sheridan. New York: Vintage.

———. 1984. *Le souci de soi.* Vol. 3 of *Histoire de la sexualité.* Paris: Gallimard.

———. 1994a. *The Birth of the Clinic: An Archaeology of Medical Perception.* 1963. Trans. A. M. Sheridan Smith. New York: Vintage.

———. 1994b. *Dits et écrits, 1954–1988.* 4 vols. Paris: Gallimard.

Fox, Christopher, Roy Porter, and Robert Wokler, eds. 1995. *Inventing Human Science: Eighteenth-Century Domains.* Berkeley: University of California Press.

Fox, Lynn H., Linda Brody, and Diane Tobin, eds. 1980. *Women and the Mathematical Mystique.* Baltimore: Johns Hopkins University Press.

Fox, Robert. 1971. *The Caloric Theory of Gases: From Lavoisier to Regnault.* Oxford: Clarendon Press.

———. 1973. "Scientific Enterprise and the Patronage of Research in France, 1800–70." *Minerva* 11:442–73. Reprinted in G. Turner, *Patronage of Science,* 9–51.

———. 1974. "The Rise and Fall of Laplacian Physics." *HSPS* 4:89–136.

———. 1976a. "La Société Zoologique de France: Ses origines et ses premières années." *Bulletin de la Société Zoologique de France* 101:1–16.

———. 1976b. "The Early History of the Société Zoologique de France." In *The Culture of Science in France, 1700–1900,* 1–16. Aldershot: Variorum; Brookfield, Vt.: Ashgate.

———. 1980. "The *Savant* Confronts His Peers: Scientific Societies in France, 1815–1914," in R. Fox and Weisz, *Organization of Science and Technology,* 241–82.

———. 1984a. "Science, the University, and the State in Nineteenth-Century France," in Geison, *Professions and the French State,* 66–145.

———. 1984b. "Presidential Address: Science, Industry, and the Social Order in Mulhouse, 1798–1871." *BJHS* 17:127–68.

———. 1997. "The University Museum and Oxford Science, 1850–1880," in Brock and Curthoys, *History of the University of Oxford,* 641–91.

Fox, Robert, and Anna Guagnini. 1998. "Laboratories, Workshops, and Sites: Concepts and Practices of Research in Industrial Europe, 1800–1914." *HSPS* 29:1:55–139.

Fox, Robert, and A. Nieto-Galàn, eds. 1999 *Natural Dyestuffs: An Industrial Culture in Europe.* Canton, Mass.: Science History.

Fox, Robert, and George Weisz. 1980a. "The Institutional Basis of French Science in the Nineteenth Century," introduction to R. Fox and Weisz, *Organization of Science and Technology,* 1–28.

———, eds. 1980b. *The Organization of Science and Technology in France, 1808–1914.* Cambridge: Cambridge University Press; Paris: Editions de la Maison des Sciences de l'Homme.

Frank, Robert G. 1987. "American Physiologists in German Laboratories, 1865–1914," in Geison, *Physiology in the American Context,* 11–46.

———. 1988. "The Telltale Heart: Physiological Instruments, Graphic Methods, and Clinical Hopes, 1854–1914," in Coleman and Holmes, *Investigative Enterprise,* 211–90.

Frankel, Eugene. 1974. "The Search for a Corpuscular Theory of Double Refraction: Malus, Laplace, and the Prize Competition of 1808." *Centaurus* 18:223–45.

———. 1977. "J. B. Biot and the Mathematization of Experimental Physics in Napoleonic France." *HSPS* 8:33–72.

Frasca-Spada, Marina, and Nick Jardine, eds. 2000. *Books and the Sciences in History.* Cambridge: Cambridge University Press.

French, John C. 1946. *A History of the University Founded by Johns Hopkins.* Baltimore: Johns Hopkins University Press.

Freudenthal, Gad, ed. 1990. *Studies on Hélène Metzger.* New York: Brill.

Freund, Ida. 1968. *The Study of Chemical Composition: An Account of Its Method and Historical Development.* Reprint. New York: Dover.

Frewer, Andreas, and Volker Roelcke, eds. 2001. *Die Institutionalisierung der Medizinhistoriographie: Entwicklungslinien vom 19. ins 20. Jahrhundert.* Stuttgart: Steiner.

Friedman, Michael. 1998. "On the Sociology of Scientific Knowledge and Its Philosophical Agenda." *SHPS* 29:239–71.

Friedman, Robert M. 1977. "The Creation of a New Science: Joseph Fourier's Analytical Theory of Heat." *HSPS* 8:73–99.

Fries, Jakob F. [1805] 1989. *Knowledge, Belief, and Aesthetic Sense.* Ed. with an introduction by Frederick Gregory. Trans. Kent Richter. Cologne: Dinter Verlag. Originally published as *Wissen, Glaube und Ahndung.*

Fritscher, Bernhard. 1991. *Vulkanismusstreit und Geochemie: Die Bedeutung der Chemie und des Experiments in der Neptunismus-Vulkanismus-Kontroverse.* Stuttgart: Steiner Verlag.

———. 1993. "Vulkane und Hochöfen: Zur Rolle metallurgischer Erfahrungen bei der Entwicklung der experimentellen Petrologie." *Beiträge zur Geschichte der Technik und Industrie* 60:27–43.

Fruton, Joseph F. 1972. *Molecules of Life: Historical Essays on the Interplay of Chemistry and Biology.* New York: Wiley.

———. 1990. *Contrasts in Scientific Styles: Research Groups in the Chemical and Biological Sciences.* Philadelphia: American Philosophical Society.

Furner, Mary. 1975. *Advocacy and Objectivity: A Crisis in the Professionalization of American Social Science.* Lexington: University Press of Kentucky.

Fye, W. Bruce. 1987. "Growth of American Physiology, 1850–1900," in Geison, *Physiology in the American Context,* 47–66.

Garber, Elizabeth. 1995. "Reading Mathematics, Constructing Physics: Fourier and His Readers, 1822–1850," in Kox and Siegel, *No Truth,* 31–54.

Garber, Elizabeth, Stephen G. Brush, and C. W. F. Everitt, eds. 1986. *Maxwell on Molecules and Gases.* Cambridge, Mass.: MIT Press.

Gascoigne, John. 1995. "The Eighteenth-Century Scientific Community: A Prosopographical Study." *SSS* 25:575–81.

Gascoigne, Robert. 1992. "The Historical Demography of the Scientific Community, 1450–1900." *SSS* 22:545–73.

Gasman, Daniel. 1971. *The Scientific Origins of National Socialism: Social Darwinism in Ernst Haeckel and the German Monist League.* New York: Science History.

———. 1998. *Haeckel's Monism and the Birth of Fascist Ideology.* New York: Peter Lang.

Gaudant, Jean. 1984. "Actualisme, antiprogressionisme, catastrophisme et créationisme dans l'oeuvre d'Alcide d'Orbigny." *RHS* 37:305–12.

Gay, Hanna, and John W. Gay. 1997. "Brothers in Science: Science and Fraternal Culture in Nineteenth-Century Britain." *HS* 35:425–53.

Gayon, Jean. 1998. *Darwinism's Struggle for Survival: Heredity and the Hypothesis of Natural Selection.* Cambridge: Cambridge University Press.

Gayrard-Valy, Yvette. 1994. *The Story of Fossils: In Search of Vanished Worlds.* 1st French ed.,

1987. Trans. I. Mark Paris. London: Thames and Hudson; New York: Harry N. Abrams.

Geiger, Roger. 1986. *To Advance Knowledge: The Growth of American Research Universities, 1900–1940.* New York: Oxford University Press.

Geikie, Archibald. 1905. *The Founders of Geology.* 2d ed. London: Macmillan.

Geison, Gerald L. 1978. *Michael Foster and the Cambridge School of Physiology: The Scientific Enterprise in Late Victorian Society.* Princeton, N.J.: Princeton University Press.

———. 1979. "Divided We Stand: Physiologists and Clinicians in the American Context," in Vogel and Rosenberg, *Therapeutic Revolution,* 67–90.

———. 1981. "Scientific Change, Emerging Specialties, and Research Schools." *HS* 19:20–40.

———. 1995. *The Private Science of Louis Pasteur.* Princeton, N.J.: Princeton University Press.

———, ed. 1984. *Professions and the French State, 1700–1900.* Philadelphia: University of Pennsylvania Press.

———. 1987. *Physiology in the American Context, 1850–1940.* Bethesda, Md.: American Physiological Society.

Geison, Gerald L., and Frederic L. Holmes, eds. 1993. *Research Schools: Historical Reappraisals. Osiris,* 2d ser., vol. 8.

Geitz, Henry, Jürgen Heideking, and Jürgen Herbst, eds. 1995. *German Influences on Education in the United States to 1917.* Cambridge: German Historical Institute, Cambridge University Press.

Genschorek, Wolfgang. 1978. *Carl Gustav Carus: Artz, Künstler, Naturforscher.* Leipzig: Hirzel Verlag.

Gericke, H. 1984. "Das Mathematische Forschungsinstitut Oberwolfach." In *Perspectives in Mathematics: Anniversary of Oberwolfach, 1984,* edited by W. Jäger, J. Moser, and R. Remmert, 23–39. Basel: Birkhäuser Verlag.

Gerschenkron, Alexander. 1952. "Economic Backwardness in Historical Perspective." In *The Progress of Underdeveloped Areas,* edited by B. F. Hoselitz, 3–29. Chicago: University of Chicago Press.

Ghiselin, Michael. 1969. *The Triumph of the Darwinian Method.* Berkeley: University of California Press.

Giddens, Anthony. 1971. *Capitalism and Modern Social Theory: An Analysis of the Writings of Marx, Durkheim, and Weber.* Cambridge: Cambridge University Press.

Gigerenzer, Gerd, Zeno Swijtink, Theodore Porter, et al. 1989. *The Empire of Chance: How Probability Changed Science and Everyday Life.* Cambridge: Cambridge University Press.

Gillespie, Neal. 1979. *Charles Darwin and the Problem of Creation.* Chicago: University of Chicago Press.

Gillies, Donald, ed. 1992. *Revolutions in Mathematics.* Oxford: Clarendon Press.

Gillispie, Charles Coulston. 1957. "The Discovery of the Leblanc Process." *Isis* 48 152–70.

———. 1959a. *Genesis and Geology: The Impact of Scientific Discoveries upon Religious Beliefs in the Decades before Darwin, 1790–1850.* Cambridge, Mass.: Harvard University Press, 1951. Reprint, New York: Harper Torchbooks.

———. 1959b. "Lamarck and Darwin in the History of Science." In *Forerunners of Darwin, 1745–1859,* edited by Bentley Glass, Owsei Temkin, and William L. Straus Jr., 265–91. Baltimore: Johns Hopkins University Press.

———. 1963. "Intellectual Factors in the Background of Analysis by Probability." In *Scientific Change,* edited by A. C. Crombie, 431–53. New York: Basic Books.

———. 1965. "Science and Technology." In *War and Peace in an Age of Upheaval, 1793–1830,* vol. 9 of *The New Cambridge Modern History,* edited by C. W. Crawley, 118–45. Cambridge: Cambridge University Press.

———, ed. 1970–80. *Dictionary of Scientific Biography.* 16 vols. New York: Charles Scribner's Sons.

Gillispie, Charles Coulston, Ivor Grattan-Guinness, and Robert Fox. 1997. *Pierre Simon Laplace, 1749–1827.* Princeton, N.J.: Princeton University Press.

Gilpin, Robert. 1968. *France in the Age of the Scientific State.* Princeton, N.J.: Princeton University Press.

Gispen, Cornelis W. R. 1983. "Selbstverständnis und Professionalisierung deutscher Ingenieure. Eine Analyse der Nachrufe." *Technikgeschichte* 50:34–61.

Gispert, Hélène. 1991. "La France mathématique: La Société mathématique de France." *Cahiers d'histoire et de philosophie des sciences,* n.s., 34:13–180.

Gispert, Hélène, and Renate Tobies. 1996. "A Comparative Study of the French and German Mathematical Societies before 1914." In *Mathematical Europe: History, Myth, Identity. L'Europe Mathématique: Histoires, Mythes, Identités,* edited by Catherine Goldstein, Jeremy Gray, and Jim Ritter, 407–30. Paris: Editions de la Maison des Sciences de l'Homme.

Giuliani, Giuseppe. 1996. *Il Nuovo Cimento: Novanti'anni di fisica in Italia, 1855–1944.* Pavia: La Goliardica Pavese.

Glass, Bentley. 1954. Review of Gillispie, *Genesis and Geology. Quarterly Review of Biology* 29:150.

Glen, Robert. 1975. Letter to the editor. *American Historical Review* 80:203–4.

Glen, William, ed. 1994. *The Mass-Extinction Debates: How Science Works in a Crisis.* Stanford, Calif.: Stanford University Press.

Goethe, Johann Wolfgang von. [1817–24] 1989. *Zur Naturwissenschaft überhaupt, besonders zur Morphologie.* Vol. 12 of *Sämtliche Werke nach Epochen seines Schaffens.* Münchener Ausgabe. Munich: Carl Hanser.

———. 1986. *Die Schriften zur Naturwissenschaft.* Vol. 9b, *Zur Morphologie.* Ed. Dorthea Kuhn. Weimar: Hermann Böhlaus Nachfolger.

Gohau, Gabriel. 1997. "Évolution des idées sur le métamorphisme et l'origine des granites." In *Le métamorphisme et la formation des granites. Évolution des idées et concepts actuels,* 13–58. Paris: Editions Nathan.

———, ed. 1997. *De la géologie à son histoire.* Paris: Ministère de l'éducation, de la Recherche et de la Technologie, Comité des Travaux Historiques et Scientifiques (Section des Sciences).

Goldman, Jay R. 1998. *The Queen of Mathematics: A Historically Motivated Guide to Number Theory.* Wellesley, Mass.: A. K. Peters.

Goldman, Lawrence. 1987. "A Peculiarity of the English: The Social Science Association and the Absence of Sociology in Nineteenth-Century Britain." *Past and Present* 114:133–71.

———. 1993. "Experts, Investigators, and the State in 1860: British Social Scientists through American Eyes." In *The State and Social Investigation in Britain and the United States,* edited by Michael J. Lacey and Mary O. Furner, 95–126. Cambridge: Cambridge University Press.

Goldstein, Daniel. 1994. " 'Yours for Science': The Smithsonian Institution's Correspondents and the Shape of Scientific Community in Nineteenth-Century America." *Isis* 85:573–99.

Goldstein, Jan. 1987. *Console and Classify: The French Psychiatric Profession in the Nineteenth Century.* Cambridge: Cambridge University Press.

———. 1994. "The Advent of Psychological Modernism in France: An Alternative Narrative," in Ross, *Modernist Impulses,* 190–209.

———. 2001. "Psychology," in T. Porter and Ross, *Cambridge History of Science.*

Golinski, Jan. 1992. *Science as Public Culture: Chemistry and Enlightenment in Britain, 1760–1820.* Cambridge: Cambridge University Press.

Good, Gregory A. 1998. "Toward a History of the Sciences of the Earth." In *Sciences of the Earth: An Encyclopedia of Events, People, and Phenomena,* 2 vols., edited by Gregory A. Good, 1: xvii–xxvi. New York: Garland Publishing.

Gooday, Graeme. 1990. "Precision Measurement and the Genesis of Physics Teaching Laboratories in Victorian Britain." *BJHS* 23:25–51.

Gould, Stephen Jay. 1965. "Is Uniformitarianism Necessary?" *American Journal of Science* 263:223–28.

———. 1977a. *Ever Since Darwin: Reflections on Natural History.* New York: Norton.

———. 1977b. *Ontogeny and Phylogeny.* Cambridge, Mass.: Harvard University Press.

———. 1981. *The Mismeasure of Man.* New York: Norton.

———. 1983. *Hen's Teeth and Horses Toes: Further Reflections on Natural History.* New York: Norton.

———. 1985. *The Flamingo's Smile: Reflections in Natural History.* New York: Norton.

———. 1987. *Time's Arrow, Time's Cycle: Myth and Metaphor in the Discovery of Geological Time.* Cambridge, Mass.: Harvard University Press.

———. 1989. *Wonderful Life: The Burgess Shale and the Nature of History.* New York: Norton.

———. 1991. *Bully for Brontosaurus: Reflections in Natural History.* New York: Norton.

———. 1993. *Eight Little Piggies: Reflections in Natural History.* New York: Norton.

———. 1995. *Dinosaur in a Haystack: Reflections in Natural History.* New York: Norton.

———. 1999. *Leonardo's Mountain of Clams and the Diet of Worms: Essays on Natural History.* New York: Norton.

Goupil, M., ed. 1992. *Lavoisier et la révolution chimique.* Palaiseau, Sabix: Ecole polytechnique.

Grabiner, Judith V. 1975. "The Mathematician, the Historian, and the History of Mathematics." *Historia Mathematica* 2:439–47. *Proceedings of the American Academy of Arts and Sciences Workshop on the Evolution of Modern Mathematics.* Ed. Garrett Birkhoff and Sue Ann Garwood.

———. 1981. *The Origins of Cauchy's Rigorous Calculus.* Cambridge, Mass.: MIT Press.

Gradmann, Christoph. 1993. "Naturwissenschaft, Kulturgeschichte und Bildungsbegriff bei Emil du Bois-Reymond." *Tractrix* 5:1–16.

———. 1998. "Leben in der Medizin: Zur Aktualität von Biographie und Prosopographie in der Medizingeschichte." In *Medizingeschichte: Aufgaben, Probleme, Perspektiven,* edited by Norbert Paul and Thomas Schlich, 243–65. Frankfurt am Main: Campus.

———. 2001. *Medizin und Mikrobiologie in Deutschland, 1870–1910: Untersuchungen zur Bakteriologie Robert Kochs.* Medical Habilitationsschrift, University of Heidelberg.

Graham, Loren. 1993. *Science in Russia and the Soviet Union: A Short History.* Cambridge: Cambridge University Press.

Graham, Loren, Wolf Lepenies, and Peter Weingart, eds. 1983. *Functions and Uses of Disciplinary Histories.* Dordrecht: Reidel.

Granshaw, Lindsay, and Roy Porter, eds. 1989. *The Hospital in History.* London: Routledge.

Grassé, Pierre-P. 1944. " 'La Biologie': Texte inédit de Lamarck." *La revue scientifique* 82:267–76.

Grattan-Guinness, Ivor. 1970. *The Development of the Foundations of Mathematical Analysis from Euler to Riemann.* Cambridge, Mass.: MIT Press.

———. 1971. "Towards a Biography of Georg Cantor." *AS* 27:345–91.

———. 1977. "History of Science Journals: 'To Be Useful, and to the Living'?." *AS* 34:193–202.

———. 1981. "Mathematical Physics in France, 1800–1840: Knowledge, Activity, and Historiography." In *Mathematical Perspectives: Essays on Mathematics and Its Historical Development,* edited by Joseph W. Dauben, 95–138. New York: Academic Press.

———. 1988. "*Grandes Ecoles, Petite Université:* Some Puzzled Remarks on Higher Education in Mathematics in France, 1795–1840." *History of Universities* 7:197–225.

———. 1990a. *Convolutions in French Mathematics, 1800–1840: From the Calculus and Mechanics to Mathematical Analysis and Mathematical Physics.* 3 vols. Basel: Birkhäuser.

———. 1990b. "Does History of Science Treat of the History of Science? The Case of Mathematics." *HS* 28:149–73.

———. 1992. "Scientific Revolutions as Convolutions? A Sceptical Enquiry," in Demidov et al., *Amphora,* 279–87.

———. 1996. Review of Parshall and Rowe, *Emergence of the American Mathematical Research Community. Isis* 87:187–88.

———. 1997. *The Rainbow of Mathematics: The Fontana History of the Mathematical Sciences.* London: Fontana. Also published as *The Rainbow of Mathematics: The Norton History of Mathematical Sciences* (New York: W. W. Norton, 1998).

Grattan-Guinness, Ivor, with J. R. Ravetz. 1972. *Joseph Fourier, 1768–1830: A Survey of His Life and Work, Based on a Critical Edition of His Monograph on the Propagation of Heat, Presented to the Institut de France in 1807.* Cambridge, Mass.: MIT Press.

Grau, Conrad. 1988. *Berühmte Wissenschaftsakademien: Von ihrem Entstehen und ihrem weltweiten Erfolg.* Thun: Deutsch.

———. 1993. *Die Preussische Akademie der Wissenschaften zu Berlin: Eine deutsche Gelehrtengesellschaft in drei Jahrhunderten.* Heidelberg: Spektrum Akademischer Verlag.

Gray, Jeremy. 1979. "Non-Euclidean Geometry: A Re-interpretation." *Historia Mathematica* 6:236–58.

———. 1989. *Ideas of Space: Euclidean, Non-Euclidean and Relativistic.* Oxford: Clarendon Press.

Greene, John C. 1959. *The Death of Adam: Evolution and Its Impact on Western Thought.* Ames: Iowa State University Press.

———. 1961. *Darwin and the Modern World View.* Baton Rouge: Louisiana State University Press.

———. 1981. *Science, Ideology, and World View: Essays in the History of Evolutionary Ideas.* Berkeley: University of California Press.

————. 1984. *American Science in the Age of Jefferson*. Ames: Iowa State University Press.

Greene, Mott T. 1982. *Geology in the Nineteenth Century: Changing Views of a Changing World*. Ithaca, N.Y.: Cornell University Press.

————. 1985. "History of Geology." *Osiris*, 2d ser., 1:97–116.

Gregory, Frederick. 1977. *Scientific Materialism in Nineteenth-Century Germany*. Dordrecht: Reidel.

————. 1989. "Kant, Schelling and the Administration of Science in the Romantic Era." *Osiris*, 2d ser., 5:17–35.

————. 1990. "Theology and the Sciences in the German Romantic Period," in Cunningham and Jardine, *Romanticism and the Sciences*, 69–81.

————. 1992a. "Hat Müller die Naturphilosophie wirklich aufgegeben," in Hagner and Wahrig-Schmidt, *Johannes Müller*, 143–54.

————. 1992b. *Nature Lost? Natural Science and the German Theological Traditions of the Nineteenth Century*. Cambridge, Mass.: Harvard University Press.

Gregory, W., ed. 1938. *International Congresses and Conferences, 1840–1937: A Union List of Their Publications Available in Libraries of the United States and Canada*. New York: H. H. Wilson.

Grinstein, Louise S., and Paul J. Campbell, eds. 1987. *Women of Mathematics: A Biobibliographic Sourcebook*. New York: Greenwood Press.

Grison E., M. Goupil, and P. Bret, eds. 1994. *A Scientific Correspondence during the Chemical Revolution: Louis Bernard Guyton-de-Morveau and Richard Kirwan, 1782–1802*. Berkeley: Berkeley Papers in History of Science.

Grmek, Mirko D. 1991. "Claude Bernard entre le matérialisme et le vitalisme: La nécessité et la liberté dans les phénomènes de la vie." In *La nécessité de Claude Bernard*, edited by Jacques Michel, 117–139. Paris: Méridiens Klincksieck.

Gross, Paul, and Norman Levitt. 1994. *Higher Superstition: The Academic Left and its Quarrels with Science*. Baltimore: Johns Hopkins University Press.

Gruber, Howard E. 1981. *Darwin on Man: A Psychological Study of Scientific Creativity*. 1974. 2d ed. New York: Dutton.

Guagnini, Anna. 1988. "Higher Education and the Engineering Profession in Italy: The Scuole of Milan and Turin, 1859–1914." *Minerva* 4:26:512–48.

Guerlac, Henry. 1976. "Chemistry as a Branch of Physics: Laplace's Collaboration with Lavoisier." *HSPS* 7:193–276.

Gunga, Hanns-Christian. 1989. *Leben und Werk des Berliner Physiologen Nathan Zuntz (1847–1920): Unter besonderer Berücksichtigung seiner Bedeutung für die Frühgeschichte der Höhenphysiologie und Luftfahtmedizin*. Husum: Matthiesen.

Guntau, Martin. 1978. "The Emergence of Geology as a Scientific Discipline." *HS* 16:280–90.

————. 1984. *Die Genesis der Geologie als Wissenschaft: Studie zu den kognitiven Prozessen und gesellschaftlichen Bedingungen bei der Herausbildung der Geologie als naturwissenschaftliche Disziplin an der Wende vom 18. zum 19. Jahrhundert*. Berlin: Akademie Verlag.

————. 1991. "Geologische Institutionen und staatliche Initiativen in der Geschichte," in Büttner and Kohler, *Geosciences / Geowissenschaften*, 229–40.

————. 1999. "Zur Geologie und Mineralogie an deutschen Bergakademien im 18. und 19. Jahrhundert." In *Proceedings of the Fourth Heritage Symposium, Banská Štiavnica, Slovakia, 7–11 September, 1998*, 105–10. Banská Štiavnica Museum. 1999.

Guralnick, Stanley M. 1975. *Science and the Ante-bellum American College.* Philadelphia: American Philosophical Society. Memoirs of the American Philosophical Society, vol. 109.

———. 1979. "The American Scientist in Higher Education, 1820–1910." In *The Sciences in the American Context: New Perspectives,* edited by Nathan Reingold, 99–141. Washington: Smithsonian Institute Press.

Gustafson, Robert. 1995. *James Woodrow (1828–1907): Scientist, Theologian, Intellectual Leader.* Lewiston, Pa.: Mellen Press.

Gustin, Bernard. 1975. The Emergence of the German Chemical Profession, 1790–1867. Ph.D. diss., University of Chicago.

Haber, Lutz F. 1958. *The Chemical Industry during the Nineteenth Century: A Study of the Economic Aspect of Applied Chemistry in Europe and North America.* Oxford: Clarendon Press.

———. 1986. *The Poisonous Cloud: Chemical Warfare in the First World War.* Oxford: Clarendon Press.

Habermas, Jürgen. 1989. *The Structural Transformation of the Public Sphere.* Trans. Thomas Burger. Cambridge, Mass.: MIT Press.

Hacking, Ian. 1983. *Representing and Intervening: Introductory Topics in the Philosophy of Natural Science.* Cambridge: Cambridge University Press.

———. 1986. "Making Up People." In *Reconstructing Individualism,* edited by Thomas C. Heller, 222–36. Stanford, Calif.: Stanford University Press.

———. 1990. *The Taming of Chance.* Cambridge: Cambridge University Press.

Hadamard, Jacques. 1949. *The Psychology of Invention in the Mathematical Field.* Princeton, N.J.: Princeton University Press.

Hagner, Michael. 1995. "Aspects of Brain Localization in Late Nineteenth-Century Germany," in Debru, *Essays in the History of the Physiological Sciences,* 73–88.

———. 1997. *Homo Cerebralis: Der Wandel vom Seelenorgan zum Gehirn.* Berlin: Berlin Verlag.

———. 2001. "Psychophysiologie und Selbsterfahrung: Metamorphosen des Schwindels und der Aufmerksamkeit im 19. Jahrhundert." In *Aufmerksamkeiten: Archäologie der literarischen Kommunikation VII,* edited by Jan Assmann and Aleida Assmann 241–63. Munich: Fink.

———, ed. 1999. *Ecce Cortex: Beiträge zur Geschichte des modernen Gehirns.* Göttingen: Wallstein.

Hagner, Michael, and Wahrig-Schmidt, Bettina, eds. 1993. *Johannes Müller und die Philosophie.* Berlin: Akademie Verlag.

Hahn, Roger. 1971. *The Anatomy of a Scientific Institution: The Paris Academy of Sciences, 1666–1803.* Berkeley: University of California Press.

———. 1975. "Scientific Research as an Occupation in Eighteenth-Century Paris." *Minerva* 13:501–13.

———. 1976. "Scientific Careers in Eighteenth-Century France." In *The Emergence of Science in Western Europe,* edited by Maurice Crosland, 127–38. New York: Science History.

———. 1995. "Lavoisier et ses collaborateurs: Une équipe au travail," in Demeulenaere-Douyère, *Il y a deux cents Lavoisier,* 55–64.

Halévy, Elie. 1928. *The Growth of Philosophic Radicalism.* New York: Macmillan.

Hall, A. Rupert. 1962. "The Changing Technical Act." *TC* 3:501–15.

———. 1974. "What Did the Industrial Revolution in Britain Owe to Science?" In *Historical Perspectives: Studies in English Thought and Society in Honour of J. H. Plumb,* edited by Neil McKendrick, 129–51. London: Europa Publications.

Hall, Marie Boas. 1984. *All Scientists Now: The Royal Society in the Nineteenth Century.* Cambridge: Cambridge University Press.

Hallett, Michael. 1984. *Cantorian Set Theory and Limitation of Size.* Oxford: Clarendon Press.

Hamlin, Christopher. 1990. *A Science of Impurity: Water Analysis in Nineteenth–Century Britain.* London: Adam-Hilger.

———. 1998. *Public Health and Social Justice in the Age of Chadwick: Britain, 1800–1854.* New York: Cambridge University Press.

Hankins, Thomas L. 1979. "In Defence of Biography: The Use of Biography in the History of Science." *History of Science* 17:1–16.

———. 1980. *Sir William Rowan Hamilton.* Baltimore: Johns Hopkins University Press.

Hannaway, Owen. 1969. Review of Crosland, *Society of Arcueil. Isis* 60:4:578–81.

———. 1975. *The Chemists and the Word: The Didactic Origins of Chemistry.* Baltimore: Johns Hopkins University Press.

———. 1976. "The German Model of Chemical Education in America: Ira Remsen at Johns Hopkins (1876–1913)." *Ambix* 23:145–64.

Hänseroth, Thomas, and Klaus Mauersberger. 1996. "Das Dresdner Konzept zur Genese technikwissenschaftlicher Disziplinen." *Dresdner Beiträge zur Geschichte der Technikwissenschaften* 24:20–45.

Hanson, Norwood Russell. 1965. *Patterns of Discovery.* Cambridge: Cambridge University Press.

Hård, Mikael. 1994. *Machines Are Frozen Spirit: The Scientification of Refrigeration and Brewing in the Nineteenth Century—A Weberian Interpretation.* Boulder, Colo.: Westview Press.

Hård, Mikael, and Andreas Knie. 1999. "The Grammar of Technology: German and French Diesel Engineering, 1920–1940." *TC* 40:26–46.

Harman, Peter M. 1982. *Energy, Force, and Matter: The Conceptual Development of Nineteenth-Century Physics.* Cambridge: Cambridge University Press.

———, ed. 1985. *Wranglers and Physicists: Studies on Cambridge Physics in the Nineteenth Century.* Manchester: Manchester University Press.

Harnack, Adolf. 1900. *Geschichte der Königlich Preußischen Akademie der Wissenschaften zu Berlin.* 3 vols. Berlin.

Harrington, Anne. 1987. *Medicine, Mind and the Double Brain.* Princeton, N.J.: Princeton University Press.

Harris, John R. 1998. *Industrial Espionage and Technology Transfer: Britain and France in the Eighteenth Century.* Aldershot: Ashgate.

Hartkopf, Werner. 1992. *Die Berliner Akademie der Wissenschaften: Ihre Mitglieder und Preisträger, 1700–1900.* Berlin: Akademie Verlag.

Hartkopf, Werner, and Gert Wangermann. 1991. *Dokumente zur Geschichte der Berliner Akademie der Wissenschaften von 1700 bis 1990.* Berlin: Spektrum Akademischer Verlag.

Haskell, Thomas. 1977. *The Birth of Professional Social Science: The American Social Science Association and the Nineteenth-Century Crisis of Authority.* Urbana: University of Illinois Press.

Hasler, Ueli. 1982. *Beherrschte Natur: Die Anpassung der Theologie an die bürgerliche Naturauffassung im 19. Jahrhundert.* Bern: Peter Lang.

Haupt, Bettina. 1984. *Deutschsprachige Chemielehrbücher (1775–1850).* Stuttgart: Deutscher Apotheker Verlag.

Haüy, R. J. 1803. *Traité élémentaire de physique.* 2 vols. Paris. De l'Impr. de Delance et Lesueur.

Hawkins, Hugh. 1960. *Pioneer: A History of the Johns Hopkins University, 1874–1899.* Ithaca, N.Y.: Cornell University Press.

Hawkins, Thomas. 1970. *Lebesgue's Theory of Integration: Its Origins and Development.* Madison: University of Wisconsin Press; New York: Chelsea.

———. 2000. *The Emergence of the Theory of Lie Groups: An Essay in the History of Mathematics, 1869–1926.* New York: Springer.

Hayek, Friedrich. 1952. *The Counter-Revolution of Science.* London: Free Press.

Hays, J. N. 1974. "Science in the City: The London Institution, 1819–1840." *BJHS* 7:146–62.

———. 1982. "London Institution." *AS* 39:229–74.

———. 1983. "The London Lecturing Empire, 1800–1850," in Inkster and Morrell, *Metropolis and Province,* 91–119.

Heilbron, J. L. 1979. *Electricity in the Seventeenth and Eighteenth Centuries: A Study of Early Modern Phyiscs.* Berkeley: University of California Press.

———. 1993. "Weighing Imponderables and Other Quantitative Science around 1800." *HSPS,* supp. to vol. 24, pt. 1: 1–337.

I Ieilbron, Johan. 1995. *Thc Rise of Social Theory.* Trans. Sheila Gogol. Minneapolis: University of Minnesota Press.

Heilbron, Johan, Lars Magnusson, and Björn Wittrock, eds. 1998. *The Rise of the Social Sciences and the Formation of Modernity: Conceptual Change in Context, 1750–1850.* Dordrecht: Kluwer.

Heischkel, Edith. 1949. "Die Geschichte der Medizingeschichtsschreibung." In *Einführung in die Medizinhistorik,* edited by Walter Artelt, 202–37. Stuttgart: Enke.

Helmholtz, Hermann von. [1869] 1896. "Über das Ziel und die Fortschritte der Naturwissenschaft. Eröffnungsrede für die Naturforscherversammlung zu Innsbruck," in Helmholtz, *Vorträge und Reden,* 1:367–98.

———. [1886] 1896. "Antwortrede gehalten beim Empfang der Graefe-Medaille zu Heidelberg," in Helmholtz, *Vorträge und Reden,* 2:311–20.

———. 1896. *Vorträge und Reden.* 2 vols. 4th ed. Braunschweig: Vieweg.

Henderson, William. 1954. *Britain and Industrial Europe, 1750–1870.* Liverpool: Liverpool University Press.

Henkin, Leo. 1940. *Darwinism in the English Novel, 1860–1910: The Impact of Evolution on Victorian Fiction.* New York: Corporate. Reprint, New York: Russell & Russell, 1963.

Henrion, Claudia, ed. 1997. *Women in Mathematics: The Addition of Difference.* Bloomington: University of Indiana Press.

Herivel, J. 1975. *Joseph Fourier: The Man and the Physicist.* Oxford: Clarendon Press.

Herries Davies, Gordon L. 1995. *North from the Hook: 150 Years of the Geological Survey of Ireland.* Dublin: Geological Survey of Ireland.

Hess, Volker. 1997. "Die Entdeckung des Krankenhauses als wissenschaftlicher Raum: Ein neues Selbstverständnis der Klinik 1800–1850." *Historia Hospitalium* 20:88–108.

———. 2000. *Der wohltemperierte Mensch: Wissenschaft und Alltag des Fiebermessens (1850–1900).* Frankfurt am Main: Campus.

————, ed. 1997. *Normierung der Gesundheit: Messende Verfahren der Medizin als kulturelle Praxis um 1900*. Husum: Matthiesen.

Hiebert, Erwin. 1962. *The Historical Roots of the Principle of Conservation of Energy*. Madison: University of Wisconsin Press.

————. 1966. "The Uses and Abuses of Thermodynamics in Religion." *Daedalus* 95: 1046–80.

Higham, Norman. 1963. *A Very Scientific Gentleman: The Major Achievements of Henry Clifton Sorby*. Oxford: Pergamon; New York: Macmillan.

Hilton, Boyd. 1988. *The Age of Atonement: The Influence of Evangelicalism on Social and Economic Thought, 1785–1865*. Oxford: Clarendon Press.

Himmelfarb, Gertrude. 1959. *Darwin and the Darwinian Revolution*. New York: W. W. Norton.

Hodge, Charles. 1994. *What Is Darwinism, and Other Writings on Science and Religion*. Ed. with an introduction by Mark Noll and David Livingstone. Reprint, Grand Rapids, Mich.: Baker Book House.

Hodgkin, Luke. 1981. "Mathematics and Revolution from Lacroix to Cauchy," in Bos, Mehrtens, and Schneider, *Social History*, 50–71.

Hölder, Helmut. 1960. *Geologie und Paläontologie in Texten und ihrer Geschichte*. Freiburg: Verlag Karl Alber.

Holmes, Frederic L. 1963a. "Elementary Analysis and the Origin of Physiological Chemistry." *Isis* 54:50–81.

————. 1963b. "The Milieu Intérieur and the Cell Theory." *BHM* 37:315–35.

————. 1974. *Claude Bernard and Animal Chemistry*. Cambridge, Mass.: Harvard University Press.

————. 1985. *Lavoisier and the Chemistry of Life: An Exploration of Scientific Creativity*. Madison: University of Wisconsin Press.

————. 1989a. "The Complementarity of Teaching and Research in Liebig's Laboratory," in Olesko, *Science in Germany*, 121–64.

————. 1989b. *Eighteenth-Century Chemistry as an Investigative Enterprise*. Office for the History of Science and Technology, University of California, Berkeley.

————. 1991–93. *Hans Krebs*. 2 vols. New York: Oxford University Press.

————. 1992. *Between Biology and Medicine: The Formation of Intermediary Metabolism*. Office for the History of Science and Technology, University of California, Berkeley.

————. 1995. "The Boundaries of Lavoisier's Chemical Revolution." *RHS* 48:9–47.

————. 1997. "Claude Bernard and the Vitalism of His Time," in Cimino and Duchesneau, *Vitalisms*, 281–95.

————. 1998. *Antoine Lavoisier: The Next Crucial Year*. Princeton, N.J.: Princeton University Press.

Holmes, Frederic L., and Trevor H. Levere, eds. 2000. *Instruments and Experimentation in the History of Chemistry*. Cambridge Mass.: MIT Press.

Holton, Gerald. 1978. *The Scientific Imagination: Case Studies*. Cambridge: Cambridge University Press.

Homburg, E., H. Schröter, A. Travis, and R. Halleux, eds. 1998. *The Chemical Industry in Europe, 1850–1914: Industrial Growth, Pollution, and Professionalization*. Dordrecht: Kluwer Academic.

Home, R. H. 1983. "Poisson's Memoirs on Electricity: Academic Politics and a New Style in Physics." *BJHS* 16:239–60.

Hong, Sungook. 1994. "Controversy over Voltaic Contact Phenomena, 1862–1900." *AHES* 47:233–89.

———. 1995a. "Efficiency and Authority in the 'Open versus Closed' Transformer Controversy." *AS* 52:49–76.

———. 1995b. "Forging Scientific Electrical Engineering: John Ambrose Fleming and the Ferranti Effect." *Isis* 86:30–51.

Hooykaas, Reijer. 1952. Review of Gillispie, *Genesis and Geology*. *Archives internationales d'Histoire des sciences* 5 (No. 18):125–28.

———. 1963. *Natural Law and Divine Miracle: The Principle of Uniformity in Geology, Biology and Theology*. Leiden: E. J. Brill.

———. 1970. *Catastrophism in Geology: Its Scientific Character in Relation to Actualism and Uniformitarianism*. Amsterdam: North Holland.

Horwich, Paul, ed. 1993. *World Changes: Thomas Kuhn and the Nature of Science*. Cambridge, Mass.: MIT Press.

Hovencamp, Herbert. 1978. *Science and Religion in America: 1800–1860*. Philadelphia: University of Pennsylvania Press.

Howard, Thomas Albert. 1999. *Religion and the Rise of Historicism: W. M. L. de Wette, Jacob Burckhardt, and the Theological Origins of Nineteenth-Century Historical Consciousness*. Cambridge: Cambridge University Press.

Howarth, Richard J. 1996. "Sources for a History of the Ternary Diagram." *BJHS* 29: 337–56.

———. 1998. "Graphical Methods in Mineralogy and Igneous Petrology (1800–1935)." In *Toward a History of Mineralogy, Petrology, and Geochemistry: Proceedings of the International Symposium on the History of Mineralogy, Petrology, and Geochemistry, Munich, March 8–9, 1996*, edited by Bernhard Fritscher and F. Henderson, 281–307. Munich: Institut für Geschichte der Wissenschaften.

———. 1999. "Measurement, Portrayal and Analysis of Orientation Data and the Origins of Early Modern Structural Geology (1670–1967)." *Proceedings of the Geologists' Association* 110:273–309.

———. 2002. "From Graphical Display to Dynamic Model: Mathematical Geology in the Earth Sciences in the Nineteenth and Twentieth Centuries," in Oldroyd, *Earth Inside and Out*, 59–97.

Høyrup, Else. 1978. *Women and Mathematics, Science, and Engineering: A Bibliography*. Roskilde, Denmark: Roskilde University Library.

Hudson, John. 1992. *The History of Chemistry*. London: MacMillan.

Huerkamp, Claudia. 1985. *Der Aufstieg der Ärzte im 19. Jahrhundert: Vom gelehrten Stand zum professionellen Experten; das Beispiel Preußens*. Göttingen: Vandenhoeck & Ruprecht.

Hufbauer, Karl. 1982. *The Formation of the German Chemical Community (1720–1795)*. Berkeley: University of California Press.

Huggett, Richard. 1989. *Cataclysms and Earth History: The Development of Diluvialism*. Oxford: Clarendon Press.

Hughes, H. Stuart. 1977. *Consciousness and Society: The Reorientation of European Social Thought, 1860–1930*. 1958. Reprint, New York: Vintage.

Hughes, Thomas P. 1989. *American Genesis: A Century of Invention and Technological Enthusiasm, 1870–1970*. New York: Viking.

Hull, David. 1973. *Darwin and His Critics: The Reception of Darwin's Theory of Evolution by the Scientific Community*. Cambridge, Mass.: Harvard University Press.

————. 1988. *Science as a Process: An Evolutionary Account of the Social and Conceptual Development of Science*. Chicago: University of Chicago Press.

Hull, David, and Michael Ruse, eds. 1998. *The Philosophy of Biology*. Oxford: Oxford University Press.

Humboldt, Alexander von. 1797. *Versuche über die gereizte Muskel- und Nervenfaser nebst den chemischen Process des Lebens in der Thier- und Pflanzenwelt*. 2 vols. Berlin: Heinrich August Rottmann.

————. 1991. *Reise in die Äquinoktial-Gegenden des Neuen Kontinents*. Ed. Ottmar Ette. 2 vols. Frankfurt am Main: Insel Verlag.

Humboldt, Alexander von, and Aimé Bonpland. 1818–29. *Personal Narrative of Travels to the Equinoctial Regions of the New Continent, during the Years 1799–1804*. 7 vols. London: Longman, Hurst, Rees, Orme, and Brown.

Hunt, Bruce J. 1983. " 'Practice vs. Theory': The British Electrical Debate, 1888–1891." *Isis* 74:341–55.

————. 1991. *The Maxwellians*. Ithaca, N.Y.: Cornell University Press.

————. 1996. "Scientists, Engineers, and Wildman Whitehouse: Measurement and Credibility in Early Cable Telegraphy." *BJHS* 29:155–69.

Hurwic, Anna. 1995. *Pierre Curie*. Paris: Flammarion.

Hutchins, Roger. 1994. "John Phillips, 'Geologist-Astronomer,' and the Origins of the Oxford University Observatory." *History of Universities* 13:193–249.

Hutchison, T. W. 1953. *A Review of Economic Doctrines, 1870–1929*. Oxford: Oxford University Press.

Hutton, James. 1788. "Theory of the Earth; or, An Investigation of the Laws Observable in the Composition, Dissolution, and Restoration of the Land upon the Globe." *Transactions of the Royal Society of Edinburgh* 1:209–304.

Huxley, Thomas Henry. [1861–94] 1997. *The Major Prose of Thomas Henry Huxley*. Ed. Alan P. Barr. Athens: University of Georgia Press.

Ihde, Aaron. 1964. *The Development of Modern Chemistry*. New York: Harper & Row.

Ihde, Don. 1990. *Technology and the Life World: From Garden to Earth*. Bloomington: Indiana University Press.

Illich, Ivan. 1977. *Limits to Medicine: The Expropriation of Health*. Harmondsworth: Penguin.

Imbrie, John, and Katherine Palmer Imbrie. 1979. *Ice Ages: Solving the Mystery*. London: Macmillan.

Infeld, Leopold. 1948. *Whom the Gods Love: The Story of Evariste Galois*. New York: Whittlesey House.

Inkster, Ian. 1975. "Science and the Mechanics' Institutes, 1820–1850: The Case of Sheffield." *AS* 32:451–74.

————. 1976. "The Social Context of an Educational Movement: A Revisionist Approach to the English Mechanics' Institutes, 1820–1850." *Oxford Review of Education* 2:277–307.

————. 1988. "Cultural Enterprise: Science, Steam, Intellect and Social Class in Rochdale *circa* 1833–1900." *SSS* 18:291–330.

————. 1991. *Science and Technology in History: An Approach to Industrial Development*. New Brunswick, N.J.: Rutgers University Press.

Inkster, Ian, and Jack Morrell, eds. 1983. *Metropolis and Province: Science in British Culture 1750–1850*. London: Hutchinson.

Irvine, William. 1955. *Apes, Angels, and Victorians*. New York: McGraw-Hill.

Israel, Giorgio, and Bruna Ingrao. 1990. *The Invisible Hand: Economic Equilibrium in the History of Science*. Cambridge, Mass.: MIT Press.

Ivanov, B. I., and V. V. Cheshev. 1982. *Entstehung und Entwicklung der technischen Wissenschaften*. Leipzig: VEB Fachbuchverlag. Originally published as *Stanovlenie i razvitie tekhnicheskikh nauk* (Leningrad: Nauka, 1977).

Ivanov, B. I., V. V. Cheshev, and O. M. Volossevich. 1980. "Die Besonderheiten der Entstehung und Entwicklung der technischen Wissenschaften." In Autorenkollektiv, *Spezifik der Technischen Wissenschaften*, 61–162. Leipzig: VEB Fachbuch-Verlag.

Jacob, Margaret C. 1988. *The Cultural Meaning of the Scientific Revolution*. Philadelphia: Temple University Press.

———. 1997. *Scientific Culture and the Making of the Industrial West*. Oxford: Oxford University Press.

Jacques, Jean. 1987. *Marcellin Berthelot: Autopsie d'un mythe*. Paris: Belin.

———. 1989. "Lavoisier et ses historiens français." *Revue des questions scientifiques* 160, no. 2:169–89.

Jacyna, L. Stephen. 1984. "The Romantic Programme and the Reception of Cell Theory in Britain." *Journal of the History of Biology* 17:13–48.

———. 1988. "The Laboratory and the Clinic: The Impact of Pathology on Surgical Diagnosis in the Glasgow Western Infirmary, 1875–1910." *BHM* 52:384–406.

Jagniaux, Raoul. 1891. *Histoire de la chimie*. 2 vols. Paris: Librairie Polytechnique, Baudry et Cie.

Jahn, Ilse, Erika Krause, Rolf Löther, et al., eds. 1998. *Geschichte der Biologie*. 3d ed. Jena: Gustav Fischer.

Jahnke, Hans Niels, and Michael Otte. 1981. "On 'Science as a Language,'" in Jahnke and Otte, *Epistemological and Social Problems*, 75–89.

———, eds. 1981. *Epistemological and Social Problems of the Sciences in the Early Nineteenth Century*. Dordrecht: Reidel.

Jaki, Stanley L. 1984. *Uneasy Genius: The Life and Work of Pierre Duhem*. The Hague: Martinus Nijhoff.

———. 1991. *Scientist and Catholic: An Essay on Pierre Duhem*. Front Royal, Va.: Christendom Press.

James, Frank A. J. L., ed. 1989. *The Development of the Laboratory: Essays on the Place of Experiment in Industrial Civilisation*. Houndmills: Macmillan.

James, Patricia. 1979. *Population Malthus: His Life and Times*. London: Routledge and Kegan Paul.

James, William. 1985. *The Varieties of Religious Experience*. Cambridge, Mass.: Harvard University Press.

Janich, Peter, and Nikolas Psarros, eds. 1998. *The Autonomy of Chemistry*. Würzburg: Königshausen & Neumann.

Jardine, Nicholas. 1991. *The Scenes of Inquiry: On the Reality of Questions in the Sciences*. Oxford: Clarendon Press.

———. 1999. "Inner History; or, How to End Enlightenment." In *The Sciences in Enlightened Europe*, edited by William Clark, Jan Golinski, and Simon Schaffer, 477–94. Chicago: University of Chicago Press.

Jardine, Nicholas, J. A. Secord, and E. C. Spary, eds. 1996. *Cultures of Natural History*. Cambridge: Cambridge University Press.

Jensen, J. Vernon. 1970. "The X-Club: Fraternity of Victorian Scientists." *BJHS* 3:63–72.

———. 1972. "Interrelationships within the Victorian 'X Club.'" *Dalhousie Review* 51:539–52.

Jensen, William. 1993. "History of Chemistry and the Chemical Community: Bridging the Gap," in Mauskopf, *Chemical Sciences,* 262–76.

Jeremy, David. 1981. *Transatlantic Industrial Revolution: The Diffusion of Textile Technologies between Britain and America, 1790–1830.* Oxford: Blackwell.

Jewson, Nicholas. 1976. "The Disappearance of the Sick-man from Medical Cosmology, 1770–1870." *Sociology* 10:225–44.

Johnson, Dale. 1977. "Prelude to Dimension Theory: The Geometrical Investigations of Bolzano." *AHES* 17:261–95.

———. 1979, 1981. "The Problem of the Invariance of Dimension in the Growth of Modern Topology, Part I." *AHES* 20:97–188; 25:85–267.

Johnson, Jeffrey Allan. 1985. "Academic Chemistry in Imperial Germany." *Isis* 76:500–24.

———. 1989. "Hierarchy and Creativity in Chemistry, 1871–1914," in Olesko, *Science in Germany,* 214–40.

———. 1990. *The Kaiser's Chemists: Science and Modernization in Imperial Germany.* Chapel Hill: University of North Carolina Press.

Jones, Greta. 1980. *Social Darwinism and English Thought: The Interaction Between Biological and Social Theory.* Atlanta Highlands, N.J.: Humanities Press.

Jordan, D. W. 1982. "The Adoption of Self-Induction by Telephony, 1886–1889." *AS* 39:433–61.

Jordanova, Ludmilla. 1989. *Sexual Visions: Images of Gender in Science and Medicine between the Eighteenth and the Twentieth Centuries.* Madison: University of Wisconsin Press.

Jordanova, Ludmilla, and Roy Porter, eds. 1979. *Images of the Earth: Essays in the History of the Environmental Sciences.* Chalfont St. Giles: British Society for the History of Science. 2d ed., 1997.

Jorland, Gerard. 1995. *Les paradoxes du capital.* Paris: O. Jacob.

Jungnickel, Christa, and Russell McCormmach. 1986. *Intellectual Mastery of Nature: Theoretical Physics from Ohm to Einstein.* Vol. 1, *The Torch of Mathematics, 1800–1870.* Vol. 2, *The Now Mighty Theoretical Physics, 1870–1925.* Chicago: University of Chicago Press.

Jütte, Robert. 1996. *Geschichte der Alternativen Medizin.* Munich: Beck.

Kadish, Alon. 1982. *The Oxford Economists in the Late Nineteenth Century.* Oxford: Clarendon Press.

Kaiser, Christopher B. 1997. *Creational Theology and the History of Physical Science: The Creationist Tradition from Basil to Bohr.* Leiden: Brill.

Kanz, Kai Torsten. 1997. *Nationalismus und internationale Zusammenarbeit in den Naturwissenschaften: Die deutsch-französischen Wissenschaftsbeziehungen zwischen Revolution und Restauration, 1789–1832.* Stuttgart: Franz Steiner.

———, ed. 1994. *Philosophie des organischen in der Goethezeit: Studien zu Werk und Wirkung des Naturforschers Carl Friedrich Kielmeyer (1765–1844).* Stuttgart: Franz Steiner Verlag.

Kapoor, S. C. 1969. "Dumas and Organic Classification." *Ambix* 16:1–65.

Karady, Victor. 1980. "Educational Qualifications and University Careers in Science in Nineteenth-Century France," in R. Fox and Weisz, *Organization of Science and Technology,* 95–124.

Kargon, Robert H. 1986. "Henry Rowland and the Physics Discipline in America." *Vistas in Astronomy* 29:131–36.

Katz, Victor J. 1993. *A History of Mathematics.* New York: HarperCollins College Publisher.

Kaufmann, Doris. 1995. *Aufklärung, bürgerliche Selbsterfahrung und die "Erfindung" der Psychiatrie in Deutschland, 1770–1850.* Göttingen: Vandenhoeck & Ruprecht.

Keillar, Ian, and John S. Smith, eds. 1995. *George Gordon: Man of Science.* Centre for Scottish Studies, University of Aberdeen.

Kelley, Donald R. 1990. *The Human Measure: Social Thought in the Western Legal Tradition.* Cambridge, Mass.: Harvard University Press.

Kendrick, Thomas D. 1956. *The Lisbon Earthquake.* London: Methuen.

Kennedy, Hubert. 1978. "The Case for James Mills Peirce." *Journal of Homosexuality* 4:179–84.

Kent, Christopher. 1978. *Brains and Numbers: Elitism, Comtism, and Democracy in Mid-Victorian England.* Toronto: University of Toronto Press.

Kevles, Daniel J. 1968. "Testing the Army's Intelligence: Psychology and the Military in World War I." *Journal of American History* 55:565–81.

———. 1978. *The Physicists: The History of a Scientific Community in America.* New York: Alfred A. Knopf.

———. 1979. "The Physics, Mathematics, and Chemistry Communities: A Comparative Analysis," in Oleson and Voss, *Organization of Knowledge,* 139–72.

Kielmeyer, Carl Friedrich. [1795] 1993. *Ueber die Verhältnisse der organischen Kräfte unter einander in der Reihe der verschiedenen Organisationen.* With an introduction by Kai Torsten Kanz. Marburg an der Lahn: Basilisken-Presse.

Kim, Dong-Won. 1991. The Emergence of the Cavendish School: An Early History of the Cavendish Laboratory, 1871–1900. Ph.D. diss., Harvard University.

———. 1995. "J. J. Thomson and the Emergence of the Cavendish School, 1885–1990." *BJHS* 28:191–226.

Kim, Mi Gyung. 1992. "The Layers of Chemical Language." *HS* 30:69–96; 397–437.

———. 1996. "Constructing Symbolic Spaces: Chemical Molecules in the Académie des Sciences." *Ambix* 43:1–31.

Kim, Stephen. 1996. *John Tyndall's Transcendental Materialism and the Conflict between Religion and Science in Victorian England.* Lewiston, Pa.: Mellen University Press.

Kipnis, Naum S. 1991. *History of the Principle of Interference of Light.* Basel: Birkhauser.

Kitcher, Philip. 1987. "Mathematical Naturalism," in Aspray and Kitcher, *New Perspectives,* 293–325.

Klaver, Jan M. Ivo. 1997. *Geology and Religious Sentiment: The Effect of Geological Discoveries on English Society and Literature between 1829 and 1859.* Leiden: Brill.

Klein, Felix. 1894. *The Evanston Colloquium: Lectures on Mathematics Delivered from Aug. 28 to Sept. 9, 1893 before Members of the Congress of Mathematics Held in Connection with the World's Fair in Chicago at Northwestern University, Evanston, Ill., by Felix Klein. Reported by Alexander Ziwet.* New York: Macmillan.

———. 1896. "The Arithmetizing of Mathematics." *Bulletin of the American Mathematical Society* 2:241–49.

———. 1926. *Vorlesungen über die Entwicklung der Mathematik im 19. Jahrhundert.* Ed. R. Courant and O. Neugebauer. Berlin: Springer.

Klein, Judy. 1997. *Statistical Visions in Time: A History of Time-Series Analysis.* Cambridge: Cambridge University Press.

Klein, Martin J. 1970. *Paul Ehrenfest: The Making of a Theoretical Physicist.* Amsterdam: North-Holland.

Klein, Ursula, ed. 2001. *Tools and Modes of Representation in the Laboratory Sciences.* Dordrecht: Kluwer Academic.

Kline, Morris. 1990. *Mathematical Thought from Ancient to Modern Times.* 1972. Reprint, New York: Oxford University Press.

Kline, Ronald. 1987. "Science and Engineering Theory in the Invention and Development of the Induction Motor, 1880–1900." *TC* 28:283–313.

———. 1992. *Steinmetz: Engineer and Socialist.* Baltimore: Johns Hopkins University Press.

Klonk, Charlotte. 1996. *Science and the Perception of Nature: British Landscape Art in the Late Eighteenth and Early Nineteenth Centuries.* New Haven: Yale University Press for the Paul Mellon Centre for Studies in British Art.

Kloppenberg, James. 1986. *Uncertain Victory: Social Democracy and Progressivism in European and American Thought, 1870–1920.* New York: Oxford University Press.

Klosterman, Leo J. 1985. "A Research School of Chemistry in Nineteenth-Century Chemistry: Jean-Baptiste Dumas and His Research Students." *AS* 42:1–80.

Knell, Simon J. 2000. *The Culture of English Geology, 1815–1851: A Science Revealed through Its Collecting.* Aldershot: Ashgate.

Knie, Andreas. 1991. *Diesel, Karriere einer Technik: Genese und Formierungsprozesse im Motorenbau.* Berlin: Edition Sigma.

Knight, David M. 1978. *The Transcendental Part of Chemistry.* Folkestone: Dawson.

———. 1986. *The Age of Science: The Scientific World-View in the Nineteenth Century.* Oxford: Blackwell.

———. 1992a. *Humphry Davy: Science and Power.* Oxford: Blackwell.

———. 1992b. *Ideas in Chemistry.* London: Athlone Press.

Knight, David M., and Helge Kragh, eds. 1998. *The Making of the Chemist: The Social History of Chemistry in Europe, 1789–1914.* Cambridge: Cambridge University Press.

Knobloch, Eberhard, and David E. Rowe, eds. 1994. *The History of Modern Mathematics. Images, Ideas, and Communities.* Boston: Academic Press.

Knorr-Cetina, Karin. 1981. *The Manufacture of Knowledge: An Essay on the Construction and Contextual Nature of Science.* Oxford: Pergamon Press.

Koblitz, Ann Hibner. 1983. *Sofia Kovalevskaia: Scientist, Writer, Revolutionary.* Boston: Birkhäuser.

Kocka, Jürgen. 1987. "Bürgertum und Bürgerlichkeit als Probleme der deutschen Geschichte vom späten 18. bis zum frühen 20. Jahrhundert." In *Bürger und Bürgerlichkeit im 19. Jahrhundert,* edited by Jürgen Kocka, 21–63. Göttingen: Vandenhoeck & Ruprecht.

Kocka, Jürgen, Rainer Hohfeld, and Peter T. Walther, eds. 1999. *Die Königlich Preußische Akademie der Wissenschaften zu Berlin im Kaiserreich.* Berlin: Akademie Verlag.

Kohler, Robert E. 1982. *From Medical Chemistry to Biochemistry: The Making of A Biomedical Discipline.* Cambridge: Cambridge University Press.

———. 1990. "The Ph.D. Machine: Building on the Collegiate Base." *Isis* 81:638–62.

Kohlstedt, Sally Gregory. 1976a. *The Formation of the American Scientific Community: The*

American Association for the Advancement of Science, 1848–60. Urbana: University of Illinois Press.

———. 1976b. "The Nineteenth-Century Amateur Tradition: The Case of the Boston Society of Natural History." In *Science and Its Public*, edited by Gerald Holton and William Blanpied, 173–90. Dordrecht: Reidel.

———. 1990. "Parlors, Primers, and Public Schooling: Education for Science in Nineteenth-Century America." *Isis* 81:425–55.

Kohlstedt, Sally Gregory, and Margaret W. Rossiter, eds. 1985. *Historical Writing on American Science. Osiris*, 2d ser., vol. 1.

Kohn, David. 1980. "Theories to Work By: Rejected Theories, Reproduction, and Darwin's Path to Natural Selection." *Studies in History of Biology* 4:67–170.

———, ed. 1985. *The Darwinian Heritage*. Princeton, N.J.: Princeton University Press.

Koizumi, Kenkichiro. 1975. "The Emergence of Japan's First Physicists, 1868–1900." *HSPS* 6:3–108.

Kolmogorov, A. N., and A. P. Yushkevich, eds. 1992–98. *Mathematics of the Nineteenth Century*. Trans. Roger Cooke. 3 vols. Vol. 1, *Mathematical Logic, Algebra, Number Theory and Probability Theory*. Vol. 2, *Geometry and Analytic Function Theory*. Vol. 3, *Function Theory According to Chebychev, Ordinary Differential Equations, Calculus of Variations, Theory of Finite Differences*. Basel: Birkhäuser.

König, Wolfgang. 1993. "Technical Education and Industrial Performance in Germany: A Triumph of Heterogeneity." In *Education, Technology and Industrial Performance in Europe, 1850–1939*, edited by Robert Fox and Anna Guagnini, 65–87. Cambridge: Cambridge University Press.

———. 1996a. "Science-based Industry or Industry-based Science? Electrical Engineering in Germany before World War I." *TC* 37:70–101.

———. 1996b. *Technikwissenschaften: Die Entstehung der Elektrotechnik aus Industrie und Wissenschaft zwischen 1880 und 1914*. Chur, Switzerland: Facultas.

———. 1999. *Künstler und Strichezieher: Konstruktions- und Technikkulturen im deutschen, britischen, amerikanischen und französischen Maschinenbau zwischen 1850 und 1930*. Frankfurt am Main: Suhrkamp.

Koot, Gerard. 1987. *English Historical Economics, 1870–1926: The Rise of Economic History and Neomercantilism*. Cambridge: Cambridge University Press.

Koselleck, Reinhard. 1979. *Vergangene Zukunft: Zur Semantik geschichtlicher Zeiten*. Frankfurt am Main: Suhrkamp.

Kox, A. J., and Daniel M. Siegel, eds. 1995. *No Truth Except in the Details: Essays in Honor of Martin J. Klein*. Dordrecht: Kluwer Academic.

Koyré, Alexandre. 1939. *Etudes Galiléennes*. Paris: Hermann & Cie.

Kozák, Jan, and Marie-Claude Thompson. 1991. *Historical Earthquakes in Europe*. Zurich: Swiss Reinsurance Company.

Kragh, Helge. 1987. *An Introduction to the Historiography of Science*. Cambridge: Cambridge University Press.

———. 1993. "Between Physics and Chemistry: Helmholtz's Route to a Theory of Chemical Thermodynamics," in Cahan, *Hermann von Helmholtz*, 432–60.

Kranakis, Eda. 1992. "Hybrid Careers and the Interaction of Science and Technology." In *Technological Development and Science in the Industrial Age: New Perspectives on the*

Science-Technology Relationship, edited by Peter Kroes and Martijn Bakker, 177–204. Dordrecht: Kluwer.

Kremer, Richard L. 1992. "Building Institutes for Physiology in Prussia, 1836–1846: Contexts, Interests and Rhetoric," in Cunningham and Williams, *Laboratory Revolution*, 72–109.

Krüger, Lorenz, Lorraine Daston, and Michael Heidelberger, eds. 1987. *The Probabilistic Revolution*. Vol. 1, *Ideas in History.* Cambridge, Mass.: MIT Press.

Krüger, Lorenz, Gerd Gigerenzer, and Mary S. Morgan, eds. 1987. *The Probabilistic Revolution*. Vol. 2, *Ideas in the Sciences.* Cambridge, Mass.: MIT Press.

Kuhn, Thomas. [1961] 1977. "The Function of Measurement in Modern Physical Science." In idem *The Essential Tension: Selected Studies in Scientific Tradition and Change*, 178–224. Chicago: University of Chicago Press.

———. [1962] 1970. *The Structure of Scientific Revolutions.* 2d rev. ed. Chicago: University of Chicago Press.

———. 1972. "Scientific Growth: Reflections on Ben-David's 'Scientific Role." *Minerva* 10:166–78.

———. 1976. "Mathematical versus Experimental Traditions in the Development of Physical Science." *Journal of Interdisciplinary History* 7:1–31.

———. 1978. *Black-Body Theory and the Quantum Discontinuity, 1894–1912.* New York: Oxford University Press.

Kuklick, Henrika. 1991. *The Savage Within: The Social History of British Anthropology, 1885–1945.* Cambridge: Cambridge University Press.

Kümmel, Werner Friedrich. 2001. "'Dem Arzt nötig oder nützlich'? Legitimierungsstrategien der Medizingeschichte im 19. Jahrhundert," in Frewer and Roelcke, *Institutionalisierung*, 75–89.

Kuper, Adam. 1988. *The Invention of Primitive Society: Transformation of an Illusion.* London: Routledge.

Kupsch, Walter O., and William A. S. Sarjeant, eds. 1979. *History of Concepts in Precambrian Geology: Papers Presented at a Symposium Organized by the International Committee on the History of Geological Sciences, Held in Montreal, August 23–24, 1972, and Additional Papers.* Special Paper no. 19. Saskatoon: Geological Association of Canada.

Kurzer, Frederick. 2001. "Chemistry and Chemists at the London Institution, 1808–1912." *AS* 58:163–201.

La Berge, Ann. 1994. "Medical Microscopy in Paris, 1830–1855," in La Berge and Feingold, *French Medical Culture*, 296–326.

———. 1998. "Dichotomy or Integration? Medical Microscopy and the Paris Clinical Tradition." In *Constructing Paris Medicine*, edited by Caroline Hannaway and Ann La Berge, 274–311. Amsterdam: Rodopi.

———. 1999. "The History of Science and the History of Microscopy." *Perspectives on Science* 7:111–42.

La Berge, Ann, and Mordechai Feingold, eds. 1994. *French Medical Culture in the Nineteenth Century.* Amsterdam: Rodopi.

Labisch, Alfons. 2001. "Von Sprengels 'pragmatischer Medizingeschichte' zu Kochs 'psychischem Apriori': Geschichte *der* Medizin und Geschichte *in der* Medizin," in Frewer and Roelcke, *Institutionalisierung*, 235–54.

Lachmund, Jens. 1997. *Der abgehorchte Körper: Zur historischen Soziologie der medizinischen Untersuchung*. Opladen: Westdeutscher Verlag.

———. 1999. "Making Sense of Sound: Auscultation and Lung Sound Codification in Nineteenth-Century French and German Medicine." *Science, Technology, & Human Values* 24:419–50.

Lachmund, Jens, and Gunnar Stollberg. 1995. *Patientenwelten: Krankheit und Medizin vom späten 18. bis zum frühen 20. Jahrhundert im Spiegel von Autobiographien*. Opladen: Leske & Budrich.

Lacroix, Silvestre François. [1802] 1816. *An Elementary Treatise on the Differential and Integral Calculus*. Cambridge: Printed by J. Smith for J. Deighton and Sons.

Laennec, R. T. H. [1819] 1961. *A Treatise on the Diseases of the Chest*. New York: Hafner.

Laidler, Keith J. 1985. "Chemical Kinetics and the Origins of Physical Chemistry." *AHES* 32:47–75.

———. 1993. *The World of Physical Chemistry*. Oxford: Oxford University Press.

Laitko, Hubert, et al. 1987. *Wissenschaft in Berlin: Von den Anfängen bis zum Neubeginn nach 1945*. Berlin: Dietz.

Lamarck, Jean-Baptiste de. 1801. "Discours d'Ouverture [1800]," in *Système des animaux sans vertèbre*. Paris: Lamarck et Deterville.

———. 1802. *Hydrogéologie*. Paris: Chez l'auteur au Muséum d'histoire naturelle.

Lammel, Hans-Uwe. 1990. *Nosologische und therapeutische Konzeptionen in der romantischen Medizin*. Husum: Matthiesen.

Lankford, John. 1997. *American Astronomy: Community, Careers, and Power, 1859–1940*. Chicago: University of Chicago Press.

Lapo, Andrei V. 1987. *Traces of Bygone Biospheres*. Trans. V. A. Purto. Moscow: Mir.

Laqueur, Thomas. 1990. *Making Sex: Body and Gender from the Greeks to Freud*. Cambridge, Mass.: Harvard University Press.

Larson, James. 1994. *Interpreting Nature: The Science of Living Form from Linnaeus to Kant*. Baltimore: Johns Hopkins University Press.

Laszlo, Alejandra C. 1987. "Physiology of the Future: Institutional Styles at Columbia and Harvard," in Geison, *Physiology in the American Context*, 67–96.

Latour, Bruno. 1987. *Science in Action: How to Follow Scientists and Engineers through Society*. Milton Keynes: Open University Press; Cambridge, Mass.: Harvard University Press.

Laudan, Rachel. 1977. "Ideas and Organizations in British Geology: A Case Study of Institutional History." *Isis* 68:527–38.

———. 1982. "The Role of Methodology in Lyell's Geology." *SHPS* 13:215–50.

———. 1987. *From Mineralogy to Geology: The Foundations of a Science, 1650–1830*. Chicago: University of Chicago Press.

———. 1993. "Histories of the Sciences and Their Uses: A Review to 1913." *HS* 31:1–34.

———. 1995. "Natural Alliance or Forced Marriage? Changing Relations between the Histories of Science and Technology." *TC*, supp. to vol. 36, no. 2:S17–S28.

Laurent, Goulven. 1987. *Paléontologie et évolution en France 1800–1860: Une histoire des idées de Cuvier et Lamarck à Darwin*. Paris: Comité des Travaux Historiques et Scientifiques.

———. 2000. "Paléontologie(s) et évolution au début du XIXe siècle: Cuvier et Lamarck." *Asclepio* 52:133–212.

———. 2001. *La naissance du transformisme: Lamarck entre Linné et Darwin.* Paris: Vuibert / Adapt.

Laurent, John. 1984. "Science, Society and Politics in Late Nineteenth-Century England: A Further Look at Mechanics' Institutes." *SSS* 14:585–619.

Lavine, Shaughan. 1994. *Understanding the Infinite.* Cambridge, Mass.: Harvard University Press.

Lawrence, Christopher. 1985a. "Incommunicable Knowledge: Science, Technology and the Clinical Art in Britain, 1850–1914." *Journal of Contemporary History* 20:503–20.

———. 1985b. "Moderns and Ancients: The 'New Cardiology' in Britain, 1880–1930." In *The Emergence of Modern Cardiology,* edited by William F. Bynum, Christopher Lawrence, and Vivian Nutton, 1–43. *Medical History,* supp. 5. London: Wellcome Institute.

Lawrence, Christopher, and Steven Shapin, eds. 1998. *Science Incarnate: Historical Embodiments of Natural Knowledge.* Chicago: University of Chicago Press.

Lawrence, Philip. 1977. "Heaven and Earth: The Relation of the Nebular Hypothesis to Geology." In *Cosmology, History, and Theology,* edited by Wolfgang Yourgrau and Allen D. Breck, 253–81. New York: Plenum.

Layton, Edwin T., Jr. 1971. "Mirror-Image Twins: The Communities of Science and Technology in Nineteenth-Century America." *TC* 12:562–80.

———. 1974. "Technology as Knowledge." *TC* 15:31–41.

———. 1976. "American Ideologies of Science and Engineering." *TC* 17:688–701.

———. 1979. "Scientific Technology, 1845–1900: The Hydraulic Turbine and the Origins of American Industrial Research." *TC* 20:64–89.

Lazarsfeld, Paul. 1961. "Notes on the History of Quantification in Sociology: Trends, Sources, and Problems." In *Quantification: A History of the Meaning of Measurement in the Natural and Social Sciences,* 277–333. Indianapolis, Ind.: Bobbs-Merrill.

Le Grand, Homer E., ed. 1990. *Experimental Inquiries: Historical, Philosophical and Social Studies of Experimentation in Science.* Dordrecht: Kluwer Academic.

Lejeune, Dominique. 1993. *Les sociétés de géographie en France et l'expansion coloniale au XIXe siècle.* Paris: Albin Michel.

LeMahieu, Daniel. 1976. *The Mind of William Paley.* Lincoln: University of Nebraska Press.

Lemaine, Gerard, Roy Macleod, Michael Mulkay, and Peter Weingart, eds. 1976. *Perspectives on the Emergence of Scientific Disciplines.* The Hague: Mouton.

Lenoir, Timothy. [1982] 1989. *The Strategy of Life: Teleology and Mechanics in Nineteenth-Century German Biology.* Chicago: University of Chicago Press.

———. 1988. "Science for the Clinic: Science Policy and the Formation of Carl Ludwig's Institute in Leipzig," in Coleman and Holmes, *Investigative Enterprise,* 139–78.

———. 1997. *Instituting Science: The Cultural Production of Scientific Disciplines.* Stanford, Calif.: Stanford University Press.

———, ed. 1998. *Inscribing Science: Scientific Texts and the Materiality of Communication.* Stanford, Calif.: Stanford University Press.

Lepenies, Wolf. 1976. *Das Ende der Naturgeschichte: Wandel kultureller Selbstverständlichkeiten in den Wissenschaften des Lebens des 18. und 19. Jahrhunderts.* Munich: Hanser.

———. 1988. *Between Literature and Science: The Rise of Sociology.* Trans. R. J. Hollingdale. Cambridge: Cambridge University Press; Paris: Editions de la Maison des Sciences de l'Homme.

Lervig, Philip. 1972. "On the Structure of Carnot's Theory of Heat." *AHES* 9:222–39.

Lesch, John E. 1984. *Science and Medicine in France: The Emergence of Experimental Physiology, 1790–1855.* Cambridge, Mass.: Harvard University Press.

Lesky, Erna. 1965. *Die Wiener medizinische Schule im 19. Jahrhundert.* Graz: Böhlau.

Levere, Trevor. 1981. *Poetry Realized in Nature: Samuel Taylor Coleridge and Early Nineteenth-Century Science.* Cambridge: Cambridge University Press.

———. 1996. "Romanticism, Natural Philosophy, and the Sciences: A Review and Bibliographical Essay." *Perspectives on Science* 4:463–88.

———. 2001. *Transforming Matter: A History of Chemistry, from Alchemy to the Buckyball.* Baltimore: Johns Hopkins University Press.

Levine, Donald N. 1995. *Visions of the Sociological Tradition.* Chicago: University of Chicago Press.

Levine, George. 1988. *Darwin and the Novelists: Patterns of Science in Victorian Fiction.* Cambridge, Mass.: Harvard University Press.

Lewenstein Bruce W. 1989. "To Improve Our Knowledge in Nature and Arts: A History of Chemical Education in the United States." *Journal of Chemical Education* 66:37–44.

Lewis, Albert C. 1990. Review of Richards, *Mathematical Visions. Historia Mathematica* 17:272–78.

Lewis, Cherry L. E., and Simon J. Knell, eds. 2001. *The Age of the Earth: From 4004 BC to AD 2002.* Special Publication no. 190. London: Geological Society.

Leyden, Ernst von. 1903. "Die deutsche Klinik am Beginn des 20. Jahrhunderts." In *Die deutsche Klinik am Eingange des zwanzigsten Jahrhunderts in akademischen Vorlesungen,* vol. 1, *Allgemeine Pathologie und Therapie,* edited by Ernst von Leyden and Felix Klemperer, 1–18. Berlin: Urban & Schwarzenberg.

Li Yan, and Du Shiran. 1987. *Chinese Mathematics: A Concise History.* Trans. J. N. Crossley and A. W.-C. Lun. Oxford. Clarendon Press.

Liebersohn, Harry. 1998. *Aristocratic Encounters: European Travelers and American Indians.* New York: Cambridge University Press.

Lightman, Bernard. 1987. *The Origins of Agnosticism: Victorian Unbelief and the Limits of Knowledge.* Baltimore: Johns Hopkins University Press.

———. 1997. "The Voices of Nature: Popularizing Victorian Science." In *Victorian Science in Context,* edited by Bernard Lightman, 187–211. Chicago: University of Chicago Press.

Lightman, Bernard, and Richard Helmstadter, eds. 1990. *Victorian Faith in Crisis: Essays on Continuity and Change in Nineteenth-Century Religious Belief.* Stanford, Calif.: Stanford University Press.

Limoges, Camille. 1980. "The Development of the Muséum d'Histoire Naturelle of Paris, 1800–1914," in R. Fox and Weisz, *Organization of Science and Technology,* 211–40.

Lindberg, David C., and Ronald L. Numbers. 1986. "Beyond War and Peace: A Reappraisal of the Encounter between Christianity and Science." *Church History* 55:338–54.

———, eds. 1986. *God and Nature: Historical Essays on the Encounter between Christianity and Science.* Berkeley: University of California Press.

Lindberg, David C., and Robert S. Westman, eds. 1990. *Reappraisals of the Scientific Revolution.* Cambridge: Cambridge University Press.

Lindenfeld, David. 1997. *The Practical Imagination: The German Sciences of State in the Nineteenth Century.* Chicago: University of Chicago Press.

Livesay, Harold C. 1975. *Andrew Carnegie and the Rise of Big Business.* Boston: Little, Brown.

Livingstone, David N. 1987. *Darwin's Forgotten Defenders: The Encounter between Evangelical Theology and Evolutionary Thought.* Grand Rapids, Mich.: W. B. Eerdmans.

———. 1992. *Preadamite Theory and the Marriage of Science and Religion.* Philadelphia: American Philosophical Society.

———. 1993. *The Geographical Tradition: Episodes in the History of a Contested Enterprise.* Oxford: Blackwell.

Livingstone, David N., D. G. Hart, and Mark A. Noll, eds. 1998. *Evangelicals and Science in Historical Perspective.* New York: Oxford University Press.

Loetz, Franziska. 1994. " 'Medikalisierung' in Frankreich, Großbritannien und Deutschland, 1750–1850: Ansätze, Ergebnisse und Perspektiven der Forschung." In *Das europäische Gesundheitssystem: Gemeinsamkeiten und Unterschiede in historischer Perspektive,* edited by Wolfgang U. Eckart and Robert Jütte, 123–61. Stuttgart: Steiner.

Lohff, Brigitte. 1978. "Johannes Müllers Rezeption der Zellenlehre in seinem 'Handbuch der Physiologie des Menschen.' " *Medizinhistorisches Journal* 12:247–58.

———. 1990. *Die Suche nach der Wissenschaftlichkeit der Physiologie in der Zeit der Romantik.* Stuttgart: Gustav Fischer.

Lombardo-Radice, Lucio. 1972. "Dai 'Manoscritti Matematici' di K. Marx." *Critica Marxista-Quaderni* 6:273–77.

Longwell, Chester R., and Richard F. Flint. 1955. *Introduction to Physical Geology.* New York: John Wiley & Sons.

Lorey, Wilhelm. 1916. *Das Studium der Mathematik an den deutschen Universitäten seit Anfang des 19. Jahrhunderts.* Leipzig: Teubner.

Lovejoy, Arthur O. 1960. *The Great Chain of Being.* 1936. Reprint, New York: Harper and Row.

Lowenberg, Bert. 1935. "The Controversy over Evolution in New England, 1859–1873." *New England Quarterly* 8:232–57.

Lowie, Robert H. 1938. *The History of Ethnological Theory.* New York: Rinehard.

Lugg, Andrew. 1978. "Overdetermined Problems in Science." *SHPS* 9:1–18.

Lundgreen, Peter. 1979. "Natur- und Technikwissenschaften an deutschen Hochschulen, 1870–1970: Einige quantitative Entwicklungen." In *Wissenschaft und Gesellschaft: Beiträge zur Geschichte der Technischen Universität Berlin, 1879–1979,* 2 vols., edited by Reinhard Rürup, 1:209–30. Berlin: Springer.

———. 1980. "The Organization of Science and Technology in France: A German Perspective," in R. Fox and Weisz, *Organization of Science and Technology,* 311–32.

———. 1990. "Engineering Education in Europe and the U.S.A., 1750–1930: The Rise to Dominance of School Culture and the Engineering Professions." *AS* 47:33–75.

Lützen, Jesper. 1990. *Joseph Liouville, 1809–1882: Master of Pure and Applied Mathematics.* New York: Springer-Verlag.

Lyell, Charles. 1990. *Principles of Geology: First Edition* [facsimile]. Vol. 1. Chicago: University of Chicago Press.

Lyons, Henry. 1944. *The Royal Society, 1660–1940: A History of Its Administration, under Its Charters.* Cambridge: Cambridge University Press.

MacLeod, Roy M. 1969. "A Victorian Scientific Network: The X-Club." *Notes and Records of the Royal Society* 24:305–22.

———. 1970a. "Science and the Civil List." *Technology and Society* 6:47–55.

————. 1970b. "The X-Club: A Social Network of Science in Late-Victorian England." *Notes and Records of the Royal Society* 24:305–23.

————. 1971a. "Of Medals and Men: A Reward System in Victorian Science, 1826–1914." *Notes and Records of the Royal Society* 26:81–105.

————. 1971b. "The Royal Society and the Government Grant: Notes on the Administration of Scientific Research, 1849–1914." *Historical Journal* 14:323–58.

————. 1971c. "The Support of Victorian Science: The Endowment of Research Movement in Great Britain, 1868–1900." *Minerva* 9:197–230.

————. 1972. "Resources of Science in Victorian England: The Endowment of Science Movement, 1868–1900." In *Science and Society, 1600–1900*, edited by Peter Mathias, 111–66. Cambridge: Cambridge University Press.

————. 1976. "Science and the Treasury: Principles, Personalities and Policies, 1870–85," in G. Turner, *Patronage of Science*, 115–72.

————. 1983. "Whigs and Savants: Reflections on the Reform Movement in the Royal Society, 1830–48," in Inkster and Morrell, *Metropolis and Province*, 55–90.

————. 1993. "The Chemists Go to War: The Mobilization of Civilian Chemists and the British War Effort, 1914–1918." *AS* 50:455–81.

————. 1996. *Public Science and Public Policy in Victorian England.* Aldershot: Variorum; Brookfield, Vt.: Ashgate.

MacLeod, Roy M., and Peter Collins, eds. 1981. *The Parliament of Science: The British Association for the Advancement of Science 1831–1981.* Northwood: Science Reviews.

MacLeod, Roy M., and Russell Moseley. 1980. "The 'Naturals' and Victorian Cambridge: The Anatomy of an Elite, 1851–1914." *Oxford Review of Education* 6:177–95.

Macnair, Peter, and Frederick Mort, eds. 1908. *History of the Geological Society of Glasgow, 1858–1908, with Biographical Notices of Prominent Members.* Glasgow: Geological Society of Glasgow.

Magendie, François. 1816. *Précis élémentaire de la physiologie.* Vol. 1. Paris: Méquignon-Marvis.

Maienschein, Jane. 1988. "Whitman at Chicago: Establishing a Chicago Style of Biology?" in Rainger, Benson, and Maienschein, *American Development of Biology*, 151–82.

————. 1991. *Transforming Traditions in American Biology, 1880–1915.* Baltimore: Johns Hopkins University Press.

Maienschein, Jane, and Michael Ruse, eds. 1999. *Biology and the Foundation of Ethics.* Cambridge: Cambridge University Press.

Malet, Antoni. 1991. "The Ecole Normale and the Education of the Scientific Elite in Nineteenth-Century France, with a Study of the *Annales Scientifiques de L'ENS*." *Asclepio* 43:1:163–87.

Malley, Marjorie. 1976. From Hyperphosphoresence to Nuclear Decay: A History of the Early Years of Radioactivity (1896–1914). Ph.D. diss., University of California, Berkeley.

Maloney, John. 1985. *Marshall, Orthodoxy, and the Professionalization of Economics.* Cambridge: Cambridge University Press.

Manier, Edward. 1978. *The Young Darwin and His Cultural Circle: A Study of Influences Which Helped Shape the Language and Logic of the First Drafts of the Theory of Natural Selection.* Dordrecht: Reidel.

Mann, Gunter, and Franz Dumont, eds. 1985. *Samuel Thomas Soemmerring und die Gelehrten der Goethezeit.* Stuttgart: Gustav Fischer.

Manten, A. A. 1966. "Historical Foundations of Chemical Geology and Geochemistry." *Chemical Geology* 1:5–31.

Manuel, Frank, and Fritzie Manuel. 1979. *Utopian Thought in the Western World.* Cambridge, Mass.: Harvard University Press.

Marcuse, Herbert. 1941. *Reason and Revolution: Hegel and the Rise of Social Theory.* New York: Oxford University Press.

Martin, Gerald P. R. 1965. "Albert Oppel: Zum 100. Todestage des Begründers der zonalen Stratigraphie." *Jahresheft des Vereins für Vaterländische Naturkunde in Württemberg* 120:185–93.

Martin, Luther H., Huck Gutman, and Patrick H. Hutton, eds. 1988. *Technologies of the Self.* Amherst: University of Massachusetts Press.

Martzloff, Jean-Claude. 1997. *A History of Chinese Mathematics.* Trans. Stephen S. Wilson. Berlin: Springer-Verlag. Originally published as *Histoire des mathématiques chinoises* (Paris: Masson, 1987).

Marvin, Ursula B. 1986. "Meteorites, the Moon, and the History of Geology." *Journal of Geological Education* 34:140–65.

———. 1996. "Ernst Florens Friedrich Chladni (1756–1827) and the Origins of Modern Meteorite Research." *Meteoritics and Planetary Research* 31:545–88.

Marx, Karl. 1983. *Mathematical Manuscripts of Karl Marx.* Clapham: New Park. Originally published, in German and Russian translation, as *Matematicheskie Rukopisi,* ed. S. A. Yanovskaya (Moscow: Nauka, 1968); and as *Mathematische Manuskripte,* ed. Wolfgang Endemann (Kronberg: Scriptor Verlag, 1974).

Mason, Stephen F. 1962. *A History of the Sciences.* Rev. ed. New York: Macmillan.

Matthews, J. Rosser. 1995. *Quantification and the Quest for Medical Certainty.* Princeton, N.J.: Princeton University Press.

Mauersberger, Klaus. 1980. "Die Herausbildung der technischen Mechanik und ihr Anteil bei der Verwissenschaftlichung des Maschinenwesens." *Dresdner Beiträge zur Geschichte der Technikwissenschaften* 2:1–52.

———. 1998. "Die 'Maschinenelemente' Carl von Bachs: Standardwerk des Maschinenbaus." In *Carl Julius von Bach (1847–1931): Pionier-Gestalter-Forscher-Lehrer-Visionär,* edited by Friedrich Naumann, 155–67. Stuttgart: Wittwer.

Maulitz, Russell C. 1979. "Physiologist versus Bacteriologist: The Ideology of Science in Clinical Medicine," in Vogel and Rosenberg, *Therapeutic Revolution,* 91–108.

———. 1987. *Morbid Appearances: The Anatomy of Pathology in the Early Nineteenth Century.* Cambridge: Cambridge University Press.

Mauskopf, Seymour H. 1976. "Crystals and Compounds. Molecular Structure and Composition in Nineteenth-Century French Science." *Transactions of the American Philosophical Society* 66 pt. 3.

———. 1988. "Gunpowder and the Chemical Revolution." *Osiris,* 2d ser., 4:93–118.

———. 1995. "Lavoisier and the Improvement of Gunpowder Production." *RHS* 48:95–121.

———, ed. 1993. *Chemical Sciences in the Modern World.* Philadelphia: University of Pennsylvania Press.

May, Kenneth O. 1966. "Quantitative Growth of the Mathematical Literature." *Science* 154:1672–73.

———. 1968. "Growth and Quality of the Mathematical Literature." *Isis* 59:363–71.

———. 1973. *Bibliography and Research Manual of the History of Mathematics.* Toronto: University of Toronto Press.

Mayr, Ernst. 1982. *The Growth of Biological Thought: Diversity, Evolution, Inheritance*. Cambridge, Mass.: Harvard University Press.

———. 1990. "When Is Historiography Whiggish?" *Journal of the History of Ideas* 51:301–309.

Maz'ia, Vladimir, and Tatyana Shaposhnikova. 1998. *Jacques Hadamard: A Universal Mathematician*. Providence, R.I.: American Mathematical Society.

Mazzolini, Renato G. 1985. "Nationale Wissenschaftsakademien im Europa des 19. Jahrhunderts." In *Nationale Grenzen und internationaler Austausch: Studien zum Kultur- und Wissenschaftstransfer in Europa*, edited by Lothar Jordan and Berndt Kortländer, 245–60. Tübingen: Max Niemeyer.

McClellan, James E., III. 1985. *Science Reorganized: Scientific Societies in the Eighteenth Century*. New York: Columbia University Press.

McClelland, Charles E. 1980. *State, Society, and University in Germany, 1700–1914*. Cambridge: Cambridge University Press.

McCloskey, Donald. 1994. "1780–1860: A Survey," in Floud and McCloskey, *Economic History of Britain*, 242–70.

McConnell, Anita. 1986. *Geophysics and Geomagnetism: Catalogue of the Science Museum Collection*. London: Her Majesty's Stationery Office.

McCormmach, Russell. 1971. Editor's foreword. *HSPS* 3: ix–xxiv.

———. 1976. Editor's foreword. *HSPS* 7: xi–xxxv.

McCosh, F. W. J. 1984. *Boussingault: Chemist and Agriculturist*. Dordrecht: Reidel.

McEvoy, John G. 1978–79. "Joseph Priestley, 'Aerial Philosopher': Metaphysics and Methodology in Priestley's Chemical Thought from 1772 to 1781." *Ambix* 25:1–55, 93–116, 153–175; 26:16–38.

———. 1997. "Positivism, Whiggism, and the Chemical Revolution." *HS* 35:1–33.

McEvoy, John G., and J. E. McGuire. 1975. "God and Nature: Priestley's Way of Rational Dissent." *HSPS* 6:325–404.

McKeown, Thomas. 1979. *The Role of Medicine: Dream, Mirage or Nemesis*. Princeton, N.J.: Princeton University Press.

McMullin, Ernan. 1987. "Scientific Controversy and Its Termination." In *Scientific Controversies: Case Studies in the Resolution and Closure of Disputes in Science and Technology*, edited by H. Tristram Engelhardt and Arthur L. Caplan, 49–91. Cambridge: Cambridge University Press.

Mehrtens, Herbert. 1979. *Die Entstehung der Verbandstheorie*. Hildesheim: Gerstenberg Verlag.

———. 1990. *Moderne, Sprache, Mathematik: Eine Geschichte des Streits um die Grundlagen der Disziplin und des Subjekts formaler Systeme*. Frankfurt am Main: Suhrkamp.

Meinel, Christoph. 1978. *Die Chemie an der Universität Marburg seit Beginn des 19. Jahrhunderts: Ein Beitrag zu ihrer Entwicklung als Hochschulfach*. Marburg: Elwert.

———. 1983. "Theory or Practice? The Eighteenth-Century Debate on the Scientific Status of Chemistry." *Ambix* 30:121–32.

———. 1992. "August Wilhelm Hofmann: 'Reigning-Chemist-in-Chief.'" *Angewandte Chemie* (international edition in English) 31:1265–82.

———, ed. 1999. *Research Laboratories and the Teaching of Chemistry*. Canton, Mass.: Science History.

Meinel, Christoph, and H. Scholz, eds. 1992. *Die Allianz von Wissenschaft und Industrie: August Wilhelm Hofmann (1818–1892)—Zeit, Werk und Wirkung*. Weinheim: VCH.

Meleshchenko, Yuri S. 1968. "Kharakter i osobennocti nauchno-tekhnicheskoy revolyut-sii." *Voprosy Filosofii* 22:13–24.

Melhado, Evan. 1981. *Jacob Berzelius: The Emergence of His Chemical System*. Stockholm: Almquist and Wicksell; Madison: University of Wisconsin Press.

———. 1985. "Chemistry, Physics, and the Chemical Revolution." *Isis* 76:195–211.

———. 1990. "On the Historiography of Science: A Reply to Perrin." *Isis* 81:273–76.

———. 1996. "Scientific Biography and Scientific Revolution." *Isis* 87:688–94.

Ménard, Claude. 1978. *La formation d'une rationalité économique: A. A. Cournot*. Paris: Flammarion.

Mendelsohn, Everett. 1963. "Cell Theory and the Development of General Physiology." *Archives internationales d'histoire des sciences* 16:419–29.

———. 1964. "The Emergence of Science as a Profession in Nineteenth-Century Europe." In *The Management of Scientists*, edited by Karl Hill, 3–48. Boston: Beacon.

———. 1965. "Physical Models and Physiological Concepts: Explanation in Nineteenth-Century Biology." *BJHS* 2:201–19.

———. 1966. "The Context of Nineteenth-Century Science." In *The Golden Age of Science: Thirty Portraits of the Giants of Nineteenth-Century Science by Their Scientific Contemporaries*, edited by Bessie Zaban Jones, xiii–xxviii. New York: Simon and Schuster (with the Smithsonian Institution, Washington, D.C.).

Merrill, George P. 1924. *The First One Hundred Years of American Geology*. New Haven: Yale University Press; London: Oxford University Press.

Merriman, Mansfield, and Robert S. Woodward. 1896. *Higher Mathematics: A Textbook for Classical and Engineering Colleges*. New York: J. Wiley and Sons.

Merton, Robert K. 1970. *Science, Technology and Society in Seventeenth-Century England*. Reprint, New York: H. Fertig. Originally published as vol. 1 of *Osiris* (1938).

Merz, John Theodore. 1904–12. *A History of European Thought in the Nineteenth Century*. 4 vols. Edinburgh: Blackwood. 1965. Reprint, New York: Dover.

———. 1915. *Religion and Science: A Philosophical Essay*. Edinburgh: Blackwood.

Meschkowski, Herbert. 1967. *Probleme des Unendlichen: Werk und Leben Georg Cantors*. Braunschweig: Vieweg.

———. 1983. *Georg Cantor: Leben, Werk und Wirkung*. Mannheim: Bibliographisches Institut.

Meschkowski, Herbert, and Winfried Nilson, eds. 1991. *Georg Cantor: Briefe*. Berlin: Springer-Verlag.

Mespoulet, Martine. 2001. *Statistique et révolution en Russie: Un compromis impossible (1880–1930)*. Rennes: Presses Universitaires de Rennes.

Metzger, Hélène. 1987. *La méthode philosophique en histoire des sciences: Textes 1914–39*. Paris: Fayard.

Mierzecki, Roman. 1991. *Historical Development of Chemical Concepts*. Warsaw: PWN Polish Scientific, Kluwer Academic.

Miller, David P. 1983. "Between Hostile Camps: Sir Humphry Davy's Presidency of the Royal Society, 1820–1827." *BJHS* 16:1–47.

———. 2002. "The 'Sobel Effect'." *Metascience* 11:185–200.

Mirowski, Philip. 1989. *More Heat than Light: Economics as Social Physics, Physics as Nature's Economics*. Cambridge: Cambridge University Press.

Mohr, Paul. 1999. *A Bibliography of the Discovery of the Geology of the East African Rift Sys-

tem (1830–1950). Sydney: International Commission on the History of Geological Sciences (INHIGEO).

Mokyr, Joel. 2000. "Knowledge, Technology, and Economic Growth during the Industrial Revolution." In *Productivity, Technology and Economic Growth*, edited by Bart Van Ark, Simon K. Kuipers, and Gerard Kuper, 253–92. The Hague: Kluwer Academic.

Montgomery, James. 1840. *A Practical Detail of the Cotton Manufacture of the United States of America, Compared with That of Great Britain*. Glasgow: J. Niven. Also published as *Die Baumwollen-Manufaktur der Vereinigten Staaten von Nordamerika zusammengehalten mit der von Großbritannien*, trans. Friedrich Georg Wieck (Leipzig: Binder, 1841).

Montgomery, Scott L. 1996. "The Eye and the Rock: Art, Observation and the Naturalistic Drawing of Earth Strata." *ESH* 15:3–24.

Moore, James R. 1979. *The Post-Darwinian Controversies: A Study of the Protestant Struggle to Come to Terms with Darwin in Great Britain and America, 1870–1900*. Cambridge: Cambridge University Press.

———. 1989. "Of Love and Death: Why Darwin 'Gave Up Christianity,'" in idem, *History, Humanity, and Evolution: Essays for John C. Greene*, 195–229. Cambridge: Cambridge University Press.

———. 1994. *The Darwin Legend*. Grand Rapids, Mich.: Baker Book House.

Morawski, Jill, ed. 1988. *The Rise of Experimentation in American Psychology*. New Haven: Yale University Press.

Morello, Nicoletta, ed. 1998. *Volcanoes in History: Proceedings of the Twentieth INHIGEO [International Commission on the History of Geological Sciences] Symposium, Napoli–Eolie–Catania (Italy)*. Genoa: Brigati.

Morgan, Mary. 1990. *The History of Econometric Ideas*. Cambridge: Cambridge University Press.

———. 1993. "Competing Notions of 'Competition' in Late Nineteenth-Century Economics." *History of Political Economy* 25:563–604.

Morrell, J. B. 1971a. "Individualism and the Structure of British Science in 1830." *HSPS* 3:183–204.

———. 1971b. "The University of Edinburgh in the Late Eighteenth Century: Its Scientific Eminence and Academic Structure." *Isis* 62:158–71.

———. 1972a. "The Chemist Breeders: The Research Schools of Justus Liebig and Thomas Thomson." *Ambix* 19:1–43.

———. 1972b. "Science and Scottish University Reform: Edinburgh in 1826." *BJHS* 6:39–56.

———. 1973. "The Patronage of Mid-Victorian Science in the University of Edinburgh." *SSS* 3:353–88. Reprinted in G. Turner, *Patronage of Science*, 53–93.

———. 1976. "London Institutions and Lyell's Career, 1820–1841." *BJHS* 9:132–46.

———. 1985. "Bradford Science, 1800–1850." *BJHS* 18:1–23.

———. 1988. "Early Yorkshire Geological and Polytechnic Society." *AS* 45:153–67.

———. 1990. "Professionalisation," in Olby et al., *Companion*, 980–89.

———. 1992. "Research in Physics at the Clarendon Laboratory, Oxford, 1919–1939." *HSPS* 22:2:263–307.

Morrell, J. B., and Arnold Thackray. 1981. *Gentlemen of Science: Early Years of the British Association for the Advancement of Science*. Oxford: Clarendon Press.

Morus, Iwan, Simon Schaffer, and Jim Secord. 1992. "Scientific London." In *London:*

World City, 1800–1840, edited by Celina Fox, 129–42. New Haven: Yale University Press.

Moscucci, Ornella. 1990. *The Science of Woman.* Cambridge: Cambridge University Press.

Moseley, Russell. 1976. Science, Government, and Industrial Research: The Origins and Development of the National Physical Laboratory, 1900–1975. Ph.D. diss., University of Sussex.

————. 1978. "The Origins and Early Years of the National Physical Laboratory: A Chapter in the Pre-History of British Science Policy." *Minerva* 16:222–50.

Moulin, Anne Marie. 1994. "Bacteriological Research and Medical Practice in and out the Pastorian School," in La Berge and Feingold, *French Medical Culture,* 327–49.

Mowery, David, and Nathan Rosenberg. 1989. *Technology and the Pursuit of Economic Growth.* Cambridge: Cambridge University Press.

Müller, D. W., J. A. McKenzie, and H. Weissert, eds. 1991. *Controversies in Modern Geology: Evolution of Geological Theories in Sedimentology, Earth History and Tectonics.* London: Academic Press.

Müller-Sievers, Helmut. 1997. *Self-Generation: Biology, Philosophy, and Literature around 1800.* Stanford, Calif.: Stanford University Press.

Multhauf, Robert P., and Gregory Good. 1987. *A Brief History of Geomagnetism and a Catalog of the Collections of the National Museum of American History.* Washington, D.C.: Smithsonian Institution Press.

Murken, Axel Hinrich. 1988. *Vom Armenhospital zum Grossklinikum: Die Geschichte des Krankenhauses vom 18. Jahrhundert bis zur Gegenwart.* Cologne: DuMont.

Musson, A. E., and Eric Robinson. 1969. *Science and Technology in the Industrial Revolution.* Manchester: Manchester University Press. 1989. Reprint, New York: Gordon and Breach.

Nash, Leonard K. 1956. "The Origin of Dalton's Atomic Theory." *Isis* 47:101–16.

National Academy of Sciences. 1913. *A History of the National Academy of Sciences, 1863–1913.* Washington, D.C.: Lord Baltimore Press.

Navarro, Luis. 1998. "Gibbs, Einstein and the Foundations of Statistical Mechanics." *AHES* 53:147–80.

Nazzaro, Antonio. 1997. *Il Vesuvio: Storia e teorie vulcanologiche.* Naples: Liguori Editore.

Needham, Joseph. 1959. *Science and Civilisation in China.* Vol. 3, *Mathematics and the Sciences of the Heavens and the Earth.* Cambridge: Cambridge University Press.

————, ed. 1925. *Science, Religion and Reality.* New York: Macmillan.

Neuenschwander, Erwin. 1996. *Riemann's Einführung in die Funktionentheorie: Eine quellenkritische Edition seiner Vorlesungen mit einer Bibliographie zur Wirkungsgeschichte der Riemannschen Funktionentheorie.* Göttingen: Vandenhoeck & Ruprecht.

Newell, Virginia K., Joella H. Gipson, L. Waldo Rich, and Beauregard Stubblefield, eds. 1980. *Black Mathematicians and Their Works.* Ardmore, Pa.: Dorrance.

Nieto-Galàn, A. 1994. Ciència a Catalunya a l'inici del segle XIX: Teoria i aplicacions tècniques a l'escola de quimica de Barcelona sota la direccióde Francec carbonell i Bravo (1805–1822). Ph.D. diss., Universitat de Barcelona.

————. 2001. *Colouring Textiles: A History of Natural Dyestuffs in Industrial Europe.* Dordrecht: Kluwer Academic.

Nieuwenkamp, W. 1975. "Trends in Nineteenth-Century Petrology." *Janus* 62:235–69.

Nisbet, Robert A. 1966. *The Sociological Tradition.* New York: Basic Books.

Noble, David. 1992. *World without Women: The Christian Clerical Culture of Modern Science.* New York: Knopf.

Nordenskiöld, Erik. [1920–24] 1936. *History of Biology: A Survey.* New York: Tudor Publishing.

Nový, Luboš. 1973. *Origins of Modern Algebra.* Prague: Academia Publishing House.

Numbers, Ronald L. 1976. *Prophetess of Health: A Study of Ellen G. White.* New York: Harper & Row.

———. 1977. *Creation by Natural Law: Laplace's Nebular Hypothesis in American Thought.* Seattle: University of Washington Press.

———. 1985. "Science and Religion," in Kohlstedt and Rossiter, *Historical Writing,* 59–80.

———, ed. 1979. *The Education of American Physicians.* Berkeley: University of California Press.

Numbers, Ronald L., and Janet S. Numbers. 1989. "Science in the Old South: A Reappraisal," in Numbers and Savitt, *Science and Medicine,* 9–35.

Numbers, Ronald L., and Todd L. Savitt, eds. 1989. *Science and Medicine in the Old South.* Baton Rouge: Louisiana State University Press.

Numbers, Ronald L., and John Stenhouse, eds. 1999. *Disseminating Darwinism: The Role of Place, Race, Religion, and Gender.* Cambridge: Cambridge University Press.

Nye, Mary Jo. 1983. "Recent Sources and Problems in the History of French Science." *HSPS* 13:401–15.

———. 1984. "Scientific Decline: Is Quantitative Evaluation Enough?" *Isis* 75:697–708.

———. 1986. *Science in the Provinces: Scientific Communities and Provincial Leadership in France, 1860–1930.* Berkeley: University of California Press.

———. 1993a. *From Chemical Philosophy to Theoretical Chemistry: Dynamics of Matter and Dynamics of Disciplines, 1800–1950.* Berkeley: University of California Press.

———. 1993b. "National Styles? French and English Chemistry in the Nineteenth and Early Twentieth Centuries." *Osiris,* 2d ser., 8:30–49.

———. 1996. *Before Big Science: The Pursuit of Modern Chemistry and Physics, 1800–1940.* New York: Twayne.

Nyhart, Lynn K. 1987. "The Disciplinary Breakdown of German Morphology, 1870–1900." *Isis* 78:365–89.

———. 1995. *Biology Takes Form: Animal Morphology and the German Universities, 1800–1900.* Chicago: University of Chicago Press.

———. 1996. "Natural History and the 'New' Biology," in Jardine, Secord, and Spary, *Cultures of Natural History,* 426–43.

———. 1998. "Civic and Economic Zoology in Nineteenth-Century Germany: The 'Living Communities' of Karl Möbius." *Isis* 89:605–30.

Oberschall, Anthony. 1965. *Empirical Social Research in Germany, 1848–1914.* Paris: Mouton.

O'Brien, Patrick K., Trevor Griffiths, and Philip Hunt. 1996. "Technological Change during the First Industrial Revolution: The Paradigm Case of Textiles, 1688–1851." In *Technological Change: Methods and Themes in the History of Technology,* edited by Robert Fox, 155–176. Amsterdam: Harwood.

O'Brien, Patrick K., and Çaglar Keyder. 1999. *Economic Growth in Britain and France, 1780–1914.* London: Routledge.

O'Donnell, John M. 1979. "The Crisis of Experimentalism in the 1920s." *American Psychologist* 34:289–95.

O'Hara, Robert J. 1991. "Representations of the Natural System in the Nineteenth Century." *Biology & Philosophy* 6:255–74.

———. 1992. "Telling the Tree: Narrative Representations and the Study of Evolutionary History." *Biology & Philosophy* 7:135–60.

Olby, Robert. [1966] 1985. *Origins of Mendelism*. 2d ed. Chicago: University of Chicago Press.

Olby, Robert, G. N. Cantor, J. R. R. Christie, and M. J. S. Hodge, eds. 1990. *Companion to the History of Modern Science*. London: Routledge.

Oldroyd, David R. 1979. "Historicism and the Rise of Historical Geology." *HS* 17:191–213, 227–57.

———. 1990. *The Highlands Controversy: Constructing Geological Knowledge through Fieldwork in Nineteenth-Century Britain*. Chicago: University of Chicago Press.

———. 1991–94. "The Archaean Controversy in Britain." Pt. 1, "The Rocks of St David's"; pt. 2, "The Malverns and Shropshire"; pt. 3, "The Rocks of Anglesey and Caernarvonshire"; pt. 4 "Some General Theoretical and Social Issues." *AS* 48:407–52; 49:401–60; 50:523–84; 51:571–92.

———. 1996a. "Sir Archibald Geikie (1835–1924) and the 'Highlands Controversy': New Archival Sources for the History of British Geology in the Nineteenth Century." *ESH* 15:141–50.

———. 1996b. *Thinking about the Earth: A History of Ideas in Geology*. London: Athlone Press; Cambridge, Mass.: Harvard University Press.

———. 1999. "The Use of Non-Written Sources in the Study of the History of Geology: Pros and Cons in the Light of the Views of Collingwood and Foucault." *AS* 56:395–415.

———, ed. 2002. *The Earth Inside and Out: Some Major Contributions to Geology in the Twentieth Century*. Special Publication no. 192. London: Geological Society.

Oldroyd, David R., and Beryl Hamilton. 1997. "Geikie and Judd, and Controversies about the Igneous Rocks of the Scottish Hebrides: Theory, Practice, and Power in the Geological Community." *AS* 54:221–68.

Oldroyd, David R., and Charlotte Klonk. 1998. "Picturing the Phenomena." *Metascience* 7:117–31.

Oldroyd, David R., and Jing-Yi Yang. 1996. "On Being the First Western Geologist in China: The Work of Raphael Pumpelly." *AS* 53:107–36.

Olesko, Kathryn M. 1988. "Commentary: On Institutes, Investigations, and Scientific Training," in Coleman and Holmes, *Investigative Enterprise*, 295–332.

———. 1991. *Physics as a Calling: Discipline and Practice in the Königsberg Seminar for Physics*. Ithaca, N.Y.: Cornell University Press.

———. 1993. "Tacit Knowledge and School Formation," in Geison and Holmes, *Research Schools*, 16–29.

———. 1994. "Civic Culture and Calling in the Königsberg Period." In *Universalgenie Helmholtz: Rückblick nach 100 Jahren*, edited by Lorenz Krüger, 22–42. Berlin: Akademie Verlag.

———, ed. 1989. *Science in Germany: The Interaction of Institutional and Intellectual Issues*. *Osiris*, 2d ser., vol. 5.

Oleson, Alexandra, and Sanborn Brown, eds. 1976. *The Pursuit of Knowledge in the Early American Republic: American Scientific and Learned Societies from Colonial Times to the Civil War*. Baltimore: Johns Hopkins University Press.

Oleson, Alexandra, and John Voss, eds. 1979. *The Organization of Knowledge in Modern America, 1860–1920.* Baltimore: Johns Hopkins University Press.

Omalius d'Halloy, Jean-Baptiste-Julien d'. 1845–46. "Note sur la succession des êtres vivants." *Bulletin de la Société Géologique de France,* n.s., 3:531–35.

Orange, Derek. 1983. "Rational Dissent and Provincial Science: William Turner and the Newcastle Literary and Philosophical Society," in Inkster and Morrell, *Metropolis and Province,* 205–30.

Ore, Øystein. 1974. *Niels Henrik Abel: Mathematician Extraordinary.* New York: Chelsea.

Oreskes, Naomi. 1999. *The Rejection of Continental Drift: Theory and Method in American Earth Science.* New York: Oxford University Press.

Ospovat, Dov. 1981. *The Development of Darwin's Theory: Natural History, Natural Theology, and Natural Selection, 1838–1859.* Cambridge: Cambridge University Press.

Osterbrock, Donald E. 1997. *Yerkes Observatory, 1892–1950: The Birth, Near Death, and Resurrection of a Scientific Institution.* Chicago: University of Chicago Press.

———. 1999. "AAS Meetings before There Was an AAS: The Pre-History of the Society," in DeVorkin, *American Astronomical Society's First Century,* 3–19.

Österreichische Geologische Gesellschaft. 1981. *Eduard Suess, Forscher und Politiker: 20.8.1831–26.4.1914. Im Gedanken zum 150. Geburtstag.* Vienna: Österreichische Geologische Gesellschaft.

Outram, Dorinda. 1980. "Politics and Vocation: French Science, 1793–1830." *BJHS* 13:27–43.

———. 1984. *Georges Cuvier: Vocation, Science, and Authority in Post-Revolutionary France.* Dover, N.H: Manchester University Press.

———. 1990. "Science and Political Ideology, 1790–1848," in Olby et al., *Companion,* 1008–23.

———. 1996. "New Spaces in Natural History," in Jardine, Secord, and Spary, *Cultures of Natural History,* 249–65.

Owen, Richard. 1842. "Report on British Fossil Reptiles, Part II." *Report of the British Association for the Advancement of Science.* London: John Murray. Pp. 60–204.

Owens, Larry. 1985. "Pure and Sound Government: Laboratories, Playing Fields, and Gymnasia in the Nineteenth-Century Search for Order." *Isis* 76:182–94.

Palsky, Gilles. 1996. *Des chiffres et des cartes: Naissance et développement de la cartographie quantitative française au XIXe siècle.* Paris: Comité des travaux historiques et scientifiques.

Parshall, Karen Hunger. 1984. "Eliakim Hastings Moore and the Founding of a Mathematical Community in America, 1892–1902." *AS* 41:313–33.

———. 1988. "America's First School of Mathematical Research: James Joseph Sylvester at the Johns Hopkins University, 1876–1883." *AHES* 38:153–96.

———. 1998. *James Joseph Sylvester: Life and Work in Letters.* Oxford: Clarendon Press.

Parshall, Karen Hunger, and David E. Rowe. 1988. "American Mathematics Comes of Age: 1875–1900," in Duren, Askey, Edwards, and Merzbach, *Century of Mathematics,* 3:3–28.

———. 1994. *The Emergence of the American Mathematical Research Community, 1876–1900: James Joseph Sylvester, Felix Klein, and Eliakim Hastings Moore.* Providence, R.I.: American Mathematical Society.

Parsons, Talcott. [1937] 1947. *The Structure of Social Action.* 2d ed. New York: Free Press.

Partington, J. R. 1964. *A History of Chemistry.* 4 vols. London: MacMillan.

Pasveer, Bernike. 1992. Shadows of Knowledge: Making a Representing Practice in Medicine—X-Ray Pictures and Pulmonary Tuberculosis, 1895–1940. Ph.D. diss., University of Amsterdam.

Patriarca, Silvana. 1996. *Numbers and Nationhood: Writing Statistics in Nineteenth-Century Italy.* Cambridge: Cambridge University Press.

Paul, Erich Robert. 1992. *Science, Religion, and Mormon Cosmology.* Urbana: University of Illinois Press.

Paul, Harry. 1972. "The Issue of Decline in Nineteenth-Century French Science." *French Historical Studies* 7:416–50.

———. 1974a. "Religion and Darwinism: Varieties of Catholic Reaction." In *The Comparative Reception of Darwinism,* edited by Thomas Glick, 403–36. Austin: University of Texas Press.

———. 1974b. "La science française de la seconde partie du XIXe siècle vue par les auteurs anglais et américains." *RHS* 27:147–63.

———. 1979. *The Edge of Contingency: French Catholic Reaction to Scientific Change from Darwin to Duhem.* Gainesville: University Presses of Florida.

———. 1985. *From Knowledge to Power: The Rise of the Science Empire in France, 1860–1939.* Cambridge: Cambridge University Press.

Paulinyi, Akos. 1982. "Der Technologietransfer für die Metallbearbeitung und die preußische Gewerbeförderung (1820–1850)." *Schriften des Vereins für Sozialpolitik, Gesellschaft für Wirtschafts- und Sozialwissenschaften,* n.s., 125:99–141.

———. 1986. "Revolution and Technology." In *Revolution in History,* edited by Roy S. Porter and Mikulas Teich, 261–89. Cambridge: Cambridge University Press.

———. 1991. "Die Umwälzung der Technik in der Industriellen Revolution zwischen 1750 und 1840." In *Mechanisierung und Maschinisierung 1600 bis 1840,* edited by Akos Paulinyi and Ulrich Troitzsch, 271–495. Propyläen Technikgeschichte, no. 3. Berlin: Propyläen.

Pauly, Philipp J. 1984. "The Appearance of Academic Biology in Late Nineteenth-Century America." *Journal of the History of Biology* 17:369–97.

———. 1987. "General Physiology and the Discipline of Physiology, 1890–1935," in Geison, *Physiology in the American Context,* 195–207.

———. 1990. *Controlling Life: Jacques Loeb and the Engineering Ideal in Biology.* 1987. Oxford: Oxford University Press.

———. 2000. *Biologists and the Promise of American Life: From Meriwether Lewis to Alfred Kinsey.* Princeton, N.J.: Princeton University Press.

Pearson, Paul N. 1996. "Charles Darwin on the Origin and Diversity of Igneous Rocks." *ESH* 15:49–67.

Peel, J. D. Y. 1971. *Herbert Spencer: The Evolution of a Sociologist.* New York: Basic Books.

Penck, Albrecht, and Eduard Brückner. 1901–9. *Die Alpen im Eiszeitalter.* 3 vols. Leipzig: Tauschner.

Penn, Granville. 1840. *Conversations on Geology: Comprising a Familiar Explanation of the Huttonian and Wernerian Systems; the Mosaic Geology, as Explained by Mr. Granville Penn; and the Late Discoveries of Professor Buckland, Humboldt, Dr. Macculloch, and Others.* London: J. W. Southgate and Son.

Pera, Marcello. 1992. *The Ambiguous Frog: The Galvani-Volta Controversy on Animal Electricity.* Trans. Jonathan Mandelbaum. Princeton, N.J.: Princeton University Press.

Perrin, Carleton E. 1986. "Of Theory Shifts and Industrial Innovations: The Relations of J. A. Chaptal and A. L. Lavoisier." *AS* 43:511–43.

———. 1990. "Chemistry as a Peer of Physics: A Response to Donovan and Melhado on Lavoisier." *Isis* 81:259–70.

Persons, Stow. 1950. "Evolution and Theology in America." In *Evolutionary Thought in America,* edited by Stow Persons, 422–53. New Haven: Yale University Press.

Phillips, Esther, ed. 1987. *Studies in the History of Mathematics.* Studies in Mathematics, vol. 26. Providence, R.I.: Mathematical Association of America.

Phillips, Melba. 1990. "The American Physical Society: A Survey of Its First 50 Years." *American Journal of Physics* 58:3:219–30.

Pick, Daniel C. 1989. *Faces of Degeneration: Aspects of a European Disorder, c. 1848–1918.* Cambridge: Cambridge University Press.

Pickering, Andrew. 1984. *Constructing Quarks: A Sociological History of Particle Physics.* Chicago: University of Chicago Press.

Pickering, Mary. 1993. *Auguste Comte: An Intellectual Biography.* Vol. 1. Cambridge: Cambridge University Press.

Pickstone, John. 1994. "Objects and Objectives: Notes on the Material Cultures of Medicine." In *Technologies of Modern Medicine,* edited by Ghislaine Lawrence, 13–22. London: Science Museum.

———. 1999. "The Development and Present State of History of Medicine in Britain." *Dynamis* 19:457–86.

———. 2000. *Ways of Knowing: A New History of Science, Technology, and Medicine.* Manchester: Manchester University Press.

Pinch, Trevor. 1990. "The Sociology of the Scientific Community," in Olby et al., *Companion,* 87–99.

Pirsson, Louis V. 1920. "Biographical Memoir of James Dwight Dana, 1813–1895." *Biographical Memoirs, National Academy of Sciences,* no. 9, 41–95.

Pittenger, Mark. 1993. *American Socialists and Evolutionary Thought, 1870–1920.* Madison: University of Wisconsin Press.

Plekhanov, Georgii Valentinovich. 1934. *Essays in the History of Materialism.* Trans. by Ralph Fox. London: John Lane.

Pointon, Marcia. 1978. "Geology and Landscape Painting in Nineteenth-Century England," in Jordanova and Porter, *Images of the Earth,* 84–116.

Poirier, Jean-Pierre. 1993. *Antoine Lavoisier, 1743–1794.* Paris: Pygmalion. English trans.: *Lavoisier: Chemist, Biologist, Economist* (Philadelphia: University of Pennsylvania Press, 1996).

Polanyi, Michael. 1983. *The Tacit Dimension.* Gloucester, Mass.: Smith.

Polya, George. 1954. *Induction and Analysis.* Vol. 1 of *Mathematics and Plausible Reasoning.* Princeton, N.J.: Princeton University Press.

Popper, Karl. 1974. "Darwinism as a Metaphysical Research Programme." In *The Philosophy of Karl Popper,* edited by Paul Schilpp, 481–98. LaSalle, Ill.: Open Court.

Porter, Dorothy, and Roy Porter. 1989. *Patient's Progress: Doctors and Doctoring in Eighteenth-Century England.* Cambridge: Polity Press.

Porter, Roy S. 1973. "The Industrial Revolution and the Rise of the Science of Geology." In *Changing Perspectives in the History of Science,* edited by Mikulas Teich and Robert Young, 320–43. London: Heinemann.

————. 1976. "Charles Lyell and the Principles of the History of Geology." *BJHS* 9:91–103.

————. 1977. *The Making of Geology: Earth Science in Britain, 1660–1815.* Cambridge: Cambridge University Press.

————. 1980. "Science, Provincial Culture, and Public Opinion in Enlightenment England." *British Journal for Eighteenth-Century Studies* 3:20–46.

————. 1985. "The Patient's View. Doing Medical History from Below." *History and Society* 14:175–98.

————. 1997. *The Greatest Benefit to Mankind: A Medical History of Humanity from Antiquity to the Present.* London: HarperCollins.

Porter, Roy S., and Mikulas Teich, eds. 1988. *Romanticism in National Context.* Cambridge: Cambridge University Press.

Porter, Theodore M. 1986. *The Rise of Statistical Thinking, 1820–1900.* Princeton, N.J.: Princeton University Press.

————. 1987. "Lawless Society: Social Science and the Reinterpretation of Statistics in Germany, 1850–1880," in Krüger, Daston, and Heidelberger, *Probabilistic Revolution,* 1:351–75.

————. 1994. "Rigor and Practicality: Rival Ideals of Quantification in Nineteenth-Century Economics." In *Natural Images in Economic Thought: "Markets Read in Tooth and Claw,"* edited by Philip Mirowski, 128–70. New York: Cambridge University Press.

————. 1995. *Trust in Numbers: The Pursuit of Objectivity in Science and Public Life.* Princeton, N.J.: Princeton University Press.

Porter, Theodore M., and Dorothy Ross, eds. 2002. *The Cambridge History of Science.* Vol. 7, *Modern Social Sciences.* New York: Cambridge University Press.

Powell, John Wesley. 1875. *Exploration of the Colorado River of the West and Its Tributaries in 1869, 1870, 1871 and 1872.* Washington, D.C.: U.S. Government Printing Office.

Price, Derek de Solla. 1984. "Of Sealing Wax and String." *Natural History* 93:48–56.

Proctor, Robert. 1991. *Value-Free Science: Purity and Power in Modern Knowledge.* Cambridge, Mass.: Harvard University Press.

Psarros, Nikos. 1995. "The Constructive Approach to the Philosophy of Chemistry." *Epistemologia* 18:27–38.

Psarros, Nikos, K. Ruthenberg, and J. Schummer, eds. 1996. *Philosophie der Chemie.* Würzburg: Königshausen & Neumann.

Purkert, Walter, and Hans Joachim Ilgauds. 1985. *Georg Cantor.* Leipzig: B. G. Teubner.

————. 1987. *Georg Cantor, 1845–1918.* Basel: Birkhäuser.

Purkyně, Jan Evangelista. 1844. "Mikroskop: Anwendung und Gebrauch bei physiologischen Untersuchungen." In *Handwörterbuch der Physiologie mit Rücksicht auf physiologische Pathologie,* 4 vols., edited by Rudolph Wagner, 2:411–41. Braunschweig: Vieweg.

Purrington, Robert D. 1997. *Physics in the Nineteenth Century.* New Brunswick, N.J.: Rutgers University Press.

Pyatt, Edward. 1983. *The National Physical Laboratory: A History.* Bristol: Adam Hilger.

Pycior, Helena M. 1987. "Marie Curie's Anti-Natural Path: Time Only for Science and Family." In *Intimate Lives: Women in Science,* edited by Pnina G. Abir-Am and Dorinda Outram, 191–215. New Brunswick, N.J.: Rutgers University Press.

————. 1996. "Pierre Curie and 'His Eminent Collaborator Mme Curie.'" In *Creative*

Couples in Science, edited by Helena M. Pycior, Nancy G. Slack, and P. G. Abir–Am, 39–56. New Brunswick, N.J.: Rutgers University Press.

Pyenson, Lewis. 1977. " 'Who the Guys Were': Prosopography in the History of Science." *HS* 15:155–88.

Pyenson, Lewis, and Douglas Skopp. 1977. "Educating Physicists in Germany *circa* 1900." *SSS* 7:329–66.

Quinn, Susan. 1995. *Marie Curie: A Life.* New York: Simon and Schuster.

Rabinbach, Anson. 1990. *The Human Motor: Energy, Fatigue, and the Origins of Modernity.* New York: Basic Books. Reprint, Berkeley: University of California Press, 1992.

Rae, Ian D. 1995. "Chemical Organisations in Australia and New Zealand." *Ambix* 42: 28–49.

Rainger, Ronald. 1985. "Paleontology and Philosophy: A Critique." *Journal of the History of Biology* 18:267–87.

———. 1991. *An Agenda for Antiquity: Henry Fairfield Osborn and Vertebrate Paleontology at the American Museum of Natural History, 1890–1935.* Tuscaloosa: University of Alabama Press.

Rainger, Ronald, Keith R. Benson, and Jane Maienschein, eds. 1988. *The American Development of Biology.* Philadelphia: University of Philadelphia Press.

Ramberg, Peter. 2000. "The Death of Vitalism and the Birth of Organic Chemistry: Wöhler's Urea Synthesis and the Disciplinary Identity of Organic Chemistry." *Ambix* 47:170–95.

Ramsay, Matthew. 1984. "The Politics of Professional Monopoly in Nineteenth-Century Medicine: The French Model and Its Rivals," in Geison, *Professions and the French State,* 225–305.

Raulff, Ulrich. 1999. *Der unsichtbare Augenblick: Zeitkonzepte in der Geschichte.* Göttingen: Wallstein.

Ravikovich, A. I. 1969. *Development of the Main Theoretical Tendencies in Geology of the Nineteenth Century* (in Russian). Moscow: Academy of Sciences of the USSR.

Read, Herbert H. 1957. *The Granite Controversy: Geological Addresses Illustrating the Evolution of a Disputant.* New York: Interscience.

Redlich, Fritz. 1968. "Frühindustrielle Unternehmer und ihre Probleme im Lichte ihrer Selbstzeugnisse." In *Wirtschafts- und Sozialgeschichtliche Probleme der frühen Industrialisierung,* edited by Wolfram Fischer, 339–412. Berlin: Colloquium.

Redondi, Pietro. 1988. "Nation et entreprise: La Société d'encouragement pour l'industrie nationale, 1801–1815." *History and Technology* 5:193–222.

Rehbock, Philip. 1983. *The Philosophical Naturalists: Themes in Early Nineteenth-Century British Biology.* Madison: University of Wisconsin Press.

Reich, Leonard S. 1985. *The Making of American Industrial Research: Science and Business at GE and Bell, 1876–1926.* Cambridge: Cambridge University Press.

Reid, Constance. 1993. *The Search for E. T. Bell.* Washington, D.C.: Mathematical Association of America.

Reid, Robert. 1974. *Marie Curie.* New York: New American Library.

Reil, Johann Christian. [1807] 1910. "Entwurf zur Organisation einer wissenschaftlich-medizinischen Schule." In *Geschichte der Königlichen Friedrich-Wilhelms-Universität zu Berlin,* 4 vols. in 5., 4:50–67. Halle a d.S.: Buchhandlung des Waisenhauses.

Reill, Peter. 1994. "Science and the Construction of the Cultural Sciences in Late Enlightenment Germany: The Case of Wilhelm von Humboldt." *History and Theory* 33:345–66.

Reingold, Nathan. 1976. "Definitions and Speculations: The Professionalization of Science in America in the Nineteenth Century," in Oleson and Brown, *Pursuit of Knowledge*, 33–69. Reprinted in Reingold, *Science, American Style*, 24–53.

———. 1991. "Graduate School and Doctoral Degree: European Models and American Realities." In idem, *Science, American Style*, 171–89. New Brunswick, N.J.: Rutgers University Press.

Reingold, Nathan, and Arthur Molella, eds. 1976. "The Interaction of Science and Technology in the Industrial Age." *TC* 17:4.

Reinhardt, Carsten. 1997. *Forschung in der chemischen Industrie: Die Entwicklung synthetischer Farbstoffe bei BASF und Hoechst, 1863 bis 1914*. Freiberg: TU Bergakademie.

———. 1998. "An Instrument of Corporate Strategy: The Central Research Laboratory at BASF, 1868–1890," in Homburg et al., *Chemical Industry in Europe*, 239–60.

Reiser, Stanley J. 1978. *Medicine and the Reign of Technology*. Cambridge: Cambridge University Press.

Renn, Jurgen. 1997. "Einstein's Controversy with Drude and the Origin of Statistical Mechanics: A New Glimpse from the 'Love Letters.' " *AHES* 51:315–54.

Reuleaux, Franz. 1875. *Theoretische Kinematik, Grundzüge einer Theorie des Maschinenwesens*. Vol. 1. Braunschweig: Vieweg.

Revel, Jacques. 1991. "Knowledge of the Territory." *Science in Context* 4:133–62.

Rheinberger, Hans-Jörg. 1987. "Zum Organismusbild der Physiologie im 19. Jahrhundert: Johannes Müller, Ernst Brücke, Claude Bernard." *Medizinhistorisches Journal* 22:342–51.

Rheinberger, Hans-Jörg, and Michael Hagner, eds. 1993. *Die Experimentalisierung des Lebens: Experimentalsysteme in den biologischen Wissenschaften 1850/1950*. Berlin: Akademie Verlag.

Rice, Adrian C., and Robin J. Wilson. 1998. "From National to International Society: The London Mathematical Society, 1867–1900." *Historia Mathematica* 25:185–217.

Rice, Adrian C., Robin J. Wilson, and J. Helen Gardner. 1995. "From Student Club to National Society: The Founding of the London Mathematical Society in 1865." *Historia Mathematica* 22:402–21.

Richards, Joan. 1988. *Mathematical Visions: The Pursuit of Geometry in Victorian England*. Boston: Academic Press.

Richards, Robert J. 1987. *Darwin and the Emergence of Evolutionary Theories of Mind and Behavior*. Chicago: University of Chicago Press.

———. 1992. *The Meaning of Evolution: The Morphological Construction and the Ideological Deconstruction of Darwin's Theory*. Chicago: University of Chicago Press.

———. 1998. "Rhapsodies on a Cat-Piano, or Johann Christian Reil and the Foundations of Romantic Psychiatry." *Critical Inquiry* 24:700–36.

———. 1999a. "Darwin's Romantic Biology, the Foundation of His Evolutionary Ethics," in Maienschein and Ruse, *Biology and the Foundation of Ethics*, 113–53.

———. 1999b. "The Epistemology of Historical Interpretation: Progressivity and Recapitulation in Darwin's Theory." In *Epistemology and Biology*, edited by Richard Creath and Jane Maienschein, 35–60. Cambridge: Cambridge University Press.

———. 2000. "Kant and Blumenbach on the *Bildungstrieb*: A Historical Misunderstanding." *Studies in History and Philosophy of Biology and Biomedical Sciences* 31C:11–32.

———. 2002. *The Romantic Conception of Life: Science and Philosophy in the Age of Goethe.* Chicago: University of Chicago Press.

Richthofen, Ferdinand von. 1868. "Principles of the Natural System of Volcanic Rocks." *Memoirs Presented to the California Academy of Sciences,* no. 1, 1–98.

Riese, Walther. 1962. "The Impact of Romanticism on the Experimental Method." *Studies in Romanticism* 2:12–22.

Ringer, Fritz. 1969. *The Decline of the German Mandarins: The German Academic Community, 1890–1933.* Cambridge, Mass.: Harvard University Press.

———. 1979. *Education and Society in Modern Europe.* Bloomington: Indiana University Press.

Risse, Günter B. 1972. "Kant, Schelling and the Early Search for a Philosophical 'Science' of Medicine in Germany." *Journal of the History of Medicine and Allied Sciences* 27:145–58.

———. 1999. *Mending Bodies, Saving Souls: A History of Hospitals.* New York: Oxford University Press.

———. 2000. "Reflected Experience in Medicine, Science, and Technology: The Example of Hospital History." *Nachrichtenblatt der Deutschen Gesellschaft für Geschichte der Medizin, Naturwissenschaft und Technik* 50:200–15.

Roberts, Gerrylynn K. 1976. "The Establishment of the Royal College of Chemistry: An Investigation of the Social Context of Early-Victorian Chemistry." *HSPS* 7:437–86.

———. 1980. "The Liberally-Educated Chemist: Chemistry in the Cambridge Natural Science Tripos, 1857–1914." *HSPS* 11:157–83.

Roberts, Jon. 1988. *Darwinism and the Divine in America: Protestant Intellectuals and Organic Evolution, 1859–1900.* Madison: University of Wisconsin Press.

Roberts, Lissa. 1991. "A Word and the World: The Significance of Naming the Calorimeter." *Isis* 82:199 22.

Roberts, Windsor Hall. 1938. *The Reaction of American Protestant Churches to the Darwinian Philosophy, 1860–1900.* Chicago: University of Chicago Libraries.

Rocke, Alan. J. 1984. *Chemical Atomism in the Nineteenth Century: From Dalton to Cannizzaro.* Columbus: Ohio State University Press.

———. 1990. "Between Two Stools: Kopp, Kolbe, and the History of Chemistry." *Bulletin for the History of Chemistry* 7:19–24.

———. 1993a. "Group Research in German Chemistry: Kolbe's Marburg and Leipzig Institutes." *Osiris,* 2d ser., 8:53–79.

———. 1993b. *The Quiet Revolution: Hermann Kolbe and the Science of Organic Chemistry.* Berkeley: University of California Press.

———. 1994. "History of Science: Adolphe Wurtz and the Renovation of the Academic Professions in France." *Ambix* 14:20–32.

———. 2001. *Nationalizing Science: Adolphe Wurtz and the Battle for French Chemistry.* Cambridge, Mass.: MIT Press.

Roderick, Gordon W., and Michael D. Stephens. 1972. *Scientific and Technical Education in Nineteenth-Century England.* New York: Barnes and Noble.

———. 1974. "Scientific Studies and Scientific Manpower in the English Civic Universities, 1870–1914." *SSS* 4:41–63.

Rodgers, Daniel. 1998. *Atlantic Crossings: Social Politics in a Progressive Age.* Cambridge, Mass.: Harvard University Press.

Romo, José, and Manuel G. Doncel. 1994. "Faraday's Initial Mistake Concerning the Direction of Induced Currents, and the Manuscript of Series I of His *Researches.*" *AHES* 47:291–385.

Roose, Theodor Georg. 1797. *Grundzüge der Lehre von der Lebenskraft.* Braunschweig: Christian Friedrich Thomas.

Rose, E. P. F., and C. P. Nathanail, eds. 2000. *Geology and Warfare: Examples of the Influence of Terrain and Geologists on Military Operations.* London: Geological Society.

Rosenberg, Charles E. 1964. "On the Study of American Biology and Medicine: Some Justifications." *BHM* 38:364–76.

———. 1979. "Toward an Ecology of Knowledge: On Discipline, Context and History," in Oleson and Voss, *Organization of Knowledge,* 440–55.

———. 1987. *The Care of Strangers: The Rise of America's Hospital System.* New York: Basic Books.

———. 1992. *Explaining Epidemics and Other Studies in the History of Medicine.* Cambridge: Cambridge University Press.

Ross, Dorothy. 1991. *The Origins of American Social Science.* Cambridge: Cambridge University Press.

———, ed. 1994. *Modernist Impulses in the Human Sciences.* Baltimore: Johns Hopkins University Press.

Ross, Sydney. 1962. "Scientist: The Story of A Word." *AS* 18:65–85.

Rossiter, Margaret. 1982. *Women Scientists in America: Struggles and Strategies to 1940.* Baltimore: Johns Hopkins University Press.

Roth, Justus L. A. 1871. "Über die Lehre vom Metamorphismus und die Entstehung der Krystallinischen Schieffer." *Abhandlungen der Königlichen Akademie der Wissenschaften zu Berlin.* No vol. no., 151–232.

Rothenberg, Marc, and Thomas R. Williams. 1999. "Amateurs and the Society during the Formative Years," in DeVorkin, *American Astronomical Society's First Century,* 40–52.

Rothman, T. 1982. "Genius and Biographers: The Fictionalization of Evariste Galois." *American Mathematical Monthly* 89:84–106.

Rothschuh, Karl E. [1953] 1973. *History of Physiology.* Trans. Günter B. Risse. Huntington: Krieger.

———. 1976. "Die Bedeutung apparativer Hilfsmittel für die Entwicklung der biologischen Wissenschaften im 19. Jahrhundert." In *Naturwissenschaft, Technik und Wirtschaft im 19. Jahrhundert,* 2 pts., edited by Wilhelm Treue and Kurt Mauel, pt. 1, 161–81. Göttingen: Vandenhoeck & Ruprecht.

Rotman, Brian. 1998. "The Technology of Mathematical Persuasion," in Lenoir, *Inscribing Science,* 55–69.

Rousseau, G. S., and Roy Porter, eds. 1980. *The Ferment of Knowledge: Studies in the Historiography of Eighteenth-Century Science.* Cambridge: Cambridge University Press.

Rowe, David E. 1989. "Interview with Dirk Jan Struik." *Mathematical Intelligencer* 11: 14–26.

———. 1994. Review of Katz, *History of Mathematics. Isis* 85:125.

———. 1996. "New Trends and Old Images in the History of Mathematics." In *Vita Mathematica: Historical Research and Integration with Teaching,* edited by Ronald Calinger, 3–16. Washington, D.C.: Mathematical Association of America.

———. 1997. "Perspective on Hilbert." *Perspectives on Science: Historical, Philosophical, Social* 5:533–70.

Rowe, David E., and John McCleary, eds. 1989. *The History of Modern Mathematics.* Vol. 1, *Ideas and Their Reception.* Vol. 2, *Institutions and Applications.* Boston: Academic Press.

Rudwick, Martin J. S. 1963. "The Foundation of the Geological Society of London: Its Scheme for Co-operative Research and Its Struggle for Independence." *BJHS* 1:325–55.

———. 1969. "Lyell on Etna, and the Antiquity of the Earth." In *Toward a History of Geology,* edited by Cecil J. Schneer, 288–304. Cambridge, Mass.: MIT Press.

———. 1970. "The Strategy of Lyell's *Principles of Geology.*" *Isis* 61:4–33.

———. 1971. "Uniformity and Progression: Reflections on the Structure of Geological Theory in the Age of Lyell." In *Perspectives in the History of Science and Technology,* edited by Duane H. D. Roller, 209–27. Norman: Oklahoma University Press.

———. 1972. *The Meaning of Fossils: Episodes in the History of Palaeontology.* London: Mac-Donald; New York: American Elsevier.

———. 1974a. "Darwin and Glen Roy: A 'Great Failure' in Scientific Method?" *SHPS* 5:97–185.

———. 1974b. "Poulett Scrope on the Volcanoes of the Auvergne: Lyellian Time and Political Economy." *BJHS* 7:205–42.

———. 1975. "Caricature as a Source for the History of Science: De la Beche's Anti-Lyellian Sketches of 1831." *Isis* 66:534–60.

———. 1976. "The Emergence of a Visual Language for Geological Science." *HS* 14:149–95.

———. 1977. "Historical Analogies in the Work of Charles Lyell." *Janus* 64:89–107.

———. 1978. "Charles Lyell's Dream of a Statistical Palaeontology." *Palaeontology* 21:225–44.

———. 1979. "Transposed Concepts from the Human Sciences in the Early Work of Charles Lyell," in Jordanova and Porter, *Images of the Earth,* 77–91.

———. 1981. "Senses of the Natural World and Senses of God: Another Look at the Historical Relation of Science and Religion." In *The Sciences and Theology in the Twentieth Century,* edited by Arthur Peacock, 241–61. Notre Dame, Ind.: University of Notre Dame Press.

———. 1982. "Charles Darwin in London: The Integration of Public and Private Science." *Isis* 73:186–206.

———. 1985a. *The Great Devonian Controversy: The Shaping of Scientific Knowledge among Gentlemanly Specialists.* Chicago: University of Chicago Press.

———. 1985b. *The Meaning of Fossils: Episodes in the History of Palaeontology.* 2d ed. Chicago: University of Chicago Press.

———. 1990. Introduction, in Lyell, *Principles of Geology,* vii–lviii.

———. 1992. *Scenes from Deep Time: Early Pictorial Representations of the Prehistoric World.* Chicago: University of Chicago Press.

———. 1996. "Cuvier and Brongniart, William Smith, and the Reconstruction of Geohistory." *ESH* 15:25–36.

———. 1997a. *Georges Cuvier, Fossil Bones, and Geological Catastrophes: New Translations and Interpretations of the Primary Texts.* Chicago: University of Chicago Press.

———. 1997b. "Smith, Cuvier et Brongniart et la reconstitution de la géohistoire," in Gohau, *De la géologie,* 119–28.

———. 2001. Jean-André de Luc and Nature's Chronology," in Lewis and Knell, *Age of the Earth,* 51–60.

Rudwick, Martin J.S., and David R. Oldroyd. 1998. "Martin Rudwick, Historian of Geology: Interviewed by David Oldroyd." *Metascience* 7:167–80.

Rupke, Nicolaas. 1983a. *The Great Chain of History: William Buckland and the English School of Geology, 1814–1849.* Oxford: Clarendon Press; New York: Oxford University Press.

———. 1983b. "The Study of Fossils in the Romantic Philosophy of History and Nature." *HS* 21:389–413.

———. 1994a. "C. C. Gillispie's *Genesis and Geology.*" *Isis* 85:261–70.

———. 1994b. *Richard Owen: Victorian Naturalist.* New Haven: Yale University Press.

———. 1996. "Eurocentric Ideology of Continental Drift." *HS* 34:251–72.

———. 1997. "Oxford's Scientific Awakening and the Role of Geology," in Brock and Curthoys, *History of the University of Oxford,* 542–62.

———. 1998. " 'The End of History' in the Early Picturing of Geological Time." *HS* 36:61–90.

———, ed. 1987. *Vivisection in Historical Perspective.* London: Croom Helm.

Ruse, Michael. 1975. "The Relationship between Science and Religion in Britain, 1830–1870." *Church History* 44:505–22.

———. 1979. *The Darwinian Revolution: Science Red in Tooth and Claw.* Chicago: University of Chicago Press.

———. 1981. "Karl Popper and Evolutionary Biology." In idem, *Is Science Sexist?: And Other Problems in the Biomedical Sciences,* 65–84. Dordrecht: D. Reidel.

———. 1986. *Taking Darwin Seriously.* Oxford: Blackwell.

———. 1995. *Evolutionary Naturalism.* London: Routledge.

———. 1996. *Monad to Man: The Concept of Progress in Evolutionary Biology.* Cambridge, Mass.: Harvard University Press.

Ruske, Walter. 1967. *100 Jahre Deutsche Chemische Gesellschaft.* Weinheim / Bergstr.: Verlag Chemie.

Russell, Colin A. 1983. *Science and Social Change, 1700–1900.* Hong Kong: Macmillan.

———. 1985. *Recent Developments in the History of Chemistry.* London: Royal Society of Chemistry.

———. 1988. "Rude and Disgraceful Beginnings: A View of History of Chemistry from the Nineteenth Century." *BJHS* 21:273–94.

———. 1993. Review of Livingstone, *Preadamite Theory. AS* 51:554–55.

———. 1996. *Edward Frankland: Chemistry, Controversy and Conspiracy in Victorian England.* Cambridge: Cambridge University Press.

Russell, Colin A., N. G. Coley, and G. K. Roberts. 1977. *Chemistry as a Profession: The Origins and Rise of the Royal Institute of Chemistry.* Milton Keynes: Open University Press.

Russell, E. S. [1916] 1982. *Form and Function: A Contribution to the History of Animal Morphology.* Chicago: University of Chicago Press.

Sack, Dorothy. 1989. "Reconstructing the Chronology of Lake Bonneville: An Historical Review," in Tinkler, *History of Geomorphology,* 223–56.

Sadoun-Goupil, Michelle. 1977. *Le Chimiste C. L. Berthollet, 1748–1822: Sa vie, son oeuvre.* Paris: Vrin.

Samelson, Franz J. 1979. "Putting Psychology on the Map: Ideology and Intelligence Testing." In *Psychology in Social Context,* edited by Allan Buss, 103–68. New York: Irvington.

Samuel, Raphael. 1977. "The Workshop of the World." *History Workshop* 3:6–72.

Sanderson, Michael. 1972. *The Universities and British Industry, 1850–1970*. London: Routledge & Kegan.

Sandmann, Jürgen. 1991. *Der Bruch mit der humanitären Tradition: Die Biologisierung der Ethik bei Ernst Haeckel und anderen Darwinisten seiner Zeit*. Stuttgart: Gustav Fischer Verlag.

Sarasin, Philipp. 2001. *Reizbare Maschinen: Eine Geschichte des Körpers, 1765–1914*. Frankfurt am Main: Suhrkamp.

Sarasin, Philipp, and Jakob Tanner, eds. 1998. *Physiologie und industrielle Gesellschaft: Studien zur Verwissenschaftlichung des Körpers im 19. und 20. Jahrhundert*. Frankfurt am Main: Suhrkamp.

Sarjeant, William A. S. 1980. *Geologists and the History of Geology: An International Bibliography from the Origins to 1978*. 5 vols. New York: Arno Press.

———. 1987. *Geologists and the History of Geology: An International Bibliography from the Origins to 1978. Supplement 1979–1984*. 2 vols. Malabar, Fla.: R. E. Krieger.

———. 1996. *Geologists and the History of Geology: An International Bibliography from the Origins to 1978. Supplement 2, 1985–1993 and Additions*. 3 vols. Malabar, Fla.: R. E. Krieger.

Sarton, George. 1936. *The Study of the History of Mathematics*. Cambridge, Mass.: Harvard University Press.

———. 1952. "Auguste Comte, Historian of Science." *Osiris* 10:358–57.

Savoye, Antoine. 1994. *Les débuts de la sociologie empirique*. Paris: Méridiens Klincksieck.

Schabas, Margaret. 1991. *A World Ruled by Number: William Stanley Jevons and the Rise of Mathematical Economics*. Princeton, N.J.: Princeton University Press.

Schaffer, Simon. 1986. "Scientific Discoveries and the End of Natural Philosophy." *SSS* 16:387–420.

———. 1988. "Astronomers Mark Time: Discipline and the Personal Equation." *Science in Context* 2:115–45.

———. 1989. "Glass Works: Newton's Prisms and the Uses of Experiment." In *The Uses of Experiment: Studies in the Natural Sciences*, edited by David Gooding, Trevor Pinch, and Simon Schaffer, 67–104. Cambridge: Cambridge University Press.

———. 1990. "Genius in Romantic Natural Philosophy," in Cunningham and Jardine, *Romanticism and the Sciences*, 82–98.

———. 1992. "Late Victorian Metrology and Its Instrumentation: A Manufactory of Ohms." In *Invisible Connections: Instruments, Institutions, and Science*, edited by Robert Bud and Susan E. Cozzens, 23–58. Bellingham, Wash.: SPIE Optical Engineering Press.

———. 1994. "Machine Philosophy: Demonstration Devices in Georgian Mechanics." *Osiris*, 2d ser., 9:157–82.

———. 1998. "Physics Laboratories and the Victorian Country House," in C. Smith and Agar, *Making Space for Science*, 149–80.

———. 1999. "OK Computer," in Hagner, *Ecce Cortex*, 254–85.

Scharlau, Winfried. 1981. "The Origins of Pure Mathematics," in Jahnke and Otte, *Epistemological and Social Problems*, 331–47.

Scheidecker, Myriam. 1997a. "Baudrimont (1806–1880): Les liens entre sa chimie et sa philosophie." *Archives internationales d'histoire des sciences* 47:26–56.

———. 1997b. "L'hypothèse d'Avogadro (1811) et d'Ampère (1814): La distinction atome / molécule et la théorie chimique." *RHS* 50:158–94.

Scheidecker, Myriam, and Robert Locqueneux. 1995. "La théorie mathématique de la combinaison chimique d'Ampère." *RHS* 48:309–52.

Schelling, Friedrich Wilhelm Joseph. [1797] 1927–28. *Ideen zu einer Philosophie der Natur*, in *Schellings Werke*, 1:77–350.

———. [1798] 1857. *Von der Weltseele, eine Hypothese der höheren Physik zur Erklärung des allgemeinen Organismus*, in Schelling, *Sämmtliche Werke*, 1:2:345–583.

———. [1800] 1927–28. *System der transscendentalen Philosophie*, in *Schellings Werke*, 2:327–734.

———. 1805. "Vorrede." *Jahrbücher der Medicin als Wissenschaft* 1: v–xx.

———. 1857. *Sämmtliche Werke*. Ed. K. F. A. Schelling. Pt. 1. 10 vols. Stuttgart: Cotta.

———. 1927–28. *Schellings Werke*. Edited by Manfred Schröter. 6 vols. Munich: Beck.

Schipperges, Heinrich, ed. 1968. *Die Versammlung Deutscher Naturforscher und Ärzte im 19. Jahrhundert*. Stuttgart: A. W. Gentner.

Schleiden, Matthias Jakob. [1842] 1849. "On the Use of the Microscope," in Schleiden, *Principles of Scientific Botany*, 575–91.

———. 1849. *Principles of Scientific Botany; or, Botany as an Inductive Science*. Trans. Edwin Lankester. London: Longman.

Schlich, Thomas, and Christoph Gradmann, eds. 1999. *Strategien der Kausalität: Konzepte der Krankheitsverursachung im 19. und 20. Jahrhundert*. Pfaffenweiler: Centaurus.

Schmid, Günther. 1935. "Über die Herkunft der Ausdrücke Morphologie und Biologie." *Nova Acta Leopoldina*, n.s., 2:597–620.

Schmiedebach, Heinz-Peter. 1993. "Pathologie bei Virchow und Traube: Experimentalstrategien in unterschiedlichem Kontext," in Rheinberger and Hagner, *Die Experimentalisierung des Lebens*, 116–34.

Schneer, Cecil J. 1978. "The Great Taconic Controversy." *Isis* 69:439–50.

Schnitter, Claude. 1996. "Le développement du Muséum national d'histoire naturelle de Paris au cours de la seconde moitié du XIXe siècle; 'se transformer ou perir.'" *RHS* 49:1:53–97.

Schofield, Robert E. 1963a. "Histories of Scientific Societies: Needs and Opportunities for Research." *HS* 2:70–83.

———. 1963b. *The Lunar Society of Birmingham: A Social History of Provincial Science and Industry in Eighteenth-Century England*. Oxford: Clarendon Press.

Scholl, Lars U. 1975. *Ingenieure in der Frühindustrialisierung: Staatliche und private Techniker im Königreich Hannover und an der Ruhr (1815–1873)*. Göttingen: Vandenhoeck & Ruprecht.

Scholz, Erhard. 1989. *Symmetrie, Gruppe, Dualität: Beziehungen zwischen theoretischer Mathematik und Anwendungen in Kristallographie und Baustatik des 19. Jahrhunderts*. Basel: Birkhäuser Verlag.

———. 1990. *Geschichte der Algebra: Eine Einführung*. Mannheim: B. I.-Wissenschafts Verlag.

Schönbein, Christian F. 1840. "On the Causes of the Change of Colour Which Takes Place in Certain Substances under the Influence of Heat." *Annals of Electricity, Magnetism & Chemistry* 5:224–36. Also published in *Annalen der Physik und Chemie* (1838).

Schorske, Carl. 1955. *German Social Democracy, 1905–1917: The Development of the Great Schism*. Cambridge, Mass.: Harvard University Press.

———. 1980. *Fin-de-Siècle Vienna: Politics and Culture*. New York: Knopf.

Schreier, Wolfgang, and Martin Franke (with Annett Fiedler). 1995. "Geschichte der Physikalischen Gesellschaft zu Berlin 1845–1900." *Physikalische Blätter* 51, no. 1:F9–F59.

Schroeder-Gudehus, Brigitte. 1966. "Caractéristiques des relations scientifiques internationales, 1870–1914." *Journal of World History* 10:161–77.

———. 1978. "Tendances de centralisation dans l'organisation de la coopération scientifique internationale," in Forbes, *Human Implications*, 150–62.

———. 1990. "Nationalism and Internationalism," in Olby et al., *Companion*, 909–17.

Schubring, Gert. 1981. "The Conception of Pure Mathematics as an Instrument in the Professionalization of Mathematics," in Bos, Mehrtens, and Schneider, *Social History*, 111–34.

———. 1985a. "Die Entwicklung des Mathematischen Seminars der Universität Bonn, 1864–1921." *Jahresberichte der Deutschen Mathematiker-Vereinigung* 87:139–63.

———. 1985b. "Das mathematische Seminar der Universität Münster, 1831 / 75 bis 1951." *Sudhoffs Archiv* 69:154–91.

———. 1989. "The Rise and Decline of the Bonn Natural Sciences Seminar," in Olesko, *Science in Germany*, 57–93.

Schultz, Susan. 1983. "The Debate over Multiple Glaciation in the United States: T. C. Chamberlin and G. F. Wright, 1889–1894." *ESH* 2:122–29.

Schummer, Joachim. 1998. "Physical Chemistry: Neither Fish nor Fowl?" in Janich and Psarros, *Autonomy of Chemistry*, 135–48.

Schumpeter, Joseph A. 1954. *A History of Economic Analysis*. New York: Oxford University Press.

Schuster, John A., and Graeme Wachirs. 1990. "Natural Philosophy, Experiment and Discourse: Beyond the Kuhn / Bachelard Problematic," in Le Grand, *Experimental Inquiries*, 1–48.

Schütt, Hans-Werner. 1996. *Eilhard Mitscherlich: Prince of Prussian Chemistry*. Philadelphia: Chemical Heritage Foundation.

"Science and Engineering." 1961. Special issue of *Technology and Culture*. *TC* 2.

Scranton, Philip. 1998. *Endless Novelty: Specialty Production and American Industrialization, 1865–1925*. Princeton, N.J.: Princeton University Press.

Secord, James A. 1986. *Controversy in Victorian Geology: The Cambrian–Silurian Dispute*. Princeton, N.J.: Princeton University Press.

———. 2001. *Victorian Sensation: The Extraordinary Publication, Reception, and Secret Authorship of Vestiges of the Natural History of Creation*. Chicago: University of Chicago Press.

Sedgwick, Adam. 1835. "Remarks on the Structure of Large Mineral Masses, and Especially on the Chemical Changes Produced in the Aggregation of Stratified Rocks during Different Periods after Their Deposition." *Transactions of the Geological Society of London*, 2d ser., 3:461–86 and plate.

Seidelman, Raymond. 1985. *Disenchanted Realists: Political Science and the American Crisis*. Albany: State University of New York Press.

Şengör, A. M. Celâl. 1982a. "Classical Theories of Orogenesis." In *Orogeny*, edited by Akiho Miyashiro, Keiiti Aki, and A. M. Celâl Şengör, 1–48. Chichester: John Wiley & Sons.

———. 1982b. "Eduard Suess' Relations to the Pre-1950 Schools of Thought in Global Tectonics." *Geologische Rundschau* 71:381–420.

———. 1991. "Timing of Orogenic Events: A Persistent Geological Controversy," in Müller, McKenzie, and Weissert, *Controversies in Modern Geology*, 405–73.

———. 1999. "Continental Interiors and Cratons: Any Relation?" *Tectonophysics* 305:1–42.

———. 2001. *Is the Present the Key to the Past or the Past the Key to the Present? James Hutton and Adam Smith Versus Abraham Gottlob Werner and Karl Marx in Interpreting History.* Special Paper no. 355. Boulder: Geological Society of America.

Servos, John W. 1980. "The Industrial Relations of Science: Chemistry at MIT." *Isis* 71:531–49.

———. 1990. *Physical Chemistry from Ostwald to Pauling: The Making of a Science in America.* Princeton, N.J.: Princeton University Press.

———. 1993. "Research Schools and Their Histories," in Geison and Holmes, *Research Schools*, 3–15.

Shakow, David, and David Rapoport. 1964. *The Influence of Freud on American Psychology.* New York: International Universities Press.

Shapin, Steven. 1972. "The Pottery Philosophical Society, 1819–35: An Examination of the Cultural Uses of Provincial Science." *SSS* 2:311–36.

———. 1974. "Property, Patronage, and the Politics of Science: The Founding of the Royal Society of Edinburgh." *BJHS* 7:1–41.

———. 1975. "Phrenological Knowledge and the Social Structure of Early Nineteenth-Century Edinburgh." *AS* 32:219–43.

———. 1990. "Science and the Public," in Olby et al., *Companion*, 990–1007.

———. 1994. *A Social History of Truth: Civility and Science in Seventeenth-Century England.* Chicago: University of Chicago Press.

Shapin, Steven, and Barry Barnes. 1977. "Science, Nature, and Control: Interpreting Mechanics' Institutes." *SSS* 7:31–74.

Shapin, Steven, and Simon Schaffer. 1985. *Leviathan and the Air-Pump: Hobbes, Boyle, and the Experimental Life.* Princeton, N.J.: Princeton University Press.

Shapin, Steven, and Arnold Thackray. 1974. "Prosopography as a Research Tool in History of Science: The British Scientific Community, 1700–1900." *HS* 12:1–28.

Shapiro, Alan. 1996. "The Gradual Acceptance of Newton's Theory of Light and Color, 1672–1727." *Perspectives on Science* 4:59–140.

Sharlin, Harold I. 1967. *The Convergent Century: The Unification of Science in the Nineteenth Century.* London: Abelard-Schuman.

Shils, Edward. 1978. "The Order of Learning in the United States from 1865 to 1920: The Ascendancy of the Universities." *Minerva* 16:159–95. Reprinted in Oleson and Voss, *Organization of Knowledge*, 19–47.

Shinn, Terry. 1979. "The French Science Faculty System, 1808–1914: Institutional Change and Research Potential in Mathematics and the Physical Sciences." *HSPS* 10:271–332.

———. 1980a. "From 'Corps' to 'Profession': The Emergence and Definition of Industrial Engineering in Modern France," in R. Fox and Weisz, *Organization of Science and Technology*, 183–208.

———. 1980b. *Savoir scientifique et pouvoir sociale: l'Ecole polytechnique 1794–1914.* Paris: Presses de la Fondation nationale des Sciences politiques.

———. 1992a. "Science, Tocqueville, and the State: The Organization of Knowledge in Modern France." *Social Research* 59:3:533–66.

———. 1992b. "Specialized Institutions: *Grandes Ecoles.*" In *The Encyclopedia of Higher Education*, 4 vols., edited by Burton Clark and Guy Neave. Vol. 2, *Analytical Perspectives*, 1225–29. Oxford: Pergamon.

Shipley, Brian C. 2001. " 'Had Lord Kelvin a Right?': John Perry, Natural Selection and the Age of the Earth, 1894–1895," in Lewis and Knell, *Age of the Earth*, 91–105.

Shorter, Edward. 1993. "The History of the Doctor-Patient Relationship," in Bynum and Porter, *Companion Encyclopedia*, 783–800.

Shortland, Michael E. 1996. *Hugh Miller and the Controversies of Victorian Science.* Oxford: Clarendon Press.

———, ed. 1995. *Hugh Miller's Memoir: From Stonemason to Geologist.* Edinburgh: Edinburgh University Press.

Shortland, Michael E., and Richard Yeo. 1996. Introduction, in Shortland and Yeo, *Telling Lives*, 1–44.

———, eds. 1996. *Telling Lives in Science: Essays on Scientific Biography.* Cambridge: Cambridge University Press.

Showalter, Elaine. 1987. *The Female Malady: Women, Madness and Culture, 1830–1980.* London: Virago.

Sibum, Heinz Otto. 1995. "Reworking the Mechanical Value of Heat: Instruments of Precision and Gestures of Accuracy in Early Victorian England." *SHPS* 26:73–106.

Siegel, Daniel M. 1986. "The Origin of Displacement Current." *HSPS* 17:99–146.

———. 1991. *Innovation in Maxwell's Electromagnetic Theory: Molecular Vortices, Displacement Current, and Light.* Cambridge: Cambridge University Press.

Sigurdsson, Haraldur. 1999. *Melting the Earth: The History of Ideas on Volcanic Eruptions.* New York: Oxford University Press.

Sigurdsson, Skuli. 1992. Review of Mehrtens, *Moderne, Sprache, Mathematik. Isis* 83:156–58.

Silliman, Robert H. 1974. "Fresnel and the Emergence of Physics as a Discipline." *HSPS* 4:137–62.

Silverberg, Helene, ed. 1998. *Gender and American Social Science.* Princeton, N.J.: Princeton University Press.

Simpson, James. 1971. *Landmarks in the Struggle between Science and Religion.* 1926. Reprint, London: Kennikat Press.

Singer, Charles. 1925. "Historical Relations of Religion and Science," in Needham, *Science, Religion and Reality*, 85–148.

Skinner, Quentin. 1969. "Meaning and Understanding in the History of Ideas." *History and Theory* 8:3–53.

Skolnick, Herman, and Kenneth M. Reese, eds. 1976. *A Century of Chemistry: The Role of Chemists and the American Chemical Society.* Washington, D.C.: American Chemical Society.

Smith, Crosbie. 1976a. " 'Mechanical Philosophy' and the Emergence of Physics in Britain, 1800–1850." *AS* 33:3–29.

———. 1976b. "Natural Philosophy and Thermodynamics: William Thomson and the 'Dynamical Theory of Heat.' " *BJHS* 9:293–320.

———. 1985. "Geologists and Mathematicians: The Rise of Physical Geology," in Harman, *Wranglers and Physicists*, 49–83.

———. 1998a. " 'Nowhere but in a Great Town': William Thomson's Spiral of Classroom Credibility," in C. Smith and Agar, *Making Space for Science*, 118–46.

———. 1998b. *The Science of Energy: A Cultural History of Energy Physics in Victorian Britain.* Chicago: University of Chicago Press.

Smith, Crosbie, and Jon Agar, eds. 1998. *Making Space for Science: Territorial Themes in the Shaping of Knowledge.* London: Macmillan.

Smith, Crosbie, and M. Norton Wise. 1989. *Energy and Empire: A Biographical Study of Lord Kelvin.* Cambridge: Cambridge University Press.

Smith, David Eugene. 1896. "History of Modern Mathematics," in Merriman and Woodward, *Higher Mathematics,* 508–70.

Smith, J. G. 1979. *The Origins and Early Development of the Heavy Chemical Industry in France.* Oxford: Clarendon Press.

Smith, Roger. 1997. *The Fontana History of the Human Sciences.* London: Fontana. Also published as *The Norton History of the Human Sciences* (New York: Norton, 1997).

Sokal, Alan. 1996a. "A Physicist Experiments with Cultural Studies." *Lingua Franca,* May / June, 62–4.

———. 1996b. "Transgressing the Boundaries: Toward a Transformative Hermeneutics of Quantum Gravity." *Social Text* 46 / 47:217–52.

Sokal, Michael M., ed. 1987. *Psychological Testing and American Society.* New Brunswick, N.J.: Rutgers University Press.

Sonnemann, Rolf. 1980. Foreword. *Dresdner Beiträge zur Geschichte der Technikwissenschaften* 1:4–9.

Sonnemann, Rolf, and Klaus Krug, eds. 1987. *Technology and Technical Sciences in History / Technik und Technikwissenschaften in der Geschichte: Proceedings of the ICOHTEC-Symposium, Dresden, 25.–29. August 1986.* Berlin: Deutscher Verlag der Wissenschaften.

Sorby, Henry C. 1857. "On the Microscopical Structure of Crystals, Indicating the Origin of Minerals and Rocks." *Quarterly Journal of the Geological Society of London* 14:453–500.

Stanziani, Alessandro. 1998. *L'économie en revolution: Le cas russe, 1870–1930.* Paris: Albin Michel.

Star, Susan Leigh. 1989. *Regions of the Mind: Brain Research and the Quest for Scientific Certainty.* Stanford: Stanford University Press.

Staudenmaier, John M., S.J. 1985. *Technology's Storytellers: Reweaving the Human Fabric.* Cambridge, Mass.: MIT Press.

Steinberg, Holger. 2001. *Kraepelin in Leipzig: Eine Begegnung von Psychiatrie und Psychologie.* Bonn: Edition Das Narrenschiff.

Steinle, Friedrich. 1996. "Work, Finish, Publish? The Formation of the Second Series of Faraday's Experimental Research in Electricity." *Physis* 33:141–220.

Stephens, Lester D. 1982. *Joseph LeConte: Gentle Prophet of Evolution.* Baton Rouge: Louisiana State University Press.

———. 1989. "Scientific Societies in the Old South: The Elliott Society and the New Orleans Academy of Sciences," in Numbers and Savitt, *Science and Medicine,* 55–78.

Stephens, Michael D., and Gordon W. Roderick. 1972. "Science, the Working Classes, and Mechanics' Institutes." *AS* 29:349–60.

Stern, Leo, et al. 1975. *Die Berliner Akademie der Wissenschaften in der Zeit des Imperialismus.* Vol. 1, *Von den neunziger Jahren des 19. Jahrhunderts bis zur Großen Sozialistischen Oktoberrevolution.* Berlin: Akademie-Verlag.

Stewart, Larry. 1992. *The Rise of Public Science: Rhetoric, Technology, and Natural Philosophy in Newtonian Britain, 1600–1750.* Cambridge: Cambridge University Press.

Stichweh, Rudolf. 1984. *Zur Entstehung des modernen Systems wissenschaftlicher Disziplinen: Physik in Deutschland, 1740–1890.* Frankfurt am Main: Suhrkamp.

———. 1992. "The Sociology of Scientific Disciplines: On the Genesis and Stability of the Disciplinary Structure of Modern Science." *Science in Context* 5:3–15.

Stigler, George. 1972. "The Adoption of the Marginal Utility Theory." *History of Political Economy* 4:571–86.

Stigler, Stephen. 1986. *The History of Statistics: The Measurement of Uncertainty before 1900.* Cambridge, Mass.: Harvard University Press.

———. 1999. *Statistics on the Table: The History of Statistical Concepts and Methods.* Cambridge, Mass.: Harvard University Press.

Stiling, Rodney L. 1995. Review of Conser, *God and the Natural World. Isis* 86:340.

Stingelin, Martin. 1992. " 'Moral und Physiologie': Nietzsches Grenzverkehr zwischen den Diskursen." In *Technopathologien,* edited by Bernhard Dotzler, 41–57. Munich: Fink.

Stocking, George. 1968. *Race, Culture, and Evolution: Essays in the History of Anthropology.* New York: Free Press.

———. 1987. *Victorian Anthropology.* New York: Free Press.

———. 1992. *The Ethnographer's Magic and Other Essays in the History of Anthropology.* Madison: University of Wisconsin Press.

———. 1995. *After Tylor: British Social Anthropology, 1888–1951.* Madison: University of Wisconsin Press.

Stoltzenberg, Dietrich. 1994. *Fritz Haber: Chemiker, Nobelpreisträger, Deutscher Jude.* Weinheim: VCH.

Stone, Lawrence. 1971. "Prosopography." *Daedalus* 100:46–79.

Strickland, Stuart. 1995. "Galvanic Disciplines: The Boundaries, Objects, and Identities of Experimental Science in the Era of Romanticism." *HS* 33:449–68.

———. 1997. *The Ideology of Self-Knowledge and the Practice of Self-Experimentation.* Preprint 65. Max Planck Institute for the History of Science, Berlin.

Struik, Dirk J. [1948] 1967. *A Concise History of Mathematics.* 3d ed. New York: Dover.

———. 1981. "Mathematics in the Early Part of the Nineteenth Century," in Bos, Mehrtens, and Schneider, *Social History,* 6–20.

Sturdy, Steve, and Roger Cooter. 1998. "Science, Scientific Management, and the Transformation of Medicine in Britain, c. 1870–1950." *HS* 36:421–66.

Suess, Eduard. 1875. *Die Entstehung der Alpen.* Vienna: Wilhelm Braumüller.

Sulloway, Frank. 1977. *Freud, Biologist of the Mind: Beyond the Psychoanalytic Legend.* New York: Basic Books.

Sutton, Geoffrey. 1981. "The Politics of Science in Early Napoleonic France: The Case of the Voltaic Pile." *HSPS* 11:329–66.

Sviedrys, Romualdas. 1970. "The Rise of Physical Science at Victorian Cambridge." *HSPS* 2:127–51 (including a commentary by Arnold Thackray, 145–49).

———. 1976. "The Rise of Physics Laboratories in Britain." *HSPS* 7:405–36.

Sweeting, G. S., ed. 1958. *The Geologists' Association 1858–1958: A History of the First Hundred Years.* London: Benham and Company Ltd. for the Geologists' Association.

Swetlitz, Marc. 1995. "The Response of American Reform Rabbis to Evolutionary Theory, 1864–1888." In *Interaction of Scientific and Jewish Culture in Modern Times,* edited by Yakov Rabkin, 132–56. Lewiston, N.Y.: Edwin Mellen Press.

Swick, Clarence H. 1921. *Modern Methods for Measuring the Intensity of Gravity.* U.S. Coast and Geodetic Survey, Special Publication no. 69. Washington, D.C.: U.S. Government Printing Office.

Szreter, Simon. 1996. *Fertility, Class, and Gender in Britain, 1860–1940.* Cambridge: Cambridge University Press.

Taton, René. 1965. "Causes of Scientific Progress in Western Europe," in Taton, *Science in the Nineteenth Century,* 549–58.

————. 1970. "Emergence and Development of Some National Scientific Communities in the Nineteenth Century." *International Social Science Journal* 22:94–110.

————, ed. 1965. *Science in the Nineteenth Century.* Trans. A. J. Pomerans. New York: Basic Books. Originally published as *Histoire générale des sciences,* vol. 3, *La science contemporaine,* pt. 1, *Le XIXe siècle* (Paris: Presses Universitaires de France, 1961).

Taquet, Philippe, ed. 2002. *Alcide d'Orbigny: du nouveau monde au passé du monde.* Paris: Muséum Nationale d'Histoire Naturelle / Nathan.

Technology and Culture, nos. 1 (1965) and 17 (October 1976).

Temkin, Owsei. 1946a. "Materialism in French and German Physiology of the Early Nineteenth Century." *BHM* 20:322–27.

————. 1946b. "The Philosophical Background of Magendie's Physiology." *BHM* 20: 10–35.

————. 1963. "Basic Science, Medicine, and the Romantic Era." *BHM* 37:97–129.

Thackray, Arnold. 1966. "The Origins of Dalton's Chemical Atomic Theory: Daltonian Doubts Resolved." *Isis* 57:33–55.

————. 1971. "Reflections on the Decline of Science in America and on Some of Its Causes." *Science* 173:27–31.

————. 1972. *John Dalton: Critical Assessments of His Life and Science.* Cambridge, Mass.: Harvard University Press.

————. 1974. "Natural Knowledge in a Cultural Context: The Manchester Model." *American Historical Review* 79:672–709.

————. 1975. Reply to Glen, 1975. *American Historical Review* 80:204–5.

Thackray, Arnold, Jeffrey Sturchio, P. Thomas Carroll, and Robert F. Bud. 1981. *Chemistry in America, 1876–1976: An Historical Application of Science Indicators.* Boston: Reidel.

Thompson, Noel W. 1984. *The People's Science: The Popular Political Economy of Exploitation and Crisis.* Cambridge: Cambridge University Press.

Tiedemann, Friedrich. 1830. "Ansichten von der Behandlungsart der Naturwissenschaften und der Medicin." *ISIS* 23:480–95.

Tikhomirov, Vladimir V. 1960. *Geology in Russia in the First Half of the Nineteenth Century* (in Russian). Moscow: Academy of Sciences of the USSR.

Tinkler, Keith J. 1985. *A Short History of Geomorphology.* Totowa, N.J.: Barnes & Noble.

————, ed. 1989. *History of Geomorphology from Hutton to Hack.* Boston: Unwin Hyman.

Tintant, Henri. 1997. "Alcide d'Orbigny (1802–1857) ou la progression des formes," in Gohau, *De la géologie,* 163–74.

Tomes, Nancy. 1990. "The Private Side of Public Health: Sanitary Science, Domestic Hygiene, and the Germ Theory, 1870–1900." *BHM* 64:509–39.

Tomkeieff, Serge I. 1983. *Dictionary of Petrology.* Eds. E. K. Walton, B. A. O. Randall, M. H. Battey, and O. Tomkeieff. Chichester: John Wiley & Sons.

Topham, Jonathan R. 1998. "Beyond the 'Common Context': The Production and Reading of the Bridgewater Treatises." *Isis* 89:233–62.

Torrens, Hugh S. 1992. "When Did the Dinosaur Get Its Name?" *New Scientist* 134:40–44.

————. 1994. "Patronage and Problems: Banks and the Earth Sciences." In *Sir Joseph Banks: A Global Perspective,* edited by R. E. R. Banks, B. Elliott, J. G. Hawkes, D. King-Hele, and G. L. Lucas, 49–75. Kew: Royal Botanical Gardens.

————. 1997. "Some Thoughts on the Complex and Forgotten History of Mineral Exploration." *Journal of the Open University Geological Society* 17:1–12.

————. 2001. "Timeless Order: William Smith (1769–1839) and the Search for Raw Materials, 1800–1820," in Lewis and Knell, *Age of the Earth*, 61–83.

————. 2002. *The Practice of British Geology, 1750–1850.* Aldershot: Ashgate Variorum.

Toti Rigatelli, Laura. 1996. *Evariste Galois, 1811–1832.* Trans. John Denton. Basel: Birkhäuser. Originally published as *Evariste Galois, Matematica sulle Barricate* (Florence: Sansoni, 1993).

Toulmin, Stephen. 1990. *Cosmopolis: The Hidden Agenda of Modernity.* Chicago: University of Chicago Press.

Touret, Jacques L. R., and Timo G. Nijland. 2002. "Metamorphism Today: New Science, Old Problems," in Oldroyd, *Earth Inside and Out*, 113–141.

Travis, Anthony S. 1989. "Science as Receptor of Technology: Paul Ehrlich and the Synthetic Dystuffs Industry." *Science in Context* 3:2:383–408.

————. 1993. *The Rainbow Makers: The Origins of the Synthetic Dyestuffs Industry in Western Europe.* Bethlehem, Pa.: Lehigh University Press.

Travis, Anthony S., Willem J. Hornix, and Robert Bud, eds. 1992. "Organic Chemistry and High Technology, 1850–1950." Special issue of *BJHS* 25:1:4:1–167.

Trenn, Thaddeus J. 1977. *The Self-Splitting Atom: The History of the Rutherford-Soddy Collaboration.* London: Taylor & Francis.

Treviranus, Gottfried Reinhold. 1802–22. *Biologie, oder Philosophie der lebenden Natur.* 6 vols. Göttingen: Johann Friedrich Röwer.

Tribe, Keith. 1988. *Governing Economy: The Reformation of German Economic Discourse, 1750–1840.* Cambridge: Cambridge University Press.

————. 1991. "Political Economy to Economics via Commerce: The Evolution of British Academic Economics," in Wagner, Wittrock, and Whitley, *Discourses on Society*, 273–302.

Trudeau, Richard. 1987. *The Non-Euclidean Revolution.* Boston: Birkhäuser.

Truesdell, C. 1960. "A Program toward Rediscovering the Rational Mechanics in the Age of Reason." *AHES* 1:1–36.

————. 1980. *The Tragicomical History of Thermodynamics, 1822–1854.* New York: Springer-Verlag.

Trümpy, Rudolph, and Marcel Lemoine. 1998. "Marcel Bertrand (1847–1907): Les nappes de charriage et le cycle orogenic cycle." *Comptes rendus de l'Académie des Sciences, Paris: Sciences de la terre et des planètes* 327:211–24.

Tsouyopoulos, Nelly. 1982. *Andreas Röschlaub und die romantische Medizin.* Stuttgart: Gustav Fischer.

————. 1992. "Schellings Naturphilosophie: Sünde oder Inspiration für den Reformer der Physiologie," in Hagner and Wahrig-Schmidt, *Johannes Müller*, 65–84.

Tuchman, Arleen Marcia. 1988. "From the Lecture to the Laboratory: The Institutionalization of Scientific Medicine at the University of Heidelberg," in Coleman and Holmes, *Investigative Enterprise*, 65–99.

————. 1993a. "Helmholtz and the German Medical Community," in Cahan, *Hermann von Helmholtz*, 17–49.

————. 1993b. *Science, Medicine, and the State in Germany: The Case of Baden, 1815–1871.* New York: Oxford University Press.

Tunzelman, G. N. von. 1978. *Steam Power and British Industrialization to 1860.* Oxford: Oxford University Press.

————. 1994. "Technology in the Early Nineteenth Century," in Floud and McCloskey, *Economic History of Britain*, 271–99.

Turner, Frank Miller. 1974. *Between Science and Religion: The Reaction to Scientific Naturalism in Late Victorian England*. New Haven: Yale University Press.

———. 1978. "The Victorian Conflict between Science and Religion: A Professional Dimension." *Isis* 69:356–76.

———. 1993a. *Contesting Cultural Authority: Essays in Victorian Intellectual Life*. Cambridge: Cambridge University Press.

———. 1993b. Review of Cantor, *Michael Faraday*. *American Historical Review* 98:495.

Turner, G. L'E., ed. 1976. *The Patronage of Science in the Nineteenth Century*. Leyden: Noordhoff International.

Turner, James. 1985. *Without God, Without Creed: The Origins of Unbelief in America*. Baltimore: Johns Hopkins University Press.

Turner, R. Steven. 1971. "The Growth of Professorial Research in Prussia, 1818–1848: Causes and Context." *HSPS* 3:137–82.

———. 1980a. "*Bildungsbürgertum* and the Learned Professions in Prussia, 1770–1830: The Origins of a Class." *Histoire Sociale—Social History* 13:105–35.

———. 1980b. "The Prussian Universities and the Concept of Research." *Internationales Archiv für Sozialgeschichte der deutschen Literatur* 5:68–93.

———. 1981. "The Prussian Professoriate and the Research Imperative," in Jahnke and Otte, *Epistemological and Social Problems*, 104–22.

———. 1982. "Justus Liebig versus Prussian Chemistry: Reflections on Early Institute-Building in Germany." *HSPS* 13:129–62.

———. 1987a. "The Great Transition and the Social Patterns of German Science." *Minerva* 25:56–76.

———. 1987b. "Universitäten." In *Handbuch der deutschen Bildungsgeschichte*, vol. 3 (1800–1870), edited by Karl-Ernst Jeismann and Peter Lundgreen, 221–49. Munich: C. H. Beck.

———. 1991. "German Science, German Universities: Historiographical Perspectives from the 1980s." In *"Einsamkeit und Freiheit" neu besichtigt: Universitätsreformen und Disziplinenbildung in Preussen als Modell für Wissenschaftspolitik im Europa des 19. Jahrhunderts*, edited by Gert Schubring, 24–36. Stuttgart: Franz Steiner.

———. 1994. *In the Eye's Mind: Vision and the Helmholtz-Hering Controversy*. Princeton, N.J.: Princeton University Press.

Turner, Stephen. 1986. *The Search for a Methodology of Social Science: Durkheim, Weber, and the Nineteenth-Century Problem of Cause, Probability, and Action*. Dordrecht: Reidel.

Underwood, J. R., and P. L. Guth, eds. 1998. *Military Geology in War and Peace*. Boulder, Colo.: Geological Society of America.

Ure, Andrew. 1836. *The Cotton Manufacture of Great Britain Systematically Investigated*. London: C. Knight.

Van Spronsen, J. W. 1969. *The Periodic System of the Chemical Elements: A History of the First Hundred Years*. Amsterdam: Elsevier.

Van Tiggelen, Brigitte. 1998. *Un chimiste des Pays-Bas autrichiens au siècle des Lumières: Karl van Bochaute*. Ph.D. diss., Université catholique de Louvain.

Vernadsky, Vladimir I. 1997. *The Biosphere*. Trans. David B. Langmuir. New York: Springer-Verlag.

Veysey, Laurence R. 1965. *The Emergence of the American University*. Chicago: University of Chicago Press.

Vierhaus, Rudolf, and Bernhard vom Brocke, eds. 1990. *Forschung im Spannungsfeld von Politik und Gesellschaft: Geschichte und Struktur der Kaiser-Wilhelm-/Max-Planck-Gesellschaft.* Stuttgart: Deutsche Verlags-Anstalt.

Vinogradov, Aleksandr P. 1963. "Development of V. I. Vernadsky's Ideas." *Soviet Soil Science* 8:727–32.

Virchow, Rudolf. [1849] 1856. "Die Einheits-Bestrebungen in der wissenschaftlichen Medicin." In *Gesammelte Abhandlungen zur wissenschaftlichen Medicin,* 1–56. Frankfurt am Main: Meidinger.

Vogel, Morris J. 1980. *The Invention of the Modern Hospital: Boston, 1870–1930.* Chicago: University of Chicago Press.

Vogel, Morris J., and Charles Rosenberg, eds. 1979. *The Therapeutic Revolution: Essays in the Social History of American Medicine.* Philadelphia: University of Pennsylvania Press.

Vom Brocke, Bernhard. 1990. "Die Kaiser-Wilhelm-Gesellschaft im Kaiserreich: Vorgeschichte, Gründung und Entwicklung bis zum Ausbruch des Ersten Weltkriegs," in Vierhaus and vom Brocke, *Forschung im Spannungsfeld,* 17–162.

———. 1996. "Die Kaiser-Wilhelm-/Max-Planck-Gesellschaft und ihre Institute zwischen Universität und Akademie: Strukturprobleme und Historiographie." In *Die Kaiser-Wilhelm-/Max-Planck-Gesellschaft und ihre Institute: Studien zu ihrer Geschichte: Das Harnack-Prinzip,* edited by Bernhard Vom Brocke and Hubert Laitko, 1–32. Berlin: Walter de Gruyter.

Wagner, Peter. 1991. "Science of Society Lost: On the Failure to Establish Sociology in Europe during the 'Classical Period,' " in Wagner, Wittrock, and Whitley, *Discourses on Society,* 219–45.

Wagner, Peter, Bjorn Wittrock, and Richard Whitley, eds. 1991. *Discourses on Society: The Shaping of the Social Science Disciplines.* Dordrecht: Kluwer.

Wahrig-Schmidt, Bettina. 1988. " 'Irgendwie, jedenfalls physiologisch': Friedrich Nietzsche, Alexandre Herzen (fils) und Charles Féré 1888." *Nietzsche-Studien* 17:434–64.

Walther, Philipp Franz von. 1807. *Physiologie des Menschen: Mit durchgängiger Rücksicht auf die comparative Physiologie der Thiere.* Vol. 1. Landshut: Krüll.

Warner, John Harley. 1985. "Science in Medicine," in Kohlstedt and Rossiter, *Historical Writing,* 37–58.

———. [1986] 1997. *The Therapeutic Perspective: Medical Practice, Knowledge, and Identity in America 1820–1885.* Princeton, N.J.: Princeton University Press.

———. 1992. "The Fall and Rise of Professional Mystery: Epistemology, Authority and the Emergence of Laboratory Medicine in Nineteenth-Century America," in Cunningham and Williams, *Laboratory Revolution,* 110–41.

———. 1995. "The History of Science and the Sciences of Medicine." In "Constructing Knowledge in the History of Science," special issue of *Osiris* (2d ser.), edited by Arnold Thackray, 10:164–93.

———. 1998. *Against the Spirit of System: The French Impulse in Nineteenth-Century American Medicine.* Princeton, N.J.: Princeton University Press.

Wartnaby, John. 1975. "John Milne (1850–1913) and the Development of the Seismograph." *Proceedings of the XIV International Congress of History of Science* 3:109–112.

Warwick, Andrew. 1998. "A Mathematical World on Paper: Written Examinations in Early Nineteenth-Century Cambridge." *Studies in the History and Philosophy of Modern Physics* 29:295–319.

Watson, Katherine D. 1995. "The Chemist as Expert: The Consulting Career of Sir William Ramsay." *Ambix* 43:143–59.

Wegmann, Eugene. 1958. "Das Erbe Werners und Huttons." *Geologie* 7:531–59.

Weikart, Richard. 1998. *Socialist Darwinism: Evolution in German Socialist Thought from Marx to Bernstein*. San Francisco, Calif.: International Scholars Publications.

Weil, André. 1980. "History of Mathematics: Why and How." In *Proceedings of the International Congress of Mathematicians* [Helsinki, 1978], 2 vols., edited by O. Lehto, 1:227–36. Helsinki: Academia Scientiarum Fennica.

Weindling, Paul J. 1979. "Geological Controversy and Its Historiography: The Pre-History of the Geological Society of London," in Jordanova and Porter, *Images of the Earth*, 248–71.

———. 1989. *Health, Race and German Politics between National Unification and Nazism, 1870–1945*. Cambridge: Cambridge University Press.

———. 1991. *Darwinism and Social Darwinism in Imperial Germany: The Contribution of the Cell Biologist Oscar Hertwig (1849–1922)*. Stuttgart: Gustav Fischer Verlag.

Weiss, John H. 1982. *The Making of Technological Man: The Social Origins of French Engineering Education*. Cambridge, Mass.: MIT Press.

Weisz, George. 1983. *The Emergence of Modern Universities in France, 1863–1914*. Princeton, N.J.: Princeton University Press.

———. 1995. *The Medical Mandarins: The French Academy of Medicine in the Nineteenth and Early Twentieth Centuries*. Oxford: Oxford University Press.

Wendel, Günter. 1975. *Die Kaiser-Wilhelm-Gesellschaft, 1911–1914: Zur Anatomie einer imperialistischen Forschungsgesellschaft*. Berlin: Akademie-Verlag.

Wertheim, Margaret. 1995. *Pythagoras' Trousers: God, Physics, and the Gender Wars*. New York: Random House.

Westbroek, Peter. 1991. *Life as a Geological Force: Dynamics of the Earth*. New York: W. W. Norton.

Westrum, Ron. 1978. "Science and Social Intelligence about Anomalies: The Case of Meteorites." *SSS* 8:461–93.

Wetzel, Walter. 1991. *Naturwissenschaften und chemische Industrie in Deutschland: Voraussetzungen und Mechanismen ihres Aufstiegs im 19. Jahrhundert*. Stuttgart: Steiner.

Wheaton, Bruce R. 1983. *The Tiger and the Shark: Empirical Roots of Wave-Particle Dualism*. Cambridge: Cambridge University Press.

[Whewell, William.] 1832. Review of Lyell, *Principles of Geology*, vol. 2. *Quarterly Review* 47:103–132.

———. [1837] 1897. *History of the Inductive Sciences from the Earliest to the Present Time*. 2 vols. New York: D. Appleton.

———. 1857. *History of the Inductive Sciences, from the Earliest to the Present Time*. 3d ed. 3 vols. London: John W. Parker.

White, Andrew Dickson. 1876. *The Warfare of Science*. New York: D. Appleton. British ed., with foreword by John Tyndall, London: H. S. King, 1876.

———. [1896] 1960. *A History of the Warfare of Science with Theology in Christendom*. 2 vols. Reprint, New York: Dover.

White, Edward A. 1952. *Science and Religion in American Thought*. Stanford, Calif.: Stanford University Press.

Whitehead, Alfred North. [1925] 1962. *Science and the Modern World*. New York: Macmillan. Reprint, New York: New American Library.

Wiesing, Urban. 1995. *Kunst oder Wissenschaft: Konzeptionen der Medizin in der deutschen Romantik.* Stuttgart-Bad Cannstatt: Frommann-Holzboog.

Wilder, Raymond L. 1968. *Evolution of Mathematical Concepts: An Elementary Study.* New York: Wiley.

Willey, Basil. 1962. *The Eighteenth-Century Background: Studies on the Idea of Nature in the Thought of the Period.* Harmondsworth: Penguin Books, Chatto and Windus.

Williams, Elizabeth A. 1994. *The Physical and the Moral: Anthropology, Physiology, and Philosophical Medicine in France, 1750–1850.* Cambridge: Cambridge University Press.

Williams, L. Pearce. 1956. "Science, Education, and Napoleon I." *Isis* 47:369–82.

———. 1965. *Michael Faraday: A Biography.* New York: Basic Books. Reprint, New York: Da Capo, 1987.

Wilson, David B. 1974. "Kelvin's Scientific Realism: The Theological Context." *Philosophical Journal* 11:41–60.

———. 1985. "The Educational Matrix: Physics Education at Early Victorian Cambridge, Edinburgh, and Glasgow Universities," in Harman, *Wranglers and Physicists,* 12–48.

Wilson, Edward O. 1975. *Sociobiology.* Cambridge, Mass.: Harvard University Press.

Wilson, Leonard G. 1972. *Charles Lyell: The Years to 1841, The Revolution in Geology.* New Haven: Yale University Press.

———. 1980. "Geology on the Eve of Charles Lyell's First Visit to America, 1841." *Proceedings of the American Philosophical Society* 124:168–202.

Winch, Donald. 1965. *Classical Political Economy and Colonies.* London: G. Bell.

———. 1969. *Economics and Policy: A Historical Study.* New York: Walker.

———. 1996. *Riches and Poverty: An Intellectual History of Political Economy in Britain, 1750–1834.* Cambridge: Cambridge University Press.

Winsor, Mary. 1991. *Reading the Shape of Nature: Comparative Zoology at the Agassiz Museum.* Chicago: University of Chicago Press.

Winter, Alison. 1998. *Mesmerized: Powers of Mind in Victorian Britain.* Chicago: University of Chicago Press.

Wise, George. 1985. *Willis R. Whitney, General Electric, and the Origins of U.S. Industrial Research.* New York: Columbia University Press.

Wise, M. Norton. 1979. "The Mutual Embrace of Electricity and Magnetism." *Science* 203:1310–18.

———. 1981. "The Flow Analogy to Electricity and Magnetism, Part I: William Thomson's Reformulation of Action at a Distance." *AHES* 25:19–70.

———. 1987. "How Do Sums Count? On the Cultural Origins of Statistical Causality," in Krüger, Daston, and Heidelberger, *Probabilistic Revolution,* 1:395–425.

———. 1989–90. "Work and Waste: Political Economy and Natural Philosophy in Nineteenth-Century Britain." *HS* 27:263–301, 391–449; 28:221–61.

———. 1993. "Mediations: Enlightenment Balancing Acts, or the Technologies of Rationalism," in Horwich, *World Changes,* 207–56.

———, ed. 1995. *The Values of Precision.* Princeton, N.J.: Princeton University Press.

Wood, Robert Muir. 1985. *The Dark Side of the Earth.* London: George Allen & Unwin.

Woodward, Horace B. 1908. *The History of the Geological Society of London.* London: Longmans, Green.

Woodward, William R., and Mitchell G. Ash, eds. 1982. *The Problematic Science: Psychology in Nineteenth-Century Thought.* New York: Praeger.

Woronoff, Denis. 1994. *Histoire de l'industrie en France du XVIe siècle à nos jours.* Paris: Seuil.

Wotiz, John. H., ed. 1993. *The Kekulé Riddle: A Challenge for Chemists and Psychologists.* New York: Glenview Press.

Wotiz, John. H., and R. S. Rudotsy. 1983. "The Unknown Kekulé." In *Essays on the History of Organic Chemistry,* edited by John G. Traynham, 21–34. Baton Rouge: Louisiana State University Press.

Wunderlich, Carl August. 1859. *Geschichte der Medizin.* Stuttgart: Ebner und Seubert.

———. 1868. *Das Verhalten der Eigenwärme in Krankheiten.* Leipzig: Wigand.

Wussing, Hans. 1979. "Zur gesellschaftlichen Stellung der Mathematik und Naturwissenschaften in der Industriellen Revolution." In *Studien zur Geschichte der Produktivkräfte,* edited by K. Lärmer, 55–68. Berlin: Akademie-Verlag.

———. 1984. *The Genesis of the Abstract Group Concept.* Trans. Abe Shenitzer. Cambridge, Mass.: MIT Press.

Yanovskaya, C. A. (Sof'ya Aleksandrovna). 1933. "O Matematicheskikh rukopisyakh Marksa (On Marx's mathematical manuscripts)," In *Marksizm i estestvoznanie* (Marxism and natural sciences), 136–80. Moscow: Nauka.

Yavetz, Ido. 1993. "Oliver Heaviside and the Significance of the British Electrical Debate." *AS* 50:135–73.

———. 1995. *From Obscurity to Enigma: The Work of Oliver Heaviside, 1872–1889.* Basel: Birkhäuser.

Yeo, Eileen Janes. 1996. *The Contest for Social Science.* London: Rivers Oram Press.

Yochelson, Ellis L. 1993. "The Question of Primordial and Cambrian / Taconic: Barrande and Logan / Marcou." *ESH* 12:111–20.

Yoder, Hatten S. 1993. "Timetable of Petrology." *Journal of Geological Education* 41:447–89.

Young, Robert M. 1966. "Scholarship and the History of the Behavioral Sciences." *HS* 5:1–51.

———. 1969. "Malthus and the Evolutionists: The Common Context of Biological and Social Theory." *Past and Present* 43:109–45.

———. 1970a. "The Impact of Darwin on Conventional Thought." In *The Victorian Crisis of Faith,* edited by Anthony Symondson, 13–35. London: SPCK.

———. 1970b. *Mind, Brain, and Adaptation in the Nineteenth Century: Cerebral Localization and Its Biological Context from Gall to Ferrier.* Oxford: Clarendon Press.

———. 1985a. "Darwinism Is Social," in Kohn, *Darwinian Heritage,* 609–40.

———. 1985b. *Darwin's Metaphor: Nature's Place in Victorian Culture.* Cambridge: Cambridge University Press.

Zimmerman, Andrew. 2001. *Anthropology and Antihumanism in Imperial Germany.* Chicago: University of Chicago Press.

Zittel, Karl A. von. 1901. *History of Geology and Palaeontology to the End of the Nineteenth Century.* 1st German ed., 1899. Trans. Maria M. Ogilvie-Gordon. London: Walter Scott.

Zloczower, Avraham. 1960. Career Opportunities and the Growth of Scientific Discovery in Nineteenth-Century Germany. M.S. diss., Hebrew University, Jerusalem.

Zöckler, Otto. 1877, 1879. *Die Geschichte der Beziehungen zwischen Theologie und Naturwissenschaft mit besonderer Rücksicht auf Schöpfungsgeschichte.* 2 vols. Gütersloh: Bertelsmann.

Zwerling, Craig S. 1980. "The Emergence of the Ecole Normale Supérieure as a Centre of Scientific Education in the Nineteenth Century," in R. Fox and Weisz, *Organization of Science and Technology,* 31–60.

———. 1990. *The Emergence of the Ecole Normale Supérieure as a Center of Scientific Education in Nineteenth-Century France.* New York: Garland.

Zylberberg, André. 1990. *L'économie mathématique en France, 1870–1914.* Paris: Economica.

INDEX